Hadamard Matrices

Hadamard Matrices

Constructions using Number Theory and Algebra

Jennifer Seberry
Emeritus Professor of Computer Science
University of Wollongong,
Wollongong, New South Wales, Australia

Mieko Yamada
Emeritus Professor of Kanazawa University
Kanazawa, Japan

WILEY

This edition first published 2020

Registered Office
John Wiley & Sons, Inc., 111 River Street, Hoboken, NJ 07030, USA

Editorial Office
111 River Street, Hoboken, NJ 07030, USA

For details of our global editorial offices, customer services, and more information about Wiley products visit us at www.wiley.com.

Wiley also publishes its books in a variety of electronic formats and by print-on-demand. Some content that appears in standard print versions of this book may not be available in other formats.

Library of Congress Cataloging-in-Publication Data

Names: Seberry, Jennifer, 1944– author.
Title: Hadamard matrices : constructions using number theory and algebra / Jennifer Seberry, Mieko Yamada.
Description: First edition. | Hoboken, NJ : Wiley, 2020. | Includes bibliographical references and index.
Identifiers: LCCN 2020014449 (print) | LCCN 2020014450 (ebook) | ISBN 9781119520245 (cloth) | ISBN 9781119520276 (adobe pdf) | ISBN 9781119520139 (epub)
Subjects: LCSH: Hadamard matrices.
Classification: LCC QA166.4. S43 2020 (print) | LCC QA166.4 (ebook) | DDC 512.9/434–dc23
LC record available at https://lccn.loc.gov/2020014449
LC ebook record available at https://lccn.loc.gov/2020014450

Cover design: Wiley
Cover image: Wiley

Set in 9.5/12.5pt STIXTwoText by SPi Global, Pondicherry, India

10 9 8 7 6 5 4 3 2 1

MIX
Paper from responsible sources
FSC www.fsc.org FSC® C013604

This book is dedicated to all those pioneering women mathematicians who now have got full access to Erdös book and know all the answers but have been forgotten in their own countries.

Contents

List of Tables

List of Figures

Preface

Tom Storer once said of Professor Koichi Yamamoto that the guy was fantastic as he had invented so much number theory to resolve the question of the existence of (v, k, λ) difference sets.

Jennifer visited Professor Koichi Yamamoto, the supervisor of Professor Mieko Yamada at Tokyo Woman's Christian University in 1972. Jennifer visited Professor Yamada for one month at Tokyo Woman's Christian University in 1989. In 1990, Mieko visited Jennifer at ADFA@UNSW in Canberra and they worked on a number of aspects of Hadamard matrix theory including M-structures. This led to the survey Hadamard matrices, sequences, and block designs, which appeared in *Contemporary Design Theory: A Collection of Surveys*. Edited by Jeffrey H. Dinitz and Douglas R. Stinson, published by John Wiley & Sons, Inc. in 1972. This present book, *Hadamard Matrices: Constructions Using Number Theory and Algebra*, has arisen from the revision and extension of this survey.

Both Mieko and Jennifer have had a long love affair with Hadamard matrices. Because of their different backgrounds they have brought two areas of Mathematics together. We hope this makes a more interesting study.

This book is heavily based on the original survey but the number theory sections have been considerably advanced in Chapters 2, 7, 8, and 10. Chapters 3, 4, 5, 6, and 9 are essentially the same, as is the Appendix. Since the original survey was written, there have been exciting developments in the knowledge of the asymptotic existence of Hadamard matrices; Chapter 11 discusses Seberry, de Launey, and Craigen's work; Chapter 12 studies three areas which impinge on the computation of Hadamard matrices: Smith normal forms, embedding maximal determinant sub-matrices, and the growth problem.

Jennifer Seberry
Wollongong, Australia

March 2019

Acknowledgments

No book this size could be the child of one or two parents. Many people are owed a debt of deep gratitude. The 1970s students Peter Robinson (ANU, Canberra), Peter Eades (ANU, Canberra), and Warren Wolfe (Queens', Kingston) had major input; the 1980s students of Seberry, Deborah J. Street (Sydney), Warwick de Launey (Sydney), and Humphrey Gastineau-Hills (Sydney), and the 2000s students Chung Le Tran (Wollongong) and Ying Zhao (Wollongong) have contributed conspicuously to the field. Many, many colleagues such as Marilena Mitrouli (Athens), Christos Koukouvinos (NTUA, Athens), and his students Stelios Georgiou and Stella Stylianou, Hadi Kharghani (Lethbridge), Wolf Holzmann (Lethbridge), Rob Craigen (Winnipeg), Ilias Kotireas (WLU), Dragomir Đoković (Waterloo), Sarah Spence Adams (Olin, Boston), Behruz Tayfeh-Rezaie (IPM, Tehran), Nikolai Balonin, Yuri Balonin, and Michael Segeev (State University of Aerospace Instrumentation, St. Petersburgh) have helped shape and correct our work. We also appreciate M. Jimbo (Chubu University), R. Fuji-Hara (University of Tsukuba), A. Munemasa (Tohoku University), M. Harada (Tohoku University), H. Kimura, and K. Momihara (Kumamoto University) for their warm encouragement. We particularly thank Koji Momihara for checking our work in Chapter 10 carefully. The researches of Yamamoto, Whiteman, and Ito had a profound influence on Yamada's research and the writing of this book. We acknowledge that at times we might have written utter rubbish without their knowledge and thoughts. We leave writing this book knowing it is an unfinished work, knowing that we have included many errors large and small. We acknowledge the LaTeXing proofing, and exceptional research assistance given us over the past 10 years by Max Norden. Seberry came to today with the help of the University of Wollongong Library and the provision of an office as Emeritus Professor. We also thank our families for their comforting words and hearty support during writing this book.

Jennifer Seberry and Mieko Yamada

Acronyms

$A * B$	The Hadamard product of matrices A and B
AOD	Amicable orthogonal design
BH array	Baumert–Hall array
BIBD	Balanced incomplete block design
BS	Base sequences
BSC	Base sequence conjecture
CP	Completely pivoted matrices (GE with complete pivoting)
DF	Difference family
EBS	Extended building set
$gcd(a, b)$	The greatest common divisor of integers a and b
GE	Gaussian elimination
GF	Galois field
GQ type	Generalized quaternion type
GR	Galois ring
H	An Hadamard matrix of order h
$M \times N$	The Kronecker product of matrices M and N
$M \bigcirc N$	The strong Kronecker product of matrices M and N
NPAF	Nonperiodic autocorrelation function
OD	Orthogonal design
PAF	Periodic autocorrelation function
PDS	Partial difference set
PG	Projective space
remrep	A real monomial representation
SBIBD	Symmetric balanced incomplete block design
sds	Supplementary difference set
SH	Signed group Hadamard matrices
SNF	Smith Normal Form
SRG	Strongly regular graph
SW	Signed group weighing matrices
$W(n, k)$	A Weighing matrix of weight k and of order n
WL array	Welch array
Z	The rational integer ring
$Z[G]$	A group ring of a group G over Z

Introduction

One hundred years ago, in 1893, Jacques Hadamard [89] found square matrices of orders 12 and 20, with entries ± 1, which had all their rows (and columns) orthogonal. These matrices, $X = (x_{ij})$, satisfied the equality of the following inequality

$$|\det X|^2 \leq \Pi_{i=1}^{n} \sum_{j=1}^{n} |x_{ij}|^2$$

and had maximal determinant. Hadamard actually asked the question of matrices with entries on the unit disc but his name has become associated with the real matrices.

Hadamard was not the first to study these matrices for J.J. Sylvester in 1867 in his seminal paper "Thoughts on inverse orthogonal matrices, simultaneous sign-successions, and tesselated pavements in two or more colours, with application to Newton's rule, ornamental tile work, and the theory of numbers" [180] had found such matrices for all orders which are powers of 2. Nevertheless, Hadamard showed matrices with elements ± 1 and maximal determinant could exist for all orders 1, 2, and $4t$ and so the Hadamard conjecture "that there exists an *Hadamard matrix*, or square matrix with every element ± 1 and all row (column) vectors orthogonal" came from here. This survey indicates the progress that has been made in the past 100 years (Figure I.1).

J. Hadamard produced his determinant inequality while studying mathematical physics: the equality of the inequality led to Hadamard matrices. A paper by U. Scarpis 1898 [150] was the next contribution to the Hadamard story anticipating the work of R. E. A. C. Paley. However, it was not until J. Williamson's work in 1944 that the name "Hadamard" attached to Hadamard matrices.

Hadamard's inequality applies to matrices from the unit circle and matrices with entries ± 1, $\pm i$, and pairwise orthogonal rows (and columns) are called *complex Hadamard matrices* (note the scalar product is $a.b = \sum a_i b_i^*$ for complex numbers). These matrices were first studied by R.J. Turyn [190]. We believe complex Hadamard matrices exist for every order $n \equiv 0 \pmod 2$. The truth of this conjecture implies the truth of the Hadamard conjecture.

In the 1960s, the US Jet Propulsion Laboratories (JPL) was working toward building the Mariner and Voyager space probes to visit Mars and the other planets of the solar system. Those of us who saw early black and white pictures of the back of the moon remember that whole lines were missing. The first black and white television pictures from the first landing on the moon were extremely poor quality. How many of us remember that the most recent flyby of Neptune was from a space probe launched in the 1970s? We take the high quality color pictures of Jupiter, Saturn, Uranus, Neptune, and their moons for granted.

In brief, these high quality color pictures are taken by using three black and white pictures taken in turn through red, green, and blue filters. Each picture is then considered as a thousand by a thousand matrix of black and white pixels. Each picture is graded on a scale of, say, 1 to 16, according to its greyness. So white is 1 and black is 16. These grades are then used to choose a codeword in, say, an eight error correction code based on, say, the Hadamard matrix of order 32. The codeword is transmitted to Earth, error corrected, the three black and white pictures reconstructed and then a computer used to reconstruct the colored pictures.

Figure I.1 The pioneers of Hadamard matrices. Jacques Hadamard; James Joseph Sylvester; and Umberto Scarpis. Source: Wikimedia Commons and http://mathscinet.ru/

Hadamard matrices were used for these codewords for two reasons, first, error correction codes based on Hadamard matrices have maximal error correction capability for a given length of codeword and, second, the Hadamard matrices of powers of 2 are analogous to the Walsh functions, thus all the computer processing can be accomplished using additions (which are very fast and easy to implement in computer hardware) rather than multiplications (which are far slower).

It was Baumert et al. [8] working with JPL who sparked the interest in Hadamard matrices in the past 50 years. They pioneered the use of computing in the construction of Hadamard matrices. The existence of an Hadamard matrix is an NPC problem (or a problem which has computational resources exponential in the input to find the answer but easy to check the answer once it has been given).

Sylvester's original construction for Hadamard matrices is equivalent to finding Walsh functions [110] which are the discrete analog of Fourier Series.

Example: Let H be a Sylvester–Hadamard matrix of order 8 and sequency order.

$$H = \begin{bmatrix} 1 & 1 & 1 & 1 & 1 & 1 & 1 & 1 \\ 1 & 1 & 1 & 1 & -1 & -1 & -1 & -1 \\ 1 & 1 & -1 & -1 & -1 & -1 & 1 & 1 \\ 1 & 1 & -1 & -1 & 1 & 1 & -1 & -1 \\ 1 & -1 & -1 & 1 & 1 & -1 & -1 & 1 \\ 1 & -1 & -1 & 1 & -1 & 1 & 1 & -1 \\ 1 & -1 & 1 & -1 & -1 & 1 & -1 & 1 \\ 1 & -1 & 1 & -1 & 1 & -1 & 1 & -1 \end{bmatrix}$$

The Walsh function generated by H is the following:

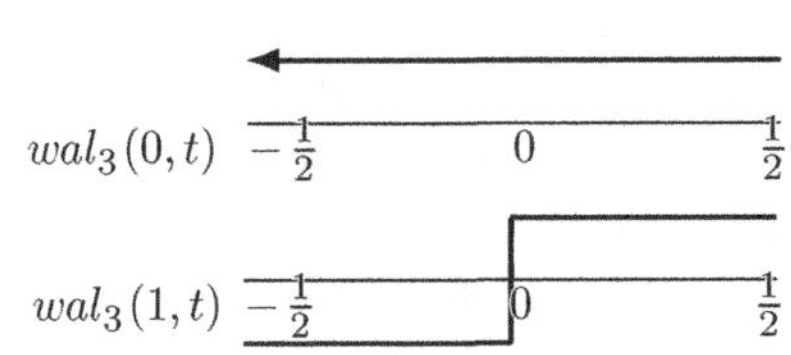

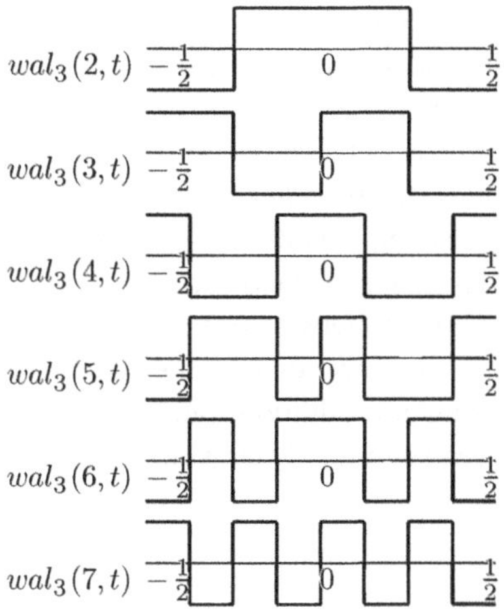

Figure I.2 gives the points of intersections of Walsh functions are identical with that of trigonometrical functions.

By mapping $w(i,t) = wal_n(i,t)$ into the interval $[-\frac{1}{2}, 0]$, then by mapping axial symmetrically into $[0, \frac{1}{2}]$, we get $w(2i,t)$ which is an even function. By operating similarly we get $w(2i-1,t)$, an odd function.

Just as any curve can be written as an infinite Fourier series

$$\sum_n a_n \sin nt + b_n \cos nt$$

the curve can be written in terms of Walsh functions

$$\sum_n a_n\, sal(i,t) + b_n\, cal(i,t) = \sum_n c_n\, wal(i,t).$$

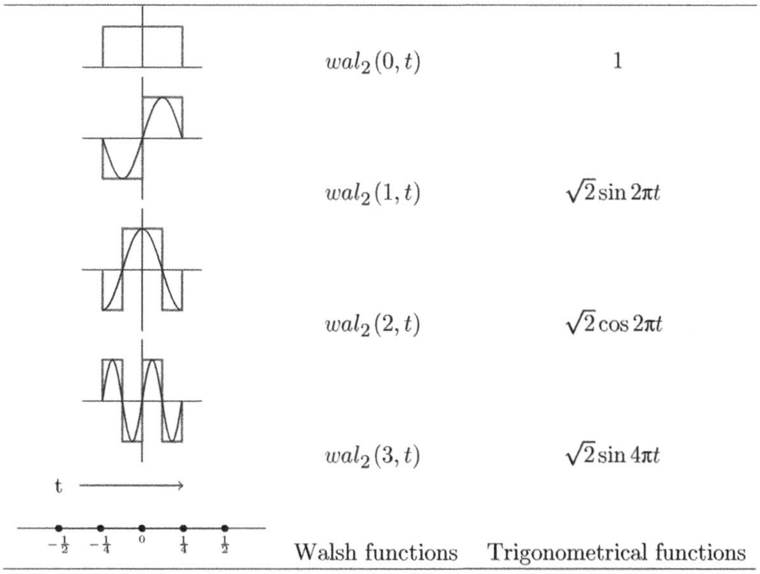

Figure I.2 Walsh functions and trigonometrical functions. Source: Seberry and Yamada [166, p. 434], Wiley.

The hardest curve to model with Fourier series is the step function $wal_2(0, t)$ and errors lead to the Gibbs phenomenon. Similarly, the hardest curve to model with Walsh functions is the basic sin $2\pi t$ or cos $2\pi t$ curve. Still, we see that we can transform from one to another.

Many problems require Fourier transforms to be taken, but Fourier transforms require many multiplications which are slow and expensive to execute. On the other hand, the fast Walsh–Hadamard transform uses only additions and subtractions (addition of the complement) and so is extensively used to transform power sequence spectrum density, band compression of television signals or facsimile signals, or image processing.

Walsh functions are easy to extend to higher dimensions (and higher dimensional Hadamard matrices) to model surfaces in three and higher dimensions – Fourier series are more difficult to extend. Walsh–Hadamard transforms in higher dimensions are also effected using only additions (and subtractions).

Constructions for Hadamard matrices can be roughly classified into three types:

- multiplication theorems
- "plug-in" methods
- direct theorems

In 1976 Jennifer Seberry Wallis in her paper "On the existence of Hadamard matrices" [213] showed that "given any odd natural number q there exists a $t \approx 2\ \ln_2(q-3)$ so that there is an Hadamard matrix of order $2^t q$ (and hence for all orders $2^s q,\ s \geq t$)." This is represented graphically in Figure I.3. In recent work many far reaching asymptotic theorems by Craigen, Kharaghani, Holzmann and Ghaderpour, de Launey, Flannery, and Horadam [32, 46, 48, 79] have substantially improved Seberry's result.

In fact, as we show in our Appendix A, Hadamard matrices are known to exist, of order $2^2 q$, for most $q < 3000$ (we have results up to 40,000 which are similar). In many other cases Hadamard matrices of order $2^3 q$ or $2^4 q$ exist. A quick look shows most of the very difficult cases are for q (prime) $\equiv 3 \pmod 4$.

Hadamard's original construction for Hadamard matrices is a "multiplication theorem" as it uses the fact that the Kronecker product of Hadamard matrices of orders $2^a m$ and $2^b n$ is an Hadamard matrix of order $2^{a+b} mn$. Our graph shows we would like to reduce this power of 2. In his book, *Hadamard Matrices and their Applications*, Agaian [1] shows how to multiply these Hadamard matrices to get an Hadamard matrix of order $2^{a+b-1} mn$ (a result

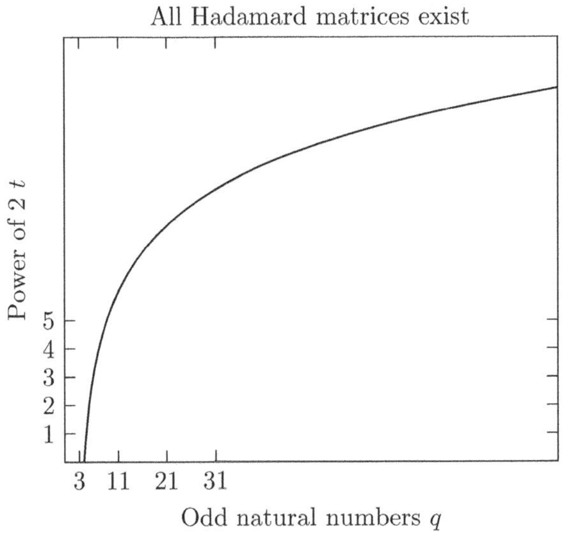

Figure I.3 Hadamard matrices of order $2^t q$. Source: Seberry and Yamada [166, p. 435], Wiley.

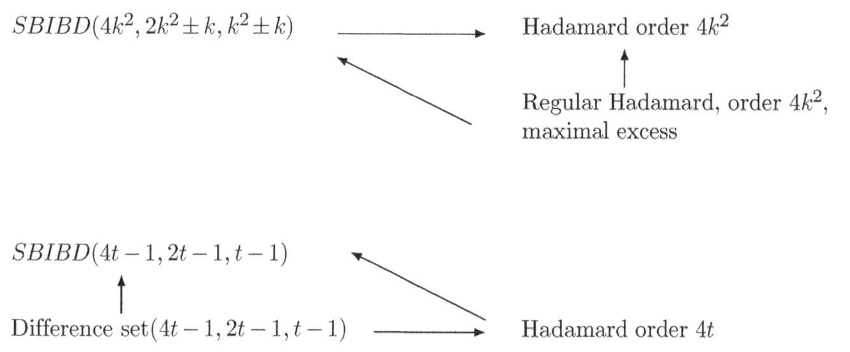

Figure I.4 Relationship between SBIBD and Hadamard matrices. Source: Seberry and Yamada [166, p. 436], Wiley.

which lowers the curve in our graph except for q, a prime). We will show how Craigen, Seberry, and Zhang [37] and de Launey [44] have further improved this result.

A paper by V. Scarpis [150] was the second addition to the story anticipating Paley's 1933 "direct" construction [144] which gives Hadamard matrices of order $\Pi_{i,j}(p_i + 1).(2(q_j + 1))$, p_i (prime power) $\equiv 3 \pmod 4$, q_j (prime power) $\equiv 1 \pmod 4$. This is extremely productive of Hadamard matrices but we note again the proliferation of powers of 2 as more products are taken.

Many people do not realize that in the same issue of the *Journal of Mathematics and Physics* as Paley's paper appears, J.A. Todd showed the equivalence of Hadamard matrices of order $4t$ and $SBIBD(4t-1, 2t-1, t-1)$. This family of SBIBD, its complementary family $SBIBD(4t-1, 2t, t)$, and the family $SBIBD(4s^2, 2s^2 \pm s, s^2 \pm s)$ are called *Hadamard designs*. The latter family satisfies the constraint $v = 4(k - \lambda)$ for $v = 4s^2$, $k = 2s^s \pm s$, and $\lambda = s^2 \pm s$ which appears in some constructions (e.g. Shrikhande [168]). Hadamard designs have the maximum number of one's in their incidence matrices of all incidence matrices of SBIBD(v, k, λ) (see Tsuzuku [187]) (Figure I.4).

In 1944 J. Williamson [223], who coined the name *Hadamard matrices*, first constructed what have come to be called *Williamson matrices*, or with a small set of conditions, *Williamson type matrices*. These matrices are used to replace the variables of a formally orthogonal matrix. We say Williamson type matrices are "plugged in" to the second matrix. Other matrices which can be "plugged in" to arrays of variables are called *suitable matrices*. Generally the arrays into which suitable matrices are plugged are *orthogonal designs* which have formally orthogonal rows (and columns) but may have variations such as Goethals–Seidel arrays, Wallis–Whiteman arrays, Spence arrays, generalized quaternion arrays, Agaian families, Kharaghani's methods, Balonin–Seberry array, regular s-sets of regular matrices which give new matrices.

This is an extremely prolific method of construction. We will discuss methods which give matrices to "plug in" and matrices to "plug into."

As a general rule if we want to check if an Hadamard matrix of a specific order $4pq$ exists we would first check if there are Williamson type matrices of order p, q, pq, then we would check if there were an $OD(4t; t, t, t, t)$ for $t = q, p, pq$. This failing, we would check the "direct" constructions. Finally we would use a "multiplication theorem." When we talk of "strength" of a construction this reflects a personal preference.

Before we proceed to more detail we consider diagrammatically some of the linkages between conjectures that will arise in this survey: the conjecture implied is "that the necessary conditions are sufficient for the existence of (say) Hadamard matrices" (see Figure I.5). (A weighing matrix, W, has elements $0, \pm 1$, is square and satisfies $WW^T = kI$.)

The hierarchy of conjectures for weighing matrices and ODs is more straightforward. Settling the OD conjecture in Table I.1 would settle the weighing matrix conjecture to its left.

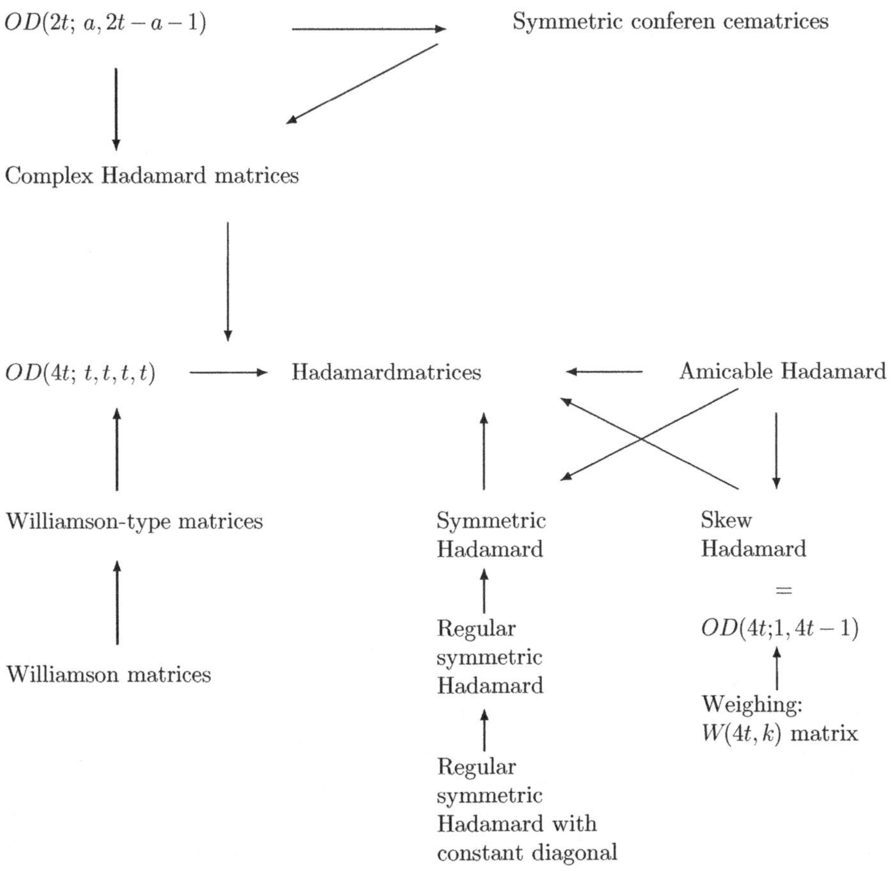

Figure I.5 Questions related to the Hadamard matrix conjecture. Source: Seberry and Yamada [166, p. 437], Wiley.

Table I.1 Weighing matrix and *OD* conjectures

	Matrices	**OD's**
Strongest	Skew-weighing	$OD(n; 1, k)$
	Weighing $W(n; k)$ n odd	
	Weighing $W(2n, k)$ n odd	$OD(2n; a, b)$ n odd
	Weighing $W(4n, k)$ n odd	$OD(4n; a, b, c, d)$ n odd
Weakest	$W(2^s n, k)$, n odd, $s \geq 3$	$OD(2^s n; u_1, u_2, ..., u_s)$ n odd

Source: Seberry and Yamada [166, table 1.1, p. 437], Wiley.

The original survey [166] emphasized those constructions, selected by us, which we believe showed the most promise toward solving the Hadamard conjecture.

Many diagrams, pictures, photos related to Hadamard matrices can be found on the website http://mathscinet.ru/catalogue/ created by Nickolay Balonin with a little help from D. Đoković, L.A. Mironovskiy, J. Seberry, and M.B. Sergeev (Figure I.6).

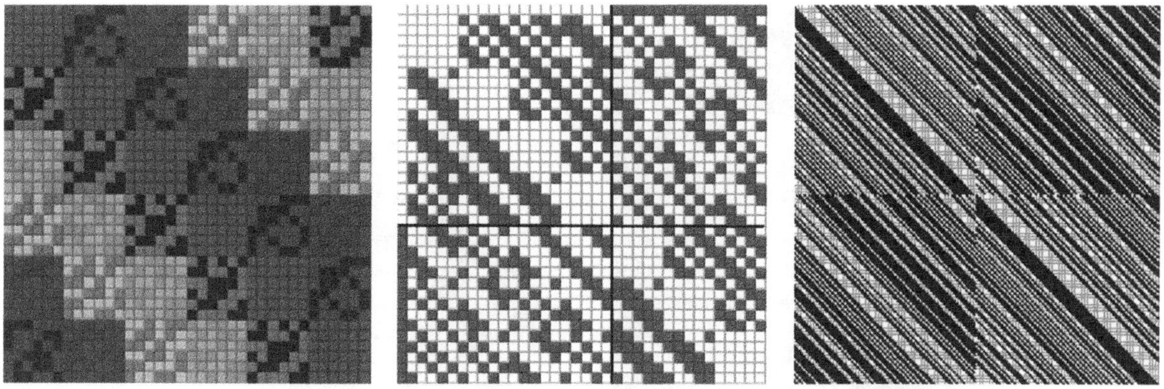

Figure I.6 Three examples of types of regular Hadamard matrices. Bush-type: cell $2m$; Vitrage: cell $2m-1$ ("Bush" with border); PerGol: cell $2m^2$. Source: http://mathscinet.ru/catalogue/regularhadamard/, Reproduced with permission of N. Balonin.

On the other hand, several arrays are related to representation theory of groups and group rings. A Williamson Hadamard matrix consists of Kronecker products of regular representation matrices of quaternion numbers and circulant matrices. A Williamson Hadamard matrix is developed to an Hadamard matrix of generalized quaternion type, which is a regular representation matrix of the element of the ring obtained from $\boldsymbol{Z}[G]$ where G is a semidirect product of a cyclic group by the generalized quaternion group. The Ito Hadamard matrix and the Williamson Hadamard matrix are recognized as Hadamard matrices of generalized quaternion type.

We extend a representation of the group ring $\boldsymbol{Z}[G]$ to an extended representation from $M_4(\boldsymbol{Z}[G])$ to $M_{4n}(\boldsymbol{Z})$ where n is the order of a group G. We obtain the Welch array of order t and the Ono–Sawade–Yamamoto array of order 9 from the ordered set $BHW(G)$. The Welch array uses the extended representation when G is a cyclic group of order t. We give an example of $t=5$. The Ono–Sawade–Yamamoto array of order 9 uses the extended representation of a direct product of two cyclic groups of order 3. The Ono–Sawade–Yamamoto array is a matrix of order 36 [166].

The signed group Hadamard matrix which Craigen introduced [30], produces an Hadamard matrix using monomial representation. Craigen obtained new results on the asymptotic existence of Hadamard matrices via signed group Hadamard matrices.

There are many notable pioneers for the number theory of this book. However, Carl Frederick Gauss, born 1777, and Carl Gustav Jacobi born 1804 are the most important for our work. Gauss is generally regarded as one of the greatest mathematicians of all time. He is referred to as *princeps mathematicorem*. In his PhD, when aged 21, he proved the fundamental theorem of algebra. Jacobi was the first to apply elliptic functions to number theory, for example proving Fermat's two-square theorem and Lagrange's four-square theorem, and similar results for six and eight squares. His other work in number theory continued the work of C. F. Gauss: new proofs of quadratic reciprocity and the introduction of the Jacobi symbol in 1837 (Figure I.7).

The direct construction of Hadamard matrices relates closely to number theory. Paley Hadamard matrix is a regular representation matrix of the element $\sum_{\alpha\in F}\psi(\alpha)\alpha \in \boldsymbol{Z}[F]$ where F is a finite field and ψ is a quadratic character of F. Then $\sum_{\alpha\in F}\eta(\alpha)\psi(\alpha)$ is a Gauss sum where η is a nontrivial additive character. The properties of a Paley core matrix are proved from the relations of Gauss sums.

Cyclotomic classes are an important tool for the construction not only of Hadamard matrices, but the other topics of combinatorics, for example, difference sets and supplementary difference sets. Cyclotomic numbers are written by Jacobi sums and conversely Jacobi sums are written by cyclotomic numbers [13]. Storer's book [179] is essential and has been a fundamental reference since first published. We will give the explicit cyclotomic numbers for $e = 2, 4, 8$ also covering the case $e = 8$ and f even not in Storer's book.

Figure I.7: Carl Frederick Gauss (Source: By Christian Albrecht Jensen, 1840 (Photo: A. Wittmann)) and Carl Gustav Jacobi. (Source: Public domain, https://commons.wikimedia.org).

An infinite family of Williamson Hadamard matrices by Whiteman [221] is constructed based on relative Gauss sums that is a ratio of Gauss sums [247]. This leads to infinite families of Hadamard matrices of generalized quaternion type constructed from relative Gauss sums.

Paley type I core matrix develops a Paley Hadamard difference set. The Stanton–Sprott difference set, which is a Paley Hadamard difference set, and its extension the Gordon–Mills–Welch difference set are associated with a relative Gauss sums.

Though it is difficult to determine the explicit values of Gauss sums and Jacobi sums, we can obtain them under the particular conditions. Index 2 Gauss sums and normalized relative Gauss sums are evaluated under the conditions of a group generated by a prime number [71, 253]. The construction theories of skew Hadamard difference sets by using index 2 Gauss sums [71, 72] and normalized relative Gauss sums [138] are new and widely applicable.

Construction of Hadamard matrices have been studied mainly over finite fields. The underlying algebraic structure in the research of combinatorics extends from finite fields to Galois rings. We obtain Menon Hadamard difference sets (which are equivalent to regular Hadamard matrices) over the Galois rings [242]. It should be noted that these difference sets have an embedding system, that is, a difference set is embedded in the ideal part of a difference set over a Galois ring of a higher characteristic. It gives one of many potential new directions to the research of Hadamard matrices.

The topics of combinatorics, relative difference sets [152], partial difference sets [146, 147], projective sets [18, 225], extended building sets [146, 147], $(q; x, y)$-partitions, [229] and planar functions [53] are also useful strong tools for the construction of Hadamard matrices. We will show how to use these tools for the construction of Hadamard matrices in this book.

1

Basic Definitions

1.1 Notations

Table 1.1 gives the notations which are used in this chapter.

1.2 Finite Fields

1.2.1 A Residue Class Ring

Let $\boldsymbol{Z}$ be the set of rational integers and (n) be a principal ideal generated by $n \in \boldsymbol{Z}$. The elements of the residue class ring $\boldsymbol{Z}/(n)$ of $\boldsymbol{Z}$ modulo (n) are

$$[0] = 0 + (n), [1] = 1 + (n), \ldots, [n-1] = n - 1 + (n).$$

Let p be a prime. $\boldsymbol{Z}/(p)$ is a field with respect to the operations

$$[a] + [b] = (a + (p)) + (b + (p)) = a + b + (p) = [a + b],$$
$$[a] \cdot [b] = (a + (p)) \cdot (b + (p)) = ab + (p) = [ab].$$

For a prime p, let $\boldsymbol{F}_p$ be the set $\{0, 1, \ldots, p-1\}$ of integers and $\varphi : \boldsymbol{Z}/(p) \longrightarrow \boldsymbol{F}_p$ be the mapping by $\varphi([a]) = a$ for $a = 0, 1, \ldots, p-1$. Since the mapping $\varphi : \boldsymbol{Z}/(p) \longrightarrow \boldsymbol{F}_p$ is an isomorphism, $\boldsymbol{F}_p$ is endowed with the field structure induced by φ.

Example 1.1: $\boldsymbol{F}_3 = \{0, 1, 2\}$. *The multiplication and the addition are described by the following:*

+	0	1	2
0	0	1	2
1	1	2	0
2	2	0	1

·	0	1	2
0	0	0	0
1	0	1	2
2	0	2	1

We have the following theorem on a simple extension. Let K be a field and $K[x]$ be a polynomial ring over K with an indeterminate x.

Theorem 1.1: *Let $\theta \in F$ be algebraic over K and let g be the minimal polynomial of θ of degree n over K. Then:*

i) $K(\theta)$ is isomorphic to $K[x]/(g)$.
ii) $[K(\theta) : K] = n$ and $\{1, \theta, \ldots, \theta^{n-1}\}$ is a basis of $K(\theta)$ over K.
iii) Every $\alpha \in K(\theta)$ is algebraic over K and its degree over K is a divisor of n.

Hadamard Matrices: Constructions using Number Theory and Algebra, First Edition. Jennifer Seberry and Mieko Yamada.

Table 1.1 Notations used in this chapter.

Notation	Description
$\boldsymbol{Z}$	The rational integer ring
(n)	A principal ideal generated by $n \in \boldsymbol{Z}$
$\boldsymbol{F}_p$, $GF(p)$	A finite field with p elements for a prime p
$\boldsymbol{F}_{p^n}$, $GF(p^n)$	A finite field with p^n elements for a prime power p^n
$\boldsymbol{F}_p(\theta)$	A simple algebraic extension by adjoining the element θ to a finite field $\boldsymbol{F}_p$
$\boldsymbol{F}_q^{\times}$, $GF(q)^{\times}$	The multiplicative group of a finite field $\boldsymbol{F}_q$ where q is a prime power
$\boldsymbol{F}_q^{+}$, $GF(q)^{+}$	The additive group of a finite field $\boldsymbol{F}_q$ where q is a prime power
$T_{F/K}$	The relative trace from F to K where K is a finite field and F is an extension of K
$N_{F/K}$	The relative norm from F to K where K is a finite field and F is an extension of K
T_F	The absolute trace of a finite field F
$[F : K]$	The degree of a finite extension F/K
$\hat{G}$	A character group of a finite group G
χ^0	A trivial character
$\boldsymbol{Z}[G]$	A group ring of a group G over $\boldsymbol{Z}$
$X^{\top}$	The transpose of the matrix X
$\overline{X}$	The complex conjugate matrix of X
$X^* = \overline{X^{\top}}$	The complex conjugate transpose matrix of X
I	The identity matrix
J	The matrix with every entry 1
$\boldsymbol{e}$	The vector with all entries 1
R	The back diagonal identity matrix
T	The basic circulant matrix
$M \times N$	The Kronecker product of the matrices M and N
Q	The Paley core
$N_A(s)$	The nonperiodic autocorrelation function
$P_A(s)$	The periodic autocorrelation function
N_X	Nonperiodic autocorrelation function of the family of the sequence X
P_X	Periodic autocorrelation function of the family of the sequence X
$\overline{x}$	Represents $-x$
A/B	The interleaving of two sequences $A = \{a_1, \dots, a_m\}$ and $B = \{b_1, \dots, b_m\}$, i.e. $\{a_1, b_1, a_2, b_2, \dots, a_m, b_m\}$
A^*	The sequence whose entries of sequence A are reversed
$\{A, B\}$	The adjoining of two sequences $\{a_1, \dots, a_m, b_1, \dots, b_m\}$
$Im(\alpha)$	The imaginary part of a complex number α
$Re(\alpha)$	The real part of a complex number α
$\sigma(H)$	The excess of an Hadamard matrix H
$\sigma(n)$	The maximum excess of Hadamard matrices of order n
S_i	The ith cyclotomic class

(*Continued*)

Table 1.1 *(Continued)*

Notation	Description
(i,j)	The cyclotomic number
$\delta_{i,j}$	Kronecker delta
$W(n,k)$	A weighing matrix of weight k and of order n
$OD(n; s_1, s_2, \dots, s_u)$	An orthogonal design of order n and type $(s_1, s_2, \dots s_u)$
$BH(4t; t, t, t, t)$	Baumert–Hall array of order $4t$
WL or $WL(t)$	Welch array of order t
$\overline{\chi}$	The conjugate character
ζ_n	A (primitive) nth root of unity
$-$	-1 when $-$ is an entry of a matrix
$n - \{v; k_1, k_2, \dots, k_n; \lambda\}$	Supplementary difference sets
$n - \{v; k; \lambda\}$	Supplementary difference sets with $k = k_1 = \cdots = k_n$
$4 - \left\{v; k_1, k_2, k_3, k_4; \sum_{i=1}^{4} k_i - v\right\}$	Hadamard supplementary difference sets
$(v, k, \lambda, \mu)PDS$	A partial difference set
$SRG(v, k, \lambda, \mu)$	A strongly regular graph
$w(H)$	The weight of an Hadamard matrix H
$(b, v, r, k, \lambda)BIBD$	A balanced incomplete block design
$(v, k, \lambda)SBIBD$	A symmetric balanced incomplete block design
$M * N$	Hadamard product of matrices M and N

This theorem says that every element of the simple algebraic extension $\boldsymbol{F}_p(\theta)$ of $\boldsymbol{F}_p$ is represented as a polynomial in θ, that is,

$$a_0 + a_1\theta + \cdots + a_{n-1}\theta^{n-1},\ a_i \in \boldsymbol{F}_p \quad \text{for} \quad 0 \le i \le n-1$$

where n is a degree of the minimal polynomial of θ over $\boldsymbol{F}_p$.

Definition 1.1: Let n be a positive integer. Assume that θ has the degree n over $\boldsymbol{F}_p$. $\boldsymbol{F}_p(\theta)$ is called a **finite field with p^n elements** or the **Galois field with p^n elements**. We denote a finite field with p^n elements by $\boldsymbol{F}_{p^n}$ or $GF(p^n)$.

Example 1.2:

1) *The polynomial $h(x) = x^2 + x + 1$ is irreducible over $\boldsymbol{F}_2$. Otherwise, $h(x) = (x - \alpha)(x - \beta)$ for some elements $\alpha, \beta \in \boldsymbol{F}_2$. We verify $h(0) \neq 0$ and $h(1) \neq 0$. Hence $h(x)$ is irreducible over $\boldsymbol{F}_2$.*
2) *The polynomial $f(x) = x^2 + x + 2$ is irreducible over $\boldsymbol{F}_3$. Let θ be a root of f. Then $\theta^2 + \theta + 2 = 0$. The other root of f is $2\theta + 2$, since $f(2\theta + 2) = (2\theta + 2)^2 + (2\theta + 2) + 2 = \theta^2 + \theta + 2 = 0$.*
   By Theorem 1.1,
   - $\boldsymbol{F}_3(\theta) \cong \boldsymbol{F}_3[x]/(f) = \{(f), 1 + (f), 2 + (f), x + (f), x + 1 + (f), x + 2 + (f), 2x + (f), 2x + 1 + (f), 2x + 2 + (f)\}$.
   - *$\boldsymbol{F}_3(\theta)$ consists of nine elements $0, 1, 2, \theta, \theta + 1, \theta + 2, 2\theta, 2\theta + 1, 2\theta + 2$. We see that $\boldsymbol{F}_3 = \{0, 1, 2\}$ is a sub-field of $\boldsymbol{F}_3(\theta)$.*

We consider $K = \boldsymbol{F}_p(\theta)$ and its simple extension $K(\delta) = (\boldsymbol{F}_p(\theta))(\delta)$ where m is the degree of δ. Then we see $K(\delta)$ is an algebraic extension and $\{1, \delta, \dots, \delta^{m-1}\}$ is a basis of $K(\delta)/K$ when replacing K by $\boldsymbol{F}_p(\theta)$ in Theorem 1.1.

1.2.2 Properties of Finite Fields

A finite field has the following properties.

Theorem 1.2:

i) A finite field has $q = p^n$ elements where p is a prime and n is a positive integer.
ii) For every prime p and every positive integer n, there exists a finite field with $q = p^n$ elements. Any finite field with p^n elements is isomorphic to the splitting field of $x^q - x$ over $\mathbf{F}_p$.
iii) For every finite field $\mathbf{F}_q$, the multiplicative group of nonzero elements of $\mathbf{F}_q$ is cyclic.

Corollary 1.1: *Let $\mathbf{F}_q$ be the finite field with $q = p^n$ elements. Then every subfield of $\mathbf{F}_q$ has order p^m, where m is a divisor of n. Conversely, if m is a divisor of n, then there is exactly one subfield of $\mathbf{F}_q$ with p^m elements.*

Example 1.3: *The subfields of the finite field $\mathbf{F}_{p^{30}}$ can be determined by listing all divisors of 30.*

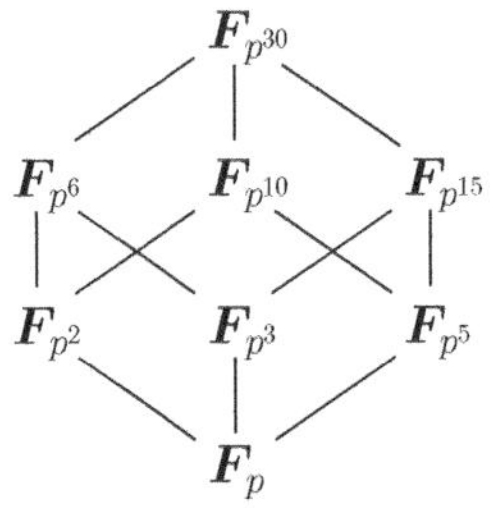

Definition 1.2: A generator of the multiplicative group $\mathbf{F}_q^\times$ of $\mathbf{F}_q$ is called a **primitive element** of $\mathbf{F}_q$.

Definition 1.3: A polynomial $f \in \mathbf{F}_q[x]$ of degree $m \geq 1$ is called a **primitive polynomial** over $\mathbf{F}_q$, if it is the minimal polynomial of a primitive element of $\mathbf{F}_q$.

Example 1.4:

a) In Example 1.2, the element θ is a generator of $\mathbf{F}_{3^2}^\times$, so is a primitive element of $\mathbf{F}_{3^2}$. Hence $x^2 + x + 2$ is a primitive polynomial.
b) The polynomial $x^2 + 1$ is irreducible over $\mathbf{F}_3$. Let η be a root of $x^2 + 1$. Then $\eta^2 = 2$, and $\eta^4 = 1$. So η is not a primitive element of $\mathbf{F}_{3^2}$. Hence the polynomial $x^2 + 1$ is not a primitive polynomial.

1.2.3 Traces and Norms

Let q be a prime power and $K = \mathbf{F}_q$ and $F = \mathbf{F}_{q^t}$. For any element $\alpha \in F$, the automorphism f of F is given by

$$\alpha^{\mathrm{f}} = \alpha^q,$$

which is called the **Frobenius automorphism**.

Definition 1.4: For $\alpha \in F$, the **relative trace** of α from F to K is defined by

$$T_{F/K}(\alpha) = \alpha + \alpha^{\mathrm{f}} + \cdots + \alpha^{\mathrm{f}^{t-1}} = \alpha + \alpha^q + \cdots + \alpha^{q^{t-1}}.$$

If K is the prime subfield of F, then $T_{F/K}(\alpha)$ is called the **absolute trace** of α and simply denoted by $T_F(\alpha)$.

From $\alpha^{q^t} = \alpha$, we have $T_{F/K}(\alpha^q) = \alpha^q + \alpha^{q^2} + \cdots + \alpha^{q^{t-1}} = T_{F/K}(\alpha)$, that is $T_{F/K}(\alpha)$ is fixed by the Frobenius automorphism. Hence $T_{F/K}(\alpha)$ is an element of K.

Example 1.5: *Let θ be a root of the polynomial x^2+x+1 over $\mathbf{F}_2$ and θ is a primitive element. Put $T = T_{\mathbf{F}_2(\theta)/\mathbf{F}_2}$. Then we have*

$$T(0) = 0+0 = 0,\ T(1) = 1+1 = 0,\ T(\theta) = \theta+\theta^2 = 1,\ T(\theta^2) = \theta^2+\theta = 1.$$

Definition 1.5: For $\alpha \in \mathbf{F}_{q^t}$, the elements $\alpha, \alpha^q, \alpha^{q^2}, \dots, \alpha^{q^{t-1}}$ are called the **conjugates of** α.

Theorem 1.3: *Let $q = p^s$ be a prime power. Let $K = \mathbf{F}_q$ and $F = \mathbf{F}_{q^t}$. Then the trace function $T_{F/K}$ satisfies the following properties:*

i) $T_{F/K}(\alpha+\beta) = T_{F/K}(\alpha) + T_{F/K}(\beta)$, $\alpha, \beta \in F$,
ii) $T_{F/K}(c\alpha) = cT_{F/K}(\alpha)$, $c \in K, \alpha \in F$,
iii) $T_{F/K}$ *is a linear transformation from F onto K,*
iv) $T_{F/K}(a) = ta$, $a \in K$.

Proof:

i) From $(\alpha+\beta)^p = \alpha^p + \beta^p$, $(\alpha+\beta)^q = \alpha^q + \beta^q$. Then

$$\begin{aligned}T_{F/K}(\alpha+\beta) &= \alpha+\beta+(\alpha+\beta)^q+\cdots+(\alpha+\beta)^{q^{t-1}} \\ &= \alpha+\beta+\alpha^q+\beta^q+\cdots+\alpha^{q^{t-1}}+\beta^{q^{t-1}} \\ &= T_{F/K}(\alpha)+T_{F/K}(\beta).\end{aligned}$$

ii) From $c^q = c$, we have $c^{q^j} = c$ for all $j \ge 0$.

$$\begin{aligned}T_{F/K}(c\alpha) &= c\alpha + c^q\alpha^q + \cdots + c^{q^{t-1}}\alpha^{q^{t-1}} \\ &= c\alpha + c\alpha^q + \cdots + c\alpha^{q^{t-1}} \\ &= cT_{F/K}(\alpha).\end{aligned}$$

iii) The properties (i), (ii) show that $T_{F/K}$ is a linear transformation from F into K. It suffices to show $T_{F/K}$ is onto. $T_{F/K}(\alpha) = 0$ if and only if α is a root of the polynomial $x^{q^{t-1}} + \cdots + x^q + x \in K[x]$ in F. Since the degree of this polynomial is q^{t-1}, it can have at most q^{t-1} roots in F. From $|F| = q^t$ and the homomorphism theorem, we see the trace function $T_{F/K}$ is onto.
iv) This follows immediately from the definition of a trace function.

□

Theorem 1.4: *Let $K = \mathbf{F}_q$ be a finite field, and let F be a finite extension of K and L be a finite extension of F. Then*

$$T_{L/K}(\alpha) = T_{F/K}\left(T_{L/F}(\alpha)\right).$$

Proof: Denote the degree of a finite extension F/K by $[F : K]$. Let $[F : K] = t$, $[L : F] = m$, so that $[L : K] = tm$. Then for $\alpha \in L$, we have

$$\begin{aligned}T_{F/K}\left(T_{L/F}(\alpha)\right) &= \sum_{i=0}^{t-1} T_{L/F}(\alpha)^{q^i} = \sum_{i=0}^{t-1}\left(\sum_{j=0}^{m-1}(\alpha^{q^{jt}})\right)^{q^i} \\ &= \sum_{i=0}^{t-1}\sum_{j=0}^{m-1}\alpha^{q^{jt+i}} = \sum_{k=0}^{tm-1}\alpha^{q^k} = T_{L/K}(\alpha).\end{aligned}$$

□

An another important function from F to K is a norm function.

Definition 1.6: Let $K = \boldsymbol{F}_q$ and $F = \boldsymbol{F}_{q^t}$. For $\alpha \in F$, the **norm** $N_{F/K}$ of α is defined by

$$N_{F/K}(\alpha) = \prod_{j=0}^{t-1} \alpha^{q^j} = \alpha \cdot \alpha^q \cdots \alpha^{q^{t-1}} = \alpha^{(q^t-1)/(q-1)}.$$

We have $N_{F/K}(\alpha^q) = N_{F/K}(\alpha)$ from $\alpha^{q^t} = \alpha$ and the definition of a norm function. Thus $N_{F/K}(\alpha)$ is fixed by the Frobenius automorphism and $N_{F/K}(\alpha) \in K$.

Theorem 1.5: *Let $K = \boldsymbol{F}_q$ and $F = \boldsymbol{F}_{q^t}$. Then the norm function $N_{F/K}$ satisfies the following properties:*

i) $N_{F/K}(\alpha\beta) = N_{F/K}(\alpha)N_{F/K}(\beta),\ \alpha, \beta \in F$,
ii) $N_{F/K}(a) = a^t,\ a \in K$,
iii) $N_{F/K}$ *is an onto mapping from* $F^\times$ *onto* $K^\times$ *where* $F^\times$ *and* $K^\times$ *are multiplicative groups of F and K, respectively.*

Proof:

i) Straightforward calculation.
ii) This follows from the definition of the norm and $a^q = a$.
iii) Since $N_{F/K}(\alpha) = 0$ if and only if $\alpha = 0$, $N_{F/K}$ is a group homomorphism from $F^\times$ into $K^\times$. The elements of the kernel of $N_{F/K}$ are the roots of the polynomial $x^{(q^t-1)/(q-1)} - 1$. Therefore the order d of the kernel satisfies $d \leq (q^t - 1)/(q - 1)$. On the other hand, the image of $N_{F/K}$ has the order $(q^t - 1)/d$ from the homomorphism theorem, so that $(q^t - 1)/d \leq q - 1$. It follows that $d = (q^t - 1)/(q - 1)$. Hence $N_{F/K}$ maps $F^\times$ onto $K^\times$.

□

Example 1.6: *Let $K = \boldsymbol{F}_3$ and θ be a root of a primitive polynomial $x^2 + x + 2$. The norms of all the elements of $F = \boldsymbol{F}_{3^2}$ are given as follows;*

$$\begin{aligned}
N_{F/K}(0) &= 0,\\
N_{F/K}(1) &= 1,\\
N_{F/K}(\theta) = N_{F/K}(\theta^3) &= 2,\\
N_{F/K}(\theta^2) = N_{F/K}(\theta^6) &= 1,\\
N_{F/K}(\theta^4) = N_{F/K}(2) &= 1,\\
N_{F/K}(\theta^5) = N_{F/K}(\theta^7) &= 2.
\end{aligned}$$

Theorem 1.6: *Let K be a finite field and F be a finite extension of K, and L be a finite extension of F. Then*

$$N_{L/K}(\alpha) = N_{F/K}\left(N_{L/F}(\alpha)\right)$$

for all $\alpha \in L$.

Proof: Let $[F : K] = t, [L : F] = m$, so that $[L : K] = tm$. For $\alpha \in L$, we have

$$\begin{aligned}
N_{F/K}\left(N_{L/F}(\alpha)\right) &= N_{F/K}\left(\alpha^{(q^{tm}-1)/(q^t-1)}\right)\\
&= \left(\alpha^{(q^{tm}-1)/(q^t-1)}\right)^{(q^t-1)/(q-1)}\\
&= \alpha^{(q^{tm}-1)/(q-1)} = N_{L/K}(\alpha).
\end{aligned}$$

□

1.2.4 Characters of Finite Fields

Let $\boldsymbol{R}$ be the real number field. A complex conjugate of a complex number $\alpha = a + bi,\ a, b \in \boldsymbol{R}, i = \sqrt{-1}$, is the number $\overline{\alpha} = a - bi$. We denote the real part a by $Re(\alpha)$ and the imaginary part b by $Im(\alpha)$.

Let G be a finite abelian group of order v. A **character** χ of G is a homomorphism from G to the complex number field $\boldsymbol{C}$. We see $\chi(1) = 1$ for the identity 1 of G as $\chi(1)\chi(1) = \chi(1)$. For $g \in G$,

$$\chi(g)^{|G|} = \chi\left(g^{|G|}\right) = \chi(1) = 1,$$

so that the value of χ is a $|G|$th root of unity. The character χ^0 with $\chi^0(g) = 1$ for all $g \in G$ is called the **trivial character**. The **conjugate character** $\overline{\chi}$ of χ is defined by $\overline{\chi}(g) = \overline{\chi(g)}$ for $g \in G$.

For characters χ_1 and χ_2, we define the product $\chi_1\chi_2$ of these characters by

$$\chi_1\chi_2(g) = \chi_1(g)\chi_2(g)$$

for all $g \in G$. Then the characters of G form a group under this multiplication.

Theorem 1.7: *Let χ be a character of G and $\hat{G}$ be a character group of G. Then*

$$\sum_{g\in G} \chi(g) = \begin{cases} 0 & \text{if } \chi \neq \chi^0, \\ |G| & \text{if } \chi = \chi^0. \end{cases}$$

Denote the identity of G by 1_G.

$$\sum_{\chi\in \hat{G}} \chi(g) = \begin{cases} 0 & \text{if } g \neq 1_G, \\ |\hat{G}| & \text{if } g = 1_G. \end{cases}$$

Proof: Suppose χ is a nontrivial character. Further suppose there exists an element a such that $\chi(a) \neq 1$. Then

$$\chi(a)\sum_{g\in G} \chi(g) = \sum_{g\in G} \chi(ag) = \sum_{g\in G} \chi(g).$$

Thus we have

$$(\chi(a) - 1)\sum_{g\in G} \chi(g) = 0,$$

so that $\sum_{g\in G} \chi(g) = 0$. If χ is a trivial character, then $\sum_{g\in G} \chi(g) = \sum_{g\in G} 1 = |G|$. Suppose $g \neq 1_G$. Further suppose there exists a character χ_1 such that $\chi_1(g) \neq 1$. Then

$$\chi_1(g)\sum_{\chi\in \hat{G}} \chi(g) = \sum_{\chi\in \hat{G}} \chi_1\chi(g) = \sum_{\chi\in \hat{G}} \chi(g),$$

so that $\sum_{\chi\in\hat{G}} \chi(g) = 0$. If $g = 1_G$, then $\sum_{\chi\in\hat{G}} \chi(g) = \sum_{\chi\in\hat{G}} 1 = |\hat{G}|$. □

Lemma 1.1: *Let q be a power of a prime p and $K = \boldsymbol{F}_q$. An additive character η_β of K for each $\beta \in K$ is given by*

$$\eta_\beta(\alpha) = \zeta_p^{T_K(\alpha\beta)}$$

where ζ_p is a primitive p-th root of unity and T_K is the absolute trace function from $\boldsymbol{F}_q$ to $\boldsymbol{F}_p$.

The **multiplicative character** χ of K is given by

$$\chi(g) = \zeta_{q-1}^s$$

where ζ_{q-1} is a primitive $(q-1)$st root of unity and s is a some integer. We extend a multiplicative character to a finite field by letting $\chi(0) = 0$. If $s = 0$, then the character is the trivial character and denoted by χ^0.

In particular, the quadratic character is frequently used.

Definition 1.7: We define the map $\chi : K^{\times} \mapsto \{1, -1\}$ by

$$\chi(a) = \begin{cases} 1 & \text{if } a \text{ is a square in } F^{\times}, \\ -1 & \text{if } a \text{ is not a square in } F^{\times}, \\ 0 & \text{if } a = 0. \end{cases}$$

The map χ is a **quadratic character** of $K^{\times}$.

Example 1.7: *Put $p = 2, q = p^2 = 2^2$, $K = \{0, 1, g, g^2\}$. Let ζ_3 be a primitive cubic root of unity. From Example 1.5, we have*

$$\begin{aligned} &\chi(g) = \zeta_3^s \quad \text{where } s \text{ is a positive integer}, \\ &\eta_g(0) = 0,\ \eta_g(1) = (-1)^{T_K(g\cdot 1)} = -1, \\ &\eta_g(g) = (-1)^{T_K(g\cdot g)} = -1, \\ &\eta_g(g^2) = (-1)^{T_K(g\cdot g^2)} = 1. \end{aligned}$$

Theorem 1.8: *The group of multiplicative characters of $\boldsymbol{F}_q$ is a cyclic group of order $q - 1$. The order of the group of additive characters of $\boldsymbol{F}_q$ is q.*

Theorem 1.9: *Let χ_1, χ_2 be multiplicative characters and $\widehat{K^{\times}}$ be the group of multiplicative characters of $K = \boldsymbol{F}_q$. Then*

$$\sum_{\alpha\in K} \chi_1(\alpha)\overline{\chi}_2(\alpha) = \begin{cases} q-1 & \text{if } \chi_1 = \chi_2, \\ 0 & \text{otherwise}, \end{cases}$$

$$\sum_{\chi_1\in\hat{K}} \chi_1(\alpha)\overline{\chi}_1(\beta) = \begin{cases} q-1 & \text{if } \alpha = \beta, \\ 0 & \text{otherwise}. \end{cases}$$

Let η_β, η_γ be additive characters and $\widehat{K^{+}}$ be the group of additive characters of K.

$$\sum_{\alpha\in K} \eta_\beta(\alpha)\overline{\eta}_\gamma(\alpha) = \begin{cases} q & \text{if } \beta = \gamma, \\ 0 & \text{otherwise}, \end{cases}$$

$$\sum_{\eta_1\in\hat{K}} \eta_\beta(\alpha)\overline{\eta}_\beta(\delta) = \begin{cases} q & \text{if } \alpha = \delta, \\ 0 & \text{otherwise}. \end{cases}$$

Proof: We transform the equations as

$$\sum_{\alpha\in K} \chi_1(\alpha)\overline{\chi}_2(\alpha) = \sum_{\alpha\in K} \chi_1\overline{\chi}_2(\alpha),$$

$$\sum_{\chi_1\in\hat{K}} \chi_1(\alpha)\overline{\chi}_1(\beta) = \sum_{\alpha\in\hat{K}} \chi_1(\alpha\beta^{-1}).$$

The assertions follow from Theorem 1.7. We obtain the equations for the case of additive characters similarly. □

1.3 Group Rings and Their Characters

Let $\boldsymbol{Z}$ be the rational integer ring and G be an additive abelian group of order v. We consider the set

$$\boldsymbol{Z}[G] = \left\{ \sum_{g\in G} a_g g \mid a_g \in \boldsymbol{Z} \right\}.$$

If we define the addition and multiplication of $\boldsymbol{Z}[G]$ as follows,

$$\sum_{g\in G} a_g g + \sum_{g\in G} b_g g = \sum_{g\in G} (a_g + b_g) g,$$

$$\sum_{g\in G} a_g g \cdot \sum_{g\in G} b_g g = \sum_{h\in G} \sum_{g+g'=h} a_g b_{g'} h,$$

then $\boldsymbol{Z}[G]$ forms a ring, which is called a **group ring** of G over $\boldsymbol{Z}$.

Let χ be a character of G. We can extend it to the group ring $\boldsymbol{Z}[G]$. If $\alpha = \sum_{g\in G} a_g g, \in \boldsymbol{Z}[G]$, we define a character χ of $\boldsymbol{Z}[G]$ by

$$\chi(\alpha) = \sum_{g\in G} a_g \chi(g).$$

The following theorem is important.

Theorem 1.10 *(Inversion formula)*: *Let $\alpha = \sum_{g\in G} a_g g \in \boldsymbol{Z}[G]$ and χ be a character of G. Denote the character group of G by $\hat{G}$. Then*

$$a_g = \frac{1}{v} \sum_{\chi\in\hat{G}} \chi(\alpha)\chi(-g).$$

If $\chi(\alpha) = 0$, then $\alpha = 0$. Thus if $\alpha, \beta \in \boldsymbol{Z}[G]$ satisfy $\chi(\alpha) = \chi(\beta)$ for all characters of G, then $\alpha = \beta$.

Proof: From the orthogonal relation in Theorem 1.7,

$$\sum_{\chi\in\hat{G}} \chi(\alpha)\chi(-g) = \sum_{\chi\in\hat{G}} \sum_{h\in G} a_h \chi(h)\chi(-g)$$
$$= \sum_{h\in G} a_h \sum_{\chi\in\hat{G}} \chi(h-g) = a_g v.$$

Thus, if $\chi(\alpha) = 0$, then $a_h = 0$ for all $h \in G$, so that $\alpha = 0$. Let $\beta = \sum_{g\in G} b_g g$. If α, β satisfy $\chi(\alpha) = \chi(\beta)$ for all characters of G, then $a_g = b_g$ for any $g \in G$. Hence $\alpha = \beta$. □

1.4 Type 1 and Type 2 Matrices

When G is an abelian group, we have two types of incidence matrices.

Definition 1.8: Let G be an additive abelian group, which are ordered in some convenient way and the ordering fixed. Let $X = \{x_i\}$ be a subset of $G, X \cap \{0\} = \emptyset$. We define the functions ψ and ϕ from G into a commutative ring such that

$$\psi(x) = \begin{cases} a & x \in X \\ b & x = 0 \\ c & x \notin X \bigcup \{0\} \end{cases}, \qquad \phi(x) = \begin{cases} d & x \in X \\ e & x = 0 \\ f & x \notin X \bigcup \{0\} \end{cases}.$$

Two matrices $M = (m_{ij})$ and $N = (n_{ij})$ defined by

$$m_{ij} = \psi(z_j - z_i) \text{ and } n_{ij} = \phi(z_j + z_i),$$

will be called **type 1 matrix of ψ on X** and **type 2 matrix of ϕ on X**, respectively.

If we put $a = 1$ and $b = c = 0$ or $d = 1$ and $e = f = 0$, then M and N will be called **type 1** $(1, 0)$ **incidence matrix** and **type 2** $(1, 0)$ **incidence matrix**, respectively. If we put $a = 1$ and $b = c = -1$ or $d = 1$ and $e = f = -1$, then M and N will be called **type 1** $(1, -1)$ **incidence matrix** and **type 2** $(1, -1)$ **incidence matrix**, respectively.

Example 1.8: *Consider the additive group $GF(3^2)$, which has elements*

$$0, 1, 2, x, x+1, x+2, 2x, 2x+1, 2x+2.$$

We assign g_i, $i = 0, \dots, 8$ in order to the above elements of $GF(3^2)$.

Define the set

$$\begin{aligned} X &= \{y | y = z^2 \text{ for some } z \in GF(3^2)\} \\ &= \{x+1, 2, 2x+2, 1\} \end{aligned}$$

using the primitive polynomial $x^2 + 2x + 2$. Now a type 1 matrix of ψ on X, $A = (a_{ij})$, is determined by a function of the type

$$\psi\,(g_j - g_i) \text{ where } \psi(x) = \begin{cases} 0 & x = 0 \\ 1 & x \in X \\ -1 & \text{otherwise.} \end{cases}$$

Then the type 1 matrix of ψ on X is

$$A = \begin{bmatrix} 0 & 1 & 1 & -1 & 1 & -1 & -1 & -1 & 1 \\ 1 & 0 & 1 & -1 & -1 & 1 & 1 & -1 & -1 \\ 1 & 1 & 0 & 1 & -1 & -1 & -1 & 1 & -1 \\ -1 & -1 & 1 & 0 & 1 & 1 & -1 & 1 & -1 \\ 1 & -1 & -1 & 1 & 0 & 1 & -1 & -1 & 1 \\ -1 & 1 & -1 & 1 & 1 & 0 & 1 & -1 & -1 \\ -1 & 1 & -1 & -1 & -1 & 1 & 0 & 1 & 1 \\ -1 & -1 & 1 & 1 & -1 & -1 & 1 & 0 & 1 \\ 1 & -1 & -1 & -1 & 1 & -1 & 1 & 1 & 0 \end{bmatrix}$$

Then keeping the same ordering as above, the type 2 matrix of ϕ on X is

$$B = \begin{bmatrix} 0 & 1 & 1 & -1 & 1 & -1 & -1 & -1 & 1 \\ 1 & 1 & 0 & 1 & -1 & -1 & -1 & 1 & -1 \\ 1 & 0 & 1 & -1 & -1 & 1 & 1 & -1 & -1 \\ -1 & 1 & -1 & -1 & -1 & 1 & 0 & 1 & 1 \\ 1 & -1 & -1 & -1 & 1 & -1 & 1 & 1 & 0 \\ -1 & -1 & 1 & 1 & -1 & -1 & 1 & 0 & 1 \\ -1 & -1 & 1 & 0 & 1 & 1 & -1 & 1 & -1 \\ -1 & 1 & -1 & 1 & 1 & 0 & 1 & -1 & -1 \\ 1 & -1 & -1 & 1 & 0 & 1 & -1 & -1 & 1 \end{bmatrix}$$

Lemma 1.2: *Suppose G is an additive abelian group of order v with elements $z_1, z_2, \dots, z_v$. Say ϕ and ψ are maps from G to a commutative ring R. Define*

$$\begin{aligned} A &= (a_{ij}), \quad a_{ij} = \phi(z_j - z_i), \\ B &= (b_{ij}), \quad b_{ij} = \psi(z_j - z_i), \\ C &= (c_{ij}), \quad c_{ij} = \mu(z_j + z_i). \end{aligned}$$

Then (independently of the ordering of elements of G except that it is fixed)

i) $C^{\top} = C$,
ii) $AB = BA$,
iii) $AC^{\top} = CA^{\top}$

where $X^{\top}$ *is the transpose matrix of the matrix* $X \in \{A, C\}$.

Proof:

i) This is trivial.
ii)

$$(AB)_{i,j} = \sum_{g \in G} \phi(g - z_i)\psi(z_j - g),$$

putting $h = z_i + z_j - g$,

$$= \sum_{h \in G} \phi(z_j - h)\psi(h - z_i)$$
$$= (BA)_{i,j}.$$

iii)

$$(AC^{\top})_{i,j} = \sum_{g \in G} \phi(g - z_i)\mu(z_j + g)$$

putting $h = z_j + g - z_i$,

$$= \sum_{h \in G} \phi(h - z_j)\mu(z_i + h)$$
$$= (CA^{\top})_{i,j}.$$

□

Lemma 1.3: *If X is a type i,* $i = 1, 2$ *matrix, then so is* $X^{\top}$.

Lemma 1.4: *(i) Let X and Y be type 2 incidence matrices generated by subsets A and B of an additive abelian group G. Suppose, further, that*

$$a \in A \Rightarrow -a \in A \quad \text{and} \quad b \in B \Rightarrow -b \in B.$$

Then,

$$XY = YX \quad \text{and} \quad XY^{\top} = YX^{\top}.$$

(ii) The same result holds if X and Y are type 1.

Proof:

i) Since X and Y are symmetric, we only have to prove that $XY^{\top} = YX^{\top}$. Suppose $X = (x_{ij})$ and $Y = (y_{ij})$ are defined by

$$x_{ij} = \phi(z_i + z_j), \qquad y_{ij} = \psi(z_i + z_j),$$

where $z_1, z_2, \ldots$ are the elements of G. Then

$$(XY^{\top})_{ij} = \sum_k \phi(z_i + z_k)\psi(z_k + z_j),$$

since $a \in A \Rightarrow -a \in A$

$$= \sum_{k} \phi(-z_i - z_k)\psi(z_k + z_j),$$

where $z_\ell = -z_k - z_i - z_j$

$$= \sum_{\ell} \phi(z_j + z_\ell)\psi(-z_\ell - z_i),$$

since $b \in B \Rightarrow -b \in B$

$$= \sum_{\ell} \phi(z_j + z_\ell)\psi(z_\ell + z_i),$$
$$= (YX^\top)_{ij}.$$

ii) The additional hypotheses on A and B force X and Y to be symmetric. The proof, then, is similar to (i), and we leave it to the reader as an easy exercise. □

Lemma 1.5: *Let $R = (r_{ij})$ be the permutation matrix of order n, defined on an additive abelian group $G = \{g_i\}$ of order n by*

$$r_{\ell,j} = \begin{cases} 1 & \text{if } g_\ell + g_j = 0, \\ 0 & \text{otherwise.} \end{cases}$$

Let M by a type 1 matrix of a subset $X = \{x_i\}$ of G. Then MR is a type 2 matrix.

Proof: Let $M = (m_{ij})$ be defined by $m_{ij} = \psi(g_j - g_i)$ where ψ maps G into a commutative ring. Then

$$(MR)_{ij} = \sum_{k} \psi(g_k - g_i)r_{kj},$$

since $r_{kj} = 1$ if $g_k + g_j = 0$,

$$\psi(g_k - g_i) = \psi(-g_j - g_i) = \psi(-1)\psi(g_j + g_i)$$

which is a type 2 matrix. □

For a cyclic group, we give definitions of some special matrices.

Definition 1.9: We let a matrix

$$T = \begin{bmatrix} 0\ 1\ 0 \cdots 0 \\ 0\ 0\ 1 \cdots 0 \\ \ddots \\ 0\ 0\ 0 \cdots 1 \\ 1\ 0\ 0 \cdots 0 \end{bmatrix}$$

of order n, which is called the **basic circulant matrix** or the **shift matrix**. We use T for the basic circulant matrix.

We see $T^\top = T^{-1}$ where T^{-1} is an inverse of a matrix T.

We define a **circulant matrix** $C = (c_{i,j})$ of order n by

$$c_{i,j} = c_{1,j-i+1}, \quad 1 \le i, j \le n.$$

Any circulant matrix with the first row $x_1, x_2, \dots, x_n$ is

$$\begin{bmatrix} x_1 & x_2 & x_3 & \cdots & x_n \\ x_n & x_1 & x_2 & \cdots & x_{n-1} \\ x_{n-1} & x_n & x_1 & \cdots & x_{n-2} \\ \vdots & \vdots & \vdots & \ddots & \vdots \\ x_2 & x_3 & x_4 & \cdots & x_1 \end{bmatrix}$$

and can be written as the polynomial of T,

$$x_1 I + x_2 T + x_3 T^2 + \cdots x_n T^{n-1}.$$

Definition 1.10: We let

$$R = \begin{bmatrix} 0 & 0 & \dots & 0 & 1 \\ 0 & 0 & \dots & 1 & 0 \\ \vdots & \vdots & \dots & 0 & \vdots \\ 0 & 1 & & \vdots & \vdots \\ 1 & 0 & \dots & 0 & 0 \end{bmatrix}$$

of order n which is called the **back diagonal identity matrix**. Then $R^2 = I$.

Definition 1.11: We define a **back circulant matrix** $B = (b_{i,j})$ of order n by

$$b_{i,j} = b_{1,i+j-1}, \quad 1 \le i,j \le n.$$

A back circulant matrix with the first row $y_1, y_2, \dots, y_n$ is

$$\begin{bmatrix} y_1 & y_2 & y_3 & \cdots & y_n \\ y_2 & y_3 & y_4 & \cdots & y_1 \\ y_3 & y_4 & y_5 & \cdots & y_2 \\ \vdots & \vdots & & \vdots & \vdots \\ y_n & y_1 & y_2 & \cdots & y_{n-1} \end{bmatrix}$$

and can be written as

$$y_n R + y_{n-1} TR + y_{n-2} T^2 R + \cdots + y_2 T^{n-2} R + y_1 T^{n-1} R.$$

Example 1.9: *$n = 4$. The basic circulant matrix and the back diagonal identity matrix B are given by*

$$T = \begin{bmatrix} 0 & 1 & 0 & 0 \\ 0 & 0 & 1 & 0 \\ 0 & 0 & 0 & 1 \\ 1 & 0 & 0 & 0 \end{bmatrix}, \quad R = \begin{bmatrix} 0 & 0 & 0 & 1 \\ 0 & 0 & 1 & 0 \\ 0 & 1 & 0 & 0 \\ 1 & 0 & 0 & 0 \end{bmatrix}.$$

The circulant matrix A and the back circulant matrix B can be written as follows:

$$A = \begin{bmatrix} 1 & 2 & 3 & 4 \\ 4 & 1 & 2 & 3 \\ 3 & 4 & 1 & 2 \\ 2 & 3 & 4 & 1 \end{bmatrix} = I + 2T + 3T^2 + 4T^3,$$

$$B = \begin{bmatrix} 1 & 2 & 3 & 4 \\ 2 & 3 & 4 & 1 \\ 3 & 4 & 1 & 2 \\ 4 & 1 & 2 & 3 \end{bmatrix} = 4R + 3TR + 2T^2R + T^3R.$$

1.5 Hadamard Matrices

1.5.1 Definition and Properties of an Hadamard Matrix

Let $X = (x_{i,j})$ be a complex matrix of order n with entries in the closed unit disk $|x_{i,j}| \leq 1$. In 1893, J. Hadamard [89] discovered that

$$|\det X| \leq n^{\frac{n}{2}}. \tag{1.1}$$

This is known as the **Hadamard determinant inequality**. The equality holds when $|x_{i,j}| = 1$ and the row vectors are pairwise orthogonal. That is $X\overline{X^{\top}} = nI$, where I is a identity matrix of order n. When the matrix X is real with maximal determinant, the entries are $+1$ or -1 and row and column vectors are orthogonal.

Definition 1.12: An **Hadamard matrix** is a square $(1, -1)$-matrix whose row vectors are orthogonal.

Lemma 1.6: *Let H be an Hadamard matrix of order h. Then:*

i) $HH^{\top} = hI_h$;
ii) $|\det H| = h^{\frac{1}{2}h}$;
iii) $HH^{\top} = H^{\top}H$;

Proof: From the definition of an Hadamard matrix, we have (*i*) and (*ii*). Since $HH^{\top} = hI$, $H^{\top} = hH^{-1}$, and $H^{\top}H = hI = HH^{\top}$. □

Definition 1.13: An Hadamard matrix H is called a **symmetric Hadamard matrix** if H is a symmetric matrix. An Hadamard matrix $M + I$ is called a **skew Hadamard matrix** if M satisfies $M^{\top} = -M$.

Example 1.10:

$$\begin{bmatrix} 1 & 1 \\ 1 & -1 \end{bmatrix}, \quad \begin{bmatrix} -1 & 1 & 1 & 1 \\ 1 & -1 & 1 & 1 \\ 1 & 1 & -1 & 1 \\ 1 & 1 & 1 & -1 \end{bmatrix}$$

are symmetric Hadamard matrices.

$$\begin{bmatrix} 1 & 1 \\ -1 & 1 \end{bmatrix} \quad \text{and} \quad \begin{bmatrix} 1 & 1 & 1 & 1 \\ -1 & -1 & 1 & 1 \\ -1 & 1 & -1 & 1 \\ -1 & 1 & 1 & -1 \end{bmatrix}$$

are skew Hadamard matrices.

Definition 1.14: Two Hadamard matrices are said to be **Hadamard equivalent** or **H-equivalent** if one can be obtained from the other by a sequence of the following operations:

i) Permute rows (or columns).
ii) Multiply any row (or column) by -1.

Definition 1.15: If a square matrix A can be obtained from a square matrix B by a sequence of the following two operations,

i) multiplying the row and the corresponding column by -1 simultaneously,
ii) interchanging two rows and the corresponding two columns simultaneously,

then A will be said to be **Seidel equivalent** to B.

Definition 1.16: A Hadamard matrix which has every element of its first row and column $+1$ is a **normalized Hadamard matrix**.

Hadamard matrices have the following properties:

Lemma 1.7 *(Hadamard matrix properties)*:

i) Let H be an Hadamard matrix of order n. If M and N are monomial matrices of order n with nonzero integers ± 1, then MHN gives an equivalent Hadamard matrix.
ii) Every Hadamard matrix is Hadamard equivalent to a normalized Hadamard matrix.
iii) An Hadamard matrix can only have order 1, 2 or $4t$, t nonnegative integer.
iv) In a normalized Hadamard matrix, any pair of rows have half their columns identical and any triple of rows has exactly one quarter of the columns identical;

Proof:

i) $(MHN)(MHN)^{\top} = MHNN^{\top}H^{\top}M^{\top} = nI.$
ii) We obtain (*ii*) from (*i*).
iii) Hadamard matrices of orders 1 and 2 are

$$[1] \quad \text{and} \quad \begin{bmatrix} 1 & 1 \\ 1 & -1 \end{bmatrix}.$$

Now suppose the order is $h > 2$. Normalize H and rearrange the first three rows to look like (using – for -1)

$$\underbrace{\begin{matrix} 1\ldots1 \\ 1\ldots1 \\ 1\ldots1 \end{matrix}}_{x} \quad \underbrace{\begin{matrix} 1\ldots1 \\ 1\ldots1 \\ -\ldots- \end{matrix}}_{y} \quad \underbrace{\begin{matrix} 1\ldots1 \\ -\ldots- \\ 1\ldots1 \end{matrix}}_{z} \quad \underbrace{\begin{matrix} 1\ldots1 \\ -\ldots- \\ -\ldots- \end{matrix}}_{w}$$

where x, y, z, and w are the number of columns of each type. Then, because the order is h

$$x + y + z + w = h$$

and taking the inner products of rows 1 and 2, 1 and 3, and 2 and 3, respectively, we get

$$x + y - z - w = 0,$$
$$x - y + z - w = 0,$$
$$x - y - z + w = 0.$$

Solving we get

$$x = y = z = w = \frac{h}{4}$$

and so $h \equiv 0 \pmod 4$.
iv) (*iv*) follows from (*iii*).

□

The assertion (*iii*) of Lemma 1.7 poses the question of when an Hadamard matrix exists. This question has not yet been solved. This has led to the following conjecture, called the Hadamard conjecture, posed probably not by Hadamard, but by Sylvester.

Conjecture 1.17: An Hadamard of order n exists whenever $n = 1,\ 2$ or n is a multiple of 4.

Wallis (Seberry) [213] proved that there exists an Hadamard matrix of order $2^t n$ for $t \geq \lfloor 2\log_2(n-3)\rfloor + 1$ where $n > 3$ is any positive integer. To prove the Hadamard conjecture, it is necessary to show $t \geq 2$. Craigen improved her result. We will show their excellent asymptotic theorems in Chapter 11.

We continue to discuss the attempts to prove this conjecture throughout the book. If the entries of X are ± 1 or $\pm i$ where $i = \sqrt{-1}$ and the rows are orthogonal, then the equality of Eq. (1.1) holds.

A complex conjugate matrix $\overline{X}$ is obtained from the matrix X by taking the complex conjugate of each entry of X.

Definition 1.18: Let X be a matrix of order n whose entries are $\pm 1, \pm i$ where $i = \sqrt{-1}$. Then if the rows (and columns) are pairwise orthogonal X called a **complex Hadamard matrix**. In other words, X is a complex Hadamard matrix if $X\overline{X}^{\top} = nI$ where $\overline{X}$ is the conjugate of X.

Definition 1.19: An Hadamard matrix H is said to be **regular** if the sum of all the entries in each row or column is a constant k. Hence $HJ = JH = kJ$ where J is a matrix with all entries +1.

Example 1.11: *Using* $\times$ *for the Kronecker product we see*

$$H = \begin{bmatrix} -1 & 1 & 1 & 1 \\ 1 & -1 & 1 & 1 \\ 1 & 1 & -1 & 1 \\ 1 & 1 & 1 & -1 \end{bmatrix} \quad \textit{and} \quad [H \times H]$$

are regular Hadamard matrices with row sums 2 and 4, respectively.

Theorem 1.11: *The order* ≥ 4 *of a regular Hadamard matrix is* $4m^2$ *where m is an positive integer.*

Proof: Let H be a regular Hadamard matrix of order $4n$ with a constant row sum t. Let $\boldsymbol{e}$ be a $1 \times n$ matrix of all 1s. The $\boldsymbol{e}H = t\boldsymbol{e}$. Thus, using $HH^{\top} = 4nI_{4n}$, we have

$$\begin{aligned} \boldsymbol{e}H(\boldsymbol{e}H)^{\top} &= t^2\boldsymbol{e}\boldsymbol{e}^{\top} = 4nt^2 \text{ and} \\ \boldsymbol{e}HH^{\top}\boldsymbol{e}^{\top} &= 4n\boldsymbol{e}\boldsymbol{e}^{\top} = 16n^2. \end{aligned}$$

Thus we see $4n = t^2$ and may be written as $4m^2$. □

An Hadamard matrix is said to be a **circulant Hadamard matrix** when the matrix is circulant.

The Hadamard matrix

$$\begin{bmatrix} - & 1 & 1 & 1 \\ 1 & - & 1 & 1 \\ 1 & 1 & - & 1 \\ 1 & 1 & 1 & - \end{bmatrix}$$

is a circulant Hadamard matrix of order 4.

We note that a circulant Hadamard matrix is regular. This has led to the following conjecture of Ryser which still has not been solved.

Conjecture 1.20 (Ryser's Conjecture): There are no circulant Hadamard matrices of order greater than 4.

We will further discuss the nonexistence of a circulant Hadamard matrix of a special form in Section 5.3.

1.5.2 Kronecker Product and the Sylvester Hadamard Matrices

Definition 1.21: If $M = (m_{ij})$ is a $m \times p$ matrix and $N = (n_{ij})$ is an $n \times q$ matrix, then the **Kronecker product** $M \times N$ is the $mn \times pq$ matrix given by

$$M \times N = \begin{bmatrix} m_{11}N & m_{12}N & \dots & m_{1p}N \\ m_{21}N & m_{22}N & \dots & m_{2p}N \\ \vdots & & & \\ m_{m1}N & m_{m2}N & \dots & m_{mp}N \end{bmatrix}$$

Lemma 1.8: *The following properties of Kronecker product follow immediately from the definition:*

i) $a(M \times N) = (aM) \times N = M \times (aN)$ *where a is a scalar,*
ii) $(M_1 + M_2) \times N = (M_1 \times N) + (M_2 \times N)$,
iii) $M \times (N_1 + N_2) = M \times N_1 + M \times N_2$,
iv) $(M_1 \times N_1)(M_2 \times N_2) = M_1M_2 \times N_1N_2$,
v) $(M \times N)^{\mathsf{T}} = M^{\mathsf{T}} \times N^{\mathsf{T}}$,
vi) $(M \times N) \times P = M \times (N \times P)$.

Example 1.12: *Let*

$$M = \begin{bmatrix} 1 & 1 \\ 1 & -1 \end{bmatrix} \text{ and } N = \begin{bmatrix} -1 & 1 & 1 & 1 \\ 1 & -1 & 1 & 1 \\ 1 & 1 & -1 & 1 \\ 1 & 1 & 1 & -1 \end{bmatrix}$$

Then

$$M \times N = \begin{bmatrix} N & N \\ N & -N \end{bmatrix} = \begin{bmatrix} -1 & 1 & 1 & 1 & -1 & 1 & 1 & 1 \\ 1 & -1 & 1 & 1 & 1 & -1 & 1 & 1 \\ 1 & 1 & -1 & 1 & 1 & 1 & -1 & 1 \\ 1 & 1 & 1 & -1 & 1 & 1 & 1 & -1 \\ -1 & 1 & 1 & 1 & 1 & -1 & -1 & -1 \\ 1 & -1 & 1 & 1 & -1 & 1 & -1 & -1 \\ 1 & 1 & -1 & 1 & -1 & -1 & 1 & -1 \\ 1 & 1 & 1 & -1 & -1 & -1 & -1 & 1 \end{bmatrix}$$

Lemma 1.9: *Let H_1 and H_2 be Hadamard matrices of orders h_1 and h_2. Then from Lemma 1.8 $H = H_1 \times N_2$ is an Hadamard matrix of order h_1h_2.*

The first occurrence of Hadamard matrices actually appears in the 1867 paper of Sylvester [180].

Theorem 1.12 ***(Sylvester's theorem):*** *There exist Hadamard matrices of order 2^t for integer $t \geq 1$.*

Proof: From Lemma 1.9, we can construct an Hadamard matrix of order 2^t from the Hadamard matrix M of order 2 in Example 1.12. □

Example 1.13:

$$M \times M = \begin{bmatrix} M & M \\ M & -M \end{bmatrix} = \begin{bmatrix} 1 & 1 & 1 & 1 \\ 1 & -1 & 1 & -1 \\ 1 & 1 & -1 & -1 \\ 1 & -1 & -1 & 1 \end{bmatrix}$$

is a Sylvester Hadamard matrix of order 4.

$$M \times M \times M = \begin{bmatrix} M \times M & M \times M \\ M \times M & -M \times M \end{bmatrix}$$

$$= \begin{bmatrix} 1 & 1 & 1 & 1 & 1 & 1 & 1 & 1 \\ 1 & -1 & 1 & -1 & 1 & -1 & 1 & -1 \\ 1 & 1 & -1 & -1 & 1 & 1 & -1 & -1 \\ 1 & -1 & -1 & 1 & 1 & -1 & -1 & 1 \\ \\ 1 & 1 & 1 & 1 & -1 & -1 & -1 & -1 \\ 1 & -1 & 1 & -1 & -1 & 1 & -1 & 1 \\ 1 & 1 & -1 & -1 & -1 & -1 & 1 & 1 \\ 1 & -1 & -1 & 1 & -1 & 1 & 1 & -1 \end{bmatrix}$$

is a Sylvester Hadamard matrix of order 8.

Theorem 1.13 ***(Day and Peterson [43], Seberry et al. [164]):*** *Every Hadamard matrix of order ≥ 4 contains a sub-matrix equivalent to*

$$\begin{bmatrix} 1 & 1 & 1 & 1 \\ 1 & - & 1 & - \\ 1 & 1 & - & - \\ 1 & - & - & 1 \end{bmatrix} \tag{1.2}$$

1.5.2.1 Remarks on Sylvester Hadamard Matrices

The first few Sylvester–Hadamard matrices of orders 2^p, $p = 1, 2, 3$ are given below. For information, we give in an additional last column the number of times the sign changes as we proceed from the first to the last element across the row:

$$S_2 = \left[\begin{array}{cc|c} 1 & 1 & 0 \\ 1 & - & 1 \end{array}\right], \qquad S_4 = \left[\begin{array}{cccc|c} 1 & 1 & 1 & 1 & 0 \\ 1 & - & 1 & - & 3 \\ 1 & 1 & - & - & 1 \\ 1 & - & - & 1 & 2 \end{array}\right],$$

$$S_8 = \left[\begin{array}{cccccccc|c} 1 & 1 & 1 & 1 & 1 & 1 & 1 & 1 & 0 \\ 1 & - & 1 & - & 1 & - & 1 & - & 7 \\ 1 & 1 & - & - & 1 & 1 & - & - & 3 \\ 1 & - & - & 1 & 1 & - & - & 1 & 4 \\ 1 & 1 & 1 & 1 & - & - & - & - & 1 \\ 1 & - & 1 & - & - & 1 & - & 1 & 6 \\ 1 & 1 & - & - & - & - & 1 & 1 & 2 \\ 1 & - & - & 1 & - & 1 & 1 & - & 5 \end{array}\right]$$

In fact we have

Lemma 1.10: *If a ± 1 Sylvester Hadamard matrix of order m, S_m, has all the sign changes 0, 1, …, $m-1$ then its Sylvester matrix S_{2m}, of order 2m, constructed via $S_m \times S_m$, ("×" is the Kronecker product), will have all the sign changes 0, 1, …, $2m-1$. The same is true for the columns. The result is true for S_2 and by induction for S_{2^t}, $t \geq z$ integer.*

This is also observed in

$$S_{16} = \left[\begin{array}{cccccccccccccccc|c} 1 & 1 & 1 & 1 & 1 & 1 & 1 & 1 & 1 & 1 & 1 & 1 & 1 & 1 & 1 & 1 & 0 \\ 1 & - & 1 & - & 1 & - & 1 & - & 1 & - & 1 & - & 1 & - & 1 & - & 15 \\ \\ 1 & 1 & - & - & 1 & 1 & - & - & 1 & 1 & - & - & 1 & 1 & - & - & 7 \\ 1 & - & - & 1 & 1 & - & - & 1 & 1 & - & - & 1 & 1 & - & - & 1 & 8 \\ \\ 1 & 1 & 1 & 1 & - & - & - & - & 1 & 1 & 1 & 1 & - & - & - & - & 3 \\ 1 & - & 1 & - & - & 1 & - & 1 & 1 & - & 1 & - & - & 1 & - & 1 & 12 \\ \\ 1 & 1 & - & - & - & - & 1 & 1 & 1 & 1 & - & - & - & - & 1 & 1 & 4 \\ 1 & - & - & 1 & - & 1 & 1 & - & 1 & - & - & 1 & - & 1 & 1 & - & 11 \\ \\ 1 & 1 & 1 & 1 & 1 & 1 & 1 & 1 & - & - & - & - & - & - & - & - & 1 \\ 1 & - & 1 & - & 1 & - & 1 & - & - & 1 & - & 1 & - & 1 & - & 1 & 14 \\ \\ 1 & 1 & - & - & 1 & 1 & - & - & - & - & 1 & 1 & - & - & 1 & 1 & 6 \\ 1 & - & - & 1 & 1 & - & - & 1 & - & 1 & 1 & - & - & 1 & 1 & - & 9 \\ \\ 1 & 1 & 1 & 1 & - & - & - & - & - & - & - & - & 1 & 1 & 1 & 1 & 2 \\ 1 & - & 1 & - & - & 1 & - & 1 & - & 1 & - & 1 & 1 & - & 1 & - & 13 \\ \\ 1 & 1 & - & - & - & - & 1 & 1 & - & - & 1 & 1 & 1 & 1 & - & - & 5 \\ 1 & - & - & 1 & - & 1 & 1 & - & - & 1 & 1 & - & 1 & - & - & 1 & 10 \end{array}\right].$$

This property is well known to users of the Walsh functions but has not been emphasized in the mathematical literature. This has prompted us to mention it explicitly here.

1.5.3 Inequivalence Classes

Hadamard equivalence and Seidel equivalence are defined in Definitions 1.14 and 1.15. M. Hall Jr. first considered the equivalence of Hadamard matrices in a 1961 JPL Research Report. He found that the Hadamard matrices of orders 1, 2, 4, 8, and 12 are unique, that there are 5 equivalence classes for order 16 and 3 equivalence classes for order 20. At the present time (2017), there are known 60 equivalence classes for order 24, 487 equivalence classes for order 28, and 13,710,027 equivalence classes for order 32.

R. Craigen verbally reported the existence of millions of equivalence classes of order 36, the strange thing about these is that every equivalence class was equivalent to a regular Hadamard matrix.

1.6 Paley Core Matrices

Let p be a prime and $q = p^t$. We recall the **quadratic character** ψ of $F = GF(q)$ defined by

$$\psi(b) = \begin{cases} 0 & \text{if } b = 0, \\ 1 & \text{if } b \text{ is a square,} \\ -1 & \text{otherwise.} \end{cases} \tag{1.3}$$

Example 1.14: *Let $q = 5$. We take 2 as a generator of $GF(5)$.*

F	0	$1 = 2^0$	$2 = 2^1$	$3 = 2^3$	$4 = 2^2$
$\psi(\alpha)$	0	1	-1	-1	1

Lemma 1.11: *Let ψ be a nontrivial quadratic character of $F = GF(q)$.*

$$\sum_{y \in F} \psi(y)\psi(y + c) = -1, \quad \text{if } c \neq 0.$$

Proof: $\psi(0)\psi(0 + c) = 0$. For $y \neq 0$, there is a unique $s \neq 0$ such that $y + c = ys$. As y ranges of the nonzero elements of $GF(q)$, s ranges over all elements of $GF(q)$ except 1. (If $s = 1$ then $c = 0$.) Hence

$$\begin{aligned} \sum_{y} \psi(y)\psi(y + c) &= \sum_{y \neq 0} \psi(y)\psi(y + c) \\ &= \sum_{y \neq 0} \psi(y^2)\psi(s) \\ &= \sum_{s \neq 1} \psi(s) \end{aligned}$$

from Theorem 1.7,

$$\begin{aligned} &= \sum_{s} \psi(s) - \psi(1) \\ &= -1. \end{aligned}$$

□

Let $a_0 . a_1, \dots, a_{q-1}$ be the elements of $F = GF(q)$ numbered such that

$$a_0 = 0, \quad a_{q-i} = -a_i, \quad i = 1, \dots, q - 1.$$

We define

$$Q = (\psi(a_i - a_j)) \tag{1.4}$$

where ψ is a nontrivial quadratic character of F defined in (1.3).

Lemma 1.12: *The diagonal entries of Q are all 0 and the other entries are in $\{1, -1\}$. The matrix Q satisfies*

i) $Q^{\top} = \begin{cases} Q & \text{if } q \equiv 1 \pmod 4, \\ -Q & \text{if } q \equiv 3 \pmod 4. \end{cases}$
ii) $JQ = QJ = 0.$
iii) $QQ^{\top} = qI - J$

where J is the matrix with all entries 1.

Proof: Since -1 is a square for $q \equiv 1 \pmod 4$ and non-square for $q \equiv 3 \pmod 4$, we have the assertion (*i*). Since exactly half of the elements of F are squares, we have (*ii*). From Lemma 1.11, we have

$$q_{i,j} = \sum_{a_t} \psi(a_i - a_t)\psi(a_j - a_t)$$

putting $g = a_i - a_t$,

$$= \sum_{g} \psi(g)\psi(g + a_j - a_i) = -1.$$

It follows the assertion (*iii*). □

Definition 1.22: The matrix Q is called the **Paley core**.

We give an another proof of Lemma 1.12 in Chapter 8 based on the properties of Gauss sums.

Example 1.15: *We let $q = 5$ in Example 1.14.*

$$Q = (\psi(a_i - a_j)) = \begin{bmatrix} 0 & 1 & -1 & -1 & 1 \\ 1 & 0 & 1 & -1 & -1 \\ -1 & 1 & 0 & 1 & -1 \\ -1 & -1 & 1 & 0 & 1 \\ 1 & -1 & -1 & 1 & 0 \end{bmatrix} = Q^{\top}.$$

We put

$$M = \begin{bmatrix} 0 & \boldsymbol{e} \\ \psi(-1)\boldsymbol{e}^{\top} & Q \end{bmatrix} \tag{1.5}$$

where $\boldsymbol{e}$ is the row vector of all 1's.

Theorem 1.14:

i) If $q \equiv 3 \pmod 4$, then there exists a skew Hadamard matrix of order $q + 1$.
ii) If $q \equiv 1 \pmod 4$, then there exists a symmetric Hadamard matrix of order $2(q + 1)$.

Proof: For $q \equiv 3 \pmod 4$, $M + I$ is a skew Hadamard matrix since $\psi(-1) = -1$. For $q \equiv 1 \pmod 4$,

$$H = \begin{bmatrix} M + I & M - I \\ M - I & -M - I \end{bmatrix}$$

is a symmetric Hadamard matrix of order $2(v + 1)$ since $\psi(-1) = 1$. □

Definition 1.23: We call the Hadamard matrix $M + I$ of order $q + 1$ **Paley type I Hadamard matrix** and H of order $2(q + 1)$ **Paley type II Hadamard matrix**.

Corollary 1.2: *There exist Hadamard matrices of order* $2^t \prod(q+1)$ *where t is a nonnegative integer and q is a prime power. When* $t = 0$, $q \equiv 3 \pmod 4$.

Definition 1.24: A matrix in the form

$$\begin{bmatrix} 1 & 1 & \dots & 1 \\ 1 & & & \\ \vdots & & A & \\ 1 & & & \end{bmatrix} \tag{1.6}$$

is said to be **bordered**. When the border is removed the matrix is said to be **trimmed**. Equation (1.5) the type I Paley matrix $M + I$ is an example.

The Paley type II matrix is an example with two row/column borders

$$\begin{bmatrix} 1 & 1 & e & e \\ 1 & - & -e & e \\ e^\top & -e^\top & Q+I & Q-I \\ e^\top & e^\top & Q-I & -Q-I \end{bmatrix} \tag{1.7}$$

We will see in Chapter 6 that the border may have 1, 2, 4, or 8 rows and columns (Theorem 6.7 and Corollary 6.12).

1.7 Amicable Hadamard Matrices

Definition 1.25: Two matrices X, Y which satisfy $XY^\top = YX^\top$ will be said to be **amicable**.

Definition 1.26: Two matrices $M = I + U$ and N will be called **(complex) amicable Hadamard matrices** if M is a (complex) skew Hadamard matrix and N a (complex) Hadamard matrix satisfying

$$U^\top = -U, \qquad N^\top = N, \quad MN^\top = NM^\top, \quad \text{if real.}$$

$$U^* = \overline{U^\top} = -U, \quad N^* = N, \quad MN^* = NM^*, \quad \text{if complex.}$$

We will only use constructions with real matrices to construct (real) amicable Hadamard matrices but it is obvious that if complex matrices are used then complex amicable Hadamard matrices can be obtained.

Lemma 1.13: *If* $M = W + I$ *and* N *are amicable Hadamard matrices then*

$$WN^\top = NW^\top.$$

Proof: Since $MN^\top = NM^\top$ we have

$$\begin{aligned} MN^\top &= (W+I)N^\top \\ &= WN^\top + N^\top \\ &= WN^\top + N \\ &= NM^\top \\ &= N(W^\top + I) \\ &= NW^\top + N \end{aligned}$$

and so

$$WN^{\top} = NW^{\top}. \qquad \square$$

Theorem 1.15 ***(Wallis [200]):*** *Suppose there are amicable Hadamard matrices of orders m and n. Then there are amicable Hadamard matrices of order mn.*

Proof: Let $I + M_1$ and N_1 be amicable Hadamard matrices of order m and $I + M_2$ and N_2 be amicable Hadamard matrices of order n. Then

$$I \times (I + M_2) + M_1 \times N_2, \qquad N_1 \times N_2$$

are the required amicable Hadamard matrices of order mn. Note $N_i^{\top} = N_i$, $i = 1, 2$.

$$\begin{aligned}(I \times (I + M_2) + M_1 \times N_2) &\times \left(I \times \left(I + M_2^{\top}\right) + M_1^{\top} \times N_2^{\top}\right) \\ &= I \times \left(I + M_2\right)\left(I + M_2^{\top}\right) + (-M_1) \times (I + M_2)N_2^{\top} \\ &\quad + M_1 \times N_2\left(I + M_2^{\top}\right) + M_1M_1^{\top} \times N_2N_2^{\top}\end{aligned}$$

since $I + M_i$ and N_i are amicable Hadamard matrices for $i = 1, 2$,

$$= I \times mI + (n-1)I \times mI = nI \times mI = nmI$$

and $(N_1 \times N_2)\left(N_1^{\top} \times N_2^{\top}\right) = nmI$.

$$\begin{aligned}\left(I \times (I + M_2) + M_1 \times N_2^{\top} - I \times I\right)^{\top} &= I \times (I - M_2) - M_1 \times N_2 - I \times I \\ &= I \times I - I \times M_2 - M_1 \times N_2 - I \times I \\ &= -I \times (I + M_2) - M_1 \times N_2 + I \times I \\ &= -\left(I \times (I + M_2) + M_1 \times N_2^{\top} - I \times I\right)\end{aligned}$$

Thus $I \times (I + M_2) + M_1 \times N_2^{\top} - I \times I$ is a skew Hadamard matrix.

$$\begin{aligned}&\left(I \times (I + M_2) + M_1 \times N_2^{\top}\right)\left(N_1^{\top} \times N_2^{\top}\right) \\ &\quad = N_1^{\top} \times (I + M_2)N_2^{\top} + M_1N_1^{\top} \times N_2N_2^{\top}\end{aligned}$$

from Lemma 1.13,

$$\begin{aligned}&= N_1 \times N_2\left(I + M_2^{\top}\right) + N_1M_1^{\top} \times N_2N_2^{\top} \\ &= (N_1 \times N_2)\left(I \times \left(I + M_2^{\top}\right) + M_1^{\top} \times N_2^{\top}\right). \qquad \square\end{aligned}$$

Corollary 1.3: *There exist amicable Hadamard matrices of order 2^t, t a nonnegative integer.*

Proof: $I + M_1 = I + M_2$ and $N_1 = N_2$ are given as

$$I + M_1 = I + M_2 = \begin{bmatrix} 1 & 1 \\ -1 & 1 \end{bmatrix} \qquad \text{and} \qquad N_2 = \begin{bmatrix} 1 & 1 \\ 1 & -1 \end{bmatrix}.$$

From the above equation, we obtain amicable Hadamard matrices of order 2^t. $\square$

Example 1.16:

$$I \times (I + M_2) + M_1 \times N_2 = \begin{bmatrix} I + M_2 & 0 \\ 0 & I + M_2 \end{bmatrix} + \begin{bmatrix} 0 & N_2 \\ -N_2 & 0 \end{bmatrix}$$

$$= \begin{bmatrix} 1 & 1 & 0 & 0 \\ -1 & 1 & 0 & 0 \\ 0 & 0 & 1 & 1 \\ 0 & 0 & -1 & 1 \end{bmatrix} + \begin{bmatrix} 0 & 0 & 1 & 1 \\ 0 & 0 & 1 & -1 \\ -1 & -1 & 0 & 0 \\ -1 & 1 & 0 & 0 \end{bmatrix}$$

$$= \begin{bmatrix} 1 & 1 & 1 & 1 \\ -1 & 1 & 1 & -1 \\ -1 & -1 & 1 & 1 \\ -1 & 1 & -1 & 1 \end{bmatrix}$$

$$N_1 \times N_2 = \begin{bmatrix} 1 & 1 & 1 & 1 \\ 1 & -1 & 1 & -1 \\ 1 & 1 & -1 & -1 \\ 1 & -1 & -1 & 1 \end{bmatrix}.$$

are amicable Hadamard matrices of order 4.

Lemma 1.14: *There are amicable Hadamard matrices of order* $q + 1$, $q \equiv 3 \pmod 4$, *q a prime power.*

Proof: Let $\boldsymbol{e}$ be the row vector of q 1's. Let Q be the Paley core defined by Eq. (1.4) for $q \equiv 3 \pmod 4$. Further let M_0 be the matrix of order $q + 1$ in Eq. (1.5).

$$M_0 = \begin{bmatrix} 0 & \boldsymbol{e} \\ -\boldsymbol{e}^\top & Q \end{bmatrix}.$$

Then M_0 has the properties

$$M_0^\top = -M_0, \quad M_0 M_0^\top = qI_{q+1}.$$

We put $h = q + 1$.

$$M = I_h + M_0$$

is a skew Hadamard matrix of order $q + 1$.

Let $U = (u_{i,j})_{0 \le i,j \le q-1}$ be the matrix of order q defined by

$$u_{i,j} = \begin{cases} 1 & \text{if } i + j = q \text{ or } i = j = 0, \\ 0 & \text{otherwise.} \end{cases}$$

Then we see $U^\top = U$ and $U^2 = 1$.

Now define N by the matrix

$$N = \begin{bmatrix} 1 & 0 \\ 0 & -U \end{bmatrix} M = \begin{bmatrix} 1 & 0 \\ 0 & -U \end{bmatrix} + \begin{bmatrix} 1 & 0 \\ 0 & -U \end{bmatrix} \begin{bmatrix} 0 & \boldsymbol{e} \\ -\boldsymbol{e}^{\top} & Q \end{bmatrix}$$

$$= \begin{bmatrix} 1 & 0 \\ 0 & -U \end{bmatrix} + \begin{bmatrix} 0 & \boldsymbol{e} \\ \boldsymbol{e}^{\top} & -UQ \end{bmatrix}$$

$$= \begin{bmatrix} 1 & \boldsymbol{e} \\ \boldsymbol{e}^{\top} & -U - UQ \end{bmatrix}.$$

We put $UQ = (c_{i,j})$. Then

$$c_{0j} = \sum_{k=0}^{q-1} u_{0k}\psi(a_k - a_j) = u_{00}\psi(a_0 - a_j) = \psi(-a_j) \text{ and}$$

$$c_{j0} = \sum_{k=0}^{q-1} u_{jk}\psi(a_k - a_0) = u_{0,q-j}\psi(a_{q-j}) = -\psi(a_j).$$

For $i \neq 0$,

$$c_{i,j} = \psi(a_{q-i} - a_j) = \psi(-a_i - a_j) = \psi(-1)\psi(a_i + a_j) = -\psi(a_i + a_j).$$

Hence UQ is symmetric, then N is also symmetric. We find

$$NN^{\top} = \begin{bmatrix} 1 & 0 \\ 0 & -U \end{bmatrix} MM^{\top} \begin{bmatrix} 1 & 0 \\ 0 & -U \end{bmatrix} = hI_h.$$

So N is a symmetric Hadamard matrix. Now

$$MN^{\top} = MM^{\top} \begin{bmatrix} 1 & 0 \\ 0 & -U \end{bmatrix} = hI_h \begin{bmatrix} 1 & 0 \\ 0 & -U \end{bmatrix} = \begin{bmatrix} 1 & 0 \\ 0 & -U \end{bmatrix} MM^{\top} = NM^{\top}.$$

Thus M and N are amicable Hadamard matrices of order $q + 1$. □

Example 1.17: *Let $q = 3$. The matrix*

$$M = \begin{bmatrix} 1 & 1 & 1 & 1 \\ -1 & 1 & -1 & 1 \\ -1 & 1 & 1 & -1 \\ -1 & -1 & 1 & 1 \end{bmatrix}$$

is a skew Hadamard matrix. Then

$$N = \begin{bmatrix} 1 & 0 \\ 0 & -U \end{bmatrix} M$$

$$= \begin{bmatrix} 1 & 0 & 0 & 0 \\ 0 & -1 & 0 & 0 \\ 0 & 0 & 0 & -1 \\ 0 & 0 & -1 & 0 \end{bmatrix} \begin{bmatrix} 1 & 1 & 1 & 1 \\ -1 & 1 & -1 & 1 \\ -1 & 1 & 1 & -1 \\ -1 & -1 & 1 & 1 \end{bmatrix} = \begin{bmatrix} 1 & 1 & 1 & 1 \\ 1 & -1 & 1 & -1 \\ 1 & 1 & -1 & -1 \\ 1 & -1 & -1 & 1 \end{bmatrix}$$

is a symmetric Hadamard matrix. We verify $NM^{\top} = MN^{\top}$. M and N are amicable Hadamard matrices of order 4.

Theorem 1.16: *If $m = 2^t \prod (q_i + 1)$ where t is a nonnegative integer and q_i is a prime power* $\equiv 3 \pmod 4$*, then there are amicable Hadamard matrices of order m.*

Proof: We obtain the assertion from Theorem 1.15, Corollary 1.3, and Lemma 1.14. □

1.8 The Additive Property and Four Plug-In Matrices

Many computer constructions for Hadamard matrices are based on finding specific orthogonal ± 1 matrices to replace the variables in some version of the **Williamson array**:

$$\begin{bmatrix} A & B & C & D \\ -B & A & D & -C \\ -C & -D & A & B \\ -D & C & -B & A \end{bmatrix}.$$

This is called the Williamson method [223].

Definition 1.27: **Williamson matrices** are four circulant symmetric ± 1 matrices of order n, A, B, C, D which satisfy Eq. (1.8)

$$AA^{\top} + BB^{\top} + CC^{\top} + DD^{\top} = kI_n \tag{1.8}$$

when $k = 4n$, which is called the **additive property**.

Definition 1.28: **Suitable weighing matrices** are four circulant symmetric $0, \pm 1$ matrices of order n, A, B, C, D which satisfy Eq. (1.8) when $k < 4n$. We say that they satisfy the **weighing additive property**. This is generalized in Definition 3.2.

Remark 1.1: Williamson's theorem uses Williamson matrices of order n in the Williamson array to form Hadamard matrices of order $4n$. See Williamson's theorem (Theorem 3.1).

Various types of suitable plug-in matrices are discussed in Section 3.3. Known Williamson and Williamson type matrices appear in Appendix A.2.

Some variants depend significantly on the structure of the plug-in matrices, this happens with circulant matrices, good matrices, propus, and luchshie matrices. Other variants depend on initial alterations in the Williamson array itself (a plug-into matrix), for example the propus array (below) and Goethals–Seidel array, Ito array, and the Balonin–Seberry array.

The Williamson, Goethals–Seidel and Propus methods give constructions for Hadamard, skew Hadamard, and symmetric Hadamard ± 1 matrices, using four specific circulant matrices A, B, C, and D, of order t see [158], satisfying the additive property (see Eq. (1.8)).

1.8.1 Computer Construction

Thus a variant of the Williamson method which allows the computer construction for skew and symmetric Hadamard matrices of order $4t$ using four circulant ± 1 matrices A, B, C, and D, of order t odd, which satisfy the additive property becomes crucial. I is the identity matrix.

Let R be the back diagonal identity matrix. Then we note that if A and B are circulant matrices of the same order t, $A(BR)^{\top} = BR(A)^{\top}$. These two circulant matrices the first, $X = A$, circulant and the second, $Y = BR$, back circulant of order t satisfy $XY^{\top} = YX^{\top}$ (see [166] and [214, p. 300]). Williamson used circulant symmetric matrices which commute and are amicable. However, in general circulant matrices commute but are not amicable. This is the key that using R unlocks. Some conditions and construction variations are discussed in Awyzio and Seberry [3].

1.8.2 Skew Hadamard Matrices

In order to construct a skew Hadamard matrix of order 36, Goethals and Seidel [81] used a variant of the Williamson array which uses four circulant matrices, A, B, C, and D, satisfying the additive property but with $(A-I)^{\mathrm{T}} = -(A-I)$. The **Goethals–Seidel array** can be written as

$$W = \begin{bmatrix} A & BR & CR & DR \\ -BR & A & D^{\mathrm{T}}R & -C^{\mathrm{T}}R \\ -CR & -D^{\mathrm{T}}R & A & B^{\mathrm{T}}R \\ -DR & C^{\mathrm{T}}R & -B^{\mathrm{T}}R & A \end{bmatrix} \quad \text{or} \quad \begin{bmatrix} A & BR & CR & DR \\ -BR & A & RD & -RC \\ -CR & -RD & A & RB \\ -DR & RC & -RB & A \end{bmatrix}.$$

This has led to the construction of many skew Hadamard matrices orders by many authors. The smallest unknown is 276 [158, p. 88].

Table 1.2 lists relevant definitions for Williamson and Goethals–Seidel arrays and skew Hadamard matrices.

1.8.3 Symmetric Hadamard Matrices

It has taken many years to find an equivalent array to construct symmetric Hadamard matrices. The first effort which parallels the Williamson construction is for **propus Hadamard matrices** [160]. These are constructed using four circulant symmetric (± 1) matrices satisfying the additive property but with $B = C$. So the propus array is

$$\begin{bmatrix} A & B & B & D \\ B & D & -A & -B \\ B & -A & -D & B \\ D & -B & B & -A \end{bmatrix}.$$

The related theorem, for reference, is

Theorem 1.17 ***(Propus symmetric Hadamard construction [160]):*** *Let A, B, C, D be four circulant symmetric ± 1 matrices of order t, with $B = C$, satisfying the additive property*

$$AA^{\mathrm{T}} + 2BB^{\mathrm{T}} + CC^{\mathrm{T}} = 4tI_t.$$

Then when these are plugged into the propus array they give a symmetric Hadamard matrix of order $4t$.

Table 1.2 Williamson, Goethals–Seidel arrays and skew Hadamard matrices.

	Four ± 1 matrices of order t, called A, B, C, D which satisfy the additive property $AA^{\mathrm{T}} + BB^{\mathrm{T}} + CC^{\mathrm{T}} + DD^{\mathrm{T}} = 4t$, will be called:
1.	*Williamson matrices* when A, B, C, D are all circulant and symmetric
2.	*Williamson-type* when A, B, C, D are all pairwise amicable and symmetric
3.	*Good* when A is circulant and skew symmetric ($(A+I)^{\mathrm{T}} = -A + I$) and B, C, D are circulant and symmetric
4.	*Four-suitable or Goethals–Seidel suitable* when A is circulant and skew symmetric ($(A+I)^{\mathrm{T}} = -A + I$) and B, C, D are all circulant
5.	*Suitable* when A, B, C, D are all circulant
	We note
(a)	The *Williamson array* uses A, B, C, D are all circulant and symmetric
(b)	The *Goethals–Seidel array* uses A, B, C, D are all circulant or amicable

Table 1.3 Balonin–Seberry array and symmetric Hadamard matrices.

	Four ± 1 matrices of order t, called A, B, C, D which satisfy the additive property $AA^\top + BB^\top + CC^\top + DD^\top = 4t$, will be called:
1.	*Williamson matrices* when A, B, C, D are all circulant and symmetric
2.	*Williamson-type* when A, B, C, D are all pairwise amicable and symmetric
3.	*Propus* when $A, B = C, D$ are all circulant and symmetric (propus matrices are also Williamson matrices)
4.	*Propus-type* when $A, B = C, D$ are all amicable and symmetric (propus-type matrices are also Williamson-type matrices)
5.	*Luchshie* when A is circulant and symmetric, $B = C, D$ are all circulant
6.	*Pirit* when $A, B = C, D$ are all circulant
	We note
(a)	The *Williamson array* uses A, B, C, D are all circulant and symmetric
(b)	The *Balonin–Seberry array (original)* [160] uses $A, B = C, D$ are all circulant and symmetric
(c)	The *Balonin–Seberry array (current)* uses A circulant and symmetric, $B = C, D$ are all circulant

Definition 1.29: Four circulant ± 1 matrices of order t, with $A^\top = A$, $B{=}C$, satisfying the additive property

$$AA^\top + 2BB^\top + CC^\top = 4tI_t,$$

will be called **luchshie** matrices. (luchshie is Russian plural for "best".) In the case where A is not necessarily symmetric we call $A, B = C, D$ **pirit** matrices (pirit is Russian for "fools gold," the mineral pyrite).

We now give a variation of the propus array which we call the Balonin–Seberry array which allows the construction of symmetric Hadamard matrices.

Definition 1.30: The **Balonin–Seberry array** uses four luchshie matrices in the following array

$$W = \begin{bmatrix} A & BR & CR & DR \\ CR & D^\top R & -A & -B^\top R \\ BR & -A & -D^\top R & C^\top R \\ DR & -C^\top R & B^\top R & -A \end{bmatrix} \quad \text{or} \quad \begin{bmatrix} A & BR & CR & DR \\ CR & RD & -A & -RB \\ BR & -A & -RD & RC \\ DR & -RC & RB & -A. \end{bmatrix}.$$

Theorem 1.18 ***(Balonin–Seberry symmetric Hadamard theorem [160]):*** *Let A, B, C, D be four luchshie matrices of order t. Then these, when plugged into the Balonin–Seberry array give a symmetric Hadamard matrix of order* $4t$.

Table 1.3 lists relevant definitions for Balonin–Seberry arrays and symmetric Hadamard matrices.

1.9 Difference Sets, Supplementary Difference Sets, and Partial Difference Sets

1.9.1 Difference Sets

Definition 1.31: Let G be a group of order v and D be a k-subset of G. Denote the identity of G by 1_G. For each $g \neq 1_G \in G$, if the number of ordered pairs (d_1, d_2), $d_1 \neq d_2$, $d_1, d_2 \in D$ such that $d_1 d_2^{-1} = g$ is equal to a constant value λ, then D is called a (v, k, λ) **difference set**.

Let $D^{-1} = \{d^{-1} | d \in D\}$. A difference set D is called **reversible** if $D^{-1} = D$.

Example 1.18: *For example the numbers* 0, 1, 3, 8, 12, 18 *modulo 31 form a* (31, 6, 1)*-difference set since*

$$\begin{array}{rrr}
1-0\equiv 1 & 12-1\equiv 11 & 8-18\equiv 21\\
3-1\equiv 2 & 12-0\equiv 12 & 3-12\equiv 22\\
3-0\equiv 3 & 0-18\equiv 13 & 0-8\equiv 23\\
12-8\equiv 4 & 1-18\equiv 14 & 1-8\equiv 24\\
8-3\equiv 5 & 18-3\equiv 15 & 12-18\equiv 25\\
18-12\equiv 6 & 3-18\equiv 16 & 3-8\equiv 26\\
8-1\equiv 7 & 18-1\equiv 17 & 8-12\equiv 27\\
8-0\equiv 8 & 18-0\equiv 18 & 0-3\equiv 28\\
12-3\equiv 9 & 0-12\equiv 19 & 1-3\equiv 29\\
18-8\equiv 10 & 1-12\equiv 20 & 0-1\equiv 30
\end{array}$$

The following lemma gives a necessary and sufficient condition of the existence of a difference set.

Lemma 1.15: *Put* $\mathcal{D}=\sum_{d\in D} d\in \mathbf{Z}[G]$, $\mathcal{D}^{-1}=\sum_{d\in D} d^{-1}\in \mathbf{Z}[G]$, *and* $\mathcal{G}=\sum_{g\in G} g\in \mathbf{Z}[G]$. *A subset D of G is a* (v,k,λ) *difference set if and only if*

$$\mathcal{D}\mathcal{D}^{-1}=(k-\lambda)1_G+\lambda\mathcal{G}.$$

Definition 1.32: Let G be an abelian group of order v and D is a (v,k,λ) difference set in G. We define

$$D+a=\{x+a\mid x\in D\}$$

of D by $a\in G$. Then $D+a$ is called a **shift** or **translate** of D. Every shift of a (v,k,λ) is itself a difference set.

Definition 1.33: Let D be a (v,k,λ) difference set in G. For an automorphism σ of G, if $\sigma(D)=D+a$ for some $a\in G$, then σ is called a **multiplier** of D. In particular, if $\sigma(D)=D$, then we say that σ fixes D.

Example 1.19: *The subset*

$$D=\{0,1,3,8,12,18\}$$

is the $(31,6,1)$ *difference set in* $\mathbf{Z}/31\mathbf{Z}$. *The automorphism* $\sigma : a\to 5a$,

$$\sigma(D)=D+28=\{0,5,9,15,28,29\}.$$

Hence σ *is a multiplier of D.*

The following theorem shows a SBIBD (see Definition 1.50) corresponds to a difference set.

Theorem 1.19: *Let G be an abelian group of order v and D be a k-subset of G. Then D is a* (v,k,λ) *difference set if and only if* $(G,\mathbb{B})$ *is a regular SBIBD where* $\mathbb{B}$ *is a set of shifts of D.*

Assume G is an abelian group.

Definition 1.34: A difference set D in additively written abelian group G is called **skew Hadamard** if and only if G is a disjoint union of D, $-D$ and $\{0\}$.

A necessary and sufficient condition can be represented in terms of characters of G.

Lemma 1.16: *Let G be an abelian group of order v. A k-subset D of G is a (v, k, λ) difference set if and only if*

$$|\chi(D)| = \sqrt{k-\lambda}$$

for every nontrivial character χ of G. A (v, k, λ) difference set is reversible if and only if $\chi(D) = \overline{\chi(D)}$.

The subset D is a skew Hadamard difference set if and only if

$$\chi(D) = \frac{1}{2}\left(-1 \pm \sqrt{-v}\right)$$

for every nontrivial character χ of G.

1.9.2 Supplementary Difference Sets

Supplementary difference sets (sds) are introduced by J. S. Wallis [205, 209]. A special case of supplementary difference sets, the complementary difference sets, such that one set is skew were introduced by Szekeres [181] and called Szekeres difference sets. We discuss Szekeres difference sets further in Chapter 9.

Definition 1.35: Let G be an additive abelian group of order v and $D_1, D_2, \ldots, D_n$ be subsets of G which contain $k_1, \ldots, k_n$ elements, respectively. For $g \neq 0$ of G, we define the number

$$\lambda_i(g) = \left|\{(d_{ij}, d_{im}) \mid g = d_{ij} - d_{im}, d_{ij}, d_{im} \in D_i\}\right| \quad \text{for } 1 \leq i \leq n.$$

If $\lambda(g) = \lambda_1(g) + \cdots + \lambda_n(g)$ has a constant value λ, then $D_1, D_2, \ldots, D_n$ are called $n - \{v; k_1, k_2, \ldots, k_n; \lambda\}$ **supplementary difference sets** or a $n - \left\{v; k_1, \ldots, k_n; \lambda\right\}$ **difference family**. If $k_1 = k_2 = \cdots = k_n = k$, we write $n - \{v; k; \lambda\}$ supplementary difference sets or a $n - \{v; k; \lambda\}$ difference family.

In particular, for the case $n = 1$, a supplementary difference set becomes a difference set in the usual sense.

Example 1.20: *Let $D_1 = \{1, 2, x+1, 2x+2\}$, $D_2 = \{0, 1, 2, x+1, 2x+2\} = D_1 \cup \{0\}$ be subsets of $GF(3^2)$ where x is a primitive element of $GF(3^2)$. Let T_1 and T_2 be the totality of differences between elements of D_1 and D_2 then*

$$\begin{aligned}
T_1 &= [1-2, 1-(x+1), 1-(2x+2), 2-1, 2-(x+1), 2-(2x+2),\\
&\qquad (x+1)-1, (x+1)-2, (x+1)-(2x+2), (2x+2)-1,\\
&\qquad (2x+2)-2, (2x+2)-(x+1)]\\
&= [2, 2x, x+2, 1, 2x+1, x, x, x+2, 2x+2, 2x+1, 2x, x+1]\\
T_2 &= [0-D_1, D_1-0, T_1]\\
&= [2, 1, 2x+2, x+1, 1, 2, x+1, 2x+2, T_1]
\end{aligned}$$

$T_1 \cup T_2 = 4(G \backslash \{0\})$. Thus D_1 and D_2 are $2-(9; 4, 5; 4)$ supplementary difference sets.

We abbreviate supplementary difference sets as **sds** and a difference family as **DF**.

Lemma 1.17: *Let $q \equiv 1 \pmod 4$ be a prime power. Let Q be a set of squares and N be a set of non-squares of $GF(q)$. Then Q and N are*

$$2 - \left\{q; \frac{1}{2}(q-1); \frac{1}{2}(q-3)\right\} \text{ sds.}$$

Proof: The equation $x - y = g$ is equivalent to the equation $xg^{-1} - yg^{-1} = 1$. Thus we consider the numbers of the solution (x, y) of the equation $x - y = 1$. Denote the number of the solutions such that $x - y = 1, x \in Q$, and $y \in Q$ by (Q, Q). We let (Q, N), (N, Q), and (N, N) similarly. Then we will show that $(Q, Q) + (N, N) = \frac{1}{2}(q-3)$.

From $x - y = -y - (-x) = 1$ and $-1 \in Q$, we have $(Q, N) = (N, Q)$. Since the equation $x - y = 1$ is equivalent to the equation $xy^{-1} - y^{-1} = 1$, $(Q, N) = (N, N)$. We have $(N, Q) + (N, N) = \frac{1}{2}(q-1)$ and $(Q, Q) + (Q, N) = \frac{1}{2}(q-3)$ by removing the case $x = 1$. Hence we have $(Q, Q) + (N, N) = \frac{1}{2}(q-3)$. Together with $|Q| = |N| = \frac{1}{2}(q-1)$, the assertion follows. □

Example 1.21: *Let* $q = 13$. $Q = \{1, 3, 4, 9, 10, 12\}$, $N = \{2, 5, 6, 7, 8, 11\}$ *are* $(13, 6, 5)$ *sds in* $GF(13)$.

Lemma 1.18: *The parameters of* $n - \{v; k_1, k_2, \dots, k_n; \lambda\}$ *supplementary difference sets satisfy*

$$\lambda(v-1) = \sum_{j-1}^{n} k_j(k_j - 1). \tag{1.9}$$

Proof: This follows immediately from the definition by counting the differences. □

We define the polynomial

$$D_i(x) = \sum_{\alpha \in D_i} x^\alpha$$

for the subset D_i, $0 \le i \le n-1$. Then the above condition is written as the polynomials

$$\sum_{i=0}^{n-1} D_i(x) D_i(x^{-1}) = \sum_{i=0}^{n-1} k_i - \lambda + \lambda J(x)$$

where $J(x) = \sum_{\alpha \in G} x^\alpha$.

Let $D_0, D_1, \dots, D_{n-1}$ be $n - \{v;\ k_0, \dots, k_{n-1};\ \lambda\}$ supplementary difference sets (DF). Let X_i be $(1, -1)$ incidence matrix of D_i for $0 \le i \le n-1$. Then we have

$$\sum_{\ell=0}^{n-1} X_\ell X_\ell^{\mathrm{T}} = 4\left(\sum_{\ell=0}^{n-1} k_\ell - \lambda\right) I + \left(nv - 4\left(\sum_{\ell=0}^{n-1} k_\ell - \lambda\right)\right) J. \tag{1.10}$$

Assume $n = 4$. Further assume $\lambda = k_0 + k_1 + k_2 + k_3 - v$. If we put $f_i(x) = 2D_i(x) - J(x)$ for $i = 0, 1, 2, 3$, then the elements of $f_i(x)$, $i = 0, 1, 2, 3$ are ± 1 and

$$\sum_{i=0}^{3} f_i(x) f_i(x^{-1}) = 4v. \tag{1.11}$$

Assume $(1, -1)$ incidence matrices X_0, X_1, X_2, X_3 of D_0, D_1, D_2, D_3 are circulant and satisfy $\sum_{\ell=0}^{3} X_\ell X_\ell^{\mathrm{T}} = 4vI$. Then it gives rise to an Hadamard matrix by plugging them into the Goethals–Seidel array. Furthermore if D_0, D_1, D_2, D_3 are all symmetric, that is $-a \in D_i$ if $a \in D_i$, $i = 0, 1, 2, 3$, then $f_i(x)$, $i = 0, 1, 2, 3$ satisfy Eq. (1.11). Thus we obtain a Williamson Hadamard matrix. If X_0 is a skew type and X_2, X_3, X_4 are symmetric, then these matrices are good matrices (see Definition 3.4).

Definition 1.36: Let D_0, D_1, D_2, D_3 be $4 - \{v;\ k_0, k_1, k_2, k_3;\ \lambda\}$ supplementary difference sets (DF) over an abelian group. If $\lambda = k_0 + k_1 + k_2 + k_3 - v$, then we obtain an Hadamard matrix of order $4v$. We call this supplementary difference sets **Hadamard supplementary difference sets** or **Hadamard difference family**.

1.9.3 Partial Difference Sets

Definition 1.37: Let G be a group of order v and D be a k-subset of G. For the elements $d_1, d_2 \in D$, $d_1 \ne d_2$, if $d_1 d_2^{-1}$ represent each non-identity element of D exactly λ times and the non-identity element of $G \backslash D$ exactly μ times, then D is called a (v, k, λ, μ) **partial difference set (PDS)**. If D does not contain the identity and $D = D^{-1} = \{g^{-1} | g \in D\}$, then we call the PDS **regular**.

A partial difference set with $\lambda = \mu$ is an ordinary difference set. Ma proved that if D is a (v, k, λ, μ) PDS with $\lambda \neq \mu$, then $D = D^{-1}$ [129].

Theorem 1.20 ***(Ma [130]):*** *Assume $\lambda \neq \mu$. Let G be a group of order v and D be a k-subset of G. Let $\mathcal{D} = \sum_{d \in D} d \in \mathbf{Z}[G]$ and $\mathcal{G} = \sum_{g \in G} g \in \mathbf{Z}[G]$. Then D is a (v, k, λ, μ) PDS if and only if*

$$\mathcal{D}^2 = \mu\mathcal{G} + (\lambda - \mu)\mathcal{D} + \gamma 1_G \tag{1.12}$$

where 1_G is the identity of G and $\gamma = k - \mu$ if $1_G \notin D$ and $\gamma = k - \lambda$ if $1_G \in D$.

Proof: See [130]. □

Suppose $\lambda \neq \mu$. we put $\beta = \lambda - \mu$ and $\Delta = \beta^2 + 4\gamma$. If D is a (v, k, λ, μ) PDS with $1_G \in D$ and $D^{-1} = D$, then $D' = D \backslash 1_G$ is also PDS with the parameters

$$(v, k - 1, \lambda - 2, \mu).$$

Therefore we can transform D to D' and discuss the regular PDS with $1_G \notin D$.

By the Eq. (1.12) in Theorem 1.20, we have $k^2 = k + \lambda k + \mu(v - k - 1)$ if $1_G \notin D$.

We have the following theorem if G is an abelian group.

Theorem 1.21 ***(Ma [130]):*** *Let G be an abelian group of order v and D be a k-subset of G with $D^{-1} = D$. Then D is a (v, k, λ, μ) partial difference set in G if and only if for any character χ of G,*

$$\chi(\mathcal{D}) = \begin{cases} k & \text{if } \chi \text{ is a trivial character of } G, \\ \frac{1}{2}\left\{\lambda - \mu \pm \sqrt{(\lambda - \mu)^2 + 4\gamma}\right\} & \text{if } \chi \text{ is a nontrivial character of } G, \end{cases} \tag{1.13}$$

where $\gamma = k - \mu$ if $1_G \notin D$ and $\gamma = k - \lambda$ if $1_G \in D$.

Example 1.22: *Let G be the additive group of $GF(13)$ and $D = \{\mathbf{0}, \mathbf{1}, \mathbf{3}, \mathbf{4}, \mathbf{9}, \mathbf{10}, \mathbf{12}\}$. We write the elements of $GF(13)$ by bold symbol to distinguish elements of $GF(13)$ and the numbers of $\mathbf{Z}$. Then*

$$\begin{aligned} \mathcal{D} &= (\mathbf{0} + \mathbf{1} + \mathbf{3} + \mathbf{4} + \mathbf{9} + \mathbf{10} + \mathbf{12})^2 \\ &= 3 \sum_{\boldsymbol{i} \in GF(13)} \boldsymbol{i} + (\mathbf{0} + \mathbf{1} + \mathbf{3} + \mathbf{4} + \mathbf{9} + \mathbf{10} + \mathbf{12}) + 3 \cdot \mathbf{0}. \end{aligned}$$

Hence D is a $(13, 7, 4, 3)$ PDS from Theorem 1.20.

We notice that $D^{-1} = D$ and $D \ni \mathbf{0}$. Let $D^- = D \backslash \mathbf{0} = \{\mathbf{1}, \mathbf{3}, \mathbf{4}, \mathbf{9}, \mathbf{10}, \mathbf{12}\}$. D^- is a $(13, 6, 2, 3)$ PDS.

We observe the connection between a strongly regular graph and a partial difference set.

Definition 1.38: A **strongly regular graph** $SRG(v, k, \lambda, \mu)$ is a k-regular graph with v vertices and has the following properties:

i) For any two adjacent vertices x and y, there are exactly λ vertices adjacent to x and y.
ii) For any two nonadjacent vertices x and y, there are exactly μ vertices adjacent to x and y.

The pentagon and the Petersen graph are famous examples of strongly regular graphs. The strongly regular graph was first studied in connection with partial geometries and symmetric association schemes of class 2.

Definition 1.39: A **Cayley graph** is defined as an undirected graph with no loops which admits an automorphism group acting regularly on the vertex set. If we identify the vertex set with the elements of the regular automorphism group G, then a Cayley graph generates the subset D such that two vertices $x, y \in G$ are adjacent if and only if $xy^{-1} \in D$.

Theorem 1.22 (*Ma [130]*): *A Cayley graph generated by subset D of the regular automorphism group G is a strongly regular if and only if D is a PDS in G with* $1_G \notin D$ *and* $D^{-1} = D$.

1.10 Sequences and Autocorrelation Function

Definition 1.40: Let $A = \{a_1, a_2, \ldots, a_n\}$ a sequence of length n. The **non periodic autocorrelation function** $N_A(s)$ is defined as

$$N_A(s) \equiv NPAF_A(s) = \sum_{i=1}^{n-s} a_i a_{i+s}, \quad s = 0, 1, \ldots, n-1, \tag{1.14}$$

The **periodic autocorrelation function** $P_A(s)$ is defined, reducing $i + s$ modulo n, as

$$P_A(s) \equiv PAF_A(s) = \sum_{i=1}^{n} a_i a_{i+s}, \quad s = 0, 1, \ldots, n-1. \tag{1.15}$$

In the case of sequences with zero nonperiodic autocorrelation function, the sequences are first padded with sufficient zeros added to the end to make their length t.

Definition 1.41: Let $X = \left\{\{a_{11}, \ldots, a_{1n}\}, \{a_{21}, \ldots, a_{2n}\}, \ldots, \{a_{m1}, \ldots, a_{mn}\}\right\}$ be m sequences of commuting variables of length n. The **nonperiodic autocorrelation function of the family of sequences** X (denoted N_X) is a function defined by

$$N_X(j) = \sum_{i=1}^{n-j} \left(a_{1,i}a_{1,i+j} + a_{2,i}a_{2,i+j} + \cdots + a_{m,i}a_{m,i+j}\right).$$

Note that if the following collection of m matrices of order n is formed,

$$\begin{bmatrix} a_{11} & a_{12} & \cdots & a_{1n} \\ & a_{11} & & a_{1,n-1} \\ & & \ddots & \\ \bigcirc & & & a_{11} \end{bmatrix}, \begin{bmatrix} a_{21} & a_{22} & \cdots & a_{2n} \\ & a_{21} & & a_{2,n-1} \\ & & \ddots & \\ \bigcirc & & & a_{21} \end{bmatrix}, \ldots, \begin{bmatrix} a_{m1} & a_{m2} & \cdots & a_{mn} \\ & a_{m1} & & a_{m,n-1} \\ & & \ddots & \\ \bigcirc & & & a_{m1} \end{bmatrix}$$

then $N_X(j)$ is simply the sum of the inner products of rows 1 and $j + 1$ of these matrices.

Definition 1.42: The **periodic autocorrelation function of the family of sequences** X (denoted P_X) is a function defined by

$$P_X(j) = \sum_{i=1}^{n} (a_{1,i}a_{1,i+j} + a_{2,i}a_{2,i+j} + \cdots + a_{m,i}a_{m,i+j}),$$

where we assume the second subscript is actually chosen from the complete set of residues $\pmod n$.

We can interpret the function P_X in the following way: form the m circulant matrices which have first rows, respectively,

$$[a_{11}a_{12}\cdots a_{1n}],\ [a_{21}a_{22}\cdots a_{2n}],\ \ldots,\ [a_{m1}a_{m2}\cdots a_{mn}]\ ;$$

then $P_X(j)$ is the sum of the inner products of rows 1 and $j+1$ of these matrices.

Definition 1.43: If X is as above with $N_X(j)=0$, $j=1,2,\ldots,n-1$, then we will call X m-**complementary sequences** of length n or X has a **zero autocorrelation** (they are also called **suitable sequences** because they work). The number of nonzero elements is called **weight.**

If $X=\{A_1,A_2,\ldots,A_m\}$ are m-complementary sequences of length n and weight $2k$ such that

$$Y=\left\{\frac{(A_1+A_2)}{2},\frac{(A_1-A_2)}{2},\ldots\frac{(A_{2i-1}+A_{2i})}{2},\frac{(A_{2i-1}-A_{2i})}{2},\ldots\right\}$$

are also m-complementary sequences (of weight k), then X will be said to be m-**complementary disjointable sequences** of length n. X will be said to be m-**complementary disjoint sequences** of length n if all $\binom{m}{2}$ pairs of sequences are disjoint, i.e. $A_i\star A_j=0$ for all i,j, where $\star$ is the Hadamard product.

Example 1.23: *For example* $\{1101\}$, $\{0010-\}$, $\{00000100-\}$, $\{0000001-\}$ *are disjoint as they have zero nonperiodic autocorrelation function and precisely one* $a_{ij}\neq 0$ *for each* j. *(Here – means "minus 1.") With padding these sequences become*

$$\{11010000000\},\{0010-000000\},\{00000100-00\},\{0000000001-\}.$$

Lemma 1.19: *Suppose* $X=\{X_1,X_2,\ldots,X_m\}$ *is a set of* $(0,1,-1)$ *sequences of length* n *for which* $N_X=0$ *or* $P_X=0$. *Further suppose the weight of* X_i *is* x_i *and the sum of the elements of* X_i *is* a_i. *Then*

$$\sum_{i=1}^{m}a_i^2=\sum_{i=1}^{m}x_i$$

Proof: Form circulant matrices Y_i for each X_i. Then

$$Y_iJ=a_iJ\quad\text{and}\quad\sum_{i=1}^{m}Y_iY_i^{\top}=\sum_{i=1}^{m}x_i$$

Now considering

$$\sum_{i=1}^{m}Y_iY_i^{\top}J=\sum_{i=1}^{m}a_i^2J=\sum_{i=1}^{m}x_iJ,$$

we have the result. □

Lemma 1.20: *Let* $A=\{a_1,a_2,\ldots,a_m\}$ *and* A^* *be* A *reversed, that is* $A^*=\{a_m,\ldots,a_2,a_1\}$. *Then*

$$N_{A^*}(s)=N_A(s).$$

Proof:

$$N_{A^*}(s)=\sum_{i=0}^{m-1-s}a_{m-1-i}a_{m-1-(i+s)}=\sum_{i=0}^{m-1-s}a_{i+s}a_i$$
$$=N_A(s).$$
□

Lemma 1.21: *Let X be a family of sequences as above. Then*

$$P_X(j) = N_X(j) + N_X(n-j), \quad j = 1, \dots, n-1.$$

Corollary 1.4: *If $N_X(j) = 0$ for all $j = 1, \dots, n-1$, then $P_X(j) = 0$ for all $j = 1, \dots, n-1$.*

Remark 1.2: $P_X(j)$ may equal 0 for all $j = 1, \dots, n-1$, even though the $N_X(j)$ do not equal zero.

1.10.1 Multiplication of NPAF Sequences

Now a few simple observations are in order, and for convenience we put them together as a lemma.

Notation 1.44: We sometimes use – for -1, and $\bar{x}$ for $-x$, and A^* to mean the order of the entries in the sequence A are reversed. We use the notation A/B to denote the interleaving of two sequences $A = \{a_1, \dots, a_m\}$ and $B = \{b_1, \dots, b_m\}$.

$$A/B = \{a_1, b_1, a_2, b_2, \dots, b_{m-1}, a_m, b_m\}$$

and $\{A, B\}$ to denote the adjoining

$$\{A, B\} = \{a_1, a_2, \dots, a_m, b_1, b_2, \dots, b_{m-1}, b_m\}.$$

One more piece of notation is in order. If g_r denotes a sequence of integers of length r, then by xg_r we mean the sequences of integers of length r obtained from g_r, by multiplying each member of g_r by x.

Lemma 1.22: *Let $X = \{A_1, A_2, \dots, A_m\}$ are m-complementary sequences of length n. Then*

i) $Y = \{A_1^*, A_2^*, \dots, A_i^*, A_{i+1}, \dots, A_m\}$ *are m-complementary sequences of length n;*
ii) $W = \{A_1, A_2, \dots, A_i, -A_{i+1}, \dots, -A_m\}$ *are m-complementary sequences of length n;*
iii) $Z = \big\{\{A_1, A_2\}, \{A_1, -A_2\}, \dots, \{A_{2i-1}, A_{2i}\}, \{A_{2i-1}, -A_{2i}\}, \dots\big\}$ *are m– (or m + 1 if m was odd when we let A_{m+1} be n zeros) complementary sequences of length 2n;*
iv) $U = \big\{\{A_1/A_2\}, \{A_1/-A_2\}, \dots, \{A_{2i-1}/A_{2i}\}, \{A_{2i-1}/-A_{2i}\}, \dots\big\}$, *are m- (or m + 1 if m was odd when we let A_{m+1} be n zeros) complementary sequences of length 2n.*

By a lengthy but straightforward calculation, it can be shown that:

Theorem 1.23: *Suppose $X = \{A_1, \dots, A_{2m}\}$ are 2m-complementary sequences of length n and weight ℓ and $Y = \{B_1, B_2\}$ are 2-complementary disjointable sequences of length t and weight 2k. Then there are 2m-complementary sequences of length nt and weight $k\ell$.*

The same result is true if X are 2m-complementary disjointable sequences of length n and weight 2ℓ and Y are 2-complementary sequences of weight k.

Proof: Using an idea of R.J. Turyn [194], we consider

$$A_{2i-1} \times \frac{(B_1 + B_2)}{2} + A_{2i} \times \frac{(B_1^* - B_2^*)}{2} \quad \text{and}$$
$$A_{2i-1} \times \frac{(B_1 - B_2)}{2} - A_{2i} \times \frac{(B_1^* + B_2^*)}{2}$$

for $i = 1, \dots, m$, which are the required sequences in the first case, and

$$\frac{(A_{2i-1} + A_{2i})}{2} \times B_1 + \frac{(A_{2i-1} - A_{2i})}{2} \times B_2^* \text{ and}$$
$$\frac{(A_{2i-1} + A_{2i})}{2} \times B_2 - \frac{(A_{2i-1} - A_{2i})}{2} \times B_1^*,$$

for $i = 1, \dots, m$, which are the required sequences for the second case. □

1.10.2 Golay Sequences

Early work of Golay [83, 84] was concerned with two $(1, -1)$ sequences with zero (nonperiodic autocorrelation function, $NPAF = 0$), but Welti [217], Tseng [185], and Tseng and Liu [186] approached the subject from the point of view of two orthonormal vectors, each corresponding to one of two orthogonal waveforms. Later work, including Turyn's [189, 193] used four or more sequences.

Definition 1.45: If $X = \{\{a_1, \dots, a_n\}, \{b_1, \dots, b_n\}\}$ are two sequences where $a_j, b_j \in \{1, -1\}$ and $N_X(j) = 0$ for $j = 1, \dots, n-1$, then the sequences in X are called **Golay complementary sequences of length** n or **Golay sequences** or **Golay pairs**. The positive integer n such that there exist Golay complementary sequences is called a **Golay number**.

Example 1.24:

$$\begin{aligned}
&n = 2 \quad 11 \textit{ and } 1- \\
&n = 10 \quad 1--1-1---1 \textit{ and } 1------11- \\
&n = 26 \quad 111--111-1-----1-11--1---- \textit{ and} \\
&\qquad\qquad ---11---1-11-1-1-11--1----.
\end{aligned}$$

We note that if X is as above and A is the circulant matrix with first row $\{a_1, \dots, a_n\}$ and B the circulant matrix with first row $\{b_1, \dots, b_n\}$, then

$$AA^T + BB^T = \sum_{i=1}^{n} (a_i^2 + b_i^2) I_n.$$

Consequently, such matrices may be used to obtain Hadamard matrices constructed from two circulants.

We would like to use Golay sequences to construct other orthogonal designs, but first we consider some of their properties.

Lemma 1.23: *Let $X = \{\{a_1, \dots, a_n\}, \{b_1, \dots, b_n\}\}$ be Golay complementary sequences of length n. Suppose k_1 of the a_i are positive and k_2 of the b_i are positive. Then*

$$n = (k_1 + k_2 - n)^2 + (k_1 - k_2)^2,$$

and n is even.

Proof: Since $P_X(j) = 0$ for all j, we may consider the two sequences as $2 - \{n; k_1, k_2; \lambda\}$ supplementary difference sets with $\lambda = k_1 + k_2 - \frac{1}{2}n$. But the parameters (counting the difference in two ways) satisfy $\lambda(n-1) = k_1(k_1 - 1) + k_2(k_2 - 1)$. On substituting λ in this equation we obtain the result of the enunciation. □

Seberry Wallis [209] introduces many properties of supplementary difference sets. Also see Sections 1.9 and 1.10.

Theorem 1.24: *Golay complementary sequences $\{\{x_1, x_2, \dots, x_n\}, \{y_1, y_2, \dots, y_n\}\}$ have the following properties:*

i) $\sum_{i=1}^{n-j}(x_ix_{i+j} + y_iy_{i+j}) = 0$ *for every* $j \neq 0, \quad j = 1, \dots, n-1,$
(where the subscripts are reduced modulo n),
i.e. $P_X = 0$.

ii) *n is even and the sum of 2 squares.*

iii) $x_{n-i+1} = e_ix_i \Leftrightarrow y_{n-i+1} = -e_iy_i$ *where* $e_i = \pm 1$.

iv) $$\left[\sum_{i\in S} x_i Re\left(\zeta^{2i+1}\right)\right]^2 + \left[\sum_{i\in D} x_i Im\left(\zeta^{2i+1}\right)\right]^2 + \left[\sum_{i\in S} y_i Im\left(\zeta^{2i+1}\right)\right]^2 + \left[\sum_{i\in D} y_i Re\left(\zeta^{2i+1}\right)\right]^2 = \frac{1}{2}n$$
where $S = \{i : 0 \le i < n, e_i = 1\}$, $D = \{i : 0 \le i < n, e_i = -1\}$, *and* ζ *is a 2n-th root of unity.*

v) *Exist for orders* $2^a10^b26^c$, a, b, c *nonnegative integers.*

vi) *Do not exist for orders 2.n (n a positive integer), when any factor of the prime decomposition of n is* $\equiv 3 \pmod 4$ *or 34, 50, or 58.*

Proof: The assertions (*i*) and (*ii*) follow from the definition and Lemma 1.23. We obtain the assertions (*iv*) and (*vi*) by Griffin [88] and Eliahou et al. [68]. □

Craigen et al. [33] and Holzmann and Kharaghani [95] has made fundamental advances in studying Golay type sequences in other contexts such as with complex entries or matrix entries. The following theorem is needed in one proof of the asymptotic existence results for Hadamard matrices:

Theorem 1.25 *(Craigen et al. [33])*: *For any positive integer n, there is a pair of Golay sequences of length* 2^n *in type 1 matrices each appearing* 2^{n-1} *times in each of the sequences.*

Proof: Let A_{n-1} and B_{n-1} be a pair of Golay sequences of length 2^{n-1} in type 1 matrices each appearing 2^{n-2} times in each of the sequences. Then $A_n = \{A_{n-1}, B_{n-1}\}$ and $B_n = \{A_{n-1}, -B_{n-1}\}$ form a Golay pair of length 2^n in type 1 matrices. □

We discuss Golay complementary sequences further in Chapter 5.
We define a complex Golay sequence which will be used in Chapter 11.

Definition 1.46: **Complex Golay sequences of length** n are a pair of sequences $\{a_1, a_2, \dots, a_n\}$ and $\{b_1, b_2, \dots, b_n\}$ where $a_j, b_j \in \{1, -1, i, -i\}$, $i = \sqrt{-1}$ and has the zero autocorrelation,

$$0 = \sum_{j=1}^{k} a_j\bar{a}_{n-k+j} + \sum_{j=1}^{k} b_j\bar{b}_{n-k+j} \quad \text{for } k = 1, 2, \dots, n-1$$

where $\overline{a_j}$ and $\overline{b_j}$ are complex conjugates of a_j and b_j, respectively.

The positive number n such that there exist complex Golay sequences of length n is called a **complex Golay number**.

1.11 Excess

We define the excess of an Hadamard matrix. We will show that Hadamard matrices with maximal excess and regular Hadamard matrices are equivalent (see [156]).

Definition 1.47: The sum of all its entries of an Hadamard matrix H is called the **excess of** H and denoted by $\sigma(H)$. The number

$$\sigma(n) = max\{\sigma(H)|H \text{ is an Hadamard matrix of order } n\}$$

is the **maximum excess**.

The number of 1's in an Hadamard matrix H is the **weight** of H and denoted by $w(H)$. The number $w(n)$ is the maximum of $w(H)$. It follows that $\sigma(H) = 2w(H) - n^2$ and $\sigma(n) = 2w(n) - n^2$.

Example 1.25: *The excess of the following Hadamard matrices*

$$H_1 = \begin{bmatrix} 1 & 1 & 1 & 1 \\ 1 & 1 & - & - \\ 1 & - & 1 & - \\ 1 & - & - & 1 \end{bmatrix}, \quad H_2 = \begin{bmatrix} - & 1 & 1 & 1 \\ 1 & - & 1 & 1 \\ 1 & 1 & - & 1 \\ 1 & 1 & 1 & - \end{bmatrix}$$

are $\sigma(H_1) = 4, \sigma(H_2) = 8$. H_2 has the maximal excess of all Hadamard matrices of order 4, then $\sigma(4) = 8$.

From Kronecker product, we have the following lemma.

Lemma 1.24: *$\sigma(mn) \geq \sigma(m)\sigma(n)$.*

The most encompassing upper bound is that of Best [14].

Theorem 1.26:

$$\sigma(n) \leq n\sqrt{n}$$

where n is a order of Hadamard matrices. The equality holds if and only if there exists a regular Hadamard matrix.

Proof: Put $H = (h_{i,j})$. Denote kth column sum $\sum_{i=1}^{n} h_{ik}$ by s_k.

$$n^2 = \sum_{i=1}^{n}\sum_{j=1}^{n}\sum_{k=1}^{n} h_{ik}h_{jk} = \sum_{k=1}^{n}\left(\sum_{i=1}^{n} h_{ik}\right)^2 = \sum_{k=1}^{n} s_k^2.$$

By Cauchy–Schwartz inequality and the above,

$$\sigma(H) = \sum_{k=1}^{n} s_k \leq \left(n\sum_{k-1}^{n} s_k^2\right)^{1/2} = \left(n \cdot n^2\right)^{1/2} = n\sqrt{n}.$$

Assume that $\sigma(n) = n\sqrt{n}$. The equality of Cauchy–Schwartz inequality holds if and only if s_k is constant. Hence H is a regular Hadamard matrix.

Conversely, assume that H is a regular Hadamard matrix. Then s_k is a constant value c for any k. Thus $n^2 = \sum_{k=1}^{n} s_k^2 = nc^2$. Therefore we have $c = \sqrt{n}$ and $\sigma(n) = nc = n\sqrt{n}$. □

In the case of $n = 4u^2$, we have $\sigma(4u^2) \leq 8u^3$.

Lower and upper bounds of the maximum excess have been given [93, 115].

Theorem 1.27: *The following are equivalent:*

i) there exists an Hadamard matrix of order $4u^2$ with maximal excess $8u^3$,
ii) there exists a regular Hadamard matrix of order $4u^2$,

iii) there is an SBIBD$(4u^2, 2u^2 + u, u^2 + u)$ *(and its complement the SBIBD* $\left(4u^2, 2u^2 - u, u^2 - u\right)$*), or* $(4u^2, 2u^2 \pm u, u^2 \pm u)$ *Menon Hadamard difference sets.*

We will define SBIBD in Section 1.12 and discuss Menon Hadamard matrices in Chapter 7.

Hadamard matrices of order $4u^2$ with maximal excess satisfying this bound have been found by Koukouvinos, Kounias, Seberry, and Yamada [114, 115, 156, 236].

An algorithm which finds an Hadamard matrix is equivalent to one which has maximal excess shows the matrix is equivalent to a regular Hadamard matrix.

1.12 Balanced Incomplete Block Designs

Definition 1.48: A **balanced incomplete block design or configuration** is a pair $(X, \mathbb{B})$ which satisfies the following properties:

i) X is a set of v elements that is called **points**,
ii) $\mathbb{B}$ is a set of b subsets of X that is called **blocks**. A block of $\mathbb{B}$ contains exactly k distinct points.
iii) Every point occurs in exactly r different blocks.
iv) Every pair of distinct points occurs together in exactly λ blocks.

We denote by a (b, v, r, k, λ)*-BIBD* or a (b, v, r, k, λ)-configuration.

Example 1.26: *For* $b = 12$, $v = 9$, $r = 4$, $k = 3$, $\lambda = 1$, *the following are blocks of* $(12, 9, 4, 3, 1)$*-BIBD:*

1 2 3	1 4 7	1 5 9	1 6 8
4 5 6	2 5 8	2 6 7	3 5 7
7 8 9	3 6 9	3 4 8	2 4 9

Counting the number of objects in two different ways we see

$$r(k-1) = \lambda(v-1) \tag{1.16}$$

and counting the number of replications of an element, we see each element is replicated r times but since each block has k elements we must have

$$bk = vr. \tag{1.17}$$

If $r = \lambda$, then every block contains X which is called the **trivial BIBD**.

The following theorem gives a necessary condition on the existence of BIBD.

Theorem 1.28 ***(Fisher's inequality):*** *A* (b, v, r, k, λ)*-BIBD exists only if* $b \geq v$.

Definition 1.49: The **complement** D' of a BIBD with parameters (b, v, r, k, λ) is obtained by replacing every blocks by its complement. Clearly if b, v, r', k', λ' are the parameters of the complementary design of a (b, v, r, k, λ)-configuration then if $0 < \lambda$ and $k < v - 1$ we have

$$\begin{cases} r' = b - r \\ k' = v - k \\ \lambda' = b - 2r + \lambda \end{cases} \tag{1.18}$$

and since $\lambda'(v-1) = r'(k'-1)$ we have $\lambda'(v-1) = (b-r)(v-k-1)$, so $0 < \lambda'$ and $k' < v - 1$.

Definition 1.50: If a (b, v, r, k, λ)-*BIBD* has $b = v$ and $r = k$, we call it a **symmetric BIBD** or **symmetric configuration** and denote it as a (v, k, λ)-***SBIBD*** or (v, k, λ)-**configuration**.

For SBIBD's, Eq. (1.16) becomes

$$k(k-1) = \lambda(v-1). \tag{1.19}$$

The most fundamental necessary condition on the existence of SBIBD is given by Bruck, Ryser, and Chowla.

Theorem 1.29 *(Bruck–Ryser–Chowla theorem [92, pp. 133–138]):* *If v, k, λ are integers for which there is a (v, k, λ)-SBIBD, then:*

i) if v is even, $k - \lambda$ is a square;
ii) if v is odd, then the Diophantine equation

$$x^2 = (k-\lambda)y^2 + (-1)^{\frac{1}{2}(v-1)}\lambda z^2 \tag{1.20}$$

has a solution in the integers not all zero.

Equation (1.19) and Theorem 1.29 give necessary conditions for the existence of a (v, k, λ)-configuration but it is not known whether or not they are sufficient.

Definition 1.51: Let $(X, \mathbb{B})$ and $(X', \mathbb{B}')$ be BIBD's. Let σ be a bijection such that $X' = \sigma(X)$ and $\mathbb{B}' = \mathbb{B}$. Then σ is a **isomorphism** of **BIBD** if and only if it satisfies $x \in B$ if and only if $\sigma(x) \in \sigma(B)$ for all point $x \in X$ and block $B \in \mathbb{B}$. Further, $(X, \mathbb{B})$ and $(X', \mathbb{B}')$ are said to be **isomorphic**.

If $X = X'$ and $\mathbb{B} = \mathbb{B}'$, then σ is called an **automorphism** of **BIBD** of $(X, \mathbb{B})$.

Definition 1.52: All automorphisms of $(X, \mathbb{B})$ form a group which is called an **automorphism group of BIBD**. If the automorphism group of $(X, \mathbb{B})$ acts regularly on X and on $\mathbb{B}$, then $(X, \mathbb{B})$ is **regular**.

Definition 1.53: The **incidence matrix** of a BIBD is a $(0,1)$-matrix of size $v \times b$ with $a_{ij} = 1$ if object i is in block j and $a_{ij} = 0$ otherwise. If $A = (a_{ij})$ is the incidence matrix of a BIBD with parameters (b, v, r, k, λ), A satisfies

$$J_{v\times v}A = kJ_{v\times b} \tag{1.21}$$

$$AA^{\top} = (r-\lambda)I_{v\times v} + \lambda J_{v\times v}. \tag{1.22}$$

Conversely, if A is a $(0,1)$-matrix satisfying Eqs. (1.21) and (1.22) then this ensures there is a BIBD with parameters (b, v, r, k, λ).

Theorem 1.30 *(Ryser [148, p. 130]):* *If A is the incidence matrix of a (v, k, λ)-SBIBD then A satisfies*

$$AA^{\top} = (k-\lambda)I + \lambda J \tag{1.23}$$

$$A^{\top}A = (k-\lambda)I + \lambda J \tag{1.24}$$

$$AJ = kJ \tag{1.25}$$

$$JA = kJ \tag{1.26}$$

where J is the matrix with all entries 1.

Theorem 1.31 *(Ryser's theorem [148, p. 369]):* *Suppose $A = (a_{ij})$ is a $v \times v$ matrix of integers and $k(k-1) = \lambda(v-1)$ and $AA^{\top} = A^{\top}A = (k-\lambda)I + \lambda J$. Then A or $-A$ is composed entirely of zeros and ones and is the incidence matrix of a SBIBD.*

Theorem 1.32: *If B is the $(1,-1)$ matrix obtained from the $(1,0)$ incidence matrix A of a (v, k, λ)-SBIBD by every entry 0 being replaced by -1, then B satisfies*

$$BB^{\top} = 4(k-\lambda)I + (v - 4(k-\lambda))J.$$

Proof: We can write $B = 2A - J$. It follows from Theorem 1.30. □

We call this matrix the $(1, -1)$ **incidence matrix** of a (v, k, λ)-*SBIBD*.

1.13 Hadamard Matrices and SBIBDs

Every Hadamard matrix H of order $4t$ is associated in a natural way with an SBIBD with parameters $(4t-1, 2t-1, t-1)$. To obtain the SBIBD we first normalize H and write the resultant matrix in the form

$$\begin{bmatrix} 1 & 1 & \dots & 1 \\ 1 & & & \\ \vdots & & A & \\ 1 & & & \end{bmatrix}.$$

Then

$$AJ = JA = -J \text{ and } AA^{\top} = 4tI - J.$$

So $B = \frac{1}{2}(A + J)$ satisfies

$$BJ = JB = (2t-1)J \text{ and } BB^{\top} = tI + (t-1)J.$$

Thus B is a $(0, 1)$-matrix satisfying the equations for the incidence matrix of an SBIBD with parameters $(4t-1, 2t-1, t-1)$. Clearly if we start with the incidence matrix B of an SBIBD with parameters $(4t-1, 2t-1, t-1)$ and

$$\begin{bmatrix} 1 & 1 & \dots & 1 \\ 1 & & & \\ \vdots & & 2B-J & \\ 1 & & & \end{bmatrix}$$

is an Hadamard matrix of order $4t$.

Theorem 1.33: *Hence Hadamard matrices can be formed from the SBIBD(v, k, λ) with parameters*

i) *SBIBD*$(4t-1, 2t-1, t-1)$, *t and integer.*
ii) *SBIBD*$\left(uv, \frac{1}{2}(uv-1), \frac{1}{4}(uv-3)\right)$ *u and v twin prime powers.*
iii) *SBIBD*$\left(4m^2, m^2 \pm m, m^2 \pm m\right)$, *m and integer.*
iv) *SBIBD*$\left(2^t - 1, 2^{t-2} - 1, 2^{t-2} - 1\right)$, *t and integer.*

Regular SBIBD corresponds to a difference set which will be shown in Section 1.9. We will discuss difference sets with the parameters (i) and (ii) in Chapter 8 and parameter (iii) in Chapter 7. (iv) follows from (i).

1.14 Cyclotomic Numbers

Let $q = ef + 1$ be an odd prime power and $K = \boldsymbol{F}_q$ be the finite field with q elements. Let g be a primitive element of K. The e**th cyclotomic classes** $S_0, S_1, \dots, S_{e-1}$ in K are defined by

$$S_i = \left\{g^{es+i} : s = 0, \dots, f-1\right\} = g^i \langle g^e \rangle$$

where $\langle g^e \rangle$ is a group generated by g^e.

Notice that S_0 is the set of eth power residues and the cyclotomic classes are pairwise disjoint and their union is $K^\times$.

Definition 1.54: For fixed i and j, $0 \le i,j \le e-1$, we define the **cyclotomic number** (i,j). That is, the number of solutions of the equation

$$x - y = 1,\ x \in S_i,\ y \in S_j.$$

We denote the cyclotomic class involves -1 by S_h. If f is even, then $h = 0$, and if f is odd, then $h = \frac{e}{2}$. The following lemma gives the basic properties of the cyclotomic numbers.

Lemma 1.25:

i) $(i,j) = (j+h, i+h)$,
ii) $(i,j) = (i-j, -j)$,
iii) $\sum_{j=0}^{e-1}(i,j) = \frac{q-1}{e} - \delta_{i,0}$,
iv) $\sum_{i=0}^{e-1}(i,j) = \frac{q-1}{e} - \delta_{j,h}$ *where* $\delta_{i,0}$, $\delta_{j,h}$ *are Kronecker deltas.*

Proof:

i) The equation $x - y = 1$ is equivalent to the equation $-y-(-x) = 1$ and $-y \in S_{j+h}$, $-x \in S_{i+h}$.
ii) The equation $x - y = 1$ is equivalent to the equation $xy^{-1} - y^{-1} = 1$ and $xy^{-1} \in S_{i-j}$, $y^{-1} \in S_{-j}$.
iii) Since the sum $\sum_{j=0}^{e-1}(i,j)$ is the number of the elements in S_i, it is equal to $f = \frac{q-1}{e}$. When $i = 0$, we remove the solution $(x,y) = (1,0)$ from $0 \notin S_j$.
iv) From (*i*) and (*iii*), we have

$$\sum_{j=0}^{e-1}(i,j) = \sum_{j=0}^{e-1}(j+h, i+h) = \frac{q-1}{e} - \delta_{j+h,0} = \frac{q-1}{e} - \delta_{j,h}.$$

□

We define the cyclotomic matrix $X = (x_{i,j})$, $i,j = 0, \ldots, e-1$ by $x_{i,j} = (i,j)$.

The case where $e = 2$
From (*iii*) in Lemma 1.25,

$$(0,0) + (0,1) = \frac{1}{2}(q-3),$$
$$(1,0) + (1,1) = \frac{1}{2}(q-1).$$

If $f = \frac{1}{2}(q-1)$ is even, then $h = 0$,

$$(0,1) = (1,0) = (1,1)$$

from (*i*) and (*ii*) in Lemma 1.25. Hence we have

$$A = (0,0) = \frac{1}{4}(q-5), \quad B = (0,1) = (1,0) = (1,1) = \frac{1}{4}(q-1).$$

For odd f, we obtain similarly

$$A = (0,0) = (0,1) = (1,1) = \frac{1}{4}(q-3), \quad B = (1,0) = \frac{1}{4}(q+1).$$

f:odd

	0	1
0	A	B
1	B	B

f:even

	0	1
0	A	A
1	B	A

Lemma 1.26: *Let q be an odd prime power. The cyclotomic numbers for $e = 2$ are given by*

$$A = \frac{1}{4}\left(q - 4 - (-1)^f\right) \text{ and } B = \frac{1}{4}\left(q - (-1)^f\right)$$

The case where $e = 4$ and f is even

Then $h = 0$. From (i) and (ii),

$$A : (0,0),$$
$$B : (0,2) = (2,0) = (2,2),$$
$$C : (1,2) = (2,1) = (3,2) = (1,3) = (3,1) = (2,3),$$
$$D : (0,1) = (1,0) = (3,3),$$
$$E : (0,3) = (3,0) = (1,1).$$

$$A + D + B + E = \frac{1}{4}(q-5).$$
$$D + E + 2C = \frac{1}{4}(q-1).$$
$$2(B + C) = \frac{1}{4}(q-1).$$

	0	1	2	3
0	A	D	B	E
1	D	E	C	C
2	B	C	B	C
3	E	C	C	D

The case where $e = 4$ and f is odd

Then $h = 2$ and

$$A : (2,0),$$
$$B : (0,0) = (2,2) = (0,2),$$
$$C : (0,1) = (3,2) = (3,3) = (1,2) = (1,1) = (0,3),$$
$$D : (1,3) = (2,1) = (3,0),$$
$$E : (1,0) = (2,3) = (3,1).$$

$$A + D + B + E = \frac{1}{4}(q-1) = \frac{1}{4}(q-5) + \frac{1}{2}h.$$
$$D + E + 2C = \frac{1}{4}(q-1).$$
$$2(B + C) = \frac{1}{4}(q-5) = \frac{1}{4}(q-1) - \frac{1}{2}h.$$

	0	1	2	3
0	B	C	B	C
1	E	C	C	D
2	A	D	B	E
3	D	E	C	C

Lemma 1.27: *Let $q = a^2 + 4b^2$, $a \equiv -1 \pmod 4$. The cyclotomic numbers for $e = 4$ are given by*

$$A = \frac{1}{16}(q - 11 + 6a + 6h), \qquad D = \frac{1}{16}(q - 3 - 2a + 8b + 2h),$$
$$B = \frac{1}{16}(q - 3 - 2a - 2h), \qquad E = \frac{1}{16}(q - 3 - 2a - 8b + 2h),$$
$$C = \frac{1}{16}(q + 1 + 2a - 2h),$$

where $h = 0$ iff is even and $h = 2$ iff is odd. The sign of b depends on the choice of a primitive element of K.

Proof: See [13]. □

We will show that b is determined by a and a primitive element in Lemma 2.3.

The case where $e = 8$ and f is odd

From Lemma 1.25, we have the following array.

$$\begin{array}{c|cccccccc} & 0 & 1 & 2 & 3 & 4 & 5 & 6 & 7 \\ \hline 0 & A & I & N & J & A & I & N & J \\ 1 & B & J & O & O & I & H & M & K \\ 2 & C & K & N & O & N & M & G & L \\ 3 & D & L & M & I & J & K & L & F \\ 4 & E & F & G & H & A & B & C & D \\ 5 & F & D & L & M & I & J & K & L \\ 6 & G & L & C & K & N & O & N & M \\ 7 & H & M & K & B & J & O & O & I \end{array} \tag{1.27}$$

$$2A + 2I + 2J + 2N = f - 1 = \frac{1}{8}(q - 9).$$
$$B + H + I + J + K + M + 2O = f = \frac{1}{8}(q - 1).$$
$$C + G + K + L + M + 2N + O = f = \frac{1}{8}(q - 1).$$
$$D + F + I + J + K + 2L + M = f = \frac{1}{8}(q - 1).$$
$$A + B + C + D + E + F + G + H = f = \frac{1}{8}(q - 1).$$

Lemma 1.28 *(Storer [179, p. 79]):* *Let $q = p^t = 8f + 1$ be a prime power. The cyclotomic numbers for $e = 8, f$ odd, are given by the following:*

I. If 2 is a fourth power in G	*II. If 2 is not a fourth power in G*
$64A = q - 15 + 2a$	$64A = q - 15 + 10a + 8c$
$64B = q + 1 - 2a + 16b + 4c$	$64B = q + 1 - 2a + 4c + 16d$
$64C = q + 1 - 6a - 16b - 8c$	$64C = q + 1 + 2a + 16b$
$64D = q + 1 - 2a - 16b + 4c$	$64D = q + 1 - 2a + 4c + 16d$
$64E = q + 1 + 18a$	$64E = q + 1 - 6a - 24c$
$64F = q + 1 - 2a + 16b + 4c$	$64F = q + 1 - 2a + 4c - 16d$
$64G = q + 1 - 6a + 16b - 8c$	$64G = q + 1 + 2a - 16b$
$64H = q + 1 - 2a - 16b + 4c$	$64H = q + 1 - 2a + 4c - 16d$
$64I = q - 7 - 2a - 4c$	$64I = q - 7 - 2a + 16b - 4c$
$64J = q - 7 - 2a - 4c$	$64J = q - 7 - 2a - 16b - 4c$
$64K = q + 1 + 6a - 4c - 16d$	$64K = q + 1 - 2a + 4c$
$64L = q + 1 - 2a + 4c$	$64L = q + 1 + 6a - 4c$
$64M = q + 1 + 6a - 4c + 16d$	$64M = q + 1 - 2a + 4c$
$64N = q - 7 + 2a + 8c$	$64N = q - 7 - 6a$
$64O = q + 1 - 2a + 4c$	$64O = q + 1 + 6a - 4c$

where a, b, c, and d are specified by:

i) If $p \equiv 1 \pmod 4$, then a and b satisfy

$$q = a^2 + 4b^2,\ a \equiv -1 \pmod 4$$

and the sign of b depends on a and a primitive element.
If $p \equiv 3 \pmod 4$, then

$$a^2 = q,\ a \equiv -1 \pmod 4,\ b = 0.$$

ii) If $p \equiv 1$ *or* $3 \pmod 8$, *then c and d satisfy*

$$q = c^2 + 2d^2, \ c \equiv -1 \pmod 4$$

and the sign of d depends on c and a primitive element.
If $p \equiv 5$ *or* $7 \pmod 8$, *then*

$$c^2 = q, \ c \equiv -1 \pmod 4, \ d = 0.$$

The case where $e = 8$ and f is even

Similarly we have the following array.

$$\begin{array}{c|cccccccc} & 0 & 1 & 2 & 3 & 4 & 5 & 6 & 7 \\ \hline 0 & A & B & C & D & E & F & G & H \\ 1 & B & H & I & J & K & L & M & I \\ 2 & C & I & G & M & N & O & N & J \\ 3 & D & J & M & F & L & O & O & K \\ 4 & E & K & N & L & E & K & N & L \\ 5 & F & L & O & O & K & D & J & M \\ 6 & G & M & N & O & N & J & C & I \\ 7 & H & I & J & K & L & M & I & B \end{array} \tag{1.28}$$

$$A + B + C + D + E + F + G + H = f - 1 = \frac{1}{8}(q - 9).$$
$$B + H + J + K + L + M + 2I = f = \frac{1}{8}(q - 1).$$
$$C + I + J + G + M + 2N + O = f = \frac{1}{8}(q - 1).$$
$$D + J + K + F + L + M + 2O = f = \frac{1}{8}(q - 1).$$
$$2E + 2K + 2N + 2L = f = \frac{1}{8}(q - 1).$$

Lemma 1.29: *Let* $q = p^t = 8f + 1$ *be a prime power. The cyclotomic numbers for* $e = 8, f$ *even, are given as follows:*

I. If 2 is a fourth power in GF(q)	*II. If 2 not a fourth power in GF(q)*
$64A = q - 23 + 18a + 24c$	$64A = q - 23 - 6a$
$64B = q - 7 - 2a + 8b - 4c + 16d$	$64B = q - 7 - 2a - 4c$
$64C = q - 7 - 6a + 16b$	$64C = q - 7 + 2a - 16b + 8c$
$64D = q - 7 - 2a - 8b - 4c + 16d$	$64D = q - 7 - 2a - 4c$
$64E = q - 7 + 2a - 8c$	$64E = q - 7 + 10a$
$64F = q - 7 - 2a + 16b - 4c - 16d$	$64F = q - 7 - 2a - 4c$
$64G = q - 7 - 6a - 16b$	$64G = q - 7 + 2a + 16b + 8c$
$64H = q - 7 - 2a - 16b - 4c - 16d$	$64H = q - 7 - 2a - 4c$
$64I = q + 1 - 2a + 4c$	$64I = q + 1 + 6a - 4c$
$64J = q + 1 + 6a - 4c$	$64J = q + 1 - 2a + 4c - 16d$
$64K = q + 1 - 2a + 4c$	$64K = q + 1 - 2a + 16b + 4c$
$64L = q + 1 - 2a + 4c$	$64L = q + 1 - 2a - 16b + 4c$
$64M = q + 1 + 6a - 4c$	$64M = q + 1 - 2a + 4c + 16d$
$64N = q + 1 + 2a$	$64N = q + 1 - 6a - 8c$
$64O = q + 1 - 2a + 4c$	$64O = q + 1 + 6a - 4c$

where a, b, c, and d are specified as in Lemma 1.28.

Remark 1.3: The cyclotomic numbers are determined by Jacobi sums. The integers a, b, c, and d are real parts and imaginary parts such that

$$J(\chi^2, \psi) = a + bi \text{ and } J(\chi, \psi) = c + d\sqrt{2}i$$

where χ is a primitive character of order 8, $\psi = \chi^4$, and $i = \sqrt{-1}$.

We will discuss the relation between cyclotomic numbers and Jacobi sums in Chapter 2.

1.15 Orthogonal Designs and Weighing Matrices

Definition 1.55: A **weighing matrix** $W(n,k)$ of weight k and of order n is a $0, \pm 1$ matrix such that $W(n,k)W(n,k)^{\top} = kI$.

In particular, $W(n, n-1)$ is called a **conference matrix**, that is, a conference matrix C of order n is $0, \pm 1$ matrix with zero diagonal satisfying $CC^{\top} = (n-1)I$.

If $n = k$, then $W(n,n)$ is an Hadamard matrix.

An orthogonal design is a generalization of an Hadamard matrix, and we use it in proving the asymptotic Theorem 11.1. First we collect a few preliminary results.

The most common structure matrices are "plugged into" is an orthogonal design.

Definition 1.56: An **orthogonal design of order** n **and type** $(s_1, \dots, s_u)$, s_i positive integers, is an $n \times n$ matrix X, with entries $\{0, \pm x_1, \dots, \pm x_u\}$ (the x_i commuting indeterminates) satisfying

$$XX^{\top} = \left(\sum_{i=1}^{u} s_i x_i^2 \right) I_n. \tag{1.29}$$

We write this as $OD(n; s_1, s_2, \dots, s_u)$.

Alternatively, each row of X has s_i entries of the indeterminate $\pm x_i$ and the distinct rows are orthogonal under the Euclidean inner product. We may view X as a matrix with entries in the field of fractions of the integral domain $\mathbf{Z}[x_1, \dots, x_u]$, ($\mathbf{Z}$ the rational integers), and then if we let $f = \Sigma_{i=1}^{u} s_i x_i^2$, X is an invertible matrix with inverse $\frac{1}{f}X^{\top}$ from $XX^{\top} = fI$, thus, we have $X^{\top}X = fI_n$ and so our alternative definition that the row vectors are orthogonal applies equally well to the column vectors of X.

Example 1.27: *Some small orthogonal designs are shown in Table 1.4. Notice that (ii) is the Hadamard equivalent of the Williamson array.*

Table 1.4 Orthogonal designs.

(i)	(ii)	(iii)	(iv)
$\begin{bmatrix} x & y \\ y & -x \end{bmatrix}$	$\begin{bmatrix} a & -b & -c & -d \\ b & a & -d & c \\ c & d & a & -b \\ d & -c & b & a \end{bmatrix}$	$\begin{bmatrix} a & b & b & d \\ -b & a & d & -b \\ -b & -d & a & b \\ -d & b & -b & a \end{bmatrix}$	$\begin{bmatrix} a & 0 & -c & 0 \\ 0 & a & 0 & c \\ c & 0 & a & 0 \\ 0 & -c & 0 & a \end{bmatrix}$
$OD(2;1,1)$	$OD(4;1,1,1,1)$	$OD(4;1,1,2)$	$OD(4;1,1)$

Source: Seberry and Yamada [166, figure 4.1, p. 458], Wiley.

An orthogonal design in which each of the entries is replaced by +1 or −1 is weighing matrix $W(n, \sum_{i=1}^{n} s_i)$.

An orthogonal design with no zeros and in which each of the entries is replaced by +1 or −1 is an **Hadamard matrix**.

In particular,

Definition 1.57: An orthogonal design $OD(4t; t, t, t, t)$ is called a **Baumert–Hall array** of order t or $BH(4t; t, t, t, t)$.

Definition 1.58: **A Welch array** of order t, denoted as **WL** or **WL(t)**, is a Baumert–Hall array constructed from 16 circulant or type 1 matrices. It is a matrix of size $4t$. A **Ono–Sawade–Yamamoto array** is a $BH(36; 9, 9, 9, 9)$ or equivalently a $WL(9)$ array constructed from 16 type 1 matrices.

1.16 *T*-matrices, *T*-sequences, and Turyn Sequences

Definition 1.59: A set of four T-**matrices**, T_i, $i = 1, \dots, 4$ of order t are four circulant or type one matrices which have entries 0, +1, or −1 and which satisfy

i) $T_i * T_j = 0, i \neq j$, ($*$ the Hadamard product)
ii) $\sum_{i=1}^{4} T_i$ is a $(1, -1)$ matrix,
iii) $\sum_{i=1}^{4} T_i T_i^T = tI_t$,
iv) $t = t_1^2 + t_2^2 + t_3^2 + t_4^2$ where t_i is the row(column) sum of T_i.

T-matrices are known (see Cohen et al. [20] for a survey) (71 occurs in [117], 73 and 79 occur in [254]) for many orders including:
1, ..., ..., 82, 84, ..., 88, 90, ..., 96, 98, ..., 102, 104, ..., 106, 108, 110, ..., 112, 114, ..., 126, 128, ..., 130, 132, 136, 138, 140, ..., 148, 150, 152, ..., 156, 158, ..., 162, 164, ..., 166, 168, ..., 172, 174, ..., 178, 180, 182, 184, ..., 190, 192, 194, ..., 196, 198, 200, ..., 210, ...

T-matrices of order t give Hadamard matrices of order $4t$. Only a few of these orders (73 and 79 in [254] and 107 in [109, p. 5]) have been discovered in the past 25 years. They pose a valuable research problem.

Definition 1.60: A set of four sequences

$$A = \left\{ \{a_{11}, \dots, a_{1n}\}, \{a_{21}, \dots, a_{2n}\}, \{a_{31}, \dots, a_{3n}\}, \{a_{41}, \dots, a_{4n}\} \right\}$$

of length n, with entries $0, 1, -1$ so that exactly one of $\{a_{1j}, a_{2j}, a_{3j}, a_{4j}\}$ is ± 1 (three are zero) for $j = 1, \dots, n$ and with zero nonperiodic autocorrelation function, that is, $N_A(j) = 0$ for $j = 1, \dots, n-1$, where

$$N_A(j) = \sum_{i=1}^{n-j} \left(a_{1i}, a_{1,i+j} + a_{2i}a_{2,i+j} + a_{3i}a_{3,i+j} + a_{4i}a_{4,i+j}\right),$$

are called T-**sequences**.

Example 1.28: *Consider*

$$T = \left\{ \{10000\}, \{01100\}, \{0001-\}, \{00000\} \right\}.$$

The sequences are disjoint, as the i-th entry is nonzero in one and only one of the four sequences. The total weight is 5, and $N_T = 0$.

Another example is obtained by using the Golay sequences

$$X = 1 - -1 - 1 - - - 1 \quad \text{and} \quad Y = 1 - - - - - -11 - .$$

Let 0_{10} be the vector of 10 zeros; then

$$T = \left\{ \{1, 0_{10}\}, \{0, \frac{1}{2}(X+Y)\}, \{0, \frac{1}{2}(X-Y)\}, \{0, 0_{10}\} \right\}$$

are T sequences of length 11.

T-matrices are a slightly weaker condition than T-sequences being defined on finite abelian groups rather than the infinite cyclic group. They are known for a few important small orders, for example 61 and 67 [100, 149], for which no T-sequences are yet known. Sequences are discussed extensively in Chapter 5. They are also known for even orders $2t$ for which no T-sequences of length t are known [113]. These will give Hadamard matrices of order $8t$.

1.16.1 Turyn Sequences

Turyn sequences are binary sequence with $NPAF = 0$ and a powerful tool to construct T-sequences, T-matrices, and Baumert–Hall array and leads to the construction of Hadamard matrices of Goethals–Seidel type which will be discussed in Chapter 3.

Definition 1.61: A sequence $X = \{x_1, x_2, \dots, x_n\}$ is **symmetric** if $X = X^*$ where $X^* = \{x_n, x_{n-1}, \dots, x_1\}$. A sequence X is **skew** if n is even and $X^* = -X$.

Definition 1.62: Four binary sequences $A = \{X, U, V, W\}$ are called **Turyn sequences** if the following conditions are satisfied:

(*i*) Sequences X and U are of length n and V and W are of length $n-1$ (the four sequences are of weight $n, n, n-1$, and $n-1$).

(*ii*) $NPAF_A = 0$.

(*iii*) If n is even, then $X^* = X$ and $V^* = -V$
and if n is odd, then $X^* = -X$ and $V^* = V$.

Example 1.29: *(Turyn sequences)* $n = 4 : \{\{11 - -\}, \{11 - 1\}, \{111\}, \{1 - 1\}\}$
$n = 5 : \{\{11 - 11\}, \{1111-\}, \{11 - -\}, \{11 - -\}\}$

We will discuss Turyn sequences in more detail in Section 5.7 and give all that are known and conjectured to exist. Turyn sequences will be used to construct base sequences and other $NPAF = 0$ sequences.

2

Gauss Sums, Jacobi Sums, and Relative Gauss Sums

2.1 Notations

Table 2.1 gives the notations which are used in this chapter.

2.2 Gauss Sums

Definition 2.1: Let $q = p^s$. We put $K = \boldsymbol{F}_q$ and $F = \boldsymbol{F}_{q^t}$. Let χ be a multiplicative character and η_β be an additive character of K. The **Gauss sum** $g(\chi, \eta_\beta)$ is defined by

$$g(\chi, \eta_\beta) = \sum_{\alpha \in K} \chi(\alpha)\eta_\beta(\alpha) = \sum_{\alpha \in K} \chi(\alpha)\zeta_p^{T_K \alpha\beta}$$

where T_K is the absolute trace.

Let ζ_m be a primitive mth root of unity. Denote a cyclotomic field obtained by adjoining ζ_m to the rational number field $\boldsymbol{Q}$ by $\boldsymbol{Q}(\zeta_m)$. The Gauss sum is an element in $\boldsymbol{Q}(\zeta_m, \zeta_p)$ where m is the order of χ.

The following theorem gives the properties of Gauss sums.

Theorem 2.1: *Suppose* $\chi \neq \chi^0$.

i) $g(\chi, \eta_\beta) = \overline{\chi}(\beta) g(\chi, \eta_1)$.
ii) $g(\chi^0, \eta_1) = -1$.
iii) $g(\chi^p, \eta_1) = g(\chi, \eta_1)$ *where p is the characteristic of* $\boldsymbol{F}_q$.
iv) $g(\chi, \eta_1)\overline{g(\chi, \eta_1)} = q$.
v) We let $g(\chi_1) = g(\chi_1, \eta_1)$. *For* $\chi_1, \chi_2, \chi_1\chi_2 \neq \chi^0$, *we have*

$$\frac{g(\chi_1)g(\chi_2)}{g(\chi_1\chi_2)} = \sum_{\alpha \in K} \chi_1(\alpha)\chi_2(1-\alpha).$$

vi) Let m be the order of χ. *We define the automorphism* $\sigma_{a,b}$ *of* $\boldsymbol{Q}(\zeta_m, \zeta_p)$ *by*

$$\sigma_{a,b}(\zeta_m) = \zeta_m^a,\ \sigma_{a,b}(\zeta_p) = \zeta_p^b.$$

Then

$$\sigma_{a,b}(g(\chi,\ \eta_1)) = \chi^{-a}(b) g(\chi^a,\ \eta_1).$$

Hadamard Matrices: Constructions using Number Theory and Algebra, First Edition. Jennifer Seberry and Mieko Yamada.

Table 2.1 Notations used in this chapter.

ζ_m	A primitive mth root of unity
$\boldsymbol{Q}(\zeta_m)$	A cyclotomic field obtained by adjoining ζ_m to the rational number field $\boldsymbol{Q}$
$g(\chi, \eta_\beta)$	The Gauss sum for a multiplicative character χ and an additive character η_β
$J(\chi_1, \chi_2)$	The Jacobi sum for multiplicative characters χ_1 and χ_2
S_0	The set of eth power residues
S_i	The eth cyclotomic class, $S_i = g^i S_0$ where g is a primitive element of a finite field
(i, j)	The cyclotomic number
$\vartheta_{F/K}(\chi_F) = \vartheta(\chi_F, \chi_K)$	The relative Gauss sum associated with the character χ_F of F induces the character χ_K in K
$\widetilde{\vartheta}_{F/K}(\chi_F) = \frac{g_F(\chi_F)}{p^{\frac{f'-f}{2}} g_K(\chi_K)}$	The normalized relative Gauss sum associated with the character χ_F of F
ω	Teichmüller character
$\mathfrak{P}, \wp, P$	A prime ideal lying above p
$\langle t \rangle$	A fractional part of a real number t

Proof: See [13, 124, 126, 127, 246]. □

Gauss sums may be determined explicitly for special characters. The following theorem gives the explicit formula of the Gauss sum for the quadratic character.

Theorem 2.2: *Let $q = p^s$ where p is an odd prime. Let χ be the quadratic character and $\eta = \eta_1$ be the additive character of $\boldsymbol{F}_q$. Then*

$$g(\chi, \eta) = \begin{cases} (-1)^{s-1}\sqrt{q} & \text{if } p \equiv 1 \pmod 4, \\ (-1)^{s-1} i^s \sqrt{q} & \text{if } p \equiv 3 \pmod 4, \end{cases}$$

where i is a primitive 4th root of unity.

Example 2.1: *Let $p = 3$, $t = 2$, and $K = \boldsymbol{F}_{3^2}$. Let χ be the primitive quadratic character and η be the additive character of K. Denote a primitive cubic root of unity by ω. Let ξ be a primitive element of K with the primitive polynomial $x^2 + 2x + 2$.*

ℓ	0	1	2	3	4	5	6	7
$T_K(\xi^\ell)$	2	1	0	1	1	2	0	2

$$g(\chi, \eta) = \sum_{\ell=0}^{7} \chi\left(\xi^\ell\right) \omega^{T_K(\xi^\ell)} = 3 = (-1)^{t-1} i^t \sqrt{q},$$

where $i = \sqrt{-1}$.

Definition 2.2: Let χ be a multiplicative character of $K = \boldsymbol{F}_q$ and F be an extension of K. The **lift** χ' of χ from K to F is given by

$$\chi'(\alpha) = \chi\left(N_{F/K}\alpha\right), \quad \alpha \in F.$$

Let η be an additive character K. The **lift** η' of η from K to F is given by

$$\eta'(\alpha) = \eta\left(T_{F/K}\alpha\right), \quad \alpha \in F.$$

The basic properties of the lift from K to F are given in the following theorem.

Theorem 2.3 ***(Berndt et al. [13, Theorem 11.4.4]):*** *Let χ be a multiplicative character of K and χ' be the lift of χ from K to F. Then*

i) χ' is a character of F,
ii) χ' and χ have the same order,
iii) a character ψ of F equals the lift χ' of some character χ of K if and only if the order of ψ divides $q-1$,
iv) if $\chi_1' = \chi_2'$, then $\chi_1 = \chi_2$.

For the lifts χ' and η' from K to F, we define the lift of the Gauss sum.

Definition 2.3: Let χ be a multiplicative character of K and η be an additive character of K. Let χ' be a lift of χ and η' be the lift of η from K to F. Then the **lift of the Gauss sum** from K to F is defined by

$$g_F(\chi',\eta') = \sum_{\alpha\in F} \chi'(\alpha)\eta'(\alpha).$$

The following two theorems give an important relationship between the Gauss sums.

Theorem 2.4 ***(Davenport–Hasse theorem on the lifted Gauss sum):*** *Let χ be a nontrivial multiplicative character and η be a nontrivial additive character of K. Let χ' and η' be the lift of χ and η from K to F, respectively. Then*

$$g_F\left(\chi',\eta'\right) = (-1)^{t-1} g_K\left(\chi,\eta\right)^t.$$

Proof: See [13, 127]. □

Theorem 2.5 ***(Davenport–Hasse product formula):*** *Let χ be a character of $K^\times$ of order $m > 1$. We put $g(\psi^m) = g(\psi^m, \eta_1)$. For every character ψ of $K^\times$, if $\psi^m \neq \psi^0$, then*

$$g(\psi^m) = \psi(m^m)\frac{\prod_{\chi^m=1} g(\chi\psi)}{\prod_{\chi^m=1} g(\chi)}.$$

If χ is a quadratic character and $\psi^2 \neq \psi^0$, then

$$g(\psi^2) = \psi(4)\frac{g(\psi)g(\chi\psi)}{g(\chi)}.$$

Proof: See [13, 124]. □

2.3 Jacobi Sums

Definition 2.4: Let χ_1, χ_2 be multiplicative characters of K. Then the sum

$$J(\chi_1,\chi_2) = \sum_{\alpha\in K} \chi_1(\alpha)\chi_2(1-\alpha)$$

in (v) of Theorem 2.1, is called the **Jacobi sum** for multiplicative characters χ_1 and χ_2. The **order of Jacobi sum** $J(\chi_1,\chi_2)$ is the least common multiple of the orders of χ_1 and χ_2.

The properties of Jacobi sums are given in the following theorem.

Theorem 2.6: *For multiplicative characters χ_1, χ_2, and χ_3, Jacobi sums satisfy the following properties:*

i) $J(\chi_1, \chi_2) = J(\chi_2, \chi_1)$.
ii) $J(\chi_1, \chi_2) = \chi_1(-1)J(\chi_1, \overline{\chi_1\chi_2})$.
iii) $J(\chi^0, \chi^0) = q - 2$, $J(\chi_1, \chi^0) = -1$ *for* $\chi_1 \neq \chi^0$.
iv) $J(\chi_1, \overline{\chi}_1) = -\chi_1(-1)$ *for* $\chi_1 \neq \chi^0$.
v) $J(\chi_1, \chi_2)J(\chi_1\chi_2, \chi_3) = J(\chi_1, \chi_2\chi_3)J(\chi_2, \chi_3)$.
vi) *If q is odd and* $\chi_1^2 \neq \chi^0$, *then*

$$J(\chi_1, \chi_1) = \overline{\chi}_1(4)J(\chi_1, \varphi)$$

where φ is a quadratic character of K.

vii) *For an automorphism* $\sigma : \zeta_{q-1} \to \zeta_{q-1}^c$, $(c, q-1) = 1$,

$$J(\chi_1^\sigma, \chi_2^\sigma) = J(\chi_1, \chi_2)^\sigma.$$

viii) *If* $\chi_1, \chi_2, \chi_1\chi_2 \neq \chi^0$, *then*

$$J(\chi_1, \chi_2)\overline{J(\chi_1, \chi_2)} = q.$$

Proof: See [13, 246]. □

2.3.1 Congruence Relations

The following congruence relations are helpful for the evaluation of Jacobi sums.

Theorem 2.7: *Let χ_j be a character of the order $k_j > 1$ for $j = 1, 2$. Denote a primitive k_j-th root of unity by ζ_{k_j} for $j = 1, 2$. Then*

$$J(\chi_1, \chi_2) \equiv -q \pmod{(1 - \zeta_{k_1})(1 - \zeta_{k_2})}.$$

Proof:

$$\begin{aligned} J(\chi_1, \chi_2) &= \sum_{\alpha \in K}(1 - \chi_1(\alpha))(1 - \chi_2(1 - \alpha)) - q \\ &\equiv -q \pmod{(1 - \zeta_{k_1})(1 - \zeta_{k_2})}. \end{aligned}$$

□

Corollary 2.1: *Let φ be a quadratic character of K. Then we have*

$$J(\chi_1, \varphi) \equiv -q \pmod{2(1 - \zeta_{k_1})}.$$

2.3.2 Jacobi Sums of Order 4

Let $q = 4f + 1$ be an odd prime power and $\chi = \chi_4$ be a primitive biquadratic character.

We put $\varpi = J(\chi, \chi^2)$. From Theorem 2.6, we have

$$J(\chi, \chi) = \chi(-1)J(\chi, \chi^2),\ J(\chi^2, \chi^2) = \chi^2(-1) = -1,\ J(\chi, \chi^3) = -\chi(-1).$$

Let $\epsilon = \chi(-1) = (-1)^f$. We obtain Table 2.2.

Lemma 2.1: *Let $\varpi = J(\chi, \chi^2) = a + bi, i = \sqrt{-1}$. Then*

Table 2.2 $J(\chi_4^{\ell}, \chi_4^m)$ for $0 \le \ell, m \le 3$.

	χ^0	χ	χ^2	χ^3
χ^0	$q-2$	-1	-1	-1
χ	-1	$\epsilon\varpi$	ϖ	$-\epsilon$
χ^2	-1	ϖ	-1	$\overline{\varpi}$
χ^3	-1	$-\epsilon$	$\overline{\varpi}$	$\epsilon\overline{\varpi}$

i) if $q \equiv 1 \pmod 8$, *then* $a \equiv -1 \pmod 4$ *and* $b \equiv 0 \pmod 4$,
ii) if $q \equiv 5 \pmod 8$, *then* $a \equiv 1 \pmod 4$ *and* $b \equiv 2 \pmod 4$.

Proof: From Corollary 2.1,

$$\varpi = J(\chi, \chi^2) \equiv -q \equiv -1 \qquad (\text{mod } 2(1-i)).$$

We obtain

$$\varpi \equiv -1 \pmod 4 \quad \text{or} \quad \varpi \equiv -1 + 2(1-i) \pmod 4.$$

The former case yields $q \equiv 1 \pmod 8$, $a \equiv -1 \pmod 4$, and $b \equiv 0 \pmod 4$. The latter case yields $q \equiv 5 \pmod 8$, $a \equiv 1 \pmod 4$, and $b \equiv 2 \pmod 4$. □

Let $\psi = \chi^2$. We can write $\varpi = J(\chi, \psi) = a + 2bi$.

Lemma 2.2: *Let* $q = 4f + 1$ *be a prime power and* $K = GF(q)$. *We put* $\varpi = J(\chi, \psi) = a + 2bi, a, b \in \mathbf{Z}$. *Then*

$$a = \frac{1}{2}(-1)^f \sum_{\alpha \in K^\times} \psi(\alpha)\psi\left(1 + \alpha^2\right)$$

and

$$b = \frac{1}{4}(-1)^{f+1} \sum_{\alpha \in K^\times} \psi(\alpha)\psi(u^{-1} + \alpha^2)$$

where u is any element such that $\chi(u) = i$.

Proof: We have

$$\begin{aligned} 2a &= J(\chi, \psi) + \overline{J(\chi, \psi)} \\ &= \sum_{\alpha \in K^\times} \chi(\alpha)\psi(1-\alpha) + \sum_{\alpha \in K^\times} \chi^3(\alpha)\psi(1-\alpha) \end{aligned}$$

repalacing α by $-\beta$ and from $\chi(-1) = (-1)^f$,

$$= 2(-1)^f \left\{ \sum_{\chi(\beta)=1} \psi(1+\beta) - \sum_{\chi(\beta)=-1} \psi(1+\beta) \right\}.$$

$$\sum_{\chi(\beta)=1} \psi(1+\beta) = \frac{1}{4} \sum_{\beta \in K^\times} \psi\left(1 + \beta^4\right)$$

and

$$\sum_{\chi(\beta)=-1} \psi(1+\beta) = \frac{1}{2}\sum_{\beta\in K^\times} \psi(1+\beta^2) - \frac{1}{4}\sum_{\beta\in K^\times} \psi\left(1+\beta^4\right),$$

we have

$$2a = (-1)^f \left\{ \sum_{\beta\in K^\times} \psi\left(1+\beta^4\right) - \sum_{\beta\in K^\times} \psi\left(1+\beta^2\right) \right\}.$$

On the other hand,

$$\begin{aligned}
&\sum_{\alpha\in K^\times} \psi(\alpha)\psi\left(1+\alpha^2\right) + \sum_{\alpha\in K^\times} \psi\left(1+\alpha^2\right) \\
&= \sum_{\alpha\in K^\times} (\psi(\alpha)+1)\psi\left(1+\alpha^2\right) = 2\sum_{\psi(\alpha)=1} \psi\left(1+\alpha^2\right) = \sum_{\alpha\in K^\times} \psi\left(1+\alpha^4\right).
\end{aligned}$$

Then we have the assertion. We have

$$\begin{aligned}
4bi &= J(\chi,\psi) - \overline{J(\chi,\psi)} \\
&= 2(-1)^f \left\{ \sum_{\chi(\beta)=i} \psi(1+\beta) - \sum_{\chi(\beta)=-i} \psi(1+\beta) \right\}.
\end{aligned}$$

Let u be an element of K such that $\chi(u) = i$.

$$\begin{aligned}
&\sum_{\chi(\beta)=i} \psi(1+\beta) = \sum_{\chi(\beta)=1} \psi(1+u\beta) = \sum_{\chi(\beta)=1} \psi(u)\psi(u^{-1}+\beta) \\
&= \frac{1}{4}\sum_{\beta\in K^\times} \psi(u^{-1}+\beta^4).
\end{aligned}$$

Thus

$$\begin{aligned}
4b &= 2(-1)^f \left\{ \sum_{\chi(\beta)=i} \psi(1+\beta) - \sum_{\psi(\beta)=-1} \psi(1+\beta) + \sum_{\chi(\beta)=-i} \psi(1+\beta) \right\} \\
&= (-1)^{f+1} \left\{ \sum_{\beta\in K^\times} \psi(u^{-1}+\beta^4) - \sum_{\beta\in K^\times} \psi(u^{-1}+\beta^2) \right\}.
\end{aligned}$$

Similarly we have,

$$b = \frac{1}{4}(-1)^{f+1} \sum_{\beta\in K^\times} \psi(\beta)\psi(u^{-1}+\beta^4).$$

□

The following theorem is called Lucas' theorem.

Theorem 2.8 (Lucas' theorem): *Let p be a prime and m, n be positive integers. The p-adic expansions of m and n are given by*

$$\begin{aligned}
m &= m_0 + m_1 p + \cdots + m_{s-1}p^{s-1}, \\
n &= n_0 + n_1 p + \cdots + n_{s-1}p^{s-1}.
\end{aligned}$$

Then

$$\binom{m}{n} \equiv \prod_{i=0}^{s-1} \binom{m_i}{n_i} \pmod{p}.$$

Proof: Since $(1+x)^{p^r} \equiv 1 + x^{p^r} \pmod p$, we have

$$\begin{aligned}(1+x)^m &= (1+x)^{m_0+m_1p+\cdots+m_{s-1}p^{s-1}} \\ &\equiv (1+x)^{m_0}\cdots(1+x^{p^r})^{m_r}\cdots(1+x^{p^{s-1}})^{m_{s-1}} \pmod p.\end{aligned}$$

The coefficient of x^n on the left is $\binom{m}{n}$. From $x^n = x^{n_0}\cdots(x^{p^r})^{n_r}\cdots(x^{p^{s-1}})^{n_{s-1}}$, the coefficient of x^n on the right is $\prod_{i=0}^{s-1}\binom{m_i}{n_i}$. Then, we have the theorem. □

Theorem 2.9: *Let $q = p^s$ and put $J(\chi,\psi) = a + 2bi$. If $p \equiv 1 \pmod 4$, then*

$$2a \equiv (-1)^{f+1}\binom{\frac{1}{2}(q-1)}{\frac{1}{4}(q-1)} \equiv (-1)^{f+1}\binom{\frac{1}{2}(p-1)}{\frac{1}{4}(p-1)}^s \pmod p.$$

If $p \equiv 3 \pmod 4$, then

$$a \equiv 0 \pmod p.$$

Proof: From Lemma 2.2,

$$\begin{aligned}\sum_{\alpha\in K^\times}\psi(\alpha)\psi\left(1+\alpha^2\right) &= \sum_{\alpha\in K^\times}\left(\alpha+\alpha^3\right)^{\frac{1}{2}(q-1)} \\ &= \sum_{\alpha\in K^\times}\sum_{k=0}^{\frac{1}{2}(q-1)}\binom{\frac{1}{2}(q-1)}{k}\alpha^{2k+\frac{1}{2}(q-1)}.\end{aligned}$$

Since

$$\sum_{\alpha\in K^\times}\alpha^{2k+\frac{1}{2}(q-1)} = \begin{cases} q-1 & \text{if } 2k+\frac{1}{2}(q-1)\equiv 0 \pmod{q-1}, \\ 0 & \text{otherwise,}\end{cases}$$

we obtain

$$2a \equiv (-1)^f\binom{\frac{1}{2}(q-1)}{\frac{1}{4}(q-1)}(-1) \equiv (-1)^{f+1}\binom{\frac{1}{2}(q-1)}{\frac{1}{4}(q-1)} \pmod q.$$

Assume that $p \equiv 1 \pmod 4$. We have the p-adic expansions

$$\begin{aligned}\frac{1}{2}(q-1) &= \frac{1}{2}(p-1)(1+p+\cdots+p^{s-1}), \\ \frac{1}{4}(q-1) &= \frac{1}{4}(p-1)(1+p+\cdots+p^{s-1}).\end{aligned}$$

Thus

$$2a \equiv (-1)^{f+1}\binom{\frac{1}{2}(q-1)}{\frac{1}{4}(q-1)} \equiv (-1)^{f+1}\binom{\frac{1}{2}(p-1)}{\frac{1}{4}(p-1)}^s \pmod p$$

from Lucas' theorem.

Assume that $p \equiv 3 \pmod 4$. Then s is even. The p-adic expansion of $\frac{1}{4}(q-1)$ is

$$\frac{1}{4}(q-1) = \sum_{t=0}^{\frac{s}{2}-1}\frac{1}{4}(3p-1)p^{2t} + \sum_{t=1}^{\frac{s}{2}}\frac{1}{4}(p-3)p^{2t-1}.$$

From Lucas's theorem, we have

$$\binom{\frac{q-1}{2}}{\frac{q-1}{4}} = \binom{\frac{p-1}{2}}{\frac{3p-1}{4}}^{\frac{s}{2}} \binom{\frac{p-1}{2}}{\frac{p-3}{4}}^{\frac{s}{2}}.$$

However, $\frac{1}{4}(3p-1) > \frac{1}{2}(p-1)$. Then $a \equiv 0 \pmod{p}$. □

The sign of b is determined by the primitive element.

Lemma 2.3 *(Berndt et al. [13 theorem 3.2.2]):* *Let g be a fixed primitive element of $K = GF(q)$. Let $J(\chi, \psi) = a + 2bi$. Then*

$$2b \equiv ag^{\frac{q-1}{4}} \pmod{p}.$$

Proof: Let ζ_{q-1} be a primitive $q-1^{\text{st}}$ root of unity and $\wp$ be a prime ideal in $\mathbf{Q}(\zeta_{q-1})$ lying above p. Denote the integer ring of $\mathbf{Q}(\zeta_{q-1})$ by $\mathcal{O}$. Then $K = GF(q) \cong \mathcal{O}/\wp$. A nonzero residue class $\alpha + \wp$ in $\mathcal{O}/\wp$ contains just one power ζ_{q-1}^x of ζ_{q-1}. That is $\zeta_{q-1}^x \equiv g^x \pmod{\wp}$.

From Theorem 2.6, $J(\chi, \psi) \equiv 0 \pmod{\wp}$. Hence

$$0 \equiv J(\chi, \psi) = a + 2bi \equiv a + 2bg^{\frac{q-1}{4}} \pmod{\wp}.$$

Thus we have $2b \equiv ag^{\frac{q-1}{4}} \pmod{p}$. □

From Theorem 2.6, we have the expression of q as a sum of 2 squares such that $q = J(\chi, \psi)\overline{J(\chi, \psi)} = a^2 + 4b^2$. Conversely, if we know the expression of q as a sum of 2 squares, $q^2 = a^2 + 4b^2$, we determine $J(\chi, \psi)$ from Lemma 2.1, Theorem 2.9, and Lemma 2.3 by choosing a primitive element.

Example 2.2:

1) *Let $q = 13 = 4 \cdot 3 + 1, f = 3$. We choose 2 as a primitive element of $GF(13)$. Let χ be a primitive biquadratic character and ψ be a quadratic character of $GF(13)$. Put $J(\chi, \psi) = a + 2bi$. The prime number 13 is expressed as $13 = 3^2 + 4^2$. From Lemma 2.1, $a \equiv 1 \pmod{4}$, then $a = -3$. Actually,*

$$2a \equiv (-1)^{3+1}\binom{\frac{q-1}{2}}{\frac{q-1}{4}} = \binom{6}{3} \equiv -6 \pmod{13}$$

from Theorem 2.9, then $a = -3$. From Lemma 2.3, $2b \equiv -3 \cdot 2^{\frac{13-1}{4}} \equiv 2 \pmod{13}, b = 1$. Thus we have $J(\chi, \psi) = -3 + 2i$.

2) *Let $q = 17 = 4 \cdot 4 + 1$. We choose 3 as a primitive element of $GF(17)$. $17 = 1^2 + 4^2$. From Lemma 2.1, $a = -1$. We also know it by*

$$2a \equiv (-1)^{4+1}\binom{\frac{q-1}{2}}{\frac{q-1}{4}} = -\binom{8}{4} \equiv -2 \pmod{17}.$$

From Lemma 2.3, $2b \equiv -1 \cdot 3^{\frac{17-1}{4}} \equiv 4 \pmod{17}, b = 2$. Thus

$$J(\chi, \psi) = -1 + 4i.$$

Table 2.3 $J(\chi_8^\ell, \chi_8^m)$ for $0 \le \ell, m \le 7$.

	χ^0	χ	χ^2	χ^3	χ^4	χ^5	χ^6	χ^7
χ^0	$q-2$	-1	-1	-1	-1	-1	-1	-1
χ	-1	$\rho\kappa$	$\epsilon\rho\varpi$	$\epsilon\kappa$	κ	$\rho\varpi$	$\epsilon\rho\kappa$	$-\epsilon$
χ^2	-1	$\epsilon\rho\varpi$	ϖ	$\epsilon\rho\kappa$	ϖ	$\epsilon\rho\varpi$	-1	$\epsilon\rho\overline{\kappa}$
χ^3	-1	$\epsilon\kappa$	$\epsilon\rho\kappa$	$\rho\kappa$	κ	$-\epsilon$	$\epsilon\rho\overline{\varpi}$	$\rho\overline{\varpi}$
χ^4	-1	κ	ϖ	κ	-1	$\overline{\kappa}$	$\overline{\varpi}$	$\overline{\kappa}$
χ^5	-1	$\rho\varpi$	$\epsilon\rho\varpi$	$-\epsilon$	$\overline{\kappa}$	$\rho\overline{\kappa}$	$\epsilon\rho\overline{\kappa}$	$\epsilon\overline{\kappa}$
χ^6	-1	$\epsilon\rho\kappa$	-1	$\epsilon\rho\overline{\varpi}$	$\overline{\varpi}$	$\epsilon\rho\overline{\kappa}$	$\overline{\varpi}$	$\epsilon\rho\overline{\varpi}$
χ^7	-1	$-\epsilon$	$\epsilon\rho\overline{\kappa}$	$\rho\overline{\varpi}$	$\overline{\kappa}$	$\epsilon\overline{\kappa}$	$\epsilon\rho\overline{\varpi}$	$\rho\overline{\kappa}$

2.3.3 Jacobi Sums of Order 8

Let $q = 8f + 1$ be a prime power. We denote a primitive character of order 8 of $K = GF(q)$ by χ and a quadratic character of K by $\psi = \chi^4$. Let $\zeta = e^{\frac{2\pi i}{8}}$ be a primitive eighth root of unity.

We put $\kappa = J(\chi, \psi)$, $\varpi = J(\chi^2, \psi)$, $\rho = \chi(4) = \pm 1$, and $\epsilon = \chi(-1) = (-1)^{\frac{q-1}{8}} = (-1)^f$. From Theorem 2.6, we have Table 2.3.

From $\zeta = \frac{1+i}{\sqrt{2}}$, we know $\mathbf{Q}(\zeta) = \mathbf{Q}(\sqrt{2}, i)$. We note $\zeta + \zeta^3 = \sqrt{2}i$ and $\zeta - \zeta^3 = \sqrt{2}$. From Theorem 2.6, we have $J(\chi, \psi) = \psi(-1)J(\psi, \overline{\psi\chi}) = J(\chi^3, \psi)$. It means that $J(\chi, \psi)$ is fixed by the automorphism $\sigma_3 : \zeta \to \zeta^3$ of $\mathbf{Q}(\zeta)$. $\mathbf{Q}(\sqrt{2}i)$ is fixed by σ_3. Then $J(\chi, \psi) \in \mathbf{Q}(\sqrt{2}i)$. Thus we can write $J(\chi, \psi) = c + d\sqrt{2}i = c + (\zeta + \zeta^3)d$.

Lemma 2.4 *(Berndt et al. [13]):* *Let $J(\chi, \psi) = c + d\sqrt{2}i$. Then*

$$c \equiv -1 \pmod 4.$$

Proof: From Corollary 2.1,

$$J(\chi, \psi) = c + d\sqrt{2}i \equiv -q \pmod{2(1-\zeta)}.$$

Then

$$\begin{aligned} J(\chi, \psi)^{\sigma_3} &= c + (\zeta + \zeta^3)d \equiv -q \pmod{2(1-\zeta^3)}, \\ J(\chi, \psi)^{\sigma_5} &= c - (\zeta + \zeta^3)d \equiv -q \pmod{2(1-\zeta^5)}, \\ J(\chi, \psi)^{\sigma_7} &= c - (\zeta + \zeta^3)d \equiv -q \pmod{2(1-\zeta^7)}. \end{aligned}$$

Then

$$\begin{aligned} &(c + q + (\zeta + \zeta^3)d)^2(c + q - (\zeta + \zeta^3)d)^2 \\ &= ((c+q)^2 - (\zeta + \zeta^3)^2 d^2)^2 = ((c+q)^2 + 2d^2)^2 \equiv 0 \pmod{32} \end{aligned}$$

and so

$$(c+q)^2 + 2d^2 = q(q + 2c + 1) \equiv 0 \pmod 8.$$

Since $q \equiv 1 \pmod 8$, $c \equiv -\frac{1}{2}(q+1) \equiv -1 \pmod 4$. □

Lemma 2.5: *Let* $J(\chi,\psi) = c + d\sqrt{2}i$ *and* $K = GF(q)$. *Then*

$$c = \frac{1}{4}(-1)^f \sum_{\alpha\in K^\times} \psi(\alpha)\psi\left(1+\alpha^4\right).$$

Proof: We have

$$\begin{aligned}4c &= J(\chi,\psi) + J\left(\chi^3,\psi\right) + J\left(\chi^5,\psi\right) + J\left(\chi^7,\psi\right)\\ &= \sum_{\alpha\in K^\times} \chi(\alpha)\psi(1-\alpha) + \sum_{\alpha\in K^\times} \chi^3(\alpha)\psi(1-\alpha) + \sum_{\alpha\in K^\times} \chi^5(\alpha)\psi(1-\alpha)\\ &+ \sum_{\alpha\in K^\times} \chi^7(\alpha)\psi(1-\alpha)\end{aligned}$$

replacing $-\alpha$ by β and from $\chi(-1) = (-1)^f$

$$\begin{aligned}&= 4(-1)^f \left\{ \sum_{\chi(\beta)=1} \psi(1+\beta) - \sum_{\chi(\beta)=-1} \psi(1+\beta) \right\}\\ &= (-1)^f \left\{ \frac{4}{8} \sum_{\alpha\in K^\times} \psi\left(1+\alpha^8\right) - 4\left(\frac{1}{4} \sum_{\alpha\in K^\times} \psi\left(1+\alpha^4\right) - \frac{1}{8} \sum_{\alpha\in K^\times} \psi\left(1+\alpha^8\right) \right) \right\}\\ &= (-1)^f \left\{ \sum_{\alpha\in K^\times} \psi\left(1+\alpha^8\right) - \sum_{\alpha\in K^\times} \psi\left(1+\alpha^4\right) \right\}.\end{aligned}$$

Similarly to the proof of Lemma 2.2, we have

$$\sum_{\alpha\in K^\times} \psi(\alpha)\psi\left(1+\alpha^4\right) + \sum_{\alpha\in K^\times} \psi\left(1+\alpha^4\right) = \sum_{\alpha\in K^\times} \psi\left(1+\alpha^8\right).$$

Therefore

$$4c = (-1)^f \sum_{\alpha\in K^\times} \psi(\alpha)\psi\left(1+\alpha^4\right).$$

□

Theorem 2.10: *Let* $q = p^s$ *and* $J(\chi,\psi) = c + d\sqrt{2}i$. *If* $p \equiv 1 \pmod 8$, *then*

$$2c \equiv (-1)^{f+1} \binom{\frac{q-1}{2}}{\frac{q-1}{8}} \equiv (-1)^{f+1} \binom{\frac{p-1}{2}}{\frac{p-1}{8}}^s \pmod p.$$

If $p \equiv 5, 7 \pmod 8$, *then*

$$c \equiv 0 \pmod p.$$

If $p \equiv 3 \pmod 8$, *then*

$$2c \equiv (-1)^{f+1} \binom{\frac{q-1}{2}}{\frac{q-1}{8}} \equiv (-1)^{f+1} \binom{\frac{p-1}{2}}{\frac{p-3}{8}}^s \pmod p.$$

Proof: From Theorem 2.5,

$$\begin{aligned}4c &= (-1)^f \sum_{\alpha\in K^\times} \psi(\alpha)\psi(1+\alpha^4) = (-1)^f \sum_{\alpha\in K^\times} (\alpha+\alpha^5)^{\frac{q-1}{2}}\\ &= (-1)^f \sum_{k=0}^{\frac{q-1}{2}} \binom{\frac{q-1}{2}}{k} \sum_{\alpha\in K^\times} \alpha^{4k+\frac{q-1}{2}}.\end{aligned}$$

If $4k + \frac{q-1}{2} \equiv 0 \pmod{q-1}$ for $0 \le k \le \frac{q-1}{2}$, that is $k = \frac{q-1}{8}$ and $k = \frac{3(q-1)}{8}$, then $\sum_{\alpha \in K^\times} \alpha^{4k+\frac{q-1}{2}} \equiv -1 \pmod{q}$. Therefore

$$4c \equiv (-1)^f \left\{ -\binom{\frac{q-1}{2}}{\frac{q-1}{8}} - \binom{\frac{q-1}{2}}{\frac{3(q-1)}{8}} \right\} \equiv 2(-1)^{f+1} \binom{\frac{q-1}{2}}{\frac{q-1}{8}} \pmod{q},$$

$$2c \equiv (-1)^{f+1} \binom{\frac{q-1}{2}}{\frac{q-1}{8}} \pmod{q}.$$

Assume that $p \equiv 1 \pmod{8}$. The p-adic expansion of $\frac{q-1}{8}$ is

$$\frac{q-1}{8} = \frac{p-1}{8}(1 + p + \cdots + p^{s-1}).$$

Thus

$$2c \equiv (-1)^{f+1} \binom{\frac{p-1}{2}}{\frac{p-1}{8}}^s \pmod{p}.$$

Assume that $p \equiv 5 \pmod{8}$. Then we know s is even. The p-adic expansion of $\frac{q-1}{8}$ is

$$\frac{q-1}{8} = \sum_{t=0}^{\frac{s}{2}-1} \frac{5p-1}{8} p^{2t} + \sum_{t=1}^{\frac{s}{2}} \frac{p-5}{8} p^{2t-1}.$$

Hence

$$\binom{\frac{q-1}{2}}{\frac{q-1}{8}} \equiv \binom{\frac{p-1}{2}}{\frac{5p-1}{8}}^{\frac{s}{2}} \binom{\frac{p-1}{2}}{\frac{p-5}{8}}^{\frac{s}{2}} \pmod{p}.$$

However, $\frac{p-1}{2} < \frac{5p-1}{8}$. Then $c \equiv 0 \pmod{p}$.

Assume that $p \equiv 7 \pmod{8}$. Then s is even. The p-adic expansion of $\frac{q-1}{8}$ is

$$\frac{q-1}{8} = \sum_{t=0}^{\frac{s}{2}-1} \frac{7p-1}{8} p^{2t} + \sum_{t=1}^{\frac{s}{2}} \frac{p-7}{8} p^{2t-1}.$$

Since $\frac{p-1}{2} < \frac{7p-1}{8}$, $c \equiv 0 \pmod{p}$.

Assume that $p \equiv 3 \pmod{8}$. The p-adic expansion of $\frac{q-1}{8}$ is

$$\frac{q-1}{8} = \sum_{t=0}^{\frac{s}{2}-1} \frac{3p-1}{8} p^{2t} + \sum_{t=1}^{\frac{s}{2}} \frac{p-3}{8} p^{2t-1}.$$

Hence

$$\binom{\frac{q-1}{2}}{\frac{q-1}{8}} \equiv \binom{\frac{p-1}{2}}{\frac{3p-1}{8}}^{\frac{s}{2}} \binom{\frac{p-1}{2}}{\frac{p-3}{8}}^{\frac{s}{2}} \pmod{p}$$

$$\equiv \binom{\frac{p-1}{2}}{\frac{p-3}{8}}^s \pmod{p}$$

from Lucas' theorem. □

Lemma 2.6 (Berndt et al. [13, theorem 3.3.2]): *Let g be a fixed primitive element of $K = GF(q)$. Let $J(\chi,\psi) = c + d\sqrt{2}i$. Then*

$$2d \equiv c\left(g^f + g^{3f}\right) \qquad (\text{mod } p).$$

Proof: As the proof of Lemma 2.4, from $\zeta_{q-1}^x \equiv g^x \pmod{\wp}$, $\zeta = \zeta_{q-1}^{\frac{q-1}{8}} \equiv g^{\frac{q-1}{8}} \equiv g^f \pmod{\wp}$.

$$0 \equiv \sqrt{2}iJ(\chi,\psi) \equiv \sqrt{2}i(c + d\sqrt{2}i) \equiv c\sqrt{2}i - 2d \pmod{\wp}$$
$$2d \equiv c\sqrt{2}i \qquad \equiv c(\zeta + \zeta^3) \qquad \equiv c(g^f + g^{3f}) \pmod{\wp}.$$

The assertion follows. □

Example 2.3: *Let $q = 5^2$ and put $J(\chi^2,\psi) = a + 2bi$. From Lemma 2.1, $a \equiv -1 \pmod 4$. Then we have $5^2 = 3^2 + 4^2$ and $a = 3$. From Lemma 2.3, $2b \equiv 3g^{\frac{25-1}{4}} \equiv 6 \pmod 5$. Hence $b \equiv 3 \pmod 5$, that is $b = -2$. Thus we have $J(\chi^2,\psi) = 3 - 4i$.*

Let $J(\chi,\psi) = c + d\sqrt{2}i$. From Theorem 2.10, $c \equiv 0 \pmod 5$. Then $c = -5$ from Lemma 2.4. We obtain $J(\chi,\psi) = -5$.

2.4 Cyclotomic Numbers and Jacobi Sums

Let $S_j, 0 \le j \le e-1$ be eth cyclotomic classes. We let

$$X_i = \sum_{x\in S_i} \zeta_p^{T_K(x)} \quad \text{for } i = 0, 1, \dots, e-1$$

where ζ_p is a primitive pth root of unity and T_K is the absolute trace from K to $\boldsymbol{F}_p$. We call this sum the **Gaussian period** for eth power residues.

Lemma 2.7:

$$X_i\overline{X_j} = \sum_{m=0}^{e-1}(i-m, j-m)X_m + \tfrac{q-1}{e}\delta_{i,j}.$$

Proof:

$$X_i\overline{X_j} = \sum_{x\in S_i}\sum_{y\in S_j} \zeta_p^{T_K(x-y)}.$$

Assume that $i \ne j$. Then $z \ne 0$ and the equation $x - y = z, x \in S_i, \ y \in S_j, \ z \in S_m$ is equivalent to the equation $x - y = 1, x \in S_{i-m}, y \in S_{j-m}$. For $i = j$ we add the solutions for $z = 0$. Hence

$$\begin{aligned} X_i\overline{X_j} &= \sum_{m=0}^{e-1}(i-m, j-m)\sum_{z\in S_m}\zeta_p^{T_K(z)} + \frac{q-1}{e}\delta_{i,j} \\ &= \sum_{m=0}^{e-1}(i-m, j-m)X_m + \frac{q-1}{e}\delta_{i,j}. \end{aligned}$$

□

$X_i\overline{X_j}$ can be written as follows from Lemma 1.25,

$$X_i\overline{X_j} = \sum_{m=0}^{e-1}(i-j, m-j)\sum_{z\in S_m}\zeta_p^{T_K(z)} + \frac{q-1}{e}\delta_{i,j}.$$

Let η be an additive character and χ be a eth residue character of K such that $\chi(\xi) = \zeta_e^k$ where ξ is a primitive element of K and k is an integer. Then Gauss sum $g(\chi, \eta)$ is represented by using a Gaussian period,

$$g(\chi, \eta) = \sum_{\alpha \in K} \chi(\alpha)\eta(\alpha) = \sum_{i=0}^{e-1} \sum_{\alpha \in S_i} \zeta_e^{ik}\eta(\alpha) = \sum_{i=0}^{e-1} \zeta_e^{ik} X_i.$$

Thus the Jacobi sum can be written using cyclotomic numbers.

Theorem 2.11: *Assume that* $-1 \in S_h$. *Let* $\chi_1(\alpha) = \zeta_e^{ik}$ *and* $\chi_2(\alpha) = \zeta_e^{ik'}$ *if* $\alpha \in S_i$. *Put* $w_1 = \zeta_e^k$ *and* $w_2 = \zeta_e^{k'}$. *Assume that* $\chi_1\chi_2 \neq \chi^0$. *Then*

$$J(\chi_1, \chi_2) = \sum_{i=0}^{e-1} \sum_{j=0}^{e-1} (i+h, j) w_1^i w_2^j.$$

Proof: Notice that $\overline{X_i} = X_{i+h}$. From Lemma 2.7,

$$\begin{aligned}
g(\chi_1, \eta) g(\chi_2, \eta) &= \sum_{i=0}^{e-1} w_1^i X_i \sum_{j=0}^{e-1} w_2^j X_j = \sum_{i=0}^{e-1} \sum_{j=0}^{e-1} w_1^i w_2^j X_i \overline{X}_{j+h} \\
&= \sum_{i=0}^{e-1} \sum_{j=0}^{e-1} w_1^i w_2^j \left\{ \sum_{m=0}^{e-1} (i-m, j+h-m) X_m + \frac{q-1}{e} \delta_{i,j+h} \right\} \\
&= \sum_{m=0}^{e-1} (w_1 w_2)^m X_m \sum_{i=0}^{e-1} \sum_{j=0}^{e-1} (i-m, j+h-m) w_1^{i-m} w_2^{j-m} \\
&\quad + \frac{q-1}{e} \sum_{i=0}^{e-1} \sum_{j=0}^{e-1} \delta_{i,j+h} w_1^i w_2^j \\
&= g(\chi_1\chi_2, \eta) \sum_{i=0}^{e-1} \sum_{j=0}^{e-1} (i, j+h) w_1^i w_2^j + \frac{q-1}{e} \sum_{i=0}^{e-1} \sum_{j=0}^{e-1} \delta_{i,j+h} w_1^i w_2^j.
\end{aligned}$$

Since $\chi_1\chi_2 \neq \chi^0$,

$$\sum_{i=0}^{e-1} \sum_{j=0}^{e-1} \delta_{i,j+h} w_1^i w_2^j = w_1^h \sum_{i=0}^{e-1} (w_1 w_2)^i = 0.$$

From Theorem 2.1,

$$J(\chi_1, \chi_2) = \sum_{i=0}^{e-1} \sum_{j=0}^{e-1} (i, j+h) w_1^i w_2^j.$$

□

Example 2.4: *Assume that* $e = 4$ *and* $h = 0$. *Let* χ *be a primitive biquadratic character of* K. *We put* $\varpi = J(\chi, \chi^2)$. *From the equations in Lemma 1.27, we have* $A + 3B = \frac{1}{4}(q-5)$ *and*

$$A + B - D - E = A + 3B - (2C + D + E) - 2(B - C) = -1 - 2(B - C).$$

If we let $\varpi = a + 2bi$, $i = \sqrt{-1}$, *then*

$$\varpi = A + B - D - E - 2(B - C) + 2i(D - E)$$

from Theorem 2.11. Therefore,

$$a = -1 - 4(B - C),$$
$$b = D - E$$

and $q = a^2 + 4b^2$ *from Theorem 2.6. We also obtain a similar result for* $h = 2$.

Conversely, cyclotomic numbers can be written using Jacobi sums.

Theorem 2.12 *(Berndt et al. [13]):* *Let* χ *be a primitive e-th residue character and* ζ_e *be a primitive e-th root of unity. For* $s, t = 0, 1, \dots, e-1$,

$$(s,t) = \frac{1}{e^2}\sum_{i=0}^{e-1}\sum_{j=0}^{e-1}\chi^i(-1)\zeta_e^{-it-js}J(\chi^i,\chi^j).$$

Proof: Notice that $\chi(-1) = (-1)^f$.

$$\begin{aligned} S &= \sum_{i=0}^{e-1}\sum_{j=0}^{e-1}\chi^i(-1)\zeta_e^{-it-js}\sum_{\alpha\in F^\times}\chi^i(\alpha)\chi^j(1-\alpha) \\ &= \sum_{i=0}^{e-1}\sum_{j=0}^{e-1}\chi^i(\xi^{-t})\chi^j(\xi^{-s})\sum_{\alpha\in F^\times}\chi^i(-\alpha)\chi^j(1-\alpha) \\ &= \sum_{i=0}^{e-1}\sum_{j=0}^{e-1}\chi^i(\xi^{-t})\chi^j(\xi^{-s})\sum_{\beta\in F^\times}\chi^i(\beta)\chi^j(1+\beta) \\ &= \sum_{\beta\in F^\times}\sum_{i=0}^{e-1}\chi^i(\beta\xi^{-t})\sum_{j=0}^{e-1}\chi^j(\xi^{-s}(1+\beta)). \end{aligned}$$

The inner sums are

$$\sum_{i=0}^{e-1}\chi^i(\beta\xi^{-t}) = \begin{cases} e & \text{if } \beta\xi^{-t}\in S_0, \\ 0 & \text{otherwise,} \end{cases}$$

and

$$\sum_{j=0}^{e-1}\chi^j(\xi^{-s}(1+\beta)) = \begin{cases} e & \text{if } \xi^{-s}(1+\beta)\in S_0, \\ 0 & \text{otherwise.} \end{cases}$$

The number of the elements such that $\beta\xi^{-t}\in S_0$ and $\xi^{-s}(1+\beta)\in S_0$ is (s,t). Therefore $S = e^2(s,t)$ and this proves the theorem. □

If the Jacobi sums are known explicitly, we determine cyclotomic numbers from Theorem 2.12.

2.4.1 Cyclotomic Numbers for $e = 2$

Put $e = 2$. Let χ be a quadratic residue character of K. From Theorem 2.6,

$$J(\chi,\chi^0) = -1,$$
$$J(\chi,\chi) = -\chi(-1) = (-1)^{f+1},$$
$$J(\chi^0,\chi^0) = q - 2.$$

Hence we have

$$
\begin{aligned}
(0,0) &= \frac{1}{4}\sum_{i=0}^{1}\sum_{j=0}^{1}\chi^i(-1)J(\chi^i,\chi^j)\\
&= \frac{1}{4}\left\{J(\chi^0,\chi^0)+J(\chi^0,\chi)+\chi(-1)J(\chi,\chi^0)+\chi(-1)J(\chi,\chi)\right\}\\
&= \frac{1}{4}\left\{q-4-(-1)^f\right\}\\
&= \begin{cases} \frac{1}{4}(q-5) & \text{if } f \equiv 0 \pmod 2,\\ \frac{1}{4}(q-3) & \text{if } f \equiv 1 \pmod 2.\end{cases}
\end{aligned}
$$

Similarly we have

$$
(0,1) = \frac{1}{4}\left\{q-2+(-1)^f\right\} = \begin{cases} \frac{1}{4}(q-1) & \text{if } f \equiv 0 \pmod 2,\\ \frac{1}{4}(q-3) & \text{if } f \equiv 1 \pmod 2,\end{cases}
$$

$$
(1,0) = \frac{1}{4}\{q-(-1)^f\} = \begin{cases} \frac{1}{4}(q-1) & \text{if } f \equiv 0 \pmod 2,\\ \frac{1}{4}(q+1) & \text{if } f \equiv 1 \pmod 2,\end{cases}
$$

$$
(1,1) = \tfrac{1}{4}\{q-2+(-1)^f\} = \begin{cases} \frac{1}{4}(q-1) & \text{if } f \equiv 0 \pmod 2,\\ \frac{1}{4}(q-3) & \text{if } f \equiv 1 \pmod 2.\end{cases}
$$

Then we have Lemma 1.26.

2.4.2 Cyclotomic Numbers for $e = 4$

Let $q = 4f + 1$ be a prime power and χ be a primitive biquadratic character of $K = GF(q)$. Put $\varpi = J(\chi,\chi^2) = a + 2bi$ where $i = \sqrt{-1}$ and $\epsilon = \chi(-1) = (-1)^f$. From Theorem 2.12 and Table 2.2,

$$
\begin{aligned}
16(1,0) &= \sum_{\ell=0}^{3}\sum_{j=0}^{3}\chi^\ell(-1)J\left(\chi^\ell,\chi^j\right)i^{-j}\\
&= J\left(\chi^0,\chi^0\right)+J\left(\chi^0,\chi\right)i^{-1}+J\left(\chi^0,\chi^2\right)i^{-2}+J\left(\chi^0,\chi^3\right)i^{-3}\\
&\quad +\epsilon\left(J\left(\chi,\chi^0\right)+J(\chi,\chi)i^{-1}+J\left(\chi,\chi^2\right)i^{-2}+J\left(\chi,\chi^3\right)i^{-3}\right)\\
&\quad +\epsilon^2\left(J\left(\chi^2,\chi^0\right)+J\left(\chi^2,\chi\right)i^{-1}+J\left(\chi^2,\chi^2\right)i^{-2}+J\left(\chi^2,\chi^3\right)i^{-3}\right)\\
&\quad +\epsilon^3\left(J\left(\chi^3,\chi^0\right)+J\left(\chi^3,\chi\right)i^{-1}+J\left(\chi^3,\chi^2\right)i^{-2}+J\left(\chi^3,\chi^3\right)i^{-3}\right)\\
&= q-1-2\epsilon-\epsilon(\varpi+\overline{\varpi})-\left(\epsilon^2+1\right)(\varpi-\overline{\varpi})i\\
&= q-1-2\epsilon-2\epsilon a+8b\\
&= \begin{cases} q+1+2a+8b & \text{if } f \text{ is odd},\\ q-3-2a+8b & \text{if } f \text{ is even}.\end{cases}
\end{aligned}
$$

Similarly we have the other cyclotomic numbers.

Theorem 2.13: *Let $q = 4f + 1$ and $J(\chi, \chi^2) = a + 2bi$ where $i = \sqrt{-1}$. Put $\epsilon = (-1)^f$. Then*

$$\begin{aligned}
16(1,0) &= q - 1 - 2\epsilon - 2\epsilon a + 8b,\\
16(0,0) &= q - 7 + 2a - 2(1+\epsilon)(1-a),\\
16(0,3) &= q - 3 - 2a - 4(1+\epsilon)b,\\
16(1,3) &= q + 1 + 2a + 4(1-\epsilon)b,\\
16(2,0) &= q - 1 - 2\epsilon - 4a + 2\epsilon a.
\end{aligned}$$

Example 2.5:

1) *Let $q = 13 = 4 \cdot 3 + 1$. Since $q \equiv 5 \pmod 8$, $a \equiv 1 \pmod 4$ from Lemma 1.25. When we choose 2 as a primitive element, $2b \equiv 2 \pmod{13}$ from Lemma 2.3. Then $J(\chi, \psi) = -3 + 4i$ and $\epsilon = (-1)^3 = -1$.*

$$\begin{aligned}
16(1,0) = 16E &= q - 1 - 2\epsilon - 2\epsilon a + 8b = 13 - 1 - 2(-1) - 2(-1)(-3) + 8 \cdot 1\\
&= 16,\ E = 1.\\
16(0,0) = 16B &= q - 7 + 2a - 2(1+\epsilon)(1-a) = 0,\ B = 0.\\
16(0,3) = 16C &= q - 3 - 2a - 4(1+\epsilon)b = 16,\ C = 1.\\
16(1,3) = 16D &= q + 1 + 2a + 4(1-\epsilon)b = 0,\ D = 0.\\
16(2,0) = 16A &= q - 1 - 2\epsilon - 4a + 2\epsilon a = 32,\ A = 2.
\end{aligned}$$

2) *Let $q = 17 = 4 \cdot 4 + 1$. Since $q \equiv 1 \pmod 8$, $a \equiv -1 \pmod 4$ from Lemma 1.25. When we choose 3 as a primitive element, $2b = 4 \pmod{17}$. Then $J(\chi, \psi) = -1 + 4i$ and $\epsilon = (-1)^4 = 1$.*

$$\begin{aligned}
16(1,0) = 16D &= q - 1 - 2\epsilon - 2\epsilon a + 8b = 17 - 1 - 2 \cdot 1 - 2 \cdot 1 \cdot (-1) + 8 \cdot 2\\
&= 32,\ D = 2.\\
16(0,0) = 16A &= q - 7 + 2a - 2(1+\epsilon)(1-a) = 0,\ A = 0.\\
16(0,3) = 16E &= q - 3 - 2a - 4(1+\epsilon)b = 0,\ E = 1.\\
16(1,3) = 16C &= q + 1 + 2a + 4(1-\epsilon)b = 16,\ C = 1.\\
16(2,0) = 16B &= q - 1 - 2\epsilon - 4a + 2\epsilon a = 16,\ B = 1.
\end{aligned}$$

In Lemma 1.27, we assume $a \equiv -1 \pmod 4$ when f is odd in order to unify the condition of a for the both cases of f, odd or even. Hence the signs of a and b in Lemma 1.27 are different from signs of real and imaginary parts of $J(\chi, \psi)$ when f is odd. The reader needs to very carefully read Lemma 1.27 to understand this concept.

2.4.3 Cyclotomic Numbers for $e = 8$

Let $q = 8f + 1$ be a prime power and χ be a primitive character of order 8 of $K = GF(q)$. Denote a primitive quadratic character by $\psi = \chi^4$. Let $\zeta = \frac{1+i}{\sqrt{2}}$ be a primitive 8th root of unity where $i = \sqrt{-1}$. Put $\varpi = J(\chi^2, \psi) = a + 2bi$ and $\kappa = J(\chi, \psi) = c + d\sqrt{2}i$, $\rho = \chi(4) = \pm 1$ and $\epsilon = \chi(-1) = (-1)^{\frac{q-1}{8}} = (-1)^f$.

Let $A, B, C, D, E, F, G, H, I, J, K, L, M, N$, and O be cyclotomic numbers in tables 1.27 and 1.28.

From Theorem 2.12 and Table 2.3,

$$
\begin{aligned}
64A = 64(0,0) = \sum_{\ell=0}^{7}\sum_{j=0}^{7} \chi^{\ell}(-1)J(\chi^{\ell},\chi^{j})
&= q-2+7(-1)+\epsilon\{-1+\rho\kappa+\epsilon\rho\varpi+\epsilon\kappa+\kappa+\rho\varpi+\epsilon\rho\kappa-\epsilon\}\\
&+\epsilon^2\{-1+\epsilon\rho\varpi+\varpi+\epsilon\rho\kappa+\varpi+\epsilon\rho\varpi-1+\epsilon\rho\overline{\kappa}\}\\
&+\epsilon^3\{-1+\epsilon\kappa+\epsilon\rho\kappa+\rho\kappa+\kappa-\epsilon+\epsilon\rho\overline{\varpi}+\rho\overline{\varpi}\}\\
&+\epsilon^4\{-1+\kappa+\varpi+\kappa-1+\overline{\kappa}+\overline{\varpi}+\overline{\kappa}\}\\
&+\epsilon^5\{-1+\rho\varpi+\epsilon\rho\varpi-\epsilon+\overline{\kappa}+\rho\overline{\kappa}+\epsilon\rho\overline{\kappa}+\epsilon\overline{\kappa}\}\\
&+\epsilon^6\{-1+\epsilon\rho\kappa-1+\epsilon\rho\overline{\varpi}+\overline{\varpi}+\epsilon\rho\overline{\kappa}+\overline{\varpi}+\epsilon\rho\overline{\varpi}\}\\
&+\epsilon^7\{-1-\epsilon+\epsilon\rho\overline{\kappa}+\rho\overline{\varpi}+\overline{\kappa}+\epsilon\overline{\kappa}+\epsilon\rho\overline{\varpi}+\rho\overline{\kappa}\}
\end{aligned}
$$

If f is odd, then $\epsilon = \chi(-1) = (-1)^f = -1$ and

$$
\begin{aligned}
64A &= q-15+6a+4c-4\rho a-4\rho c\\
&= \begin{cases} q-15+2a & \text{if } \rho=\chi^2(2)=1,\\ q-15+10a+8c & \text{if } \rho=\chi^2(2)=-1.\end{cases}
\end{aligned}
$$

If f is even, then $\epsilon = \chi(-1) = (-1)^f = 1$ and

$$
\begin{aligned}
64A &= q-23+6a+12c+12\rho a+12\rho c\\
&= \begin{cases} q-23+18a+24c & \text{if } \rho=1,\\ q-23-6a & \text{if } \rho=-1.\end{cases}
\end{aligned}
$$

Thus we have the following results.

Theorem 2.14: *Let $q = 8f+1$. Let $\varpi = J(\chi^2,\psi) = a+2bi$ and $\kappa = J(\chi,\psi) = c+d\sqrt{2}i = c+(\zeta+\zeta^3)d$. Put $\rho = \chi(4) = \pm1$ and $\epsilon = \chi(-1) = (-1)^{\frac{q-1}{8}} = (-1)^f$. Assume that f is odd. Then*

$$
\begin{aligned}
64A &= q-15+6a+4c-4\rho a-4\rho c\\
&= \begin{cases} q-15+2a & \text{if } \rho=\chi^2(2)=1,\\ q-15+10a+8c & \text{if } \rho=\chi^2(2)=-1.\end{cases}
\end{aligned}
$$

$$
\begin{aligned}
64B &= q+1-2a+8(1+\rho)b+4c+8(1-\rho)d\\
&= \begin{cases} q+1-2a+16b+4c & \text{if } \rho=1,\\ q+1-2a+4c+16d & \text{if } \rho=-1.\end{cases}
\end{aligned}
$$

$$
\begin{aligned}
64C &= q+1-2a-4\rho a-16\rho b-4(1+\rho)c\\
&= \begin{cases} q+1-6a-16b-8c & \text{if } \rho=1,\\ q+1+2a+16b & \text{if } \rho=-1.\end{cases}
\end{aligned}
$$

$$
\begin{aligned}
64D &= q+1-2a-8(1+\rho)b+4c+8(1-\rho)d\\
&= \begin{cases} q+1-2a-16b+4c & \text{if } \rho=1,\\ q+1-2a+4c+16d & \text{if } \rho=-1.\end{cases}
\end{aligned}
$$

$$64E = q + 1 + 6a + 12\rho a - 12(1-\rho)c = \begin{cases} q + 1 + 18a & \text{if } \rho = 1, \\ q + 1 - 6a - 24c & \text{if } \rho = -1. \end{cases}$$

$$64F = q + 1 - 2a + 8(1+\rho)b + 4c - 8(1-\rho)d = \begin{cases} q + 1 - 2a + 16b + 4c & \text{if } \rho = 1, \\ q + 1 - 2a + 4c - 16d & \text{if } \rho = -1. \end{cases}$$

$$64G = q + 1 - 2a - 4\rho a + 16\rho b - 4(1+\rho)c = \begin{cases} q + 1 - 6a + 16b - 8c & \text{if } \rho = 1, \\ q + 1 + 2a - 16b & \text{if } \rho = -1. \end{cases}$$

$$64H = q + 1 - 2a - 8(1+\rho)b + 4c - 8(1-\rho)d = \begin{cases} q + 1 - 2a - 16b + 4c & \text{if } \rho = 1, \\ q + 1 - 2a + 4c - 16d & \text{if } \rho = -1. \end{cases}$$

$$64I = q - 7 - 2a + 8(1-\rho)b - 4c = \begin{cases} q - 7 - 2a - 4c & \text{if } \rho = 1, \\ q - 7 - 2a + 16b - 4c & \text{if } \rho = -1. \end{cases}$$

$$64J = q - 7 - 2a - 8(1-\rho)b - 4c = \begin{cases} q - 7 - 2a - 4c & \text{if } \rho = 1, \\ q - 7 - 2a - 16b - 4c & \text{if } \rho = -1. \end{cases}$$

$$64K = q + 1 + 2a + 4\rho a - 4\rho c - 8(1+\rho)d = \begin{cases} q + 1 + 6a - 4c - 16d & \text{if } \rho = 1, \\ q + 1 - 2a + 4c & \text{if } \rho = -1. \end{cases}$$

$$64L = q + 1 + 2a - 4\rho a + 4\rho c = \begin{cases} q + 1 - 2a + 4c & \text{if } \rho = 1, \\ q + 1 + 6a - 4c & \text{if } \rho = -1. \end{cases}$$

$$64M = q + 1 + 2a + 4\rho a - 4\rho c + 8(1+\rho)d = \begin{cases} q + 1 + 6a - 4c + 16d & \text{if } \rho = 1, \\ q + 1 - 2a + 4c & \text{if } \rho = -1. \end{cases}$$

$$64N = q - 7 - 2a + 4\rho a + 4(1+\rho)c = \begin{cases} q - 7 + 2a + 8c & \text{if } \rho = 1, \\ q - 7 - 6a & \text{if } \rho = -1. \end{cases}$$

$$64O = q + 1 + 2a - 4\rho a + 4\rho c = \begin{cases} q + 1 - 2a + 4c & \text{if } \rho = 1, \\ q + 1 + 6a - 4c & \text{if } \rho = -1. \end{cases}$$

Example 2.6: *Let $q = 5^2 = 8 \cdot 3 + 1$ and χ be a primitive character of order 8 of $GF(5^2)$. Put $\psi = \chi^4$, a quadratic character of $GF(5^2)$. From Example 2.3, $J(\chi, \psi) = -5$ and $J(\chi^2, \psi) = 3 - 4i$. 2 is not a fourth power. Thus we have*

$$64A = q - 15 + 10a + 8c = 0, \qquad 64B = q + 1 - 2a + 4c + 16d = 0,$$

$$64C = q + 1 + 2a + 16b = 0, \qquad 64D = q + 1 - 2a + 4c + 16d = 0,$$

$$64E = q + 1 - 6a - 24c = 128, \qquad 64F = q + 1 - 2a + 4c - 16d = 0,$$

$$64G = q + 1 + 2a - 16b = 64, \qquad 64H = q + 1 - 2a + 4c - 16d = 0,$$

$$64I = q - 7 - 2a + 16b - 4c = 0, \qquad 64J = q - 7 - 2a - 16b - 4c = 64,$$

$$64K = q + 1 - 2a + 4c = 0, \qquad 64L = q + 1 + 6a - 4c = 64,$$

$$64M = q + 1 - 2a + 4c = 0, \qquad 64N = q - 7 - 6a = 0,$$

$$64O = q + 1 + 6a - 4c = 64.$$

Theorem 2.15: *Let q, ϖ, κ, ρ, and ϵ be as in Lemma 2.14. Assume that f is even. Then*

$$64A = q - 23 + 6a + 12c + 12\rho a + 12\rho c = \begin{cases} q - 23 + 18a + 24c & \text{if } \rho = 1, \\ q - 23 - 6a & \text{if } \rho = -1. \end{cases}$$

$$64B = q - 7 - 2a + 4(1+\rho)b - 4c + 8(1+\rho)d = \begin{cases} q - 7 - 2a + 8b - 4c + 16d & \text{if } \rho = 1, \\ q - 7 - 2a - 4c & \text{if } \rho = -1. \end{cases}$$

$$64C = q - 7 - 2a - 4\rho a + 16\rho b + 4(1-\rho)c = \begin{cases} q - 7 - 6a + 16b & \text{if } \rho = 1, \\ q - 7 + 2a - 16b + 8c & \text{if } \rho = -1. \end{cases}$$

$$64D = q - 7 - 2a - 4(1+\rho)b - 4c + 8(1+\rho)d = \begin{cases} q - 7 - 2a - 8b - 4c + 16d & \text{if } \rho = 1, \\ q - 7 - 2a - 4c & \text{if } \rho = -1. \end{cases}$$

$$64E = q - 7 + 6a - 4\rho a - 4(1+\rho)c = \begin{cases} q - 7 + 2a - 8c & \text{if } \rho = 1, \\ q - 7 + 10a & \text{if } \rho = -1. \end{cases}$$

$$64F = q - 7 - 2a + 8(1+\rho)b - 4c - 8(1+\rho)d = \begin{cases} q - 7 - 2a + 16b - 4c - 16d & \text{if } \rho = 1, \\ q - 7 - 2a - 4c & \text{if } \rho = -1. \end{cases}$$

$$64G = q - 7 - 2a - 4\rho a - 16\rho b + 4(1-\rho)c = \begin{cases} q - 7 - 6a - 16b & \text{if } \rho = 1, \\ q - 7 + 2a + 16b + 8c & \text{if } \rho = -1. \end{cases}$$

$$64H = q - 7 - 2a - 8(1+\rho)b - 4c - 8(1+\rho)d = \begin{cases} q - 7 - 2a - 16b - 4c - 16d & \text{if } \rho = 1, \\ q - 7 - 2a - 4c & \text{if } \rho = -1. \end{cases}$$

$$64I = q + 1 + 2a - 4\rho a + 4\rho c = \begin{cases} q + 1 - 2a + 4c & \text{if } \rho = 1, \\ q + 1 + 6a - 4c & \text{if } \rho = -1. \end{cases}$$

$$64J = q + 1 + 2a + 4\rho a - 4\rho c - 8(1 - \rho)d = \begin{cases} q + 1 + 6a - 4c & \text{if } \rho = 1, \\ q + 1 - 2a + 4c - 16d & \text{if } \rho = -1. \end{cases}$$

$$64K = q + 1 - 2a + 8(1 - \rho)b + 4c = \begin{cases} q + 1 - 2a + 4c & \text{if } \rho = 1, \\ q + 1 - 2a + 16b + 4c & \text{if } \rho = -1. \end{cases}$$

$$64L = q + 1 - 2a - 8(1 - \rho)b + 4c = \begin{cases} q + 1 - 2a + 4c & \text{if } \rho = 1, \\ q + 1 - 2a - 16b + 4c & \text{if } \rho = -1. \end{cases}$$

$$64M = q + 1 + 2a + 4\rho a - 4\rho c + 8(1 - \rho)d = \begin{cases} q + 1 + 6a - 4c & \text{if } \rho = 1, \\ q + 1 - 2a + 4c + 16d & \text{if } \rho = -1. \end{cases}$$

$$64N = q + 1 - 2a + 4\rho a - 4(1 - \rho)c = \begin{cases} q + 1 + 2a & \text{if } \rho = 1, \\ q + 1 - 6a - 8c & \text{if } \rho = -1. \end{cases}$$

$$64O = q + 1 + 2a - 4\rho a + 4\rho c = \begin{cases} q + 1 - 2a + 4c & \text{if } \rho = 1, \\ q + 1 + 6a - 4c & \text{if } \rho = -1. \end{cases}$$

Example 2.7: *Let $q = 7^2 = 8 \cdot 6 + 1$ and χ be a primitive character of order 8 of $GF(7^2)$. Put $\psi = \chi^4$, a quadratic character of $GF(7^2)$. From Lemma 2.1 and Theorem 2.9, $J(\chi^2, \psi) = 7$. From Lemma 2.4 and Theorem 2.10, $J(\chi, \psi) = 7$. 2 is a fourth power. Thus we have*

$$64A = q - 23 + 18a + 24c = 320, \qquad 64B = q - 7 - 2a + 8b - 4c + 16d = 0,$$

$$64C = q - 7 - 6a + 16b = 0, \qquad 64D = q - 7 - 2a - 8b - 4c + 16d = 0,$$

$$64E = q - 7 + 2a - 8c = 0, \qquad 64F = q - 7 - 2a + 16b - 4c - 16d = 0,$$

$$64G = q - 7 - 6a - 16b = 0, \qquad 64H = q - 7 - 2a - 16b - 4c - 16d = 0,$$

$$64I = q + 1 - 2a + 4c = 64, \qquad 64J = q + 1 + 6a - 4c = 64,$$

$$64K = q + 1 - 2a + 4c = 64, \qquad 64L = q + 1 - 2a + 4c = 64,$$

$$64M = q + 1 + 6a - 4c = 64, \qquad 64N = q + 1 + 2a = 64,$$

$$64O = q + 1 - 2a + 4c = 64.$$

We specify the real part c and the imaginary part d of $J(\chi, \psi)$ if we write $J(\chi, \psi)\overline{J(\chi, \psi)} = q = c^2 + 2d^2$ with $c \equiv -1 \pmod 4$, then we determine c and obtain d when we choose a primitive element of $GF(q)$ from Lemma 2.6. Thus we determine $J(\chi, \psi)$.

2.5 Relative Gauss Sums

Definition 2.5: Let χ_F be a nontrivial character of an extension F of K. Put $g_F(\chi_F) = g(\chi_F, \eta_1)$. Let $g_K(\chi_K)$ be the Gauss sum associated with the induced character of χ_F in K. The ratio

$$\vartheta_{F/K}(\chi_F) = \vartheta(\chi_F, \chi_K) = \frac{g_F(\chi_F)}{g_K(\chi_K)}$$

of two Gauss sums is called the **relative Gauss sum** or the **Eisenstein sum** associated with χ_F.

Denote the orders of χ_F and χ_K by m' and m, respectively. Let f' and f be positive integers such that

$$p^{f'} \equiv 1 \pmod{m'}, \quad p^f \equiv 1 \pmod{m}.$$

If $\chi_F \neq \chi_F^0$ and $\chi_K \neq \chi_K^0$, then the ratio

$$\widetilde{\vartheta}_{F/K}(\chi_F) = \frac{g_F(\chi_F)}{p^{\frac{f'-f}{2}} g_K(\chi_K)},$$

has $\left|\widetilde{\vartheta}_{F/K}(\chi_F)\right| = 1$. We call this ratio a **normalized relative Gauss sum.**

The concept of the relative Gauss sum was first introduced in the paper by Yamamoto and Yamada [247] to construct a Williamson Hadamard matrix, which will be discussed in Chapter 3. Momihara constructed a family of skew Hadamard difference sets from normalized relative Gauss sums that we will discuss in Chapter 10.

The relative Gauss sum is written in the following form.

Theorem 2.16 *(Yamamoto and Yamada [247]):* *Suppose that χ_F is a multiplicative character of F inducing a nontrivial character χ_K in K. Then the relative Gauss sum associated with χ_F can be written in the following form*

$$\vartheta_{F/K}(\chi_F) = \sum_{\alpha \mod K^\times} \chi_F(\alpha)\overline{\chi_K}(T_{F/K}\alpha)$$

where the sum is extended over a system of representatives of the quotient group $F^\times/K^\times$. Furthermore we have the norm relation

$$\vartheta_{F/K}(\chi_F)\overline{\vartheta_{F/K}(\chi_F)} = q^{t-1}.$$

Proof: Put $\chi = \chi_F$. An element of $F^\times$ is uniquely written as $a\alpha$ where $a \in K^\times$ and α runs over a system of representatives of the quotient group $F^\times/K^\times$, so that

$$\begin{aligned}
g_F(\chi_F) &= \sum_{\alpha \mod K^\times} \sum_{a \in K^\times} \chi(a\alpha)\zeta_p^{T_K(aT_{F/K}\alpha)} \\
&= \sum_{\alpha \mod K^\times} \sum_{a \in K^\times} \chi(\alpha)\chi(a)\overline{\chi}\left(T_{F/K}\alpha\right)\zeta_p^{T_K(a)} \\
&= \sum_{a \in K^\times} \chi(a)\zeta_p^{T_K(a)} \sum_{\alpha \mod K^\times} \chi(\alpha)\overline{\chi}\left(T_{F/K}\alpha\right) \\
&= g_K(\chi_K) \sum_{\alpha \mod K^\times} \chi(\alpha)\overline{\chi}(T_{F/K}\alpha).
\end{aligned}$$

From Theorem 2.1, we have

$$\vartheta_{F/K}(\chi_F)\overline{\vartheta_{F/K}(\chi_F)} = \frac{g_F(\chi_F)\overline{g_F(\chi_F)}}{g_K(\chi_K)\overline{g_K(\chi_K)}} = \frac{q^t}{q} = q^{t-1}.$$

□

Suppose $T_{F/K}(\alpha) = a \neq 0$. If we put $\beta = a^{-1}\alpha$, then $T_{F/K}(\beta) = 1$. Hence any element $\alpha \in F$ with $T_{F/K}(\alpha) \neq 0$ is represented as $\alpha = a\beta$, $a \in K$, and $T_{F/K}(\beta) = 1$. Suppose $T_{F/K}(\alpha) = 0$. There exists an element γ such that $T_{F/K}(\alpha\gamma) = a \neq 0$. Then $\alpha\gamma$ is represented uniquely as $\alpha\gamma = a\beta$, $a \in K^\times$, and $T_{F/K}(\beta) = 1$. Hence $\alpha = a\beta\gamma^{-1}$ and $T_{F/K}(\beta\gamma) = 0$ from $a \neq 0$. Thus we can take a system $\mathcal{L}$ of representatives of the quotient group $F^\times/K^\times$ and we decompose $\mathcal{L}$ in two parts as follows,

$$\mathcal{L} = \mathcal{L}_0 + \mathcal{L}_1,\ \mathcal{L}_0 = \{\beta \mid T_{F/K}\beta = 0\},\ \mathcal{L}_1 = \{\beta \mid T_{F/K}\beta = 1\}.$$

Theorem 2.17 *(Yamada [235])*: *Let χ_F be a multiplicative character of F and denote the induced character of χ_F in K by χ_K. We take the set $\mathcal{L}$ as a system of representatives of $F^\times/K^\times$. Then*

$$\sum_{\beta\in\mathcal{L}_1} \chi_F(\beta) = \begin{cases} \vartheta_{F/K}(\chi_F), & \text{if } \chi_K \neq \chi^0, \\ -\frac{1}{q}g_F(\chi_F), & \text{if } \chi_K = \chi^0 \text{ and } \chi_F \neq \chi^0. \\ q^{t-1}, & \text{if } \chi_F = \chi^0. \end{cases}$$

Proof: Put $\chi = \chi_F$. Since an element $\alpha \in F^\times$ is represented uniquely as $\alpha = a\beta$, $a \in K^\times$, and $\beta \in \mathcal{L}$. Thus we have

$$g_F(\chi) = \sum_{a\in K^\times}\sum_{\beta\in\mathcal{L}} \chi(a\beta)\zeta_p^{T_F(a\beta)} = \sum_{\beta\in\mathcal{L}} \chi(\beta) \sum_{a\in K^\times} \chi_K(a)\zeta_p^{T_K(aT_{F/K}\beta)}.$$

We distinguish two cases.

(i) If $\beta \in \mathcal{L}_1$ or $T_{F/K}(\beta) = 1$, then

$$\begin{aligned}\sum_{a\in K^\times} \chi_K(a)\zeta_p^{T_K(aT_{F/K}\beta)} &= \sum_{a\in K^\times} \overline{\chi}(T_{F/K}\beta)\chi(aT_{F/K}\beta)\zeta_p^{T_K(aT_{F/K}\beta)} \\ &= \overline{\chi}(T_{F/K}\beta)g_K(\chi_K) = g_K(\chi_K).\end{aligned}$$

(ii) If $\beta \in \mathcal{L}_0$ or $T_{F/K}(\beta) = 0$, then

$$\sum_{a\in K^\times} \chi(a)\zeta_p^{T_K(aT_{F/K}\beta)} = \sum_{a\in K^\times} \chi_K(a) = \begin{cases} 0 & \text{if } \chi_K \neq \chi^0, \\ q-1 & \text{if } \chi_K = \chi^0. \end{cases}$$

If $\chi_K \neq \chi^0$, then we obtain

$$g_F(\chi) = \sum_{\beta\in\mathcal{L}_1} \chi(\beta)g_K(\chi_K)$$

from (*i*). Assume $\chi_K = \chi^0$. Since $\overline{\chi}(T_{F/K}\beta)g_K(\chi_K) = -1$ for $\beta \in \mathcal{L}_1$, we have

$$\begin{aligned}g_F(\chi) &= (q-1)\sum_{\beta\in\mathcal{L}_0} \chi(\beta) - \sum_{\beta\in\mathcal{L}_1} \chi(\beta) \\ &= -\sum_{\beta\in\mathcal{L}} \chi(\beta) + q\sum_{\beta\in\mathcal{L}_0} \chi(\beta).\end{aligned}$$

When $\chi \neq \chi_0$, $\sum_{\beta\in\mathcal{L}} \chi(\beta) = 0$. Hence

$$g_F(\chi) = q\sum_{\beta\in\mathcal{L}_0} \chi(\beta) = -q\sum_{\beta\in\mathcal{L}_1} \chi(\beta).$$

When $\chi = \chi^0$, we have

$$g_F(\chi) = -\sum_{\beta \in \mathcal{L}} \chi(\beta) + q \sum_{\beta \in \mathcal{L}_0} \chi(\beta)$$
$$= -\frac{q^t - 1}{q - 1} + q \sum_{\beta \in \mathcal{L}_0} \chi(\beta)$$
$$= -1.$$

Then

$$\sum_{\beta \in \mathcal{L}_0} \chi(\beta) = \frac{q^{t-1} - 1}{q - 1}$$

and

$$\sum_{\beta \in \mathcal{L}_1} \chi(\beta) = q^{t-1}.$$

□

Example 2.8: *Let $K = \mathbf{F}_5, F = \mathbf{F}_{5^2}$. Let χ_4 be a primitive biquadratic character and χ_3 be a primitive cubic character of F. Set $\chi = \chi_4\chi_3$. Let η be an additive character of F. Denote the induced characters of χ and η in K by χ_K and η_K, respectively. Denote a primitive biquadratic root of unity by i and a primitive cubic root of unity by ω. Let ξ be a primitive element of F. Notice that the character χ induces a quadratic character in K.*

ℓ	$T_{F/K}\xi^\ell$
0	2
1	1
2	2
3	0
4	1
5	1
6	4
7	2
8	4
9	0
10	2
11	2
12	3
13	4
14	3
15	0
16	4
17	4
18	1
19	3
20	1
21	0
22	3
23	3

From Definition 2.5, we have

$$g_F(\chi, \eta) = \sum_{\ell=0}^{23} \chi_4\chi_3(\xi^\ell)\zeta_5^{T_{F/K}\xi^\ell}$$
$$= (-\zeta_5 - \zeta_5^4 + \zeta_5^2 + \zeta_5^3)(2 + i)$$
$$= -(2 + i)g_K(\chi_K, \eta_K).$$

Then we have

$$\vartheta_{F/K}(\chi) = \frac{g_F(\chi)}{g_K(\chi)} = -(2 + i).$$

On the other hand, from Theorem 2.16,

$$\sum_{\alpha \mod K^\times} \chi(\alpha)\overline{\chi}(T_{F/K}\alpha)$$
$$= \sum_{\ell=0}^{5} \chi_4(\xi^\ell)\chi_3(\xi^\ell)\overline{\chi}_4(T_{F/K}\xi^\ell)\overline{\chi}_3(T_{F/K}\xi^\ell)$$
$$= -(2 + i)$$

From the table, $\mathcal{L}_1 = \{\xi, \xi^4, \xi^5, \xi^{18}, \xi^{20}\}$, we verify

$$\sum_{\beta \in \mathcal{L}_1} \chi(\beta) = -(2 + i).$$

For more details and discussions of Gauss sums, Jacobi sums, and relative Gauss sums, we refer the reader to see [13, 124, 126, 127, 234, 237, 247].

2.6 Prime Ideal Factorization of Gauss Sums

2.6.1 Prime Ideal Factorization of a Prime *p*

Let $\zeta_m = e^{2\pi i/m}$ and $\boldsymbol{Q}(\zeta_m)$ be a cyclotomic field obtained by adjoining a primitive mth sum root of unity ζ_m to the rational numbers. Let φ be the Euler's function.

First we give a prime ideal factorization of a prime number p in a cyclotomic field $\boldsymbol{Q}(\zeta_m)$.

Theorem 2.18: *Let p be a prime such that $\gcd(p, m) = 1$. Let f be the smallest positive integer such that $p^f \equiv 1 \pmod{m}$, that is, f is the order of p in $\boldsymbol{Z}_m^\times = (\boldsymbol{Z}/m\boldsymbol{Z})^\times$. If we put $\varphi(m) = f \cdot g$, then*

$$p \sim \mathcal{P}_1 \mathcal{P}_2 \cdots \mathcal{P}_g$$

and $N_{Q(\zeta_m)/Q}\mathcal{P}_i = p^f$ where $\mathcal{P}_i, i = 1, 2, \dots, g$ are prime ideals in $\boldsymbol{Q}(\zeta_m)$ lying above p and $N_{Q(\zeta_m)/Q}$ is a relative norm from $\boldsymbol{Q}(\zeta_m)$ to $\boldsymbol{Q}$ of a prime ideal.

Proof: See Berndt, et al. [13]. □

Theorem 2.19: *If $m = p^h$, then $p \sim \mathcal{P}^n$, $\mathcal{P} \sim (1 - \zeta_{p^h})$ where $n = \varphi(p^h)$ and ζ_{p^h} is a primitive p^h-th root of unity.*

Proof: See [13]. □

2.6.2 Stickelberger's Theorem

We let $q = p^n, n \in \boldsymbol{Z}, n \geq 2$. Let $\wp$ be a prime ideal in $\boldsymbol{Q}(\zeta_{q-1})$ lying above the prime number p where ζ_{q-1} is a primitive $q-1$st root of unity. Denote the integer ring of $\boldsymbol{Q}(\zeta_{q-1})$ by $\mathcal{O}$. The finite field $K = \boldsymbol{F}_q$ is identified with the residue class field $\mathcal{O}/\wp$.

Denote the group of $q-1$st roots of unity by μ_{q-1}. A nonzero residue class $\alpha + \wp$ in $\mathcal{O}/\wp$ contains just one power ζ_{q-1}^x of ζ_{q-1}. Then the map $\omega : K^\times \cong \mathcal{O}/\wp \to \mu_{q-1}$ defined by

$$\omega(\alpha + \wp) = \zeta_{q-1}^x$$

is a multiplicative character of $K^\times$, which is called the **Teichmüller character** of K. The Teichmüller character generates the group of multiplicative characters of $K^\times$ and any multiplicative character is a power of ω.

Let k be an integer and $0 \leq k < q-1$. For the p-adic expansion of k, $k = k_0 + k_1 p + \cdots + k_{n-1}p^{n-1}$, $0 \leq k_i < p$, we define

$$s(k) = k_0 + k_1 + \cdots + k_{n-1},$$
$$\gamma(k) = k_0! k_1! \cdots k_{n-1}!$$

in the range $0 \leq k < q-1$.

Theorem 2.20 (Stickelberger's congruence for Gauss sums): *Let $\mathfrak{P}$ be a prime ideal in $\boldsymbol{Q}(\zeta_{q-1}, \zeta_p)$ lying above $\wp$. For an integer $k \geq 1$, we have*

$$\frac{g(\omega^{-k})}{(\zeta_p - 1)^{s(k)}} \equiv \frac{-1}{\gamma(k)} \pmod{\mathfrak{P}}.$$

In particular,

$$ord_{\mathfrak{P}} g(\omega^{-k}) = s(k).$$

Proof: See Berndt et al. [13] and Lang [124]. □

2.6.3 Prime Ideal Factorization of the Gauss Sum in $Q(\zeta_{q-1})$

The Galois group $\tilde{G} = \mathrm{Gal}(\boldsymbol{Q}(\zeta_{q-1}, \zeta_p)/\boldsymbol{Q})$ is a direct product of the Galois group G of $\boldsymbol{Q}(\zeta_{q-1})/\boldsymbol{Q}$ and the Galois group G' of $\boldsymbol{Q}(\zeta_p)/\boldsymbol{Q}$. Denote the group of ℓth root of unity by μ_ℓ. Then

$$\tilde{G} = \{\sigma_{c,d} = \sigma_{c,1} \cdot \sigma_{1,d} \quad \sigma_{c,1} : \zeta_{q-1} \to \zeta_{q-1}^c,\ \sigma_{c,1} \text{ is identity on } \mu_p,$$
$$\sigma_{1,d} : \zeta_p \to \zeta_p^d,\ \sigma_{1,d} \text{ is identity on } \mu_{q-1}\}$$

where $0 \le c \le q-2$, $0 \le d \le p-1$, $gcd(c, q-1) = gcd(d, p) = 1$. Denote the decomposition group of $\tilde{G}$, G, and G' by $\tilde{Z}$, Z, and Z', respectively. From Theorem 2.20,

$$\mathrm{ord}_{\sigma_{c,1}\mathfrak{P}} g(\omega^{-k}) = \mathrm{ord}_{\mathfrak{P}} \sigma_{c,1}(g(\omega^{-k})) = \mathrm{ord}_{\mathfrak{P}} g(\omega^{-kc}) = s(kc).$$

Since the prime number p is totally ramified in $\boldsymbol{Q}(\zeta_p)$ and $[G' : Z'] = 1$, we obtain the prime factorization of the Gauss sum $g(\omega^{-k})$ as follows,

$$g(\omega^{-k}) \sim \prod_{\sigma_{c,1}^{-1} \pmod{Z}} \prod_{\sigma_{1,d}^{-1} \pmod{Z'}} \left(\mathfrak{P}^{\sigma_{c,1}^{-1}\sigma_{1,d}^{-1}}\right)^{\mathrm{ord}_{\sigma_{c,1}^{-1}\sigma_{1,d}^{-1}\mathfrak{P}} g(\omega^{-k})}$$
$$\sim \prod_{\sigma_{c,1}^{-1} \pmod{Z}} \left(\mathfrak{P}^{\sigma_{c,1}^{-1}}\right)^{\mathrm{ord}_{\sigma_{c,1}^{-1}\mathfrak{P}} g(\omega^{-k})}$$
$$\sim \prod_{c \in \boldsymbol{Z}_{q-1}^{\times}/\langle p\rangle} \left(\mathrm{P}^{\sigma_{c,1}^{-1}}\right)^{s(kc)}$$

from Theorem 2.20.

We put $\sigma_c = \sigma_{c,1}$ and

$$\theta(k, \mathfrak{P}) = \sum_{c \in \boldsymbol{Z}_{q-1}^{\times}/\langle p\rangle} s(kc)\sigma_c^{-1}.$$

We denote a fractional part of a real number t by $\langle t \rangle$.

Lemma 2.8: *For any integer k, we have*

$$s(k) = (p-1)\sum_{i=0}^{n-1} \left\langle \frac{kp^i}{q-1} \right\rangle.$$

Proof: The relation is obvious for $k = 0$. Assume that $1 \le k < q-1$. If we write $k = k_0 + k_1 p + \cdots + k_{n-1}p^{n-1}$, $0 \le k_i \le p-1$, then we have

$$k_0 + k_1 p + \cdots + k_{n-1}p^{n-1} = (q-1)\left\langle \frac{k}{q-1} \right\rangle,$$
$$k_{n-1} + k_0 p + \cdots + k_{n-2}p^{n-1} = (q-1)\left\langle \frac{kp}{q-1} \right\rangle,$$
$$\vdots$$
$$k_1 + k_2 p + \cdots + k_0 p^{n-1} = (q-1)\left\langle \frac{kp^{n-1}}{q-1} \right\rangle.$$

The system yields

$$\frac{p^n-1}{p-1}(k_0+k_1+\cdots+k_{n-1})=(q-1)\sum_{i=0}^{n-1}\left\langle\frac{kp^i}{q-1}\right\rangle.$$

Hence we obtain the relation in the theorem. □

From Lemma 2.8, we have

$$\begin{aligned}\theta(k,\mathfrak{P}) &= (p-1)\sum_{c\in \mathbf{Z}_{q-1}^{\times}/\langle p\rangle}\sum_{i=0}^{n-1}\left\langle\frac{kcp^i}{q-1}\right\rangle\sigma_c^{-1}\\ &= (p-1)\sum_{c\in \mathbf{Z}_{q-1}^{\times}}\left\langle\frac{kc}{q-1}\right\rangle\sigma_c^{-1}.\end{aligned}$$

We know the prime ideal $\wp$ of $\mathbf{Q}(\zeta_{q-1})$ is totally ramified in $\mathbf{Q}(\zeta_{q-1},\zeta_p)$, $\wp\sim\mathfrak{P}^{p-1}$. Then we have the following theorem.

Theorem 2.21 *(Lang [124]):* *Let $\wp$ is a prime ideal in $\mathbf{Q}(\zeta_{q-1})$ lying above p and $\mathfrak{P}$ is a prime ideal in $\mathbf{Q}(\zeta_{q-1},\zeta_p)$ lying above $\wp$.*

$$\begin{aligned}g\left(\omega^{-k}\right) &\sim \mathfrak{P}^{(p-1)\sum_{c\in \mathbf{Z}_{q-1}^{\times}}\left\langle\frac{kc}{q-1}\right\rangle\sigma_c^{-1}}\\ &\sim \wp^{\sum_{c\in \mathbf{Z}_{q-1}^{\times}}\left\langle\frac{kc}{q-1}\right\rangle\sigma_c^{-1}}.\end{aligned}$$

We call $\sum_{c\in \mathbf{Z}_{q-1}^{\times}}\left\langle\frac{kc}{q-1}\right\rangle\sigma_c^{-1}$ the **Stickelberger element.**

2.6.4 Prime Ideal Factorization of the Gauss Sums in $Q(\zeta_m)$

We assume the character ω^{-k} of $\mathbf{F}_q$ has the order m, $p\nmid m$. Let f be the smallest integer such that $p^f\equiv 1 \pmod m$. Then $N_{Q(\zeta_m)/Q}\mathcal{P}=p^f$ where $\mathcal{P}$ is a prime ideal in $\mathbf{Q}(\zeta_m)$ lying above p from Theorem 2.18.

From Theorem 2.21,

$$\theta(k,\wp)=\sum_{c\in \mathbf{Z}_{q-1}^{\times}}\left\langle\frac{kc}{q-1}\right\rangle\sigma_c^{-1}=\sum_{c'\in \mathbf{Z}_m^{\times}}\ \sum_{\substack{c\in \mathbf{Z}_{q-1}^{\times}\\ c\equiv c' \pmod m}}\left\langle\frac{kc}{q-1}\right\rangle\sigma_c^{-1}.$$

An element $c\in \mathbf{Z}_{q-1}^{\times}$ such that $c\equiv c' \pmod m$, $c'\in \mathbf{Z}_m^{\times}$ can be written as $c=c'\cdot d$, $d\equiv 1 \pmod m$, $d\in \mathbf{Z}_{q-1}^{\times}$. It follows that

$$\begin{aligned}\theta(k,\wp) &= \sum_{c'\in \mathbf{Z}_m^{\times}}\ \sum_{\substack{d\in \mathbf{Z}_{q-1}^{\times}\\ d\equiv 1 \pmod m}}\left\langle\frac{kc'd}{q-1}\right\rangle\sigma_{c'd}^{-1}\\ &= \sum_{c'\in \mathbf{Z}_m^{\times}}\left\langle\frac{kc'}{q-1}\right\rangle\sigma_{c'}^{-1}\cdot\sum_{\substack{d\in \mathbf{Z}_{q-1}^{\times}\\ d\equiv 1 \pmod m}}\sigma_d^{-1}.\end{aligned}$$

From

$$\mathrm{Gal}(\mathbf{Q}(\zeta_{q-1})/\mathbf{Q}(\zeta_m))=\left\{\sigma_d:\zeta_{q-1}\to\zeta_{q-1}^d,\ d\equiv 1 \pmod m\right\}$$

and $N_{\mathbf{Q}(\zeta_m)/\mathbf{Q}}\mathcal{P} = p^f$,

$$\wp^{\sum_{d\in \mathbf{Z}_{q-1}^{\times},\ d\equiv 1 \pmod{m}} \sigma_d^{-1}} = N_{\mathbf{Q}(\zeta_{q-1})/\mathbf{Q}(\zeta_m)}\,\wp = \mathcal{P}^{\frac{n}{f}}$$

in $\mathbf{Q}(\zeta_m)$. Since ω^{-k} has the order m, $g(\omega^{-k})^m$ belongs to $\mathbf{Q}(\zeta_m)$. The prime ideal decomposition of $g(\omega^{-k})^m$ is given by the following theorem.

Theorem 2.22: *Let $q = p^n$ and let f be the smallest integer such that $p^f \equiv 1 \pmod{m}$. Assume the character ω^{-k} of $GF(q)$ has the order m, $p \nmid m$. Then*

$$g(\omega^{-k})^m \sim \mathcal{P}^{\frac{mn}{f}\sum_{c\in \mathbf{Z}_m^{\times}}\left\langle \frac{kc}{q-1}\right\rangle \sigma_c^{-1}}$$

or symbolically

$$g(\omega^{-k}) \sim \mathcal{P}^{\frac{n}{f}\sum_{c\in \mathbf{Z}_m^{\times}}\left\langle \frac{kc}{q-1}\right\rangle \sigma_c^{-1}}.$$

where $\mathcal{P}$ is a prime ideal in $\mathbf{Q}(\zeta_m)$ lying above p.

3

Plug-In Matrices

3.1 Notations

Table 3.1 gives the notations which are used in this chapter.

3.2 Williamson Type and Williamson Matrices

Williamson's famous theorem [223] is:

Theorem 3.1: *Let n be an odd positive integer. Suppose there exist four symmetric circulant* $(1, -1)$ *matrices* A, B, C, D *of order n. Further, suppose*

$$A^2 + B^2 + C^2 + D^2 = 4nI_n.$$

Then

$$W_H = \begin{bmatrix} A & B & C & D \\ -B & A & -D & C \\ -C & D & A & -B \\ -D & -C & B & A \end{bmatrix} \tag{3.1}$$

is an Hadamard matrix of order $4n$.

Definition 3.1: The matrix in (3.1) is called a **Williamson Hadamard matrix of order** $4n$ or a **Hadamard matrix of quaternion type**.

We recall the component matrices A, B, C, and D are called Williamson matrices (see Definition 1.27). The component matrices in the Williamson Hadamard matrix satisfy the additive property (1.8).

The immediate generalization is

Definition 3.2: We can generalize Definition 1.27 by considering four ± 1 matrices A, B, C, and D of order n that satisfy both

$$XY^\top = YX^\top \quad \text{for} \quad X, Y \in \{A, B, C, D\},$$

that is, A, B, C, D are pairwise amicable, and

$$AA^\top + BB^\top + CC^\top + DD^\top = 4nI$$

will be called **Williamson-type matrices**.

Hadamard Matrices: Constructions using Number Theory and Algebra, First Edition. Jennifer Seberry and Mieko Yamada.

Table 3.1 Notations used in this chapter.

$X^{\top}$	The transpose of the matrix X
$+$	$+1$ when $+$ is an element of a sequence
$-$	-1 when $-$ is an entry of a matrix or a sequence
$M \times N$	The Kronecker product
R	The back diagonal identity matrix
I	The identity matrix
A/B	The interleaving of two sequences $A = \{a_1, \dots, a_m\}$ and $B = \{b_1, \dots b_{m-1}\}$
$\{A, B\}$	The adjoining of two sequences $\{a_1, \dots, a_m\}$ and $\{b_1, \dots b_m\}$
0_l	A sequence of zeros of length l
ζ	A (primitive) root of unity
$GF(q)$	A finite field with q elements for a prime power q
$4-\left\{v; k_1, k_2, k_3, k_4; \sum_{i=1}^{4} k_i - v\right\}$	Hadamard supplementary difference sets
$\vartheta_{F/K}(\chi_F)$	The relative Gauss sum associated with the character χ_F of a finite field F
$J(\chi_1, \chi_2)$	The Jacobi sum for multiplicative characters χ_1 and χ_2

Then

$$\begin{bmatrix} A & B & C & D \\ -B & A & -D & C \\ -C & D & A & -B \\ -D & -C & B & A \end{bmatrix}$$

is an Hadamard matrix.

Thus we have:

Theorem 3.2: *Let n be an odd integer. Suppose there exists four pairwise amicable $(1, -1)$ matrices of order n, A, B, C, D satisfying the additive property $AA^{\top} + BB^{\top} + CC^{\top} + DD^{\top} = 4n$. Then the matrix in (3.1) is a Williamson type Hadamard matrix.*

We show a Williamson Hadamard matrix is associated with a quaternion group. Let A, B, C, and D be a circulant matrix of order n. Let Q be a quaternion group such that $Q = \{\pm 1, \pm i, \pm j, \pm k\}$ with the conditions $i^2 = -1, j^2 = -1, k^2 = -1, ij = -ji, jk = -kj, ki = -ik$. The right regular representation matrices $R(1), R(i), R(j), R(k)$ of $1, i, j, k$ are

$$R(1) = \begin{bmatrix} 1 & 0 & 0 & 0 \\ 0 & 1 & 0 & 0 \\ 0 & 0 & 1 & 0 \\ 0 & 0 & 0 & 1 \end{bmatrix}, \qquad R(i) = \begin{bmatrix} 0 & 1 & 0 & 0 \\ - & 0 & 0 & 0 \\ 0 & 0 & 0 & - \\ 0 & 0 & 1 & 0 \end{bmatrix},$$

$$R(j) = \begin{bmatrix} 0 & 0 & 1 & 0 \\ 0 & 0 & 0 & 1 \\ - & 0 & 0 & 0 \\ 0 & - & 0 & 0 \end{bmatrix}, \qquad R(k) = \begin{bmatrix} 0 & 0 & 0 & 1 \\ 0 & 0 & - & 0 \\ 0 & 1 & 0 & 0 \\ - & 0 & 0 & 0 \end{bmatrix}.$$

Thus the Williamson Hadamard matrix H can be written as

$$H = A \times R(1) + B \times R(i) + C \times R(j) + D \times R(k)$$

where $\times$ is the tensor (or Kronecker) product. It means that a Williamson Hadamard matrix is a right regular representation matrix of a quaternion number $\alpha = a + bi + cj + dk$ by substituting a matrix A for a, and so on. We see that a left regular representation matrix of α is Hadamard equivalent to a right regular representation matrix of α.

We use T for the basic circulant matrix (see Definition 1.9).

We define the polynomials in $\mathbf{Z}[x]/(x^n - 1)$ corresponding to the first rows of A, B, C, and D. That is,

$$f_A(x) = \sum_{m=0}^{n-1} a_m x^m, \quad f_B(x) = \sum_{m=0}^{n-1} b_m x^m,$$
$$f_C(x) = \sum_{m=0}^{n-1} c_m x^m, \quad f_D(x) = \sum_{m=0}^{n-1} d_m x^m.$$

where $(a_0, a_1, \dots, a_{n-1}), (b_0, b_1, \dots, b_{n-1}), (c_0, c_1, \dots, c_{n-1}), (d_0, d_1, \dots, d_{n-1})$ are first rows of A, B, C, D, respectively. Then $A = f_A(T), B = f_B(T), C = f_C(T)$, and $D = f_D(T)$.

The conditions for the matrices A, B, C, and D in Theorem 3.1 are equivalent to the conditions for the polynomials $f_A(x), f_B(x), f_C(x), f_D(x)$ as follows:

i) The coefficient of the polynomials $f_A(x), f_B(x), f_C(x), f_D(x)$ is ± 1;
ii) $f_A(x^{-1}) \equiv f_A(x) \pmod{x^n - 1}$. Similarly for $f_B(x), f_C(x), f_D(x)$;
iii)

$$f_A(x)^2 + f_B(x)^2 + f_C(x)^2 + f_D(x)^2 \equiv 4n \pmod{x^n - 1}. \tag{3.2}$$

The Eq. (3.2) is called the **Williamson equation**.

Lemma 3.1: *We may assume $a_0 = b_0 = c_0 = d_0 = 1$ by replacing A (or B, C, D) by $-A$ (or $-B, -C, -D$) if necessary. Then three of the coefficients $a_\ell, b_\ell, c_\ell, d_\ell$, $1 \le \ell \le n-1$, are of the same sign.*

Proof: We define the polynomials

$$g_A(x) = \sum_{a_\ell = 1} x^\ell, \quad g_B(x) = \sum_{b_\ell = 1} x^\ell,$$
$$g_C(x) = \sum_{c_\ell = 1} x^\ell, \quad g_D(x) = \sum_{d_\ell = 1} x^\ell$$

and $k_A = |\{\ell | a_\ell = 1\}|$ and k_B, k_C, k_D in the same way.
Since

$$f_A(x) = \sum_{a_\ell = 1} x^\ell - \sum_{a_\ell = -1} x^\ell$$

and

$$\sum_{a_\ell = 1} x^\ell + \sum_{a_\ell = -1} x^\ell = \sum_{\ell = 0}^{n-1} x^\ell = J(x)$$

we have

$$f_A(x) = 2g_A(x) - J(x), \qquad f_B(x) = 2g_B(x) - J(x),$$
$$f_C(x) = 2g_C(x) - J(x), \qquad f_D(x) = 2g_D(x) - J(x).$$

From the Williamson equation (3.2),

$$g_A(x)^2 + g_B(x)^2 + g_C(x)^2 + g_D(x)^2 \equiv n + (k_A + k_B + k_C + k_D - n)J(x) \pmod{x^n - 1}.$$

Since k_A, k_B, k_C, k_D are all odd, $\lambda = k_A + k_B + k_C + k_D - n$ is also odd. Then

$$g_A(x)^2 = \left(\sum_{a_\ell=1} x^\ell\right)^2 \equiv \sum_{a_\ell=1} x^{2\ell} \equiv g_A(x^2) \pmod{2}.$$

We have $g_B(x) \equiv g_B(x^2)$, $g_C(x) \equiv g_C(x^2)$, $g_D(x) \equiv g_D(x^2) \pmod{2}$ similarly. Thus

$$g_A(x^2) + g_B(x^2) + g_C(x^2) + g_D(x^2) \equiv 1 + J(x) \pmod{2, x^n - 1}.$$

Since n is odd, it is equivalent to

$$g_A(x) + g_B(x) + g_C(x) + g_D(x) \equiv 1 + J(x) \pmod{2, x^n - 1}$$

replacing x^2 by x. Hence for each $\ell = 1, \ldots, n-1$, we see that every x^ℓ occurs odd times on the left side of the above equation. □

Thus we define the polynomials in $\mathbf{Z}[x]/(x^n - 1)$,

$$\begin{aligned}
\phi_A(x) &= \tfrac{1}{2}\left\{-f_A(x) + f_B(x) + f_C(x) + f_D(x)\right\},\\
\phi_B(x) &= \tfrac{1}{2}\left\{f_A(x) - f_B(x) + f_C(x) + f_D(x)\right\},\\
\phi_C(x) &= \tfrac{1}{2}\left\{f_A(x) + f_B(x) - f_C(x) + f_D(x)\right\},\\
\phi_D(x) &= \tfrac{1}{2}\left\{f_A(x) + f_B(x) + f_C(x) - f_D(x)\right\}.
\end{aligned}$$

From Lemma 3.1, one of the coefficients of x^ℓ, $1 \le \ell \le n-1$ is ± 2 and the other three coefficients are all 0. Then $\phi_A(x)$ has the form

$$\phi_A(x) = 1 + 2\sum_{m=1}^{n-1/2} e_m\left(x^m + x^{-m}\right)$$

and $\phi_B(x), \phi_C(x), \phi_D(x)$ have the similar forms. This leads to

Theorem 3.3: *Suppose n is an odd integer. Let $\mathcal{A}, \mathcal{B}, \mathcal{C}, \mathcal{D}$ be a partition of $\left\{1, 2, \ldots, \frac{n-1}{2}\right\}$ and $e_m = \pm 1$ ($m = 1, 2, \ldots, \frac{n-1}{2}$). Further we put $u_m = x^m + x^{-m}$. If the equation*

$$\left(1 + 2\sum_{m\in\mathcal{A}} e_m u_m\right)^2 + \left(1 + 2\sum_{m\in\mathcal{B}} e_m u_m\right)^2 + \left(1 + 2\sum_{m\in\mathcal{C}} e_m u_m\right)^2 + \left(1 + 2\sum_{m\in\mathcal{D}} e_m u_m\right)^2 \equiv 4n \pmod{x^n - 1} \tag{3.3}$$

holds, then there exists an Hadamard matrix of order 4n. We call the Eq. (3.3) the ***reduced Williamson equation****.*

If we put $x = 1$, Eq. (3.3) shows that $4n$ is written as a sum of 4 squares of odd integers. Williamson gave solutions for n, odd, $n \le 13$. Baumert and Hall [9] gave all solutions for all representations of $4n$ as a sum of 4 odd squares $3 \le n \le 23$. Further results are found by Đoković [55, 57, 59], Holzmann et al. [97], Sawade [149], Spence [176], Yamada [242], Yamamoto and Yamada [247], Koukouvinos and Kounias [112], and Xia et al. [229].

Conversely, if the Eq. (3.3) in Theorem 3.3 holds and we put

$$\begin{aligned}
f_A(x) &= \tfrac{1}{2}\left\{-\phi_A(x) + \phi_B(x) + \phi_C(x) + \phi_D(x)\right\},\\
f_B(x) &= \tfrac{1}{2}\left\{\phi_A(x) - \phi_B(x) + \phi_C(x) + \phi_D(x)\right\},\\
f_C(x) &= \tfrac{1}{2}\left\{\phi_A(x) + \phi_B(x) - \phi_C(x) + \phi_D(x)\right\},\\
f_D(x) &= \tfrac{1}{2}\left\{\phi_A(x) + \phi_B(x) + \phi_C(x) - \phi_D(x)\right\},
\end{aligned}$$

then the Williamson equation (3.2) holds.

Example 3.1:

1) *Let $n = 3$. Then $4 \cdot 3 = 12 = 1^2 + 1^2 + 1^2 + 3^2$. We define the sets, $\mathcal{A}^+ = \mathcal{A}^- = \mathcal{B}^+ = \mathcal{B}^- = \mathcal{C}^+ = \mathcal{C}^- = \mathcal{D}^+ = \emptyset$ and $\mathcal{D}^- = \{1\}$. The polynomials are*

$$\phi_A(x) = \phi_B(x) = \phi_C(x) = 1, \quad \phi_D(x) = 1 - 2(x + x^2).$$

and the Williamson equation is

$$1 + 1 + 1 + (1 - 2(x + x^2))^2 \equiv 12 \pmod{x^3 - 1},$$

that is,

$$f_A(x) = f_B(x) = f_C(x) = 1 - (x + x^2), \quad f_D(x) = 1 + x + x^2$$

and Williamson matrices,

$$A = B = C = \begin{bmatrix} 1 & - & - \\ - & 1 & - \\ 1 & 1 & - \end{bmatrix}, \qquad D = \begin{bmatrix} 1 & 1 & 1 \\ 1 & 1 & 1 \\ 1 & 1 & 1 \end{bmatrix}$$

where "–" denotes –1.

2) *Let $n = 5$. $4 \cdot 5 = 20 = 1^2 + 1^2 + 3^2 + 3^2$. The sets $\mathcal{A}^+ = \mathcal{A}^- = \mathcal{B}^+ = \mathcal{B}^- = \mathcal{C}^+ = \mathcal{D}^+ = \emptyset$, $\mathcal{C}^- = \{1\}$, and $\mathcal{D}^- = \{2\}$. Then the polynomials are*

$$\phi_A(x) = \phi_B(x) = 1, \quad \phi_C(x) = 1 - 2(x + x^4), \quad \phi_D(x) = 1 - 2(x^2 + x^3),$$

and

$$\begin{aligned}
f_A(x) &= f_B(x) = 1 - (x + x^2 + x^3 + x^4),\\
f_C(x) &= 1 + (x + x^4) - (x^2 + x^3),\\
f_D(x) &= 1 - (x + x^4) + (x^2 + x^3).
\end{aligned}$$

Then the Williamson matrices are

$$A = B = \begin{bmatrix} 1 & - & - & - & - \\ - & 1 & - & - & - \\ - & - & 1 & - & - \\ - & - & - & 1 & - \\ - & - & - & - & 1 \end{bmatrix}, \quad C = \begin{bmatrix} 1 & 1 & - & - & 1 \\ 1 & 1 & 1 & - & - \\ - & 1 & 1 & - & - \\ - & - & 1 & 1 & - \\ - & - & - & 1 & 1 \end{bmatrix}, \quad D = \begin{bmatrix} 1 & - & 1 & 1 & - \\ - & 1 & - & 1 & 1 \\ 1 & - & 1 & - & 1 \\ 1 & 1 & - & 1 & - \\ - & 1 & 1 & - & 1 \end{bmatrix}.$$

3) *Let $n = 7$. Then $4 \cdot 7 = 28 = 1^2 + 3^2 + 3^2 + 3^2 = 1^2 + 1^2 + 1^2 + 5^2$.*
   a) *For the case $28 = 1^2 + 3^2 + 3^2 + 3^2$, we give $A^+ = A^- = B^+ = C^+ = D^+ = \emptyset$, $B^- = \{1\}$, $C^- = \{2\}$, $D^- = \{3\}$.*
   b) *For the case $28 = 1^2 + 1^2 + 1^2 + 5^2$, we give $A^+ = A^- = B^+ = B^- = D^- = \emptyset$, $C^+ = \{1\}$, $C^- = \{2\}$, $D^+ = \{3\}$.*

We have the following polynomials,

$$
\begin{aligned}
(a)\ & f_A(x) = 1 - (x + x^2 + x^3 + x^4 + x^5 + x^6),\\
& f_B(x) = 1 + x - x^2 - x^3 - x^4 - x^5 + x^6,\\
& f_C(x) = 1 - x + x^2 - x^3 - x^4 + x^5 - x^6,\\
& f_D(x) = 1 - x - x^2 + x^3 + x^4 - x^5 - x^6
\end{aligned}
$$

and

$$
\begin{aligned}
(b)\ & f_A(x) = 1 + x - x^2 + x^3 + x^4 - x^5 + x^6,\\
& f_B(x) = f_A(x),\\
& f_C(x) = 1 - x + x^2 + x^3 + x^4 + x^5 - x^6,\\
& f_D(x) = 1 + x - x^2 - x^3 - x^4 - x^5 + x^6.
\end{aligned}
$$

Thus we obtain Williamson matrices of order 7 *and an Hadamard matrix of order* 28.

3.3 Plug-In Matrices

When we plug Williamson matrices A, B, C, and D into the matrix W_H in (3.1), we obtain a Williamson Hadamard matrix. We call Williamson matrices A, B, C, and D **plug-in matrices** and the plugged-in array a **plug-into array**.

In this section, we give several plug-in matrices and related plug-into arrays. We first consider different combination of symmetric and skew symmetric. We will discuss plug-into array more precisely in Chapter 4.

3.3.1 The Ito Array

Definition 3.3: Assume ± 1 circulant matrices A, B, C, and D of order n satisfy the Ito additive property

$$
\begin{aligned}
& AA^\top + BB^\top + CC^\top + DD^\top = 4nI,\\
& AB^\top + CD^\top = BA^\top + DC^\top.
\end{aligned}
$$

Then

$$
\begin{bmatrix} A & B & C & D \\ B & -A & D & -C \\ C^\top & -D^\top & -A^\top & B^\top \\ D^\top & C^\top & -B^\top & -A^\top \end{bmatrix} \tag{3.4}
$$

is called **Ito's Hadamard matrix** or the **Ito array**.

Ito showed the Paley type I Hadamard matrices are equivalent to the Ito's Hadamard matrix [102].

3.3.2 Good Matrices : A Variation of Williamson Matrices

Good matrices first appeared in the PhD Thesis of Jennifer (Seberry) Wallis [203]. There the matrices, which were given no name, were given for $n = 1, \ldots, 15, 19$. In 1971 she gave good matrices for $n = 23$ [202]. Seberry Wallis gave an array, the **Seberry–Williamson array**, which is a modification of the Williamson array to make skew Hadamard matrices.

R is defined in Lemma 1.5. Written in terms of circulant and R it is

$$W_{\text{Seberry-Williamson-array}} = \begin{bmatrix} A & BR & CR & DR \\ -BR & A & DR & -CR \\ -CR & -DR & A & BR \\ -DR & CR & -BR & A \end{bmatrix}.$$

Good matrices were first used by name in [210].

Definition 3.4: Four ± 1 circulant (or type 1) matrices, A_1, A_2, A_3, and A_4, of order n will be called **good matrices** if they satisfy

i) $(A_1 - I)^\top = -(A_1 - I), A_i^\top = A_i, i = 2, 3, 4;$
ii) $\sum_{i=1}^{4} A_i A_i^\top = 4nI_n$, (the additive property (1.8)).

Substituting A_1 for A, A_2 for B, A_3 for C, and A_4 for D gives us

Theorem 3.4: *Suppose there exist good matrices of order n. Then using the Seberry–Williamson array there exists a skew Hadamard matrix of order* 4*n*.

Hunt (Unpublished material. Private communication with author Seberry, 1971) gave good matrices for $n = 1$, …, 25. Later Szekeres [183] gave a list for order $n = 1, \ldots, 31$. Djoković [57, 58] provided orders $n = 33, 35$, and 127. Then Georgiou et al. [75] provided 37, 39. Đoković (Regarding good matrices of order 41. Email communication to author Seberry, 23 July 2014) says that only one set of supplementary difference sets, (41;20, 20, 16, 16;31), for 41 remains to be searched.

We note that while there are no Williamson matrices of order 35 there are good matrices of order 35.

(Seberry) Wallis [203] gave a construction for $n = 19$ and Đoković [57] for $n = 33, 35$, and 127. The remainder were found by computer search.

3.3.3 The Goethals–Seidel Array

Part (i) of Definition 3.4 can be relaxed in an array called the **Goethals–Seidel array** (Wallis Whiteman array) which only needs four circulant ± 1 matrices satisfying the additive property (the additive property (1.8)). This variation was first given to find a skew Hadamard matrix of order 36 [81] but has proved highly useful in making new Hadamard matrices and orthogonal designs. We have

Theorem 3.5 *(Goethals and Seidel [81])*: *If A, B, C, D are square circulant ± 1 matrices of order n, and R is the back diagonal identity matrix, then if*

$$AA^\top + BB^\top + CC^\top + DD^\top = 4nI, \tag{3.5}$$

then the array

$$H = \begin{bmatrix} A & BR & CR & DR \\ -BR & A & D^\top R & -C^\top R \\ -CR & -D^\top R & A & B^\top R \\ -DR & C^\top R & -B^\top R & A \end{bmatrix} \tag{3.6}$$

is an Hadamard matrix of order 4*n*. *Furthermore if A is skew type, the Hadamard matrix is skew Hadamard.*

3.3.4 Symmetric Hadamard Variation

Seberry and Balonin [159] have recently given another variation of the Williamson array, called the **propus array**.

Theorem 3.6 *(Seberry–Balonin [159]):* *Suppose there exist four ±1 circulant (or type 1) matrices, A, $B = C$ and D, of order n, where*

$$\text{(i)}\quad A^{\top} = A,\ B^{\top} = B,\ C^{\top} = C,\ \text{and } D^{\top} = D,$$

$$\text{(ii)}\quad AA^{\top} + BB^{\top} + CC^{\top} + DD^{\top} = 4nI_n. \tag{3.7}$$

Then

$$H = \begin{bmatrix} A & B & B & D \\ B & D & -A & -B \\ B & -A & -D & B \\ D & -B & B & -A \end{bmatrix} \tag{3.8}$$

is a symmetric (propus) Hadamard matrix of order 4n.

All our constructions so far have relied on circulant or back circulant, but all the constructions are valid for type 1 and type 2 matrices. See Wallis et al. [214] and Wallis and Whiteman [216].

3.4 Eight Plug-In Matrices

In [212, lemma 5] many specific variations of the plug-into arrays $OD(8; 1, 1, 1, 1, 1, 1, 1, 1)$ are given which use eight plug-in matrices. Another variant uses the Kharaghani array (see Theorem 3.7) and amicable sets, which are a variant of eight Williamson matrices, where matrices are pairwise amicable.

Definition 3.5: Eight circulant ±1 matrices $A_1, A_2, \dots, A_8$ of order w which are symmetric and which satisfy

$$\sum_{i=1}^{8} A_i A_i^{\top} = 8wI_w$$

will be called **eight Williamson matrices**.

Eight ±1 amicable matrices $A_1, A_2, \dots, A_8$ of order w which satisfy both

$$\sum_{i=1}^{8} A_i A_i^{\top} = 8wI_w, \qquad A_j A_i^{\top} = A_i A_j^{\top}, \qquad i,j = 1, \dots, 8$$

will be called **eight Williamson-type matrices**.

3.4.1 The Kharaghani Array

Plotkin [145] showed that, if there is an Hadamard matrix of order $2t$, then there is an $OD(8t; t, t, t, t, t, t, t, t)$. In the same paper, he also constructed an $OD(24; 3, 3, 3, 3, 3, 3, 3, 3)$. This orthogonal design (OD) has appeared in [78], [161] and in [166]. It is conjectured that there is an $OD(8t; t, t, t, t, t, t, t, t)$ for each odd integer t. Until recently, no-one except the original for $t = 3$ found by Plotkin, had been constructed in the ensuing 28 years. Holzmann and Kharaghani [96] using a new method constructed many new Plotkin *OD*s of order 24 and two new Plotkin *OD*s of order 40 and 56. Actually their construction provides many new orthogonal designs in 6, 7, and 8 variables which include the Plotkin *OD*s of order 40 and 56.

Following Kharaghani [106]

Definition 3.6: A set $\{A_1, A_2, \dots, A_{2n}\}$ of square real matrices is an **amicable set** if

$$\sum_{i=1}^{n} \left(A_{\sigma(2i-1)}A_{\sigma(2i)}^{\mathrm{T}} - A_{\sigma(2i)}A_{\sigma(2i-1)}^{\mathrm{T}} \right) = 0$$

for some permutation σ of the set $\{1, 2, \dots, 2n\}$.

For simplicity, we will always take $\sigma(i) = i$ unless otherwise specified. So

$$\sum_{i=1}^{n} \left(A_{2i-1}A_{2i}^{\mathrm{T}} - A_{2i}A_{2i-1}^{\mathrm{T}} \right) = 0. \tag{3.9}$$

Clearly a set of mutually amicable matrices is amicable, but the converse is not true in general.

Theorem 3.7 ***(Kharaghani [106]):*** *Let* $\{A_i\}_{i=1}^{8}$ *be an amicable set of circulant* ± 1 *matrices of order t, satisfying the additive property, that is* $\sum_{i=1}^{8} A_iA_i^{\mathrm{T}} = 8tI_t$. *Let R be a back diagonal identity matrix. Then the Kharaghani array H gives an Hadamard matrix of order* $8t$.

$$H = \begin{bmatrix} A_1 & A_2 & A_4R & A_3R & A_6R & A_5R & A_8R & A_7R \\ -A_2 & A_1 & A_3R & -A_4R & A_5R & -A_6R & A_7R & -A_8R \\ -A_4R & -A_3R & A_1 & A_2 & -A_8^{\mathrm{T}}R & A_7^{\mathrm{T}}R & A_6^{\mathrm{T}}R & -A_5^{\mathrm{T}}R \\ -A_3R & A_4R & -A_2 & A_1 & A_7^{\mathrm{T}}R & A_8^{\mathrm{T}}R & -A_5^{\mathrm{T}}R & -A_6^{\mathrm{T}}R \\ -A_6R & -A_5R & A_8^{\mathrm{T}}R & -A_7^{\mathrm{T}}R & A_1 & A_2 & -A_4^{\mathrm{T}}R & A_3^{\mathrm{T}}R \\ -A_5R & A_6R & -A_7^{\mathrm{T}}R & -A_8^{\mathrm{T}}R & -A_2 & A_1 & A_3^{\mathrm{T}}R & A_4^{\mathrm{T}}R \\ -A_8R & -A_7R & -A_6^{\mathrm{T}}R & A_5^{\mathrm{T}}R & A_4^{\mathrm{T}}R & -A_3^{\mathrm{T}}R & A_1 & A_2 \\ -A_7R & A_8R & A_5^{\mathrm{T}}R & A_6^{\mathrm{T}}R & -A_3^{\mathrm{T}}R & -A_4^{\mathrm{T}}R & -A_2 & A_1 \end{bmatrix}$$

3.5 More *T*-sequences and *T*-matrices

In order to satisfy the conditions of Theorem 3.11, we are led to study sequences of a more restricted type. In this section we further develop the discussion on T-matrices, T-sequences, and Turyn sequences in Section 1.16.

Theorem 3.8 ***(Turyn [194]):*** *Suppose* $A = \{X, U, Y, V\}$ *are Turyn sequences of length* ℓ, *where X is skew and Y is symmetric for* ℓ *even and X is symmetric and Y is skew for* ℓ *odd. Then there are T-sequences of length* $2\ell - 1$ *and* $4\ell - 1$.

Proof: We use the notation A/B, $\{A, B\}$ as before. Let 0_t be a sequence of zeros of length t. Then

$$T_1 = \left\{ \left\{ \frac{1}{2}(X+Y), 0_{\ell-1} \right\}, \left\{ \frac{1}{2}(X-Y), 0_{\ell-1} \right\}, \left\{ 0_\ell, \frac{1}{2}(Y+V) \right\}, \left\{ 0_\ell, \frac{1}{2}(Y-V) \right\} \right\}$$

and

$$T_2 = \left\{ \{1, 0_{4\ell-2}\}, \{0, X/Y, 0_{2\ell-1}\}, \{0, 0_{2\ell-1}, U/0_{\ell-1}\}, \{0, 0_{2\ell-1}, 0_\ell/V\} \right\}$$

are the T-sequences of lengths $2\ell - 1$ and $4\ell - 1$, respectively. □

Corollary 3.1: *There are T-sequences constructed from Turyn sequences of lengths* 3, 5, 7, 9, 11, 13, 15, 19, 23, 25, 27, 29, 31, 51, 59.

Theorem 3.9: *If X and Y are Golay sequences of length r, then*

$$T = \left\{ \{1, 0_r\},\ \left\{0, \frac{1}{2}(X+Y)\right\},\ \left\{0, \frac{1}{2}(X-Y)\right\},\ \{0_r, 1\} \right\}$$

are T-sequences of length $r+1$.

Corollary 3.2 (*Turyn [194]*): *There exist T-sequences of lengths* $1 + 2^a 10^b 26^c$, *where* a, b, c *are nonnegative integers.*

Combining these last two corollaries we have

Corollary 3.3: *There exist T-sequences of lengths* $3, 5, 7, \ldots, 33, 41, 51, 53, 59, 65, 81, 101, 105, 107$.

A desire to fill the gaps in the list in Corollary 3.3 leads to the following idea.

Lemma 3.2: *Suppose* $X = \{A, B, C, D\}$ *are four complementary sequences of length* $\ell, \ell, \ell-1, \ell-1$, *respectively, and weight k. Then*

$$Y = \left\{ \{A, C\},\ \{A, -C\},\ \{B, D\},\ \{B, -D\} \right\}$$

are four-complementary sequences of length $2\ell - 1$ *and weight* $2k$. *Further, if* $\frac{1}{2}(A+B)$ *and* $\frac{1}{2}(C+D)$ *are also* $(0, 1, -1)$ *sequences, then*

$$Z = \left\{ \left\{\frac{1}{2}(A+B), 0_{\ell-1}\right\}, \left\{\frac{1}{2}(A-B), 0_{\ell-1}\right\}, \left\{0_\ell, \frac{1}{2}(C+D)\right\}, \left\{0_\ell, \frac{1}{2}(C-D)\right\} \right\}$$

are four-complementary sequences of length $2\ell - 1$ *and weight k. If* A, B, C, D *are* $(1, -1)$ *sequences, then Z consists of T-sequences of length* $2\ell - 1$.

In fact, Turyn has found sequences satisfying these conditions.

Corollary 3.4: *The following four* $(1, -1)$ *sequences are of lengths* $24, 24, 23, 23$:

$$\begin{array}{l}
1\ -1\ -1\ -1\ \ 1\ -1\ \ 1\ -1\ -1\ -1\ -1\ \ 1\ \ 1\ \ 1\ \ 1\ \ 1\ \ 1\ -1\ -1\ \ 1\ -1\ -1\ -1\ \ 1 \\
1\ -1\ -1\ \ 1\ -1\ -1\ \ 1\ -1\ \ 1\ \ 1\ \ 1\ -1\ -1\ -1\ -1\ -1\ \ 1\ -1\ -1\ -1\ \ 1\ -1\ -1\ -1 \\
1\ \ 1\ \ 1\ -1\ -1\ -1\ \ 1\ \ 1\ -1\ -1\ \ 1\ -1\ -1\ -1\ -1\ \ 1\ -1\ -1\ -1\ -1\ \ 1\ -1\ \ 1 \\
1\ \ 1\ -1\ -1\ \ 1\ -1\ \ 1\ \ 1\ -1\ \ 1\ -1\ \ 1\ \ 1\ \ 1\ -1\ \ 1\ -1\ -1\ \ 1\ -1\ -1\ -1\ \ 1
\end{array}$$

Hence there are T-sequences of length 47.

These sequences may be used in the Goethals–Seidel array to construct the Hadamard matrix of order 188 which was unknown for over 40 years. This method is far more insightful than the construction given in Hedayat and Wallis [94].

We may summarize these results in one theorem which will become extremely powerful in constructing plug-into arrays for Hadamard matrices.

Theorem 3.10: *If there exist T-sequences of length t, then*

i) there exist T-matrices of order t;
ii) there exists a Baumert–Hall array of order t;
iii) there exists an orthogonal design $OD(4t; t, t, t, t)$.

Proof: (i) follows by using the T-sequences as first rows of circulant matrices. The rest follows from Theorem 3.11. □

The following result, in a slightly different form, was also discovered by R.J. Turyn. It is one of the most useful methods for constructing $OD(4n; n, n, n, n)$, that is, matrices to "plug-into."

Theorem 3.11 ***(Cooper and J. Wallis [23]):*** *Suppose there exist circulant or type one T-matrices (T-sequences) X_i, $i = 1, \dots, 4$ of order n. Let a, b, c, d be commuting variables. Then*

$$\begin{aligned} A &= \ \ aX_1 + bX_2 + cX_3 + dX_4 \\ B &= -bX_1 + aX_2 + dX_3 - cX_4 \\ C &= -cX_1 - dX_2 + aX_3 + bX_4 \\ D &= -dX_1 + cX_2 - bX_3 + aX_4 \end{aligned}$$

can be used in the Goethals–Seidel (or Wallis–Whiteman) array to obtain an $OD(4n; n, n, n, n)$ and an Hadamard matrix of order $4n$.

Corollary 3.5: *If there is a T-matrix of order t there is an $OD(4t; t, t, t, t)$.*

The results on T-matrices and T-sequences as applied to Hadamard matrices are given in Chapter 6.

The appropriate theorem for the construction of Hadamard matrices (it is implied by Williamson, Baumert–Hall, Welch, Cooper–J.Wallis, Turyn) is:

Theorem 3.12: *Suppose there exists an $OD(4t; t, t, t, t)$ and four suitable matrices A, B, C, D of order w which are pairwise amicable, have entries +1 or −1, and which satisfy*

$$AA^{\mathrm{T}} + BB^{\mathrm{T}} + CC^{\mathrm{T}} + DD^{\mathrm{T}} = 4wI_w.$$

Then there is an Hadamard matrix of order $4wt$.

Williamson matrices (which are discussed in Section 3.2) are suitable matrices for $OD(4t; t, t, t, t)$: Williamson matrices are plugged into the OD.

Corollary 3.6: *If there are circulant or type one T-matrices of order t and there are Williamson matrices of order w, there is an Hadamard matrix of order $4tw$. Alternatively, if there is an $OD(4t; t, t, t, t)$ and Williamson matrices of order w there is an Hadamard matrix of order $4tw$.*

3.6 Construction of T-matrices of Order $6m + 1$

In the following we assume $q = 6m + 1$ $(m > 0)$ is a prime power and

$$q = a^2 + 3b^2.$$

Without loss of generality, we may assume $a, b \geq 0$. It is shown in [230] that every prime power $q \equiv 1 \pmod 6$ may be written as $q = a^2 + 3b^2$.

Let g be a generator of the multiplicative group of $GF(q)$. Let

$$C_i = \{g^{2mj+i} : j = 0, 1, 2\}, \ i = 0, \dots, 2m - 1$$

be cyclotomic classes of order $2m$.

Let X be a subset of $S = \boldsymbol{Z}/2m\boldsymbol{Z}$, $Y = S - X$, and $|X| = r \equiv m \pmod 2$. We define the elements Q_1, Q_2, Q_3, Q_4 of the group ring $\boldsymbol{Z}[GF(q)^+]$. We identify the subset C_i as the element of the group ring $\boldsymbol{Z}[GF(q)^+]$ and denote 0 by z.

$$\begin{aligned} Q_1 &= z + \sum_{i\in X} \alpha_i C_i, \\ Q_2 &= \sum_{j\in Y} \beta_j g^{2m\delta_j+j}, \\ Q_3 &= \sum_{j\in Y} \beta_j g^{2m(\delta_j+1)+j}, \\ Q_4 &= \sum_{j\in Y} \beta_j g^{2m(\delta_j+2)+j} \end{aligned} \tag{3.10}$$

where α_i, β_j are +1 or −1 such that

$$\left|1 + 3\sum_{i\in X}\alpha_i\right| = a, \quad \left|\sum_{j\in Y}\beta_j\right| = b. \tag{3.11}$$

Let λ and μ be the numbers of ones in a set $\{a_i | i \in X\}$ and a set $\{\beta_j | j \in Y\}$, respectively. If we take $\mu = m + \frac{1}{2}(b - r)$, then the second equation of (3.11) will be satisfied. If we take $\lambda = \frac{1}{2}(r + \frac{a-1}{3})$ for $a \equiv 1 \pmod 3$ or $\frac{1}{2}(r - \frac{a+1}{3})$ for $a \equiv 2 \pmod 3$, then the first equation of (3.11) will be satisfied too.

We define $Q^{-1} = \sum_{g\in GF(q)} \alpha(-g)$ for $Q = \sum_{g\in GF(q)} \alpha g$.

Definition 3.7: If the elements Q_1, Q_2, Q_3, Q_4 of $\boldsymbol{Z}[GF(q)^+]$ satisfy

$$\sum_{i=1}^{4} Q_i Q_i^{-1} = q, \tag{3.12}$$

then Q_1, Q_2, Q_3, Q_4 are said to satisfy ***T*-property**.

Clearly, such Q_i, $i = 1, 2, 3, 4$, are completely determined by these X, Y, α_i, β_j, and δ_j, where $i \in X$ and $j \in Y$.

For a subset D of $GF(q)$, we define the incidence matrix $A(D) = (a_{\gamma\epsilon})_{\gamma,\epsilon\in GF(q)}$ of order q by

$$a_{\gamma\epsilon} = \begin{cases} 1 & \text{if } \gamma - \epsilon \in D, \\ 0 & \text{otherwise.} \end{cases}$$

Then we have the $(0, \pm 1)$ matrices from (3.10),

$$\begin{aligned} T_1 &= I + \sum_{i\in X} \alpha_i A(C_i), \\ T_2 &= \sum_{j\in Y} \beta_j A(\{g^{2m\delta_j+j}\}), \\ T_3 &= \sum_{j\in Y} \beta_j A(\{g^{2m(\delta_j+1)+j}\}), \\ T_4 &= \sum_{j\in Y} \beta_j A(\{g^{2m(\delta_j+2)+j}\}). \end{aligned}$$

Theorem 3.13: *Assume that the elements Q_1, Q_2, Q_3, Q_4 of $\boldsymbol{Z}[GF(q)^+]$ satisfy equations (3.10), (3.11) and has T-property. Then the incidence matrices T_1, T_2, T_3, T_4 are T-matrices of order q.*

Proof: We see $T_i * T_j = 0, (i \neq j)$ and $\sum_{i=1}^{4} T_i$ is a $(1, -1)$ matrix. From (3.12), these matrices satisfy

$$T_1T_1^{\top} + T_2T_2^{\top} + T_3T_3^{\top} + T_4T_4^{\top} = qI.$$

Let t_i, be the row(column) sum of T_i, $i = 1, 2, 3, 4$. From 3.11,

$$t_1^2 + t_2^2 + t_3^2 + t_4^2 = \left(1 + 3\sum_{i\in X} \alpha_i\right)^2 + 3\left(\sum_{j\in Y} \beta_j\right)^2 = a^2 + 3b^2 = q.$$

Hence T_1, T_2, T_3, T_4 are T-matrices. □

In order to search Q_1, Q_2, Q_3, Q_4 satisfying conditions in Theorem 3.13, we assume that X is chosen with one of the following two patterns:

(w1) $i \in X$, if and only if $2m - i \in X$;

(w2) $X = \{0, \dots, r-1\}$.

Example 3.2: *Let $q = 19 = 4^2 + 3 \cdot 1^2$, $a = 4$, $b = 1$. Then $m = 3$. We take 2 as a primitive element. Set*

$$C_i = \{2^{6j+i} \; : \; j = 0, 1, 2\}, \; i = 0, 1, 2, 3, 4, 5.$$

- *$r = 1$ and $X = \{0\}$, $Y = \{1, 2, 3, 4, 5\}$. We have (α, β, δ) sequences of lengths 1, 5, 5 as follows:*

$$\begin{array}{ll} \alpha & +, \\ \beta & + + + - -, \\ \delta & 0\ 0\ 0\ 1\ 0. \end{array}$$

Q_1, Q_2, Q_3, Q_4 determined by the sequences above are

$$\begin{aligned} Q_1 &= z + 1 + 7 + 11, \\ Q_2 &= \sum_{j\in Y} \beta_j 2^{6\delta_j + j} \\ &= 2 + 4 + 8 + (-1) \cdot 17 + (-1) \cdot 13, \\ Q_3 &= \sum_{j\in Y} \beta_j 2^{6(\delta_j+1)+j} \\ &= 14 + 9 + 18 + (-1) \cdot 5 + (-1) \cdot 15, \\ Q_4 &= \sum_{j\in Y} \beta_j 2^{6(\delta_j+2)+j} \\ &= 3 + 6 + 12 + (-1) \cdot 16 + (-1) \cdot 10. \end{aligned}$$

- *$r = 3$ and $X = \{0, 1, 2\}$, $Y = \{3, 4, 5\}$. We have (α, β, δ) sequences of lengths 3, 3, 3 as follows:*

$$\begin{array}{ll} \alpha & + - +, \\ \beta & + - +, \\ \delta & 0\ 0\ 0. \end{array}$$

The corresponding sequences Q_1, Q_2, Q_3, Q_4 are:

$$\begin{aligned} Q_1 &= z + C_0 - C_1 + C_2 \\ &= z + 1 + 4 + 6 + 7 + 9 + 11 + (-1) \cdot 2 + (-1) \cdot 3 + (-1) \cdot 14, \\ Q_2 &= \sum_{j\in Y} \beta_j g^{6\delta_j + j} = 8 + 13 + (-1) \cdot 16, \\ Q_3 &= \sum_{j\in Y} \beta_j g^{6(\delta_j+1)+j} = 18 + 15 + (-1) \cdot 17, \\ Q_4 &= \sum_{j\in Y} \beta_j g^{6(\delta_j+2)+j} = 12 + 10 + (-1) \cdot 5. \end{aligned}$$

- *$r = 5$. There is no solution of this form.*

T-matrices of orders 73 and 79 are first found in Zuo et al. [254] by another method.

In Table 3.2, we give four sequences (X, α's, β's, δ's) of lengths m, m, m, m with the T-property and X which satisfy the condition ($w1$):

When $m = 2$ and 8, there are not four-sequences of lengths m, m, m, m with X satisfying ($w1$).

In Table 3.3 we give the (α's, β's, δ's) with the T-property such that X satisfying ($w2$), i.e. $X = \{0, \dots, r-1\}$. In this case, we omit X and write three sequences ($\alpha' s, \beta' s, \delta' s$) of lengths $r, 2m-r, 2m-r$.

To search for T-matrices of order $6m+1$ using base sequences, one must search for four-sequences of lengths $3m+1$, $3m+1$, $3m$, $3m$, respectively [111, 166]. This method makes the searching process simpler and greatly reduces the computation time.

From the searching algorithm, we have the following conclusion:

There are (α, β, δ) sequences of lengths m, m, m satisfying equation (3.12) at least for $m = 1, 2, 3, 6$, and (α, β, δ) sequences of lengths $m-2$, $m+2$, and $m+2$ satisfying equation (3.12) at least for $m = 2, 3, 4, 5, 6, 7, 8, 10, 11, 12, 13$.

For more details, see Table 3.3.

We have the following conjecture:

Conjecture 3.8: If $6m+1$ is a prime power, then there are (α, β, δ) sequences of lengths r, $2m-r$, $2m-r$ satisfying equation (3.12), where $r \equiv m (\text{mod } 2)$, $0 \le r < m$ for $m > 1$.

3.7 Williamson Hadamard Matrices and Paley Type II Hadamard Matrices

3.7.1 Whiteman's Construction

A Williamson Hadamard matrix corresponds to a complex Hadamard matrix of the form

$$\begin{bmatrix} X & Y \\ -Y^* & X^* \end{bmatrix}$$

where X, Y are symmetric circulant matrices whose entries ± 1 and $\pm i$ and $X^* = \overline{X^{\mathsf{T}}}$ is the complex conjugate transpose matrix of X. If A, B, C, and D are Williamson matrices, then X and Y can be written by

$$\begin{aligned} 2X &= A + B + i(A - B), \\ 2Y &= C + D + i(C - D). \end{aligned} \tag{3.13}$$

Turyn found that if $q \equiv 1 \pmod 4$, then the Paley core matrix can be put in the form

$$\begin{bmatrix} P & S \\ S & -P \end{bmatrix} \tag{3.14}$$

where P and S are symmetric circulant and satisfy $PP^{\mathsf{T}} + SS^{\mathsf{T}} = qI$ [192]. Then Turyn proved the following Theorem 3.15 from (3.14) and the above correspondence. Whiteman gave another proof determining the Williamson equation. We show Whiteman's construction.

Let v be a non-square element of $K = GF(q)$. Then the polynomial $h(x) = x^2 - v$ is irreducible in K. The polynomials $ax + b, a, b \in K$ form a finite field $F = GF(q^2)$. Let χ be a primitive fourth residue character of F. We notice that $\chi(g) = -1$ if $q \equiv 1 \pmod 4$ where g is a primitive element of K.

Lemma 3.3: *Let ζ be a n-th root of unity. If we let $f(\zeta) = \sum_{k=0}^{n-1} a_k \zeta^k$, then*

$$a_k = \frac{1}{n} \sum_{j=0}^{n-1} f(\zeta^j) \zeta^{-jk}.$$

Table 3.2 $(X, \alpha, \beta, \delta)$ of lengths m, m, m, m.

m	q	a	b	g	X	α	β	δ
1	7	2	1	3	$\{0\}$	$-$	$+$	0
3	19	4	1	2	$\{0, 1, 5\}$	$-++$	$+-+$	000
4	25	5	0	$x+1^a$	$\{0, 1, 4, 7\}$	$+---$	$++--$	0110
5	31	2	3	3	$\{0, 3, 4, 6, 7\}$	$-++--$	$++-++$	00112
6	37	5	2	2	$\{0, 1, 5, 6, 7, 11\}$	$-+---+$	$+----+$	010120
7	43	4	3	3	$\{0, 3, 5, 6, 8, 9, 11\}$	$+-+-+-+$	$+-----+$	0011002
10	61	7	2	2	$\{0, 1, 4, 5, 6, 10, 14, 15, 16, 19\}$	$++--+++--+$	$++------++$	0110111012
11	67	8	1	2	$\{0, 3, 6, 8, 9, 10, 12, 13, 14, 16, 19\}$	$-++------++$	$+--+-++--++$	00220011120
12	73	5	4	5	$\{0, 2, 3, 5, 7, 9, 12, 15, 17, 19, 21, 22\}$	$-+--+-+-+--+$	$-++++++--++-$	010202011020
13	79	2	5	3	$\{0, 1, 2, 3, 4, 7, 8, 18, 19, 22, 23, 24, 25\}$	$--+++----+++-$	$+-++-++--++++$	0002001221010

[1](mod $x^2 - 3$, mod 5).

Table 3.3 (α, β, δ) Sequences for $X = \{0, 1, \ldots, r-1\}$.

m	q	a	b	g	r	α	β	δ	Note
2	13	1	2	2	2	+−	++	01	
3	19	4	1	2	1	+	+++−−	00010	
4	25	5	0	$x+1$	2	−−	+++−−−	021020	(mod x^2-3, mod 5)
5	31	2	3	3	1	−	+++−−−+++	012001012	
5	31	2	3	3	3	−+−	++−−−−−	0000102	
6	37	5	2	2	2	−−	+++−−−−+++	0010011001	
6	37	5	2	2	6	+−−+−−	++−−++	100110	
6	37	5	2	2	4	+−−−	++−+−−++	00222021	
7	43	4	3	3	1	+	++++−−−−−++++	0022220012021	
7	43	4	3	3	3	+−+	+++−−+−−+++	01010012012	
7	43	4	3	3	5	−+++−	++++−−−++	020200002	
8	49	1	4	$x+2$	0	∅	−++−−++++++−−++−	0102102122212011	(mod x^2+1, mod 7)
8	49	1	4	$x+2$	6	−−−+++	+−−−+−−−−+	0110111120	(mod x^2+1, mod 7)
8	49	7	0	$x+2$	6	++−−++	++++−−−−−+	0201002222	(mod x^2+1, mod 7)
8	49	7	0	$x+2$	2	++	+++−−−−++++−−−	02010111210202	(mod x^2+1, mod 7)
10	61	7	2	2	8	+++++−−−	+++−−−++−−++	022201100101	
11	67	8	1	2	9	−−−+++−−−	+−−+−+−+−++−−	0121002000221	
12	73	5	4	5	10	−−+−++−+−−	++−+−++−−+−+++	02021100021211	
13	79	2	5	3	11	+−+++−−−+−−	+−+++−++−−++++−	000101220002211	

Furthermore we have,

$$\sum_{k=0}^{n-1} a_k \overline{a}_{k+r} = \sum_{j=0}^{n-1} |f(\zeta^j)|^2 \zeta^{jr}.$$

Proof:

$$\sum_{j=0}^{n-1} f(\zeta^j)\zeta^{-jk} = \sum_{j=0}^{n-1}\left(\sum_{i=0}^{n-1} a_i \zeta^{ij}\right)\zeta^{-jk} = \sum_{i=0}^{n-1} a_i \sum_{j=0}^{n-1} \zeta^{j(i-k)} = a_k.$$

$$\begin{aligned}\sum_{j=0}^{n-1} f(\zeta^j)\overline{f(\zeta^j)}\zeta^{jr} &= \sum_{j=0}^{n-1}\left(\sum_{i=0}^{n-1} a_k \zeta^{kj}\right)\left(\sum_{i=0}^{n-1} \overline{a}_i \zeta^{-ji}\right)\zeta^{jr}\\ &= \sum_{k=0}^{n-1}\sum_{k=0}^{n-1} a_k \overline{a}_i \sum_{j=0}^{n-1} \zeta^{j(k-i+r)} = n\sum_{k=0}^{n-1} a_k \overline{a}_{k+r}.\end{aligned}$$

□

Lemma 3.4: *Let ξ be a primitive element of $F = GF(q^2)$. Then*

$$\sum_{k=0}^{q} \chi\left(tr\left(\xi^k\right)\right)\chi\left(tr\left(\xi^{k+r}\right)\right) = \begin{cases} (-1)^j q & \text{if } r = j(q+1),\\ 0 & \text{otherwise.}\end{cases}$$

Theorem 3.14 *(Whiteman, [221]):* *Let $q \equiv 1 \pmod 4$ be an odd prime power and $n = \frac{q+1}{2}$. Let χ be a primitive fourth residue character of $F = GF(q^2)$. Put $\xi^k = ax + b, a, b \in K = GF(q)$. We define*

$$a_k = \chi(a), \quad b_k = \chi(b).$$

Then the sums

$$f(\zeta) = \sum_{k=0}^{n-1} a_{4k}\zeta^k \text{ and } \quad g(\zeta) = \sum_{k=0}^{n-1} b_{4k}\zeta^k$$

satisfy

$$f^2(\zeta) + g^2(\zeta) = q$$

where ζ is a n-th root of unity including $\zeta = 1$.

Proof: Since $\xi^{n(q-1)} = -1$, then $tr(\xi^n) = \xi^n + \xi^{nq} = 0$. So we put $\xi^n = wx, w \in K$. Then the numbers a_k and b_k satisfy the following relations,

$$\begin{aligned} b_{k+n} &= -\chi(w)a_k,\\ b_{k+2n} &= -b_k,\\ b_{k+4n} &= b_k.\end{aligned} \tag{3.15}$$

Since v is a non-square element, we get

$$a_k = (-1)^k a_{4n-k}, \quad b_k = (-1)^k b_{4n-k}$$

from $tr(\xi^k) = 2b$ and (3.15).

For $k = 4\ell$, we have

$$a_{4(n-\ell)} = a_{4\ell}, \qquad b_{4(n-\ell)} = b_{4\ell}, \quad 0 \le \ell \le n.$$

Therefore $f(\zeta)$ and $g(\zeta)$ are real. From Lemma 3.3 we obtain

$$\sum_{\ell=0}^{n-1}\left(a_{4\ell}a_{4\ell+r}+b_{4\ell}b_{4\ell+r}\right)=\frac{1}{n}\sum_{j=0}^{n-1}\left(f^2\left(\zeta^j\right)+g^2\left(\zeta^j\right)\right)\zeta^{jr}. \tag{3.16}$$

Since $r=0$ is the only value such that $4r$ divides $q+1$ in $0\le r\le n-1$ and $\chi(tr(\xi^k))=\chi(2b)$,

$$\sum_{k=0}^{q} b_k b_{k+4r}=\begin{cases} q & r=0\\ 0 & 1\le r\le n-1\end{cases} \tag{3.17}$$

from Lemma 3.4. We note

$$\boldsymbol{Z}/2n\boldsymbol{Z}=\{4\ell \pmod{2n} \mid 0\le \ell\le n-1\}\cup\{4\ell+n \pmod{2n} \mid 0\le\ell\le n-1\}.$$

From the relations (3.15),

$$\begin{aligned}\sum_{k=0}^{q} b_k b_{k+4r} &= \sum_{\ell=0}^{n-1} b_{4\ell}b_{4\ell+4r}+\sum_{\ell=0}^{n-1} b_{4\ell+n}b_{4\ell+4r+n}\\ &=\sum_{\ell=0}^{n-1}\left(b_{4\ell}b_{4\ell+4r}+a_{4\ell}a_{4\ell+4r}\right).\end{aligned}$$

Combining Eqs. (3.16) and (3.17), it follows that

$$f^2\left(\zeta^j\right)+g^2\left(\zeta^j\right)=\sum_{\ell=0}^{n-1}\left(b_{4\ell}b_{4\ell+4r}+a_{4\ell}a_{4\ell+4r}\right)\zeta^{-jr}=q$$

for $j=0,\dots,n-1$ by Lemmas 3.3 and 3.4. □

Corollary 3.7 *(Whiteman [221])*: *Let* $q=2n-1\equiv 1 \pmod 4$ *be a prime power. Put*

$$\psi_1(\zeta)=1+f(\zeta),\quad \psi_2(\zeta)=1-f(\zeta),\quad \psi_3(\zeta)=\psi_4(\zeta)=g(\zeta)$$

where $f(\zeta)$ *and* $g(\zeta)$ *are defined in Theorem 3.14. Then the identity*

$$\psi_1^2(\zeta)+\psi_2^2(\zeta)+\psi_3^2(\zeta)+\psi_4^2(\zeta)=4n$$

is satisfied for each n-th root of unity including $\zeta=1$.

Theorem 3.15 *(Turyn [192], Whiteman [221])*: *Let* $q=2n-1\equiv 1 \pmod 4$ *be a prime power. Then there exists a Williamson Hadamard matrix of order* $4n$*, in which A and B agree on the diagonal entries and* $C=D$.

From Theorem 3.12, we have,

Corollary 3.8: *Let A, B, C, and D be as in Theorem 3.15. Suppose that there exists* $OD(4t;t,t,t,t)$*. Then there exists an Hadamard matrix of order* $4tn=2t(q+1)$.

3.7.2 Williamson Equation from Relative Gauss Sums

In this section, we show that the Williamson equation in Theorem 3.15 can be constructed based on the relative Gauss sum.

Let $Q = \{\pm 1, \pm i, \pm j, \pm k\}$ with the conditions $i^2 = -1, j^2 = -1, k^2 = -1, ij = -ji, jk = -kj, ki = -ik$. Denote the rational quaternion field by $R = \mathbf{Q} + \mathbf{Q}i + \mathbf{Q}j + \mathbf{Q}k$ where $\mathbf{Q}$ is the rational number field. For $\xi = a + bi + cj + dk \in R$, we define the quaternion conjugate by $\bar{\xi} = a - bi - cj - dk$. Then we have the norm $\xi\bar{\xi} = a^2 + b^2 + c^2 + d^2$.

Assume that $q = 2n - 1 \equiv 1 \pmod 4$. Denote the primitive character of order m of $F = GF(q^2)$ by χ_m. The character $\chi_F = \chi_{4n}^t$ t : odd, of F induces the quadratic character ψ of $K = GF(q)$. Since

$$\frac{1}{4n} = \frac{\epsilon}{4} + \frac{1}{n} \cdot \frac{1 - \epsilon n}{4}, \quad \text{for } \epsilon = (-1)^{\frac{n-1}{2}} \equiv n \pmod 4,$$

we have $\chi_{4n} = \chi_4^{\epsilon} \chi_n^{\frac{1-\epsilon n}{4}}$. Let ξ be a primitive element of F. Then we can take $\{\langle \xi \rangle \pmod{K^\times}\}$ as the system of representatives of the quotient group $F^\times / K^\times$. From Theorem 2.16,

$$\vartheta_{F/K}(\chi_F) = \sum_{\ell=0}^{2n-1} \psi\left(T_{F/K}\xi^{\ell}\right) \chi_4^{t\epsilon}\left(\xi^{\ell}\right) \chi_n^{t(1-\epsilon n)/4}\left(\xi^{\ell}\right).$$

If we put $m = \frac{1}{4}(1 - \epsilon n)\ell$, then $\ell \equiv 4m \pmod{2n}$ if ℓ is even and $\ell \equiv 4m - n \pmod{2n}$ if ℓ is odd. Hence the above sum

$$\vartheta_{F/K}(\chi_F) = \psi(2) + \sum_{m=1}^{n-1} \left(\psi\left(T_{F/K}\xi^{4m}\right) - i^t \psi\left(T_{F/K}\xi^{4m-n}\right)\right) \zeta_n^{tm}.$$

We write

$$e_m = \psi\left(2T_{F/K}\xi^{4m}\right), \quad d_m = \psi\left(2T_{F/K}\xi^{4m-n}\right)$$

and define the polynomial

$$f(x) = 1 + \sum_{m=1}^{n-1} \left(e_m - id_m\right) x^m \in \mathbf{Q}(i)[x]$$

and extend the quaternion conjugate to the polynomial ring as

$$\overline{f(x)} = f_1(x) - if_2(x) \text{ for } f(x) = f_1(x) + if_2(x),\ f_1(x), f_2(x) \in \mathbf{Q}[x].$$

Then

$$\overline{f(x)} \equiv f\left(x^{n-1}\right) \pmod{x^n - 1}$$

from $e_{-m} = e_m$ and $d_{-m} = d_m$.

If $t \equiv 1 \pmod 4$, then

$$\vartheta_{F/K}(\chi_F) = \psi(2) f\left(\zeta_n^t\right)$$

and if $t \equiv 3 \pmod 4$, then

$$\overline{\vartheta_{F/K}(\chi_F)} = \psi(2) f\left(\zeta_n^t\right).$$

Hence, $\vartheta_{F/K}(\chi_F)\overline{\vartheta_{F/K}(\chi_F)} = q$ in Theorem 2.16, implies

$$f(x)\overline{f(x)} \equiv q \pmod{x^n - 1}.$$

We recognize the element $a + bi \in \mathbf{Q}(i)$ as the element of $\mathbf{Q} + \mathbf{Q}i \subset R$. Thus the element $j \in Q$ acts dihedral on $\mathbf{Q}(i)$.

Theorem 3.16 (*Yamamoto and Yamada [247]*): *Let $q = 2n - 1 \equiv 1 \pmod 4$. Let $F = GF(q^2)$ be a quadratic extension of $K = GF(q)$, ψ be the quadratic character of K, and ξ be a primitive element of F. We define*

$$z_m = \frac{1}{1-i}\left(\psi\left(2T_{F/K}\xi^{4m}\right) - i\psi\left(2T_{F/K}\xi^{4m-n}\right)\right) \quad \text{for } m = 1, 2, \ldots, n-1,$$

and the polynomial $g(x) \in \mathbb{R}[x]$ *by*

$$g(x) = \frac{1+i+j+k}{2} + \sum_{m=1}^{n-1} z_m x^m.$$

Then it holds that $2g(x) \cdot \overline{2g(x)} \equiv 4n \pmod{x^n - 1}$, *which is the reduced Williamson equation.*

Proof: From $g(x) = \frac{1+i}{2}j + \frac{1+i}{2}f(x)$,

$$\begin{aligned} g(x)\overline{g(x)} &\equiv \frac{1}{2}\left(j + f(x)\right)\left(-j + \overline{f(x)}\right) \\ &\equiv \frac{1}{2}\left(1 + f(x)\overline{f(x)} + j\overline{f(x)} - f(x)j\right) \\ &\equiv \frac{1}{2}(1+q) \equiv n \pmod{x^n - 1}. \end{aligned}$$

Therefore $2g(x)\overline{2g(x)} \equiv 4n \pmod{x^n - 1}$. We straightforwardly verify $z_{-m} = z_m$ and $z_m = \pm 1$. Then we can write

$$2g(x) = j + k + \left(1 + 2\sum_{z_m=1} u_m - 2\sum_{z_m=-1} u_m\right) + \left(1 + 2\sum_{z_m=i} u_m - 2\sum_{z_m=-i} u_m\right) i$$

where $u_m = x^m + x^{-m}$. Consequently we obtain

$$4n \equiv 1 + 1 + \left(1 + 2\sum_{z_m=1} u_m - 2\sum_{z_m=-1} u_m\right)^2 + \left(1 + 2\sum_{z_m=i} u_m - 2\sum_{z_m=-i} u_m\right)^2 \pmod{x^n - 1},$$

which is the reduced Williamson equation. □

Corollary 3.9 ***(Yamamoto–Yamada, [247]):*** *Assume that the reduced Williamson equation has the form*

$$1^2 + 1^2 + \left(\epsilon + 2\sum_{m\in A} e_m x^m\right)^2 + \left(\epsilon + 2\sum_{m\in B} e_m x^m\right)^2 \equiv 4n \pmod{x^n - 1},\ e_{-m} = e_m,$$

where $\epsilon = (-1)^{\frac{n-1}{2}}$ *and* A, B *is a partition of* $\Omega = \{1, 2, \ldots, n-1\}$. *Denote the subsets of* A *consisting of* m *with* $e_m = 1$ *or* -1 *by* A_+, A_- *respectively, and by* B_+, B_- *the similar subsets of* B. *Then*

$$A_+ = A \cap 2A,\ A_- = A \cap 2B,\ B_+ = B \cap 2B,\ B_- = B \cap 2A.$$

Moreover, if $s = |A| - |B|$, *then* $2n - 1 = r^2 + s^2$ *for an integer* $r \equiv 1 \pmod 4$ *and*

$$\begin{aligned} |A_+| &= \frac{1}{4}\{n - 1 - \epsilon + r + 2s\}, \\ |B_+| &= \frac{1}{4}\{n - 1 - \epsilon + r - 2s\}, \\ |A_-| &= |B_-| = \frac{1}{4}\{n - 1 - \epsilon + r\}. \end{aligned}$$

Remark 3.1: The numbers $|A_+|, |A_-|, |B_+|$, and $|B_-|$ are cyclotomic numbers $(0,1), (2,3), (0,3)$, and $(2,1)$ of order 4, respectively [247].

Example 3.3: *$n = 7$, $\epsilon = -1$, $28 = 1^2 + 1^2 + 1^2 + 5^2$. $q = 2n - 1 = 13 = (-2)^2 + (-3)^2$, $r = -3$, and $A = \{3, 4\}$, $B = \{1, 6, 2, 5\}$, $|A| - |B| = -2$.*

$$\begin{aligned}
A_+ &= \{3,4\} \cap \{1,6\} = \emptyset \\
A_- &= \{3,4\} \cap \{4,3,2,5\} = \{3,4\}, \\
B_+ &= \{1,6,2,5\} \cap \{4,3,2,5\} = \{2,5\}, \\
B_- &= \{1,6,2,5\} \cap \{1,6\} = \{1,6\}. \\
|A_+| &= \tfrac{1}{4}(7 - 1 - (-1) + (-3) - 2 \cdot 2) = 0, \\
|B_+| &= \tfrac{1}{4}(7 - 1 - (-1) + (-3) + 2 \cdot 2) = 2, \\
|A_-| = |B_-| &= \tfrac{1}{4}(7 - 1 + (-1) - (-3)) = 2.
\end{aligned}$$

The reduced Williamson equation is given by

$$1^2 + 1^2 + \left(-1 - 2\left(x^3 + x^4\right)\right)^2 + \left(-1 + 2\left(x^2 + x^5\right) - 2\left(x + x^6\right)\right)^2 = 28.$$

Some of the theory may be developed to the generalized quaternion type.

3.8 Hadamard Matrices of Generalized Quaternion Type

3.8.1 Definitions

We know that a Williamson matrix is a product of a right regular representation matrix of quaternion numbers and a circulant matrix. In this section we generalize this construction to a generalized quaternion group type. Paley type I and type II Hadamard matrices are recognized as Hadamard matrices of generalized quaternion type.

A generalized quaternion group Q_s of order 2^{s+2} is a group generated by the two elements ρ, j such that

$$\rho^{2^{s+1}} = 1, \quad j^2 = \rho^{2s}, \quad j\rho j^{-1} = \rho^{-1}.$$

Let G be a semidirect product of a cyclic group $\langle \zeta \rangle$ of an odd order n by the generalized group Q_s of order 2^{s+2}. That is, G is generalized by ρ, j, and ζ with the conditions

$$\rho^{2^s} = -1,\ j^2 = -1,\ j\rho j^{-1} = \rho^{-1},\ \rho\zeta\rho^{-1} = \zeta,\ j\zeta j^{-1} = \zeta^{-1},\ \zeta^n = 1.$$

We consider the ring $\mathcal{R}$ from the group ring $\boldsymbol{Z}[G]$ by identifying the elements ± 1 in the center of Q_s with ± 1 of the rational integer ring $\boldsymbol{Z}$. Put $\mathcal{H} = \left\{\rho^k \zeta^\ell, 0 \le k \le 2^s - 1, 0 \le \ell \le n - 1\right\}$ and choose the basis $\mathcal{L} = \mathcal{H} \cup \mathcal{H}j$ of $\mathcal{R}$.

An element δ in $\mathcal{R}$ takes the following form

$$\delta = \sum_{k=0}^{2N-1} \sum_{\ell=0}^{n-1} a_{k\ell} \zeta^\ell \rho^k + \sum_{k=0}^{2N-1} \sum_{\ell=0}^{n-1} b_{k\ell} \zeta^\ell \rho^k j = \alpha + \beta j \tag{3.18}$$

where $N = 2^{s-1}$ and

$$\alpha = \sum_{k=0}^{2N-1} \sum_{\ell=0}^{n-1} a_{k\ell} \zeta^\ell \rho^k, \qquad \beta = \sum_{k=0}^{2N-1} \sum_{\ell=0}^{n-1} b_{k\ell} \zeta^\ell \rho^k.$$

We define the conjugate $\overline{\delta} = \overline{\alpha} - \beta j$ of δ based on the automorphisms $\tau : \rho \to \rho^{-1}, \zeta \to \zeta^{-1}$ of G. Furthermore we define the norm $\mathcal{N}(\delta) = \delta\overline{\delta}$, so that

$$\mathcal{N}(\delta) = \alpha\overline{\alpha} + \beta\overline{\beta}, \quad \mathcal{N}(\delta\theta) = \mathcal{N}(\delta)\mathcal{N}(\theta),\ \delta, \theta \in \mathcal{R}.$$

The right regular representation matrix $R(\delta)$ is given by

$$R(\delta) = \begin{bmatrix} \mathcal{A} & \mathcal{B} \\ -\mathcal{B}^{\mathrm{T}} & \mathcal{A}^{\mathrm{T}} \end{bmatrix},$$

such that

$$\mathcal{A} = \begin{bmatrix} A_0 & A_1 & \cdots & A_{2N-1} \\ -A_{2N-1} & A_0 & \cdots & A_{2N-2} \\ -A_{2N-2} & -A_{2N-1} & \cdots & A_{2N-3} \\ \vdots & \vdots & & \vdots \\ -A_1 & -A_2 & \cdots & A_0 \end{bmatrix}, \quad \mathcal{B} = \begin{bmatrix} B_0 & b_1 & \cdots & B_{2N-1} \\ -B_{2N-1} & B_0 & \cdots & B_{2N-2} \\ -B_{2N-2} & -B_{2N-1} & \cdots & B_{2N-3} \\ \vdots & \vdots & & \vdots \\ -B_1 & -B_2 & \cdots & B_0 \end{bmatrix}$$

where $A_k = \sum_{\ell=0}^{n-1} a_{k\ell}T^{\ell}$ and $B_k = \sum_{\ell=0}^{n-1} b_{k\ell}T^{\ell}$ are the circulant matrices of order n, where T is the basic circulant matrix.

Definition 3.9: If an element δ in $\mathcal{R}$ which is given by the Eq. (3.18) satisfies

i) all the coefficients $a_{k\ell}, b_{k\ell}$ are from $\{1, -1\}$, and
ii) $\mathcal{N}(\delta) = 2^{s+1}n = 4nN$,
then the right regular matrix representation $R(\delta)$ becomes an Hadamard matrix of order $2^{s+1}n$, which is called an **Hadamard matrix of generalized quaternion type**.
Similarly if the following conditions are satisfied;
iii) $a_{kk} = 0$ and all other coefficients $a_{k\ell}, b_{k\ell}, k \neq \ell$ are from $\{1, -1\}$, and
iv) $\mathcal{N}(\delta) = 2^{s+1}n - 1 = 4nN - 1$,

then $R(\delta)$ is a conference matrix of order $2^{s+1}n$, which is called a **conference matrix of generalized quaternion type**.

We abbreviate generalized quaternion type as **GQ type** for convenience sake.
The conditions (i) and (ii) in Definition 3.9 are expressed in terms of the matrices A_k, B_k:

$$\sum_{k=0}^{2N-1} A_k A_k^{\mathrm{T}} + \sum_{k=0}^{2N-1} B_k B_k^{\mathrm{T}} = 4nNI,$$

$$\sum_{k=0}^{\ell-1} \left(A_k A_{2N-\ell+k}^{\mathrm{T}} + B_k B_{2N-\ell+k}^{\mathrm{T}}\right) - \sum_{k=0}^{2N-\ell-1} \left(A_k^{\mathrm{T}} A_{k+\ell} + B_k^{\mathrm{T}} B_{k+\ell}\right) = 0,$$

$$\text{for } 1 \leq \ell \leq 2N - 1.$$

In particular, assuming $N = 1$, the conditions become

$$A_0 A_0^{\mathrm{T}} + A_1 A_1^{\mathrm{T}} + B_0 B_0^{\mathrm{T}} + B_1 B_1^{\mathrm{T}} = 4nI,$$
$$A_0 A_1^{\mathrm{T}} - A_1 A_0^{\mathrm{T}} + B_0 B_1^{\mathrm{T}} - B_1 B_0^{\mathrm{T}} = 0.$$

In this case, $R(\delta)$ gives Ito's Hadamard matrix

$$\begin{bmatrix} A & B & C & D \\ B & -A & D & C \\ C^{\mathsf{T}} & -D^{\mathsf{T}} & -A^{\mathsf{T}} & B^{\mathsf{T}} \\ D^{\mathsf{T}} & C^{\mathsf{T}} & -B^{\mathsf{T}} & -A^{\mathsf{T}} \end{bmatrix}$$

by replacing A_0, A_1, B_0, B_1 by A, B, C, D respectively. Further if A, B, C, D are symmetric, then it gives a Williamson Hadamard matrix.

3.8.2 Paley Core Type I Matrices

If $q \equiv 3 \pmod 4$, then the Paley core type I matrix Q is recognized as a conference matrix of GQ type. Seidel-equivalence is defined in Definition 1.15.

Theorem 3.17 ***(Yamada [238]):*** *The Paley core type I matrix Q is Seidel-equivalent to a conference matrix of GQ type with the several additional properties,*

i) $\mathcal{A}$ *is skew symmetric,*
ii) $B_{2N-m-1} = -B_m^{\mathsf{T}}$ *for* $m = 0, 1, \dots, N-1$ *where* $q+1 = 2^{s+1}n$, $s \geq 1$, *n odd,* $N = 2^{s-1}$.

3.8.3 Infinite Families of Hadamard Matrices of GQ Type and Relative Gauss Sums

We construct some infinite families of Hadamard matrices of GQ type from the relative Gauss sum. Let $F = GF(q^2)$, a quadratic extension of $K = GF(q)$, and ξ be a primitive element of F. Let χ be a character of F.

Lemma 3.5 ***(Yamada [238]):*** *Let* $q+1 = 2^s n, s \geq 2$, *n odd,* ρ *be a primitive* 2^{s+1}*st root of unity, and* ζ *an arbitrary n-th root of unity. We put* $\chi = \chi_{2^{s+1}}\chi_n$ *where* $\chi_{2^{s+1}}(\xi) = \rho$ *and* $\chi_n(\xi) = \zeta$. *Then* χ *induces a quadratic character* ψ *in K.*

The relative Gauss sum $\vartheta_\chi = \vartheta_{F/K}(\chi)$ *can be written as*

$$\vartheta_\chi = \alpha + \beta\rho^n,$$
$$\alpha = \sum_{m=0}^{(q-1)/2} \psi\left(T_{F/K}\xi^{2m}\right)\rho^{2m}\zeta^{2m},$$
$$\beta = \sum_{m=0}^{(q-1)/2} \psi\left(T_{F/K}\xi^{2m+n}\right)\rho^{2m}\zeta^{2m}.$$

Proof: From Theorem 2.16,

$$\vartheta_\chi = \sum_{m=0}^{q} \chi\left(\xi^m\right)\psi\left(T_{F/K}\xi^m\right) = \sum_{m=0}^{q} \psi\left(T_{F/K}\xi^m\right)\rho^m\zeta^m.$$

According as m even or m odd, we obtain the assertion of the theorem. □

Theorem 3.18 ***(Yamada [238]):*** *Let* α *and* β *be as in Lemma 3.5. Then the right regular representation matrix of*

$$\alpha \pm i + \beta j$$

gives an Hadamard matrix of GQ type of order $2^s n$.

Proof: From Lemma 3.5, we have $\overline{\vartheta_\chi} = \alpha - \beta\rho^n$, so that $\overline{\alpha} = \alpha$, $\overline{\beta\rho^n} = -\beta\rho^n$. Therefore

$$\vartheta_\chi\overline{\vartheta_\chi} = \left(\alpha + \beta\rho^n\right)\left(\overline{\alpha} + \overline{\beta\rho^n}\right) = \alpha\overline{\alpha} + \beta\overline{\beta} = q$$

from Theorem 2.16. The coefficients of $\alpha \pm i + \beta j$ are from $\{1, -1\}$. Thus $\mathcal{N}(\alpha \pm i + \beta j) = q + 1$ and the right regular representation matrix of $\alpha \pm i + \beta j$ is an Hadamard matrix of GQ type of $2^s n$. □

In Theorem 3.15, Turyn and Whiteman showed that type II Hadamard matrix can be transformed into the Williamson Hadamard matrix. Theorem 3.16 asserts this Williamson Hadamard matrix is constructed based on the relative Gauss sum. The following theorem shows this Williamson Hadamard matrix is regarded as a GQ type.

Theorem 3.19 *(Yamada [238])*: *Assume $s = 1$, that is $q + 1 = 2n$, n odd. α and β can be written as in Lemma 3.5. Then the right regular representation matrix of*

$$\delta = (1 - j)(\vartheta_\chi + ij) = (1 - j)(\alpha - i + \beta\rho^n j)$$

is an Hadamard matrix of GQ type of order $4n$ and gives an Hadamard matrix in Theorem 3.15.

3.9 Supplementary Difference Sets and Williamson Matrices

3.9.1 Supplementary Difference Sets from Cyclotomic Classes

Let $q = ef + 1$ be an odd prime power and $F = GF(q)$ be the finite field with q elements. Let g be a primitive element of F. Let $S_0, S_1, \dots, S_{e-1}$ in F be e cyclotomic classes;

$$S_i = \left\{g^{es+i} : s = 0, \dots, f - 1\right\}.$$

Note that S_0 is the set of eth power residues. For non-empty subsets $A_0, A_1, \dots, A_{n-1}$ of $\Omega = \{0, 1, \dots, e - 1\}$, we define the subsets

$$D_i = \bigcup_{\ell \in A_i} S_\ell, \quad \text{for } i = 0, \dots, n - 1,$$

of F, which are unions of some eth cyclotomic classes. It is not necessary that the D_i are pairwise disjoint. When we put $|A_i| = u_i$, $0 \le i \le n - 1$, then $|D_i| = u_i f$. We give necessary and sufficient conditions that the subsets $D_0, \dots, D_{n-1}$ become $n - \left\{q; u_0 f, u_1 f, \dots, u_{n-1} f; \lambda\right\}$ *sds*. Many construction for sds have been found essentially using cyclotomy, see Stanton and Sprott [178], and Wallis [205, 209].

Definition 3.10: Let D be a subset of $F = GF(q)$. The function

$$\psi_D(\alpha) = \begin{cases} 1 & \text{if } \alpha \in D, \\ 0 & \text{otherwise} \end{cases}$$

is called a **characteristic function** of D.

The characteristic function φ_{S_ℓ} of an eth cyclotomic class S_ℓ is given by

$$\varphi_{S_\ell}(\alpha) = \frac{1}{e}\sum_{k=0}^{e-1} \zeta_e^{-\ell k}\chi^k(\alpha), \quad \text{for } \alpha \in F$$

where χ is a primitive eth power residue character and ζ_e is a primitive eth root of unity.

Necessary and sufficient conditions such that the subsets $D_0, D_1, \ldots, D_{n-1}$ become supplementary difference sets is given by Jacobi sums. Explicit determination of Jacobi sum is difficult for large e.

Theorem 3.20 (*Yamada [239]*): *Let $A_0, A_1, \ldots, A_{n-1}$ be non-empty subsets of $\Omega = \{0, 1, \ldots, e-1\}$ and $|A_i| = u_i$ for $i = 0, \ldots, n-1$. The subsets*

$$D_i = \bigcup_{\ell \in A_i} S_\ell$$

of F determined by A_i, $i = 0, \ldots, n-1$, become $n - \{q; u_0 f, u_1 f, \ldots, u_{n-1} f; \lambda\}$ sds if and only if the following equations are satisfied:

(i) $\sum_{i=0}^{n-1} u_i(u_i f - 1) \equiv 0 \pmod{e}$.

(ii) $\sum_{i=0}^{n-1} \sum_{m=0}^{e-1} J(\chi^m, \chi^{-t}) \omega_{i,m} \omega_{i,t-m} = 0$ *for all* $t = 1, \ldots, e-1$,

where $\omega_{i,m} = \sum_{\ell \in A_i} \zeta_e^{-\ell m}$, *$\zeta_e$ is a primitive e-th root of unity, χ is a primitive e-th power residue character, and $J(\chi^m, \chi^{-t})$ is the Jacobi sum for e-th power residue characters χ^m, χ^{-t}. If f is odd, i.e. $-1 \notin S_0$, we have only to verify equation (ii) for even t.*

Example 3.4: *We let $e = 4$ and $n = 2$. Let $C_i, i = 0, 1, 2, 3$ be cyclotomic classes of order 4. Let A_0 and A_1 be the union of cyclotomic classes of order 4. Suppose that $|A_0| = 1$ and $|A_1| = 2$. The equation (i) of Theorem 3.20 is*

$$\left(\frac{q-1}{4} - 1\right) + 2\left(\frac{2(q-1)}{4} - 1\right) \equiv 0 \pmod{4}.$$

Then $\lambda = \frac{1}{16}(5q - 17)$ and it implies $q \equiv 13 \pmod{16}$. We can assume $A_0 = C_0$ and $A_1 = C_1 \cup C_3$.

Let χ be a primitive biquadratic character of F. We write $\omega = J(\chi, \chi^2) = a + 2bi$. The equation (ii) of Theorem 3.20 is

$$-2 + \omega(1 + 2i) + \overline{\omega}(1 - 2i) = 2a - 8b - 2 = 0.$$

Namely $a = 1 + 4b$. Thus we have

$$q = a^2 + 4b^2 = 1 + 8b + 20b^2,$$
$$5q = (10b + 2)^2 + 1.$$

By replacing $10b + 2$ by z, we have $q = \frac{1}{5}(z^2 + 1) \equiv 13 \pmod{16}$ and $z \equiv 2 \pmod{5}$.

We will give numerical results for $e = 2, n = 1, 2$ and for $e = 4, n = 1, 2, 3, 4$ in Appendix B.

3.9.2 Constructions of an Hadamard 4-sds

We recall the definition of Hadamard supplementary difference sets.

Lemma 3.6: $4 - \{v; k_1, k_2, k_3, k_4; \sum k_i - v\}$ *sds can be used to form circulant (± 1) matrices of order n and hence, using the Goethals–Seidel array, an Hadamard matrix of order $4v$,*

We show Xia–Lui constructions of a supplementary difference set by straightforward calculations.

Assume that $q \equiv 1 \pmod 4$ is a prime power. Let S_i, $0 \le i \le 2q+1$, be $(2q+2)$th cyclotomic classes of $F = GF(q^2)$. We define the subset D of F by

$$D = \bigcup_{i=0}^{(q-1)/4-1} \bigcup_{u=1}^{3} S_{4i+\frac{q+1}{2}u} \bigcup S_{2q} \bigcup \left(\bigcup_{j=1}^{(q-1)/4} S_{4j-2} \right).$$

Furthermore we put

$$E_0 = D, \quad E_1 = \xi^{\frac{q+1}{2}} D, \quad E_2 = \xi^{q+1} D, \quad E_3 = \xi^{\frac{3(q+1)}{2}} D$$

where ξ is a primitive element of F. Xia and Liu [227] showed the existence of the following supplementary difference sets.

Theorem 3.21 *(Xia and Liu [227]):* *Assume that $q \equiv 1 \pmod 4$ is a prime power. Then E_0, E_1, E_2, E_3 are $4-\left\{q^2;\ \frac{1}{2}q(q-1);\ q^2-2q\right\}$ symmetric Hadamard supplementary difference sets (DF) in $F = GF(q^2)$. Hence there exists an Hadamard matrix of order $4q^2$.*

If we put $q = p^r$ where $p \equiv 3 \pmod 4$ an odd prime for a positive integer r, then r should be even to satisfy the condition $q \equiv 1 \pmod 4$. Xia gave an another construction in this case.

Theorem 3.22 *(Xia [226]):* *The subsets $E_0, E_1, E_2, \overline{E_3} = g^{\frac{3(q+1)}{2}}(F-D)$ are*

$$4-\left\{q^2;\ \tfrac{1}{2}q(q-1), \tfrac{1}{2}q(q-1), \tfrac{1}{2}q(q-1), \tfrac{1}{2}q(q+1);\ q(q-1)\right\}$$

symmetric supplementary difference sets (DF) of F when $p \equiv 3 \pmod 4$ and $q = p^r$, r even. Hence there exists an Hadamard matrix of order $4q^2$.

Supplementary difference sets in Theorems 3.21 and 3.22 are symmetric and satisfy $\lambda = k_0 + k_1 + k_2 + k_3 - v$. Hence these difference families give rise to Williamson Hadamard matrices by Theorem 3.1. Xia and Liu partly solved the case for $q \equiv 3 \pmod 4$.

Theorem 3.23 *(Xia and Liu [228]):* *If $q \equiv 3 \pmod 8$ is a prime power, then there exist*

$$4-\left\{q^2;\ \tfrac{1}{2}q(q-1);\ q^2-2q\right\} \text{ Hadamard sds.}$$

Hence there exists an Hadamard matrix of order $4q^2$.

We now consider Xiang's generalized Xia–Liu construction.

Let q be an odd prime power and $L_i, 0 \le i \le q$, be a $q+1$st cyclotomic classes.

Let A_i be a subset of $\{0, 1, \ldots, q\}$ satisfying $|A_i| = \frac{(e-1)q-1}{2e}$ and for any $x \in A_i, x \not\equiv i \pmod e$.

Denote the 2eth cyclotomic classes by C_i, $0 \le i \le 2e-1$. Assume that $q \equiv e-1 \pmod{2e}$ and $e \equiv 2 \pmod 4$. We define the subsets of $F = GF(q^2)$ by

$$D_i = C_i \bigcup \left(\bigcup_{j \in A_i} L_j \right)$$

and

$$D_{i+e} = C_{i+e} \bigcup \left(\bigcup_{j \in A_i} L_j \right)$$

for $0 \le i \le e-1$. Since $-1 \in C_0$, D_i is symmetric for $0 \le i \le 2e-1$.

Theorem 3.24 (Xiang [231]): *Assume* $q \equiv e-1 \pmod{2e}$ *and* $e \equiv 2 \pmod 4$. *Then* $D_0, D_1, \dots, D_{2e-1}$ *are* $2e - \left\{q^2; \frac{q^2-q}{2}; \frac{e}{2}q^2 - eq\right\}$ *symmetric supplementary difference sets (DF) in F.*

Especially for $e = 2$, we have

Corollary 3.10 (Xiang [231]): *In the above theorem, let* $e = 2$. *Then* D_0, D_1, D_2, D_3 *are* $4 - \left\{q^2; \frac{q^2-q}{2}; q^2 - 2q\right\}$ *symmetric Hadamard supplementary difference sets (DF). Hence there exists an Hadamard matrix of order* $4q^2$.

Example 3.5: $q = 5$. *Let* ξ *be a primitive element of* $GF(5^2)$. *We take the subsets* $A_0 = \{1\}$ *and* $A_1 = \{2\}$ *of* $\{0, 1, \dots, 5\}$. *Then*

$$\begin{aligned}
D_0 &= C_0 \cup L_1 = \{1, \xi^4, \xi^8, \xi^{12}, \xi^{16}, \xi^{20}, \xi^7, \xi^{13}, \xi^{19}, \xi\},\\
D_1 &= C_1 \cup L_2 = \{\xi, \xi^5, \xi^9, \xi^{13}, \xi^{17}, \xi^{21}, \xi^8, \xi^{14}, \xi^{20}, \xi^2\},\\
D_2 &= C_2 \cup L_1 = \{\xi^2, \xi^6, \xi^{10}, \xi^{14}, \xi^{18}, \xi^{22}, \xi^7, \xi^{13}, \xi^{19}, \xi\},\\
D_3 &= C_3 \cup L_2 = \{\xi^3, \xi^7, \xi^{11}, \xi^{15}, \xi^{19}, \xi^{23}, \xi^8, \xi^{14}, \xi^{20}, \xi^2\}
\end{aligned}$$

are $4 - \{25, 10, 15\}$ *supplementary difference sets in* $GF(5^2)$.
In the group ring notation, we verify

$$\chi(D_0)^2 + \chi(D_1)^2 + \chi(D_2)^2 + \chi(D_3)^2 = 25$$

for all nontrivial additive characters of $GF(5^2)$.

We assume that $q \equiv e-1 \pmod{2e}$ and $e \equiv 0 \pmod 4$. The subsets A_i of $\{0, 1, \dots, q\}$ are as above. We define the subsets

$$D_i = C_i \bigcup \left(\bigcup_{j \in A_i} L_j\right) \quad \text{for } 0 \le i \le e-1.$$

Since $-1 \in C_e$, the D_i are not symmetric.

Theorem 3.25 (Xiang [231]): *The subsets* $D_0, D_1, \dots, D_{e-1}$ *are*

$$e - \left\{q^2; \frac{1}{2}(q^2 - q); \frac{e}{4}(q^2 - 2q)\right\}$$

supplementary difference sets (DF) in F.

If we let $e = 4$ in the above theorem, we have the supplementary difference set with the same parameters as in Corollary 3.10.

Corollary 3.11 (Xiang [231]): *In the above theorem, let* $e = 4$. *Then* D_0, D_1, D_2, D_3 *are* $4 - \left\{q^2; \frac{q^2-q}{2}; q^2 - 2q\right\}$ *Hadamard supplementary difference sets (DF) when* $q \equiv 3 \pmod 8$. *Hence there exists an Hadamard matrix of order* $4q^2$.

Supplementary difference sets with the same parameters as Theorem 3.21 and Corollaries 3.10 and 3.11 were obtained by Yamada [241] and Xia et al. [229]. Yamada used Jacobi sums for the order $q + 1$ and Xia–Xia–Seberry used $(q; x, y)$-partitions.

Let q be an odd prime power and $e = q + 1$. Put $F = GF(q^2)$ and $K = GF(q)$.

Lemma 3.7 (Yamada [241]): *Let g be a primitive element of F. For the integers $0 \leq x, t \leq q$, we have*

$$N = |\{\delta \in K^{\times} \,|1 - g^t\delta \in g^x K^{\times}\}| = \begin{cases} 1 & \text{if } x \neq t \text{ and } t \neq 0, \\ 0 & \text{if } x = t \text{ and } t \neq 0, \\ q-2 & \text{if } x = t = 0, \\ 0 & \text{if } x \neq 0 \text{ and } t = 0. \end{cases}$$

Proof: If $t = 0$, then $x = 0$. If $x = t$, $1 - g^t\delta \in g^t K^{\times}$, then $1 \in g^t K^{\times}$, that is $t = 0$, then $N = q - 2$.
Assume that $x \neq t$, $t \neq 0$. The element $1 - g^t\delta$ is contained in the cyclotomic class $g^x K^{\times}$ for some x.
If $1 - g^t\delta \in g^x K^{\times}$ and $1 - g^t\delta' \in g^x K^{\times}$ for $\delta, \delta' \in K^{\times}$, then $\delta = \delta'$. □

Theorem 3.26 (Yamada [241]): *Put $e = q + 1$. Let χ be a primitive e-th residue character and χ^0 be the trivial character of F. For $0 \leq k, k' \leq e - 1 = q$, we have*

$$J\left(\chi^k, \chi^{k'}\right) = \begin{cases} q^2 - 2 & \text{if } k = k' = 0, \\ -1 & \text{if } k + k' = 0, \text{ or } k \neq 0 \text{ and } k' = 0, \\ q & \text{if } k + k \neq 0, k \neq 0 \text{ and } k' \neq 0. \end{cases}$$

Proof: From Theorem 2.6, we have $J(\chi^0, \chi^0) = q^2 - 2$, $J(\chi^0, \chi^k) = -1$, and $J(\chi^k, \chi^{-k}) = -\chi^k(-1) = -1$ for $k \neq 0$. Assume that $k + k \neq 0$, $k \neq 0$, and $k' \neq 0$. From Lemma 3.7,

$$\begin{aligned} J\left(\chi^k, \chi^{k'}\right) &= \sum_{t=0}^{e-1} \chi^{-k'}\left(g^t\right) \sum_{\delta \in K^{\times}} \chi^k\left(1 - g^t\delta\right) \\ &= (q-2) + \sum_{t=1}^{e-1} \chi^{k'}\left(g^t\right)\left(-\chi^k(1) - \chi^k\left(g^t\right)\right) = q. \end{aligned}$$

□

Assume $q \equiv 1 \pmod 4$. Denote the $e = q + 1$st cyclotomic classes by $S_0, S_1, \ldots, S_q$. Let

$$\begin{aligned} \Omega &= \{0, 1, \ldots, e - 1\} = \Omega_0 \cap \Omega_1, \\ \Omega_0 &= \{a \in \Omega | a \equiv 0 \pmod 2\} \end{aligned}$$

and

$$\Omega_1 = \{a \in \Omega | a \equiv 1 \pmod 2\}.$$

Assume that A_0 is a subset of Ω_1 and A_1 is a subset of Ω_0 such that $|A_0| = |A_1| = \frac{q-1}{4}$. Further we let $A_2 = A_0$ and $A_3 = A_1$. Let C_0, C_1, C_2, and C_3 be biquadratic cyclotomic classes.

Theorem 3.27 (Yamada [241]): *Assume $q \equiv 1 \pmod 4$. Then*

$$D_i = \left(\bigcup_{\ell \in A_i} S_\ell\right) \bigcup C_i \quad \text{for } i = 0, 1, 2, 3,$$

are $4 - \{q^2; \frac{1}{2}q(q-1); q(q-2)\}$ Hadamard supplementary difference sets. Hence there exists an Hadamard matrix of order $4q^2$.

Proof: Put $e = q + 1$ and $w_{i,k} = \sum_{\ell \in A_i} \zeta_e^{-\ell k}$. Let χ be a primitive $(q+1)$st residue character and ψ be a primitive biquadratic character. Then the characteristic function of $\bigcup_{\ell \in A_i} S_\ell$ is

$$f_{A_i}(\alpha) = \frac{1}{e} \sum_{k=0}^{e-1} w_{i,k} \chi^k(\alpha)$$

and the characteristic function of C_i is

$$f_{C_i}(\alpha) = \frac{1}{4} \sum_{k'=0}^{3} \zeta_4^{-ik'} \psi^{k'}(\alpha).$$

Therefore

$$D_i = \sum_{i=0}^{3} \sum_{\alpha \in F^\times} \left(f_{A_i}(\alpha) + f_{C_i}(\alpha) \right) \alpha.$$

When we verify $\sum_{i=0}^{3} D_i D_i^{-1} = 2q(q-1) + q(q-2) \sum_{\alpha \in K^\times} \alpha$, we need the values of Jacobi sums,

$$\sum_{k=0}^{e-1} w_{i,k} \chi^k(\alpha) \cdot \sum_{k'=0}^{3} \zeta_4^{-ik'} \psi^{k'}(\alpha) = 0 \text{ and for odd } k' \sum_{i=0}^{3} w_{i,k} \zeta_4^{-ik'} = 0.$$

□

3.9.3 Construction from $(q; x, y)$-Partitions

Xia–Xia–Seberry introduced the concept of $(q; x, y)$-partitions to construct a four-supplementary difference set from the cyclotomic classes. The concept $(q; x, y)$-partition is a generalization of the partition by Chen [18]. It provides a useful method for the constructing difference families, difference sets, and Williamson Hadamard matrices.

Assume that $q \equiv 1 \pmod 4$ is an odd prime power. We know that every prime power $q \equiv 1 \pmod 4$ is represented as the sum of 2 squares,

$$q = x^2 + 4y^2$$

with $x \equiv 1 \pmod 4$. Put $q = 4m + 1, m = \frac{q-1}{4}$.

Let g be a primitive element of $K = GF(q)$. Then $x^2 - g \in K[x]$ is an irreducible polynomial. The algebraic extension $K(\omega)$ of K by adjoining the root ω of $x^2 - g$ is a finite field $F = GF(q^2)$ with q^2 elements. The polynomials $\alpha\omega + \beta, \alpha, \beta \in K = GF(q)$ form a finite field $F = GF(q^2)$.

Let ξ be a primitive element of $F = GF(q^2)$ such that $\xi^{q+1} = g$ and $n = \frac{q+1}{2}$. Denote the $2(q+1)$st cyclotomic classes of F by $S_i = \xi^i S_0$. $S_0 = S$ is the set of $2(q+1)$st residues of F and the set of quadratic residues of K. $S_{q+1} = \xi^{q+1} S_0 = N$ is the set of quadratic non-residues of K.

For a cyclotomic class $S_i = \xi^i S_0, 0 \le i \le 2q+1, i \ne q+1$, write $\xi^i = \alpha\omega + \beta$. Then $\alpha \ne 0$ and

$$S_i = \xi^i S_0 = \left\{ \alpha g^{2k} \omega + \beta g^{2k} \mid k = 0, 1, \dots, \frac{q-3}{2} \right\}$$
$$= \left\{ \left(\alpha g^{2k}, \alpha^{-1}\beta \left(\alpha g^{2k}\right)\right) \mid k = 0, 1, \dots, \frac{q-3}{2} \right\}.$$

We represent S_i by $\{(\eta, r\eta), \eta \in S\}$ or $\{(\eta, r\eta), \eta \in N\}$ according as $\alpha \in S$ or $\alpha \in N$. For convenience, we denote

$$S_0 = (0, S), \quad S_{q+1} = (0, N)$$

and

$$\{(\beta, r\beta)\ \beta \in X\}, \quad X \in \{S, N\}.$$

We identify a subset X of F with an element of $\sum_{\alpha\in X}\alpha$ of $\boldsymbol{Z}[F]$. Thus we note that

$$XY = \sum_{x\in X} x \cdot \sum_{y\in Y} y = \sum_{x\in X}\sum_{y\in Y}(x+y).$$

Theorem 3.28 (Xia et al. [229]): *Let $q \equiv 1 \pmod 4$ be an odd prime power. There exist four subsets X_1, X_2, X_3, and X_4 of $GF(q)$, such that*

$$\{|X_1|, |X_2|\} = \{m+y, m-y\},\ \{|X_3|, |X_4|\} = \left\{m+\tfrac{1}{2}(1+x), m+\tfrac{1}{2}(1-x)\right\}, \tag{3.19}$$

$$X_1 + X_2 + X_3 + X_4 = GF(q), \tag{3.20}$$

$$X_1S + X_2N = \sum_{i=1}^{4}(|X_i|-1)X_i, \tag{3.21}$$

$$X_1N + X_2S = |X_2|X_1 + |X_1|X_2 + |X_4|X_3 + |X_3|X_4, \tag{3.22}$$

$$X_3S + X_4N = |X_3|X_1 + |X_4|X_2 + |X_2|X_3 + |X_1|X_4, \tag{3.23}$$

$$X_3N + X_4S = |X_4|X_1 + |X_3|X_2 + |X_1|X_3 + |X_2|X_4, \tag{3.24}$$

*for some x, y satisfying $q = x^2 + 4y^2, x \equiv 1 \pmod 4$. We call the partition satisfying (3.19)–(3.24) a **$(q; x, y)$-partition**.*

Proof: We define the subsets X_1, X_2, X_3, X_4 of $K = GF(q)$ from biquadratic cyclotomic classes C_0, C_1, C_2, C_3 and show these subsets satisfy the conditions (3.19)–(3.24). It is clear that

$$C_i = \bigcup_{j=0}^{2m} S_{4j+i}, \quad i = 0, 1, 2, 3.$$

C_0 and $C_2 = \xi^{q+1}C_0$ can be written in the forms

$$C_0 = (0, S) \cup \{(S, rS), r \in X_1\} \cup \{(N, rN), r \in X_2\}, \tag{3.25}$$

$$C_2 = (0, N) \cup \{(N, rN), r \in X_1\} \cup \{(S, rS), r \in X_2\} \tag{3.26}$$

for some subsets X_1 and X_2 of K. Similarly we have

$$C_1 = \{(S, rS), r \in X_3\} \cup \{(N, rN), r \in X_4\}, \tag{3.27}$$

$$C_3 = \{(N, rN), r \in X_3\} \cup \{(S, rS), r \in X_4\}. \tag{3.28}$$

From

$$|C_1| = |C_2| = |C_3| = |C_4| = \frac{q^2-1}{4},$$

$$|X_1| + |X_2| = 2m = \frac{q-1}{2}, \text{ and}$$

$$|X_3| + |X_4| = 2m+1 = \frac{q+1}{2}.$$

Then we have (3.20).

Let ζ be a nth root of unity. Let $f(\zeta)$, $g(\zeta)$, χ, a_k, and b_k be as in Theorem 3.14. We have

$$\begin{aligned} f(1) &= \sum_{i=0}^{n-1} a_{4i} \\ &= |\{\alpha \in S \mid \xi^{4i} = \alpha\omega + \beta \in C_0\}| - |\{\alpha \in N \mid \xi^{4i} = \alpha\omega + \beta \in C_0\}| \\ &= |(S, rS), r \in X_1| - |(N, rN), r \in X_2| \\ &= |X_1| - |X_2|. \end{aligned}$$

Since $C_1 = \xi^{\frac{q+1}{2}} C_0$ and $\omega^2 = g = \xi^{q+1}$, $\xi^{\frac{q+1}{2}}(\alpha\omega + \beta) = \alpha g + \beta\omega \in C_1$. Hence

$$g(1) = \sum_{i=0}^{n-1} b_{4i} = |(S, rS), r \in X_3| - |(N, rN), r \in X_4| = |X_3| - |X_4|.$$

From Theorem 3.14, $(|X_1| - |X_2|)^2 + (|X_3| - |X_4|)^2 = q$, $(|X_1| - |X_2|)^2 = 4y^2$ and $(|X_3| - |X_4|)^2 = x^2$. Then we obtain Eq. (3.19).

If $h \in C_0$, then $hC_0 = C_0$ and $hC_2 = C_2$. Assume that $r_0 \in X_1$. We take $h = \alpha\omega + \beta \in C_0$ such that $\alpha \in S, \alpha^{-1}\beta = -r_0$.

If $\gamma\omega + \delta \in (S, rS), r \in X_1$, then

$$\begin{aligned} h(\gamma\omega + \delta) &= (\alpha\omega + \beta)(\gamma\omega + \delta) \\ &= \gamma(\alpha r + \beta)\omega + \gamma(\alpha g + \beta r) \in \left\{\left((\alpha r + \beta)S,\ (\alpha g + \beta r)S\right), r \in X_1\right\}. \end{aligned}$$

If $\delta \in (0, S)$, then $h\delta = \delta(\alpha\omega + \beta) \in (\alpha S, \beta S)$.

$$\begin{aligned} hC_0 =& (\alpha S, \beta S) \cup \left\{\left((\alpha r + \beta)S, (\alpha g + \beta r)S\right), r \in X_1\right\} \\ & \cup \left\{\left((\alpha r + \beta)N, (\alpha g + \beta r)N\right), r \in X_2\right\}. \end{aligned}$$

We see $(\alpha S, \beta S) = (S, -r_0 S)$, $\alpha r + \beta = \alpha(r - r_0)$ and $\alpha g + \beta r = \alpha(g - r_0 r)$. Especially, $((\alpha r_0 + \beta)S, (\alpha g + \beta r_0)S) = (0, (g - r_0^2)S)$. Hence

$$\begin{aligned} hC_0 &= (0, S) \cup (S, -r_0 S) \cup \left\{\left((r - r_0)S, (g - r_0 r)S\right), r \in X_1, r \neq r_0\right\} \\ & \quad \cup \left\{\left((r - r_0)N, (g - r_0 r)N\right), r \in X_2\right\}. \end{aligned}$$

Denote $r - r_0$ by r_1 if $r \in X_1$ and by r_2 if $r \in X_2$. If $r_1 = r - r_0 \in S$, then $r_1 \in (X_1 - r_0) \cap S$. Thus

$$\begin{aligned} & \left\{\left((r - r_0)S, (g - r_0 r)S\right), r \in X_1, r \neq r_0\right\} \\ & \quad = \left\{(S, r_1^{-1}(g - r_0^2 - r_0 r_1)S), r_1 \in (X_1 - r_0) \cap S\right\}. \end{aligned}$$

For $r_2 = r - r_0 \in N$, we have similarly

$$\begin{aligned} & \left\{\left((r - r_0)N, (g - r_0 r)N\right), r \in X_2\right\} \\ & \quad = \left\{\left(S, r_2^{-1}(g - r_0^2 - r_0 r_2)N\right), r_2 \in (X_2 - r_0) \cap S\right\}. \\ hC_0 &= (0, S) \cup (S, -r_0 S) \cup \left\{(S, r_1^{-1}(g - r_0^2 - r_0 r_1)S), r_1 \in E_1\right\} \\ & \quad \cup \left\{(N, r_2^{-1}(g - r_0^2 - r_0 r_2)N), r_2 \in E_2\right\} \end{aligned}$$

where

$$\begin{aligned} E_1 &= ((X_1 - r_0) \cap S) \cup ((X_2 - r_0) \cap N), \\ E_2 &= ((X_1 - r_0) \cap N) \cup ((X_2 - r_0) \cap S). \end{aligned}$$

Comparing hC_0 and the Eq. (3.25), we have

$$|E_1| = |X_1| - 1, \quad |E_2| = |X_2|.$$

It means the coefficients of r_0 in $X_1 S + X_2 N$ and $X_1 N + X_2 S$ are $|X_1| - 1$ and $|X_2|$, respectively. One can obtain the Eqs. (3.21) through (3.24) hold similarly. □

Theorem 3.29 *(Xia et al., [229]):* *Let $q \equiv 1 \pmod 4$ be an odd prime power. Let $W = \{X_1, X_2, X_3, X_4\}$ be a $(q; x, y)$-partition of $GF(q)$, $\beta \neq 0, r \in GF(q)$. $\tilde{W} = \{\tilde{X_1}, \tilde{X_2}, \tilde{X_3}, \tilde{X_4}\}$ obtained from W under the following transformations:*

i) $\tilde{X_i} = X_i + r, i = 1, 2, 3, 4,$
ii) $\tilde{X_i} = \{\alpha^p : \alpha \in X_i\}, i = 1, 2, 3, 4,$
iii) $\tilde{X_i} = \beta X_i, i = 1, 2, 3, 4 \, for \, \beta \in S,$
iv) $\tilde{X_1} = \beta X_2, \tilde{X_2} = \beta X_1, \tilde{X_i} = \beta X_i, \; i = 3, 4 \, for \, \beta \in N.$

Then $\tilde{W}$ is also a $(q; x, y)$-partition of $GF(q)$.

The subsets C_0, C_1, C_2, C_3 of $F = GF(q^2)$ satisfying equations (3.25), (3.26), (3.27) and (3.28) with $(q; x, y)$-partition $\{X_1, X_2, X_3, X_4\}$ of $K = GF(q)$ are not necessarily cyclotomic classes, but the union of cyclotomic classes.

Example 3.6 *(Xia et al. [229]):* *Put $q = 5$, $F = GF(5^2)$, $K = GF(5)$. $g = 2, \omega^2 = 2.5 = 1^2 + 4 \cdot 1^2$, $x = 1$, $y = 1$, $m = 1$.*

$$X_1 = \{0, 1\}, \; X_2 = \emptyset, \; X_3 = \{2, 4\}, \; X_4 = \{3\}.$$
$$S = \{1, 4\}, \; N = \{2, 3\}.$$

It is easy to verify that X_1, X_2, X_3, X_4 satisfy (3.19)–(3.24). We define the subsets C_0, C_1, C_2, C_3 by (3.25), (3.26), (3.27), and (3.28). Let ξ be a primitive element of F.

$$C_0 = (0, S) \cup (S, 0) \cup (S, S) = \{1, \xi^{12}, \xi^3, \xi^{15}, \xi^{10}, \xi^{22}\},$$
$$C_1 = (S, 2S) \cup (S, 4S) \cup (N, 3N) = \{\xi^{13}, \xi, \xi^{14}, \xi^2, \xi^{11}, \xi^{23}\},$$
$$C_2 = (0, N) \cup (N, 0) \cup (N, N) = \{\xi^6, \xi^{18}, \xi^9, \xi^{21}, \xi^{16}, \xi^4\},$$
$$C_3 = (N, 2N) \cup (N, 4N) \cup (S, 3S) = \{\xi^{19}, \xi^7, \xi^{20}, \xi^8, \xi^5, \xi^{17}\}.$$

We notice that C_0 is not the set of biquadratic residues.

Theorem 3.30 *(Xia et al. [229]):* *Let $q = 4m + 1 \equiv 1 \pmod 4$ is an odd prime power. Suppose $\{X_1, X_2, X_3, X_4\}$ is a $(q; x, y)$-partition of $GF(q)$. C_0, C_1, C_2, C_3 are the subsets given as in (3.25)–(3.28), respectively. Then*

$$C_i C_j = \delta_{ij} 2m(2m + 1) + \sum_{k=0}^{3} \langle j - i, k \rangle \, C_{i+k}, \; 0 \leq i \leq j \leq 3 \tag{3.29}$$

where $\delta_{i,j} = 0$ or 1 according as $i \neq j$ or $i = j$, respectively, and $\langle i,j \rangle$ is given by the following table with $\langle i,j \rangle$ in the i-th row and j-th column, $i, j = 0, 1, 2, 3$,

	0	1	2	3
0	A	D	B	E
1	D	E	C	C
2	B	C	B	C
3	E	C	C	D

- $A = m^2 - m - 1 + 3y^2$,
- $D = m^2 + m - y^2 + ef$,
- $B = m^2 + m - y^2$,
- $E = m^2 + m - y^2 - ef$,
- $C = m^2 + y^2$,
- $e = \frac{1}{2}\left(|X_1| - |X_2|\right), \; f = |X_3| - |X_4|$.

Proof: For $\alpha = \alpha_1\omega + \alpha_2 \in F, \alpha_1, \alpha_2 \in K$, we denote an additive character by $\mu_{(\alpha_1,\alpha_2)}$. Denote the right-hand side of the Eq. (3.29) by $R_{i,j}$. Then we verify $\mu_{\alpha_1,\alpha_2}(C_1C_2) = \mu_{\alpha_1,\alpha_2}(R_{i,j})$ for all additive characters of F. For the detailed proof, see [229]. □

C_0, C_1, C_2, C_3 given as in (3.25), (3.26), (3.27), (3.28) can be written in the following form,

$$C_0 = \bigcup_{i=0}^{2m} S_{a_i}, \quad C_2 = \bigcup_{i=0}^{2m} S_{a_i+4m+2},$$
$$C_1 = \bigcup_{j=0}^{2m} S_{b_j}, \quad C_3 = \bigcup_{j=0}^{2m} S_{b_j+4m+2},$$

where $\{a_0, a_1, ..., a_{2m}, b_0, b_1, ..., b_{2m}\} \equiv \{0, 1, ..., q = 4m+1\} \pmod{4m+2}$. Take any subsets $F_1 \subset \{a_0, a_1, \ldots, a_{2m}\}$, $F_2 \subset \{b_0, b_1, \ldots, b_{2m}\}$ such that $|F_1| = |F_2| = m$. Put

$$A = \bigcup_{a_i \in F_1} (S_{a_i} \cup S_{a_i+4m+2}), \quad B = \bigcup_{b_j \in F_2} (S_{b_j} \cup S_{b_j+4m+2}), \tag{3.30}$$

$$D_0 = B \cup C_0,\ D_2 = B \cup C_2,\ D_1 = A \cup C_1,\ D_3 = A \cup C_3. \tag{3.31}$$

Theorem 3.31 *(Xia et al., [229]):* *Let $q \equiv 1 \pmod 4$ is an odd prime power. Then D_0, D_1, D_2, D_3 are*

$$4 - \left\{q^2;\ \tfrac{1}{2}q(q-1);\ q(q-2)\right\}$$

supplementary difference sets (DF) in $GF(q^2)$. There exists an Hadamard matrix of order $4q^2$.

It is trivial that A and B are symmetric, since $-1 \in S_0$. Hence D_0, D_1, D_2, D_3 are symmetric. The parameters satisfy $\lambda = k_1 + k_2 + k_3 + k_4 - v$. Then, the ± 1 incidence matrices of the supplementary difference sets in Theorem 3.31 are Williamson matrices.

If q is a complete square, then $x = \pm\sqrt{q}, y = 0$. In this case, we have a simple theorem.

Example 3.7: *Let C_0, C_1, C_2, C_3 be given as in Example 3.6. Let $S_i, 0 \le i \le 11$ be 12th cyclotomic classes. Then C_0, C_1, C_2, C_3 can be written as*

$$C_0 = S_0 \cup S_3 \cup S_{10},\ C_1 = S_1 \cup S_2 \cup S_{11},$$
$$C_2 = S_6 \cup S_9 \cup S_4, \quad C_3 = S_7 \cup S_8 \cup S_5.$$

We take $F_1 = \{0\}$ and $F_2 = \{1\}$. Thus we have $A = S_0 \cup S_6 = \left\{1, \xi^{12}, \xi^6, \xi^{18}\right\}$, $B = S_1 \cup S_7 = \left\{\xi, \xi^{13}, \xi^7, \xi^{19}\right\}$. Then the subsets

$$D_0 = C_0 \cup B = \left\{1, \xi^{12}, \xi^3, \xi^{15}, \xi^{10}, \xi^{22}, \xi, \xi^{13}, \xi^7, \xi^{19}\right\},$$
$$D_1 = C_1 \cup A = \left\{\xi^{13}, \xi, \xi^{14}, \xi^2, \xi^{11}, \xi^{23}, 1, \xi^{12}, \xi^6, \xi^{18}\right\},$$
$$D_2 = C_2 \cup B = \left\{\xi^6, \xi^{18}, \xi^9, \xi^{21}, \xi^{16}, \xi^4, \xi, \xi^{13}, \xi^7, \xi^{19}\right\},$$
$$D_3 = C_3 \cup A = \left\{\xi^{19}, \xi^7, \xi^{20}, \xi^8, \xi^5, \xi^{17}, 1, \xi^{12}, \xi^6, \xi^{18}\right\},$$

are $4 - (25, 10, 15)$ sds.

3.10 Relative Difference Sets and Williamson-Type Matrices over Abelian Groups

In this section, we will show that the existence of Williamson-type matrices over an abelian group G of order m is equivalent to the existence of relative $(4m, 2, 4m, 2m)$ difference sets in the direct product of G and the quaternion group, which was proved by Schmidt [152].

Definition 3.11: Let G be a group of order mn and N be a normal subgroup of G of order n. A k-subset R of G is called an (m, n, k, λ) **difference set in** G **relative to** N if and only if the multi-set of $r_1 r_2^{-1}$ of distinct elements $r_1, r_2 \in R$ contains each element of $G \backslash N$ exactly λ times and contains no element of N. The parameters satisfy $k(k-1) = \lambda n(m-1)$.

Lemma 3.8: *A k-subset R of a group G of order mn is a relative (m, n, k, λ) difference set in G relative to a normal subgroup N of order n if and only if*

$$RR^{-1} = k + \lambda(G - N)$$

in $\mathbf{Z}[G]$.

A (ut, u, ut, t) relative difference set is called **semi-regular relative difference set**. Ito [103] conjectured that a $(4t, 2, 4t, 2t)$ semi-regular relative difference set in dicycle groups exists for all positive integers t.

Let G (not necessary abelian) be a group of order m. A square matrix of order m is called G-**invariant** if the rows and columns of $A = (a_{g,h})$ can be indexed with elements g, h of G such that $a_{gk,hk} = a_{g,h}$ for all $g, h, k \in G$. If A is a type 1 matrix such that $a_{g,h} = \psi(gh^{-1})$, then A is G-invariant from

$$A_{gk,hk} = \psi\left(gk^{-1}(hk)^{-1}\right) = \psi\left(gh^{-1}\right) = a_{g,h}.$$

Let G be a group of order m. Let A, B, C, D be G-invariant matrices of order m with entries ± 1. If

$$H = \begin{bmatrix} A & B & C & D \\ -B & A & -D & C \\ -C & D & A & -B \\ -D & -C & B & A \end{bmatrix}$$

is an Hadamard matrix of order $4m$, then A, B, C, D are said to be **Williamson-type of order m over a group G**.

In the group ring notation, a necessary and sufficient condition such that H becomes an Hadamard matrix is

$$AA^{-1} + BB^{-1} + CC^{-1} + DD^{-1} = 4m, \tag{3.32}$$

$$XY^{-1} - X^{-1}Y + VZ^{-1} - V^{-1}Z = 0, \tag{3.33}$$

where (X, Y, V, Z) stands for (A, B, C, D) or (A, D, B, C) or (A, C, D, B).

Let $Q_8 = \left\langle x, y \mid x^4 = y^4 = 1, x^2 = y^2, y^{-1}xy = x^{-1} \right\rangle$ be the quaternion group of order 8.

Theorem 3.32 (Schmidt [152]): *Let G be an abelian group of order m. Williamson-type matrices of order m over G exists if and only if there exists a $(4m, 2, 4m, 2m)$ semi-regular relative difference set in $T = G \times Q_8$ relative to $N = \langle 1 \rangle \times \left\langle x^2 \right\rangle$.*

Proof: Let R be a $(4m, 2, 4m, 2m)$ semi-regular relative difference set in T relative to N. Let $U = G \times \langle x^2 \rangle$ and write $R = E + Fx + Ky + Lxy$ with $E, F, K, L \subset U$. From Lemma 3.8, R is a $(4m, 2, 4m, 2m)$ relative difference set in T relative to N if and only if

$$EE^{-1} + FF^{-1} + KK^{-1} + LL^{-1} = 4m + 2m(U - N), \tag{3.34}$$

$$E^{-1}F + K^{-1}L + \left(EF^{-1} + KL^{-1}\right)x^2 = 2mU, \tag{3.35}$$

$$FL^{-1} + E^{-1}K + \left(EK^{-1} + F^{-1}L\right)x^2 = 2mU, \tag{3.36}$$

$$F^{-1}K + E^{-1}L + \left(EL^{-1} + FK^{-1}\right)x^2 = 2mU. \tag{3.37}$$

We let $X = X_1 + X_2x^2, X_1, X_2 \in G$ for $X = E, F, K, L$. Since

$$|R| = |E| + |F| + |K| + |L| = 4m \text{ and } |E|^2 + |F|^2 + |K|^2 + |L|^2 = 4m^2,$$

we have $|E| = |F| = |K| = |L| = m$ and

$$X_1 + X_2 = G, \quad X_i = E_i, F_i, K_i, L_i, \ i = 1, 2.$$

Let $A = E_1 - E_2, B = F_1 - F_2, C = K_1 - K_2, D = L_1 - L_2$. Then A, B, C, D satisfy the relation (3.32) by straightforward verification. From

$$\chi\left(X_1\right) = -\chi\left(X_2\right), \ X_i = E_i, F_i, K_i, L_i, \ i = 1, 2$$

for a nontrivial character χ of G and from (3.35), we see

$$\chi(AB^{-1} - A^{-1}B + CD^{-1} - C^{-1}D) = 0$$

for all characters of G. We can show the other relations of (3.33) similarly.

Conversely we assume that A, B, C, D satisfy the relations (3.32) and (3.33). Since A, B, C, D yield the Williamson-type matrices, $X_1 + X_2 = G, X_i = E_i, F_i, K_i, L_i, i = 1, 2$. We show

$$\chi\left(EE^{-1} + FF^{-1} + KK^{-1} + LL^{-1}\right) = \chi(4m + 2m(U - N)) \tag{3.38}$$

for all character χ of U. If χ is trivial on N, then the right-hand side of (3.38) is equal to 0. From $\chi(X) = \chi(X_1 + X_2) = \chi|_G(G)$ for $X \in \{E, F, K, L\}$, the left-hand side of (3.38) is 0. Suppose χ is nontrivial on N. Then

$$\chi(E) = \chi(E_1 - E_2) = \chi|_G(A),$$

$$\chi(F) = \chi|_G(B),$$

$$\chi(K) = \chi|_G(C),$$

$$\chi(L) = \chi|_G(D)$$

where $\chi|_G$ is the character χ of U restricted to G. Hence the relation (3.34) follows. We prove the relations (3.35)–(3.37) similarly. □

Example 3.8: *Let $G = \mathbf{Z}_3 =< \alpha >$ be a cyclic group of order 3. Put $T = G \times Q_8, N = \langle 1 \rangle \times \langle x^2 \rangle$ and $U = G \times \langle x^2 \rangle$. We write $R = E + Fx + Ky + Lxy$ where*

$$E = F = K = 1 + \left(\alpha + \alpha^2\right)x^2, \quad L = 1 + \alpha + \alpha^2.$$

Then E, F, K, L satisfy (3.34)–(3.37). Hence

$$\begin{aligned} R = {} & 1 + (\alpha + \alpha^2)x^2 + (1 + (\alpha + \alpha^2)x^2)x \\ & + (1 + (\alpha + \alpha^2)x^2)y + (1 + \alpha + \alpha^2)xy \end{aligned}$$

is a $(12, 2, 12, 6)$*-semi-regular relative difference set in* $\boldsymbol{Z}_3 \times Q_8$*. We let*

$$A = B = C = E_1 - E_2 = 1 - (\alpha + \alpha^2), \quad D = 1 + \alpha + \alpha^2.$$

Then A, B, C, D *satisfy (3.32) and (3.33). Denote the* ± 1 *incidence matrices of* A, B, C, D *by* $\mathcal{A}, \mathcal{B}, \mathcal{C}, \mathcal{D}$*.*

$$\mathcal{A} = \mathcal{B} = \mathcal{C} = \begin{bmatrix} 1 & - & - \\ - & 1 & - \\ - & - & 1 \end{bmatrix}, \quad \mathcal{D} = \begin{bmatrix} 1 & 1 & 1 \\ 1 & 1 & 1 \\ 1 & 1 & 1 \end{bmatrix}$$

are G-invariant and the Williamson matrices over $\boldsymbol{Z}_3$*.*

Let G be an abelian group of order m. Denote the semidirect product of G and Q_8 by $Q_8(G)$ such that $x^{-1}gx = x, y^{-1}gy = g^{-1}$ for all $g \in G$.

Theorem 3.33 ***(Schmidt [152]):*** *A* $(4m, 2, 4m, 2m)$ *relative difference set in* $Q_8(G)$ *relative to* $\langle y^2 \rangle$ *exists if and only if there is a Hadamard matrix of the form*

$$H = \begin{bmatrix} A & B & C & D \\ -B & A & -D & C \\ -C^{\mathrm{T}} & D^{\mathrm{T}} & A^{\mathrm{T}} & -B^{\mathrm{T}} \\ -D^{\mathrm{T}} & -C^{\mathrm{T}} & B^{\mathrm{T}} & A^{\mathrm{T}} \end{bmatrix}$$

where A, B, C, D *are G-invariant matrices of order* m*.*

Remark 3.2: Horadam and de Launey discuss the study of Hadamard matrices from the perspective of co-cycles. They showed Sylvester–Hadamard, Williamson Hadamard, Paley Hadamard, Ito's Hadamard matrix, and generalized quaternion type exist based on co-cycles. See [46]. We refer the reader to the excellent book of K. Horadam [98] to explore the power and beauty of co-cyclic Hadamard matrices.

3.11 Computer Construction of Williamson Matrices

Paley's work [144] left many orders for Hadamard matrices unresolved. Later Williamson [223] gave a method that many researchers hoped would give results for all orders of Hadamard matrices. Many computer generated results are given in [8, 9]. That the Williamson method would give results for all orders of Hadamard matrices was first disproved by Đoković [58].

Williamson matrices are presently known for n, odd, $1 \leq n \leq 33, 37, 39, 41, 43, 45, 49, 51, 55, 57, 61, 63, 69, 75, 79, 81, 85, 87, 91, 97, 99$. Williamson-type matrices (see Definition 3.2) are known for orders $53, 73, 83, 89$. Orders $35, 47$, and 59 are eliminated by computer search.

Table 3.4 Existence of inequivalent Williamson matrices for small orders.

n	**1**	**3**	**5**	**7**	**9**	**11**	**13**	**15**	**17**	**19**	**21**	**23**	**25**	**27**	**29**
N	**1**	**1**	**1**	**2**	**3**	**1**	**4**	**4**	**4**	**6**	**7**	**1**	**10**	**6**	**1**
n	31	33	35	37	39	41	43	45	47	49	51	53	55	57	59
N	2	5	0	4	1	1	2	1	0	1	2	0	1	1	0

Source: Holzmann et al. [97, table 1, p. 347], Reproduced with permission of Springer Nature.

While there are no Williamson matrices of orders 47, 53, and 59, there are good matrices and luchshie matrices for these orders. Using Goethals–Seidel and Balonin–Seberry arrays, respectively, we get skew Hadamard and symmetric Hadamard matrices of orders $4 \cdot 47 = 188, 4 \cdot 53 = 212$, and $4 \cdot 59 = 236$.

The startling paper of Holzmann et al. [97] gives the existence of inequivalent Williamson matrices for small orders. See Table 3.4

Conjecture 3.12 (Holzmann, et al. [97]): There are no Williamson matrices of prime order $p \geq 43$ unless $p = \frac{1}{2}(q+1)$, $q \equiv 1 \pmod 4$ a prime power.

4

Arrays: Matrices to Plug-Into

4.1 Notations

Table 4.1 gives the notations which are used in this chapter.

4.2 Orthogonal Designs

Orthogonal designs, their quadratic forms and algebras are studied extensively in Seberry's book [158]. For easy reference and the reader's convenience we repeat Table 1.4 here as Table 4.2. We only give a small idea of their usage here.

Lemma 4.1: *Let D be an* $OD(n; u_1, u_2, \dots, u_t)$ *orthogonal design, on the t commuting variables* $x_1, x_2, \dots, x_t$. *Then the following orthogonal designs exist:*

i) $OD(n; u_1, u_2, \dots, u_i + u_j, \dots, u_t)$ *on* $t-1$ *variables;*
ii) $OD(n; u_1, \dots, u_{i-1}, u_{i+1}, \dots, u_t)$ *on* $t-1$ *variables;*
iii) $OD(2n; u_1, u_2, \dots, u_t)$ *on t variables;*
iv) $OD(2n; 2u_1, 2u_2, \dots, 2u_t)$ *on t variables;*
v) $OD(2n; u_1, u_1, u_2, \dots, u_t)$ *on* $t+1$ *variables;*
vi) $OD(2n; u_1, u_1, 2u_2, \dots, 2u_t)$ *on* $t+1$ *variables.*

Example 4.1: *Let*

$$D_1 = \begin{bmatrix} a & -b & -c & -d \\ b & a & -d & c \\ c & d & a & -b \\ d & -c & b & a \end{bmatrix}, \quad D_2 = \begin{bmatrix} a & b \\ b & -a \end{bmatrix}.$$

Then D_1 *is the* $OD(4; 1, 1, 1, 1)$ *design and* D_2 *is the* $OD(2; 1, 1)$ *design (see Table 1.4). When we let* $a = b$ *and* $c = d$ *in* D_1, *the resultant design is* $OD(4; 2, 2)$ *which is equivalent to the orthogonal design in Table 4.2 (iii). When we let* $d = 0$ *in* D_1, *the resultant design is* $OD(4; 1, 1, 1)$ *in Table 4.2 (i). Replacing the variables a and b of* $OD(2; 1, 1)$ D_2 *by*

$$\begin{bmatrix} x & 0 \\ 0 & x \end{bmatrix} \quad \text{and} \quad \begin{bmatrix} y & 0 \\ 0 & y \end{bmatrix}$$

respectively, we obtain $OD(4; 1, 1)$ *in Table 4.2 (iv). Replacing a and b by*

$$\begin{bmatrix} x & x \\ x & -x \end{bmatrix} \quad \text{and} \quad \begin{bmatrix} y & y \\ y & -y \end{bmatrix}$$

respectively, we obtain $OD(4; 2, 2)$ *in Table 4.2 (iii).*

Hadamard Matrices: Constructions using Number Theory and Algebra, First Edition. Jennifer Seberry and Mieko Yamada.

Table 4.1 Notations used in this chapter.

$x \to x^*$	An involution of $\mathbf{Z}[G]$
$OD(n; u_1, u_2, \dots, u_t)$	Orthogonal design of order n and type $(u_1, u_2, \dots, u_t)$
$OD(4t; t, t, t, t)$	Baumert–Hall array of order t
$BHW(G)$	An ordered set on a group G which is used to produce Welch (WL) and Ono–Sawade–Yamamoto arrays
$M \times N$	The Kronecker product
R	The back diagonal identity matrix
C_n	A cyclic group of order n

Table 4.2 Orthogonal designs.

$$\begin{bmatrix} a & -b & -c & 0 \\ b & a & 0 & c \\ c & 0 & a & -b \\ 0 & -c & b & a \end{bmatrix} \quad \begin{bmatrix} x & 0 & y & 0 \\ 0 & x & 0 & y \\ y & 0 & -x & 0 \\ 0 & y & 0 & -x \end{bmatrix}$$

(i) $OD(4; 1, 1, 1)$ (ii) $OD(4; 1, 1)$

$$\begin{bmatrix} x & x & y & y \\ x & -x & y & -y \\ y & y & -x & -x \\ y & -y & -x & x \end{bmatrix} \quad \begin{bmatrix} z & x & 0 & y \\ -x & z & y & 0 \\ 0 & y & -z & -x \\ y & 0 & x & -z \end{bmatrix}$$

(iii) $OD(4; 2, 2)$ (iv) $OD(4; 1, 1)$

4.2.1 Baumert–Hall Arrays and Welch Arrays

A special orthogonal design, the $OD(4t; t, t, t, t)$, is especially useful in the construction of Hadamard matrices. An $OD(12; 3, 3, 3, 3)$ was first found by L. Baumert and M. Hall Jr [9] and an OD(20; 5, 5, 5, 5) by Welch (see below). $OD(4t; t, t, t, t)$ are sometimes called Baumert–Hall arrays (see Definition 1.57) or BH of order $4t$.

Another set of matrices, of a very different kind, can be obtained by partitioning a matrix:

Definition 4.1: Let M be a matrix of order tm. Then M can be expressed as a t^2 **block** M**-structure** when M is an orthogonal matrix:

$$M = \begin{bmatrix} M_{11} & M_{12} & \cdots & M_{1t} \\ M_{21} & M_{22} & \cdots & M_{2t} \\ & & \cdots & \\ M_{t1} & M_{t2} & \cdots & M_{tt} \end{bmatrix}$$

where M_{ij} is of order m $(i,j = 1, 2, \dots, t)$.

M-structures are studied in Chapter 6 and use the internal structure of an orthogonal matrices as their tools. However a very different kind of M-structure is given in parts (v) and (vi) of Definition 4.2 below.

Definition 4.2: An M-structure comprising 16 blocks, each of which is circulant (or type 1) and of size $t \times t$ will be called a **Welch array**, $WL(t)$, of order t. An $OD(4t; t, t, t, t)$ and a Welch array where each of the 16 blocks is of size 3×3, and of type 1, will be called an **Ono–Sawade–Yamamoto array**, i.e. an $OD(36; 9, 9, 9, 9)$. Examples of these for $t = 5$ and 9 are given below in Tables 4.3 and 4.4.

Some orthogonal designs of special interest are:

i) The Williamson array – the $OD(4; 1, 1, 1, 1)$

$$R = \begin{bmatrix} A & B & C & D \\ -B & A & -D & C \\ -C & D & A & -B \\ -D & -C & B & A \end{bmatrix}, \quad L = \begin{bmatrix} A & B & C & D \\ -B & A & D & -C \\ -C & -D & A & B \\ -D & C & -B & A \end{bmatrix}$$

R is the right regular representation of quaternions and L is the left regular representation of quaternions. R is equivalent to L.

ii) The $OD(8; 1, 1, 1, 1, 1, 1, 1, 1)$

$$\left[\begin{array}{cccc|cccc} A & B & C & D & E & F & G & H \\ -B & A & D & -C & F & -E & -H & G \\ -C & -D & A & B & G & H & -E & -F \\ -D & C & -B & A & H & -G & F & -E \\ \hline -E & -F & -G & -H & A & B & C & D \\ -F & E & -H & G & -B & A & -D & C \\ -G & H & E & -F & -C & D & A & -B \\ -H & -G & F & E & -D & -C & B & A \end{array}\right]$$

The $OD(8; 1, 1, 1, 1, 1, 1, 1, 1)$ is the left regular representation of Cayley numbers.

iii) The Baumert–Hall array – the $OD(12; 3, 3, 3, 3)$

$$A(x,y,z,w) = \begin{bmatrix} y & x & x & x & -z & z & w & y & -w & w & z & -y \\ -x & y & x & -x & w & -w & z & -y & -z & z & -w & -y \\ -x & -x & y & x & w & -y & -y & w & z & z & w & -z \\ -x & x & -x & y & -w & -w & -z & w & -z & -y & -y & -z \\ -y & -y & -z & -w & z & x & x & x & -w & -w & z & -y \\ -w & -w & -z & y & -x & z & x & -x & y & y & -z & -w \\ w & -w & w & -y & -x & -x & z & x & y & -z & -y & -z \\ -w & -z & w & -z & -x & x & -x & z & -y & y & -y & w \\ -y & y & -z & -w & -z & -z & w & y & w & x & x & x \\ z & -z & -y & -w & -y & -y & -w & -z & -x & w & x & -x \\ -z & -z & y & z & -y & -w & y & -w & -x & -x & w & x \\ z & -w & -w & z & y & -y & y & z & -x & x & -x & w \end{bmatrix}$$

or alternatively we obtain the $OD(12; 3, 3, 3, 3)$ from the Cooper–J.Wallis theorem. The matrices

$$X_1 = \begin{bmatrix} 1 & 0 & 0 \\ 0 & 1 & 0 \\ 0 & 0 & 1 \end{bmatrix}, \quad X_2 = \begin{bmatrix} 0 & 1 & 0 \\ 0 & 0 & 1 \\ 1 & 0 & 0 \end{bmatrix}, \quad X_3 = \begin{bmatrix} 0 & 0 & 1 \\ 1 & 0 & 0 \\ 0 & 1 & 0 \end{bmatrix}, \quad X_4 = 0$$

are circulant T-matrices of order 3 (see Definition 1.59). From Cooper–J.Wallis Theorem 3.12, we have

$$A = \begin{bmatrix} a & b & c \\ c & a & b \\ b & c & a \end{bmatrix}, \quad B = \begin{bmatrix} -b & a & d \\ d & -b & a \\ a & d & -b \end{bmatrix},$$

$$C=\begin{bmatrix}-c & -d & a\\ a & -c & -d\\ -d & a & -c\end{bmatrix},\quad D=\begin{bmatrix}-d & c & -b\\ -b & -d & c\\ c & -b & -d\end{bmatrix}.$$

Then the Goethals–Seidel (J.Wallis–Whiteman) array

$$\begin{bmatrix}A & BR & CR & DR\\ -BR & A & -D^{\top}R & C^{\top}R\\ -CR & D^{\top}R & A & -B^{\top}R\\ -DR & -C^{\top}R & B^{\top}R & A\end{bmatrix}$$

$$=\left[\begin{array}{ccc|ccc|ccc|ccc}
a & b & c & d & a & -b & a & -d & -c & -b & c & -d\\
c & a & g & a & -b & d & -d & -c & a & c & -d & -b\\
b & c & a & -b & d & a & -c & a & -d & -d & -b & c\\ \hline
-d & -a & b & a & b & c & -c & b & d & -d & a & -c\\
-a & b & -d & c & a & b & b & d & -c & a & -c & -d\\
b & -d & -a & b & c & a & d & -c & b & -c & -d & a\\ \hline
-a & d & c & c & -b & -d & a & b & c & -a & -d & b\\
d & c & -a & -b & -d & c & c & a & b & -d & b & -a\\
c & -a & d & -d & c & -b & b & c & a & b & -a & -d\\ \hline
b & -c & d & d & -a & c & a & d & -b & a & b & c\\
-c & d & b & -a & c & d & d & -b & a & c & a & b\\
d & b & -c & c & d & -a & -b & a & d & b & c & a
\end{array}\right]$$

is the $OD(12;3,3,3,3)$.

iv) The Plotkin array – the $OD(24;3,3,3,3,3,3,3,3)$ Let $A(x,y,z,w)$ be as in (iii) and

$$B(x,y,z,w)=\begin{bmatrix}
y & x & x & x & -w & w & z & y & -z & z & w & -y\\
-x & y & x & -x & -z & z & -w & -y & w & -w & z & -y\\
-x & -x & y & x & -y & -w & y & -w & -z & -z & w & z\\
-x & x & -x & y & w & w & -z & -w & -y & z & y & z\\
 & & & & & & & & & & & \\
-w & -w & -z & -y & z & x & x & x & -y & -y & z & -w\\
y & y & -z & -w & -x & z & x & -x & -w & -w & -z & y\\
-w & w & -w & -y & -x & -x & z & x & z & y & y & z\\
z & -w & -w & z & -x & x & -x & z & y & -y & y & w\\
 & & & & & & & & & & & \\
z & -z & y & -w & y & y & w & -z & w & x & x & x\\
y & -y & -z & -w & -z & -z & -w & -y & -x & w & x & -x\\
z & z & y & -z & w & -y & -y & w & -x & -x & w & x\\
-w & -z & w & -z & -y & y & -y & z & -x & x & -x & w
\end{bmatrix}$$

$A(x,y,z,w)$ and $B(x,y,z,w)$ are both Baumert–Hall arrays, but $B(x,y,z,w)$ is not equivalent to $A(x,y,z,w)$. Then

$$\begin{bmatrix}A(x_1,x_2,x_3,x_4) & B(x_5,x_6,x_7,x_8)\\ B(-x_5,x_6,x_7,x_8) & -A(-x_1,x_2,x_3,x_4)\end{bmatrix}$$

is the required design.

v) The Welch array – the $OD(20;5,5,5,5)$ constructed from 16 block circulant matrices is given below in Table 4.3.

vi) The Ono–Sawade–Yamamoto array – the $OD(36;9,9,9,9)$ constructed from 16 type 1 matrices is given below in Table 4.4.

Table 4.3 The first known Welch array – the $OD(20; 5, 5, 5, 5)$.

−D	B	−C	−C	−B	C	A	−D	−D	−A	−B	−A	C	−C	−A	A	−B	−D	D	−B
−B	−D	B	−C	−C	−A	C	A	−D	−D	−A	−B	−A	C	−C	−B	A	−B	−D	D
−C	−B	−D	B	−C	−D	−A	C	A	−D	−C	−A	−B	−A	C	D	−B	A	−B	−D
−C	−C	−B	−D	B	−D	−D	−A	C	A	C	−C	−A	−B	−A	−D	D	−B	A	−B
B	−C	−C	−B	−D	A	−D	−D	−A	C	−A	C	−C	−A	−B	−B	−D	D	−B	A
−C	A	D	D	−A	−D	−B	−C	−C	B	−A	B	−D	D	B	−B	−A	−C	C	−A
−A	−C	A	D	D	B	−D	−B	−C	−C	B	−A	B	−D	D	−A	−B	−A	−C	C
D	−A	−C	A	D	−C	B	−D	−B	−C	D	B	−A	B	−D	C	−A	−B	−A	−C
D	D	−A	−C	A	−C	−C	B	−D	−B	−D	D	B	−A	B	−C	C	−A	−B	−A
A	D	D	−A	−C	−B	−C	−C	B	−D	B	−D	D	B	−A	−A	−C	C	−A	−B
B	−A	−C	C	−A	A	B	−D	D	B	−D	−B	C	C	B	−C	A	−D	−D	−A
−A	B	−A	−C	C	B	A	B	−D	D	B	−D	−B	C	C	−A	−C	A	−D	−D
C	−A	B	−A	−C	D	B	A	B	−D	C	B	−D	−B	C	−D	−A	−C	A	−D
−C	C	−A	B	−A	−D	D	B	A	B	C	C	B	−D	−B	−D	−D	−A	−C	A
−A	−C	C	−A	B	B	−D	D	B	A	−B	C	C	B	−D	A	−D	−D	−A	−C
−A	−B	−D	D	−B	B	−A	C	−C	−A	C	A	D	D	−A	−D	B	C	C	−B
−B	−A	−B	−D	D	−A	B	−A	C	−C	−A	C	A	D	D	−B	−D	B	C	C
D	−B	−A	−B	−D	−C	−A	B	−A	C	D	−A	C	A	D	C	−B	−D	B	C
−D	D	−B	−A	−B	C	−C	−A	B	−A	D	D	−A	C	A	C	C	−B	−D	B
−B	−D	D	−B	−A	−A	C	−C	−A	B	A	D	D	−A	C	B	C	C	−B	−D

Source: Seberry and Yamada [166, p. 448], Wiley.

vii) The Goethals–Seidel [81] (J. Wallis and Whiteman [216]) array:

$$\begin{bmatrix} A & BR & CR & DR \\ -BR & A & -D^{\top}R & C^{\top}R \\ -CR & D^{\top}R & A & -B^{\top}R \\ -DR & -C^{\top}R & B^{\top}R & A \end{bmatrix}$$

or

$$\begin{bmatrix} A & BR & CR & DR \\ -BR & A & D^{\top}R & -C^{\top}R \\ -CR & -D^{\top}R & A & B^{\top}R \\ -DR & C^{\top}R & -B^{\top}R & A \end{bmatrix}$$

where A, B, C, D are circulant (type 1) matrices satisfying the additive property (Eq. (1.8)) and R is the back diagonal identity $(0, 1)$ matrix (equivalent type 2 matrix).

Definition 4.3: **Suitable matrices** of order w for an $OD(n; s_1, s_2, \dots, s_u)$ are u pairwise amicable matrices (see Section 1.7), A_i, $i = 1, \dots, u$ which have entries $+1$ or -1 and which satisfy

$$\sum_{i=1}^{u} s_i A_i A_i^{\top} = \left(\Sigma s_i\right) w I_w. \tag{4.1}$$

Table 4.4 Ono–Sawade–Yamamoto array – the $OD(36;9,9,9,9)$.

a	a	a	b	c	d	−b	−d	−c	b	−a	a	b	c	−d	b	d	−c	c	−a	a	−b	c	d	b	−d	c	d	−a	a	b	−c	d	−b	d	c
a	a	a	d	b	c	−c	−b	−d	a	b	−a	−d	b	c	−c	b	d	a	c	−a	d	−b	c	c	b	−d	a	d	−a	d	b	−c	c	−b	d
a	a	a	c	d	b	−d	−c	−b	−a	a	b	c	−d	b	d	−c	b	−a	a	c	c	d	−b	−d	c	b	−a	a	d	−c	d	b	d	c	−b
−b	−d	−c	a	a	a	b	c	d	b	d	−c	b	−a	a	b	c	−d	b	−d	c	c	−a	a	−b	c	d	−b	d	c	d	−a	a	b	−c	d
−c	−b	−d	a	a	a	d	b	c	−c	b	d	a	b	−a	−d	b	c	c	b	−d	a	c	−a	d	−b	c	c	−b	d	a	d	−a	d	b	−c
−d	−c	−b	a	a	a	c	d	b	d	−c	b	−a	a	b	c	−d	b	−d	c	b	−a	a	c	c	d	−b	d	c	−b	−a	a	d	−c	d	b
b	c	d	−b	−d	−c	a	a	a	b	c	−d	b	d	−c	b	−a	a	−b	c	d	b	−d	c	c	−a	a	b	−c	d	−b	d	c	d	−a	a
d	b	c	−c	−b	−d	a	a	a	−d	b	c	−c	b	d	a	b	−a	d	−b	c	c	b	−d	a	c	−a	d	b	−c	c	−b	d	a	d	−a
c	d	b	−d	−c	−b	a	a	a	c	−d	b	d	−c	b	−a	a	b	c	d	−b	−d	c	b	−a	a	c	−c	d	b	d	c	−b	−a	a	d
−b	a	−a	−b	c	−d	−b	d	−c	a	a	a	b	−c	−d	−b	d	c	−d	−a	a	b	c	−d	−b	−d	−c	c	a	−a	b	c	d	−b	−d	c
−a	−b	a	−d	−b	c	−c	−b	d	a	a	a	−d	b	−c	c	−b	d	a	−d	−a	−d	b	c	−c	−b	−d	−a	c	a	d	b	c	c	−b	−d
a	−a	−b	c	−d	−b	d	−c	−b	a	a	a	−c	−d	b	d	c	−b	−a	a	−d	c	−d	b	−d	−c	−b	a	−a	c	c	d	b	−d	c	−b
−b	d	−c	−b	a	−a	−b	c	−d	−b	d	c	a	a	a	b	−c	−d	−b	−d	−c	−d	−a	a	b	c	−d	−b	−d	c	c	a	−a	b	c	d
−c	−b	d	−a	−b	a	−d	−b	c	c	−b	d	a	a	a	−d	b	−c	−c	−b	−d	a	−d	−a	−d	b	c	c	−b	−d	−a	c	a	d	b	c
d	−c	−b	a	−a	−b	c	−d	−b	d	c	−b	a	a	a	−c	−d	b	−d	−c	−b	−a	a	−d	c	−d	b	−d	c	−b	a	−a	c	c	d	b
−b	c	−d	−b	d	−c	−b	a	−a	b	−c	−d	−b	d	c	a	a	a	b	c	−d	−b	−d	−c	−d	−a	a	b	c	d	−b	−d	c	c	a	−a
−d	−b	c	−c	−b	d	−a	−b	a	−d	b	−c	c	−b	d	a	a	a	−d	b	c	−c	−b	−d	a	−d	−a	d	b	c	c	−b	−d	−a	c	a
c	−d	−b	d	−c	−b	a	−a	−b	−c	−d	b	d	c	−b	a	a	a	c	−d	b	−d	−c	−b	−a	a	−d	c	d	b	−d	c	−b	a	−a	c
−c	a	−a	−b	−c	d	b	−d	−c	d	a	−a	b	c	d	−b	d	−c	a	a	a	−b	c	−d	b	d	−c	−b	−a	a	−b	c	d	−b	−d	−c
−a	−c	a	d	−b	−c	−c	b	−d	−a	d	a	d	b	c	−c	−b	d	a	a	a	−d	−b	c	−c	b	d	a	−b	−a	d	−b	c	−c	−b	−d
a	−a	−c	−c	d	−b	−d	−c	b	a	−a	d	c	d	b	d	−c	−b	a	a	a	c	−d	−b	d	−c	b	−a	a	−b	c	d	−b	−d	−c	−b
b	−d	−c	−c	a	−a	−b	−c	d	−b	d	−c	d	a	−a	b	c	d	b	d	−c	a	a	a	−b	c	−d	−b	−d	−c	−b	−a	a	−b	c	d
−c	b	−d	−a	−c	a	d	−b	−c	−c	−b	d	−a	d	a	d	b	c	−c	b	d	a	a	a	−d	−b	c	−c	−b	−d	a	−b	−a	d	−b	c
−d	−c	b	a	−a	−c	−c	d	−b	d	−c	−b	a	−a	d	c	d	b	d	−c	b	a	a	a	c	−d	−b	−d	−c	−b	−a	a	−b	c	d	−b
−b	−c	d	b	−d	−c	−c	a	−a	b	c	d	−b	d	−c	d	a	−a	−b	c	−d	b	d	−c	a	a	a	−b	c	d	−b	−d	−c	−b	−a	a
d	−b	−c	−c	b	−d	−a	−c	a	d	b	c	−c	−b	d	−a	d	a	−d	−b	c	−c	b	d	a	a	a	d	−b	c	−c	−b	−d	a	−b	−a
−c	d	−b	−d	−c	b	a	−a	−c	c	d	b	d	−c	−b	a	−a	d	c	−d	−b	d	−c	b	a	a	a	c	d	−b	−d	−c	−b	−a	a	−b
−d	a	−a	b	−c	−d	−b	−d	c	−c	−a	a	b	−c	d	−b	−d	−c	b	a	−a	b	c	d	b	−d	−c	a	a	a	−b	−c	d	b	−d	c
−a	−d	a	−d	b	−c	c	−b	−d	a	−c	−a	d	b	−c	−c	−b	−d	−a	b	a	d	b	c	−c	b	−d	a	a	a	d	−b	−c	c	b	−d
a	−a	−d	−c	−d	b	−d	c	−b	−a	a	−c	−c	d	b	−d	−c	−b	a	−a	b	c	d	b	−d	−c	b	a	a	a	−c	d	−b	−d	c	b
−b	−d	c	−d	a	−a	b	−c	−d	−b	−d	−c	−c	−a	a	b	−c	d	b	−d	−c	b	a	−a	b	c	d	b	−d	c	a	a	a	−b	−c	d
c	−b	−d	−a	−d	a	−d	b	−c	−c	−b	−d	a	−c	−a	d	b	−c	−c	b	−d	−a	b	a	d	b	c	c	b	−d	a	a	a	d	−b	−c
−d	c	−b	a	−a	−d	−c	−d	b	−d	−c	−b	−a	a	−c	−c	d	b	−d	−c	b	a	−a	b	c	d	b	−d	c	b	a	a	a	−c	d	−b
b	−c	−d	−b	−d	c	−d	a	−a	b	−c	d	−b	−d	−c	−c	−a	a	b	c	d	b	−d	−c	b	a	−a	−b	−c	d	b	−d	c	a	a	a
−d	b	−c	c	−b	−d	−a	−d	a	d	b	−c	−c	−b	−d	a	−c	−a	d	b	c	−c	b	−d	−a	b	a	d	−b	−c	c	b	−d	a	a	a
−c	−d	b	−d	c	−b	a	−a	−d	−c	d	b	−d	−c	−b	−a	a	−c	c	d	b	−d	−c	b	a	−a	b	−c	d	−b	−d	c	b	a	a	a

Source: Seberry and Yamada [166, pp. 448–449], Wiley.

Four suitable matrices are used in Baumert–Hall and Welch arrays. For the Plotkin array, eight suitable matrices are needed. They are used in the following theorem.

Theorem 4.1 ***(Geramita and Seberry [78]):*** *Suppose there exists an orthogonal design $OD(\Sigma s_i; s_1, \ldots, s_u)$ and u suitable matrices of order m. Then there is an Hadamard matrix of order $(\Sigma s_i)m$.*

Table 4.5 The relationship between "plug-in" and "plug-into" matrices.

Matrices to "plug-in"	Matrices to "plug-into"
Hardest to find	
Williamson	$OD(4t; t, t, t, t)$ or $BH(4t; t, t, t, t)$
Williamson type	$OD(4t; t, t, t, t)$ or $BH(4t; t, t, t, t)$
Good, best, G-matrices	$OD(4t; t, t, t, t)$
Propus	Propus array
8-Williamson	$OD(8t; t, t, t, t, t, t, t, t)$
8-Williamson type	$OD(8t; t, t, t, t, t, t, t, t)$
Easiest to find	
Suitable matrices	$OD(2^t n; u_1, u_2, \dots, u_t)$
4 Circulant suitable matrices	Goethals–Seidel
4 Type 1 suitable matrices	J. Wallis–Whiteman
4 Lushshie matrices	Balonin–Seberry array
M-structures	
Amicable sets and Kharaghani matrices	

Source: Seberry and Yamada [166, table 3.1, p. 451], Wiley.

If some of the suitable matrices have entries $0, +1, -1$, then weighing matrices rather than Hadamard matrices could have been constructed.

An overview of matrices to "plug-in" and "plug-into" is given in Table 4.5.

One of the most prolific method for constructing matrices to "plug-into" uses T-matrices or T-sequences. The other, using Welch and Ono–Sawade–Yamamoto arrays, is described in the next Section 4.3.

4.3 Welch and Ono–Sawade–Yamamoto Arrays

We modify a construction of Turyn to obtain the first theorem which capitalized on Welch arrays. The $OD(4s; u_1, \dots, u_n)$ of the next theorem is a Welch or Ono–Sawade–Yamamoto array.

Theorem 4.2 ***(Seberry et al. [165, 188]):*** *Suppose there are T-matrices of orthogonal design order t. Further suppose there is an $OD(4s; u_1, \dots, u_n)$ constructed of 16 circulant (or type 1) $s \times s$ blocks on the variables $x_1, \dots, x_n$. Then there is an $OD(4st; tu_1, \dots, tu_n)$. In particular, if there is an $OD(4s; s, s, s, s)$ constructed of 16 circulant (or type 1) $s \times s$ blocks then there is an $OD(4st; st, st, st, st)$.*

Proof: We write the OD as (N_{ij}), $i, j = 1, 2, 3, 4$, where each N_{ij} is circulant (or type 1). Hence, we are considering the OD purely as an M-structure. Since we have an OD

$$N_{i1}N_{j1}^{\mathrm{T}} + N_{i2}N_{j2}^{\mathrm{T}} + N_{i3}N_{j3}^{\mathrm{T}} + N_{i4}N_{j4}^{\mathrm{T}} = \begin{cases} \sum_{k=1}^{4} u_k x_k^2 I_s, & i = j, \\ 0, & i \neq j. \end{cases}$$

Suppose the T-matrices are T_1, T_2, T_3, T_4. Then form the matrices

$$A = T_1 \times N_{11} + T_2 \times N_{21} + T_3 \times N_{31} + T_4 \times N_{41}$$
$$B = T_1 \times N_{12} + T_2 \times N_{22} + T_3 \times N_{32} + T_4 \times N_{42}$$

$$C = T_1 \times N_{13} + T_2 \times N_{23} + T_3 \times N_{33} + T_4 \times N_{43}$$
$$D = T_1 \times N_{14} + T_2 \times N_{24} + T_3 \times N_{34} + T_4 \times N_{44}.$$

Now

$$AA^{\top} + BB^{\top} + CC^{\top} + DD^{\top} = t \sum_{k=1}^{4} u_k x_k^2 I_{st},$$

and since A, B, C, D are type 1, they can be used in the J. Wallis–Whiteman generalization of the Goethals–Seidel array to obtain the result. □

We use the Welch, $WL(5)$, and the Ono–Sawade–Yamamoto array, $WL(9)$, to see

Corollary 4.1: *Suppose T-matrices exist of order t. Then there are orthogonal designs $OD(20t; 5t, 5t, 5t, 5t)$ and $OD(36t; 9t, 9t, 9t, 9t)$.*

4.4 Regular Representation of a Group and *BHW*(*G*)

Bell and Đoković in [12] give infinitely many Welch arrays, $WL(t)$, (but which they call Baumert–Hall–Welch arrays) for t even. We choose to concentrate on the case for t odd. The existence of WL arrays for odd t is an important research question (see Appendix C, Problem C.15).

Let G be a finite group of order n, written multiplicatively with identity element 1. The inversion $x \to x^{-1}$ on G extends to an involution $*$ of the group ring $\boldsymbol{Z}[G]$ as

$$\left(\sum_{x\in G} u_x x\right)^* = \sum_{x\in G} u_x x^{-1}, \quad u_x \in \boldsymbol{Z}.$$

Let $M_k(\boldsymbol{Z}[G])$ be the set of the matrices of order k with entries in $\boldsymbol{Z}[G]$. If $A \in M_k(\boldsymbol{Z}[G])$, then the matrix A^* is obtained from A by transposing A and applying $*$ to each of the entries.

Definition 4.4: For the element $z = \sum_{x\in G} u_x x$, the set $X = \{x \in G; u_x \neq 0\}$ is a subset of G and is called a **support** of z. If $X = G$, then z has **full support**. Two elements of $Z[G]$ are said to be **disjoint** if their support are disjoint sets. Two matrices $A, B \in M_k(\boldsymbol{Z}[G])$ are said to be **disjoint** if the entries $A(i,j)$ and $B(i,j)$ are disjoint for all i and j.

Definition 4.5: Let G be a finite group of order n. **An ordered set for a Welch array on a group** G, $\boldsymbol{BHW(G)}$, is the ordered set (A_1, A_2, A_3, A_4) of the matrices of order 4 over $\boldsymbol{Z}[G]$ satisfying the following conditions:

i) A_i and A_j are disjoint for $i \neq j$,
ii) $A_i A_i^* = nI, \quad (i = 1, 2, 3, 4)$,
iii) $A_i A_j^* + A_j A_i^* = 0$ for $i \neq j$.

The regular representation of G extends naturally to the representation ψ of the group ring $\boldsymbol{Z}[G]$, $\psi : \boldsymbol{Z}[G] \to M_n(\boldsymbol{Z})$. The representation ψ extends to an embedding of the ring $M_k(\boldsymbol{Z}[G])$ into $M_{nk}(\boldsymbol{Z})$. If $A \in M_k(\boldsymbol{Z}[G])$, then $\psi(A)$ is obtained from A by replacing each entry $A(i,j) \in \boldsymbol{Z}[G]$ by the matrix $\psi(A(i,j)) \in M_n(\boldsymbol{Z})$.

Theorem 4.3: *Let G be a finite group of order n and let $(A_1, A_2, A_3, A_4) \in BHW(G)$. Then the matrix*

$$A = \sum_{i=1}^{4} x_i \psi(A_i)$$

is $OD(4n; n, n, n, n)$ with variables x_1, x_2, x_3, x_4.

We note that in [12] the following pretty description of the first two known Welch array on the group G is given. We introduce the following four auxiliary examples:

$$\sigma_1 = \begin{bmatrix} 1 & 0 & 0 & 0 \\ 0 & 1 & 0 & 0 \\ 0 & 0 & 1 & 0 \\ 0 & 0 & 0 & 1 \end{bmatrix}, \qquad \sigma_2 = \begin{bmatrix} 0 & 1 & 0 & 0 \\ - & 0 & 0 & 0 \\ 0 & 0 & 0 & - \\ 0 & 0 & 1 & 0 \end{bmatrix},$$

$$\sigma_3 = \begin{bmatrix} 0 & 0 & 1 & 0 \\ 0 & 0 & 0 & 1 \\ - & 0 & 0 & 0 \\ 0 & - & 0 & 1 \end{bmatrix}, \qquad \sigma_4 = \begin{bmatrix} 0 & 0 & 0 & 1 \\ 0 & 0 & - & 0 \\ 0 & 1 & 0 & 0 \\ - & 0 & 0 & 0 \end{bmatrix},$$

We have $(\sigma_1, \sigma_2, \sigma_3, \sigma_4) \in BHW(G)$ where G is the trivial group. $\sum_{i=1}^{4} x_i\sigma_i$ is $OD(4; 1, 1, 1, 1)$.

The first (nontrivial) example of a Welch array was constructed by L.R. Welch in 1971 (see [78]). In his example $G = C_5$ is a cyclic group of order 5. Let x be a generator of C_5. Define matrices A_i by

$$A_1 = \begin{bmatrix} 0 & 1 & x-x^4 & -x-x^4 \\ 1 & 0 & x+x^4 & x^4-x \\ x^4-x & x+x^4 & 0 & 1 \\ -x-x^4 & x-x^4 & 1 & 0 \end{bmatrix},$$

$$A_2 = \begin{bmatrix} x^2+x^3 & 0 & -1 & x^3-x^2 \\ 0 & x^2+x^3 & x^2-x^3 & 1 \\ -1 & x^3-x^2 & -x^2-x^3 & 0 \\ x^2-x^3 & 1 & 0 & -x^2-x^3 \end{bmatrix},$$

and

$$A_3 = A_1\sigma_3,$$
$$A_4 = A_2\sigma_3.$$

Since σ_3 commutes with A_1 and anti-commutes with A_2, it is easy to verify that $(A_1, A_2, A_3, A_4) \in BHW(C_5)$. From Theorem 4.3, we obtain the Welch array.

The second example of a Welch array was constructed by Ono, Sawade, and Yamamoto in 1984 (see [142, 165, 166]) (T. Ono and K. Sawade. Baumert–Hall–Welch array of order 36. Personal communication, translation of original by Mieko Yamada, published originally in Japanese, 1984). In their example $G = C_3 \times C_3$ is the direct product of two cyclic groups of order 3 with generators x and y. The matrix

$$A_1 = A_1(x) = \begin{bmatrix} 1+x+x^2 & x^2-x & x^2-x & x^2-x \\ x-x^2 & 1+x+x^2 & x^2-x & x-x^2 \\ x-x^2 & x-x^2 & 1+x+x^2 & x^2-x \\ x-x^2 & x^2-x & x-x^2 & 1+x+x^2 \end{bmatrix}$$

is hermitian (i.e. $A_1^* = A_1$), satisfies the equation $A_1A_1^* = 9I_4$, and anti-commutes with σ_2, σ_3, and σ_4. Set

$$A_2 = A_1(y)\sigma_2, \quad A_3 = A_1(xy)\sigma_3, \quad A_4 = A_1(x^2y)\sigma_4.$$

Now it is easy to check that $(A_1, A_2, A_3, A_4) \in BHW(C_3 \times C_3)$. From Theorem 4.3, we obtain the Ono–Sawade–Yamamoto array.

5

Sequences

5.1 Notations

Table 5.1 gives the notations which are used in this chapter.

5.2 PAF and NPAF

We concentrate on the powerful construction techniques using disjoint orthogonal matrices and sequences with zero autocorrelation. Since we are at first concerned with orthogonal designs, we shall also consider sequences of commuting variables.

To clarify our notation we specify

1) T-matrices are disjoint $0, \pm 1$ matrices with $PAF = 0$ (see Definition 1.59);
2) T-sequences are disjoint 0 ± 1 sequences with $NPAF = 0$ (see Definition 1.60);
3) Turyn sequences $A = \{X, U, Y, V\}$ are $(1, -1)$ sequences with $NPAF = 0$ where
   a) for $\ell = 2m$, $8m - 6$ is the sum of 2 squares

$$\begin{aligned} X &= \{x_1 = 1, x_2, \ldots, x_m, -x_m, \ldots, -x_2, -x_1 = 1\} \\ U &= \{u_1 = 1, u_2, \ldots, u_m, -u_m, \ldots, -u_2, u_1 = 1\} \\ Y &= \{y_1, y_2, \ldots, y_{m-1}, y_m, y_{m-1}, \ldots, y_2, y_1\} \\ V &= \{v_1, v_2, \ldots, v_{m-1}, v_m, v_{m-1}, \ldots, v_2, v_1\} \end{aligned}$$

   the sequences are said to be skew.
   b) for $\ell = 2m + 1$, $8m + 2$ is the sum of 2 squares

$$\begin{aligned} X &= \{x_1 = 1, x_2, \ldots, x_m, x_{m+1}, x_m, \ldots, x_2, x_1 = 1\} \\ U &= \{u_1 = 1, u_2, \ldots, u_m, u_{m+1}, u_m, \ldots, u_2, -1\} \\ Y &= \{y_1, y_2, \ldots, y_m, -y_m, \ldots, -y_2, -y_1\} \\ V &= \{v_1, v_2, \ldots, v_m, -v_m, \ldots, -v_2, -v_1\} \end{aligned}$$

   the sequences are said to be symmetric (see Definitions 1.61 and 1.62). It is further developed in Corollaries 5.1 and 5.2).
4) 6-Turyn-type sequences are $(1, -1)$ sequences A, B, C, C, D, D, of lengths m, m, m, m, m, m with $NPAF = 0$ (this is further developed in Section 5.8).

Hadamard Matrices: Constructions using Number Theory and Algebra, First Edition. Jennifer Seberry and Mieko Yamada.

Table 5.1 Notations used in this chapter.

$\overline{x}$	Represents $-x$
$\boldsymbol{e}$	$1 \times n$ matrix of all 1's
X^*	The elements of a sequence X written in the reverse order
X/Y	The interleaving of two sequences $\{x_1, y_1, \dots, x_m, y_m\}$
(X, Y)	The adjoining of two sequences $\{x_1, x_2, \dots, x_m, y_1, y_2, \dots, y_m\}$
$\overline{X}$	Represents the sequence $-X = \{-x_1, -x_2, \dots, -x_m\}$ for the sequence $X = \{x_1, x_2, \dots, x_m\}$
N_X or $N_X(j)$ or $NPAF_X$	Nonperiodic autocorrelation function of sequence X
P_X or $P_X(j)$ or PAF_X	Periodic autocorrelation function of sequence X
$BS(m, n)$	Four sequences of elements $+1, -1$ of lengths m, m, n, n which have zero nonperiodic auto-correlation function: base sequences
6-Turyn Type	Sequences A, B, C, C, D, D with $NPAF = 0$ and lengths m, m, n, n, n, n
$BH(4n)$	Baumert–Hall array of $4n$, i.e. the $OD(4n; n, n, n, n)$
WL or $WL(t)$	A $BH(4t; t, t, t, t)$ constructed using 16 circulant or type 1 matrices
y	A Yang number
β	Is an integer in $A \cup B$ of Proposition 5.1
$\kappa = KK$	A Koukouvinos–Kounias number
$\boldsymbol{Z}$	A rational integer ring
xg_r	The sequence of integers of length r obtained from g_r by multiplying each member of g_r by x
$C(A, B)$	A matrix C with submatrices A and B
$C'(A, B) = C(C_1, C_2)$	A matrix C' with submatrices C_1 and C_2

5.3 Suitable Single Sequences

We first observe that there is a circulant Hadamard matrix of order 4. We have

$$\begin{bmatrix} - & 1 & 1 & 1 \\ 1 & - & 1 & 1 \\ 1 & 1 & - & 1 \\ 1 & 1 & 1 & - \end{bmatrix}$$

based on the single sequence – 1 1 1. Ryser's conjecture (1.20) is that this is the only circulant Hadamard matrix.

5.3.1 Thoughts on the Nonexistence of Circulant Hadamard Matrices for Orders >4

The Ryser conjecture which is still unproven is "that there are no circulant Hadamard matrices of order greater than four" has been the subject of much recent research. In the following we give some thoughts on the question.

A circulant Hadamard matrix, C written $C(1, -1)$, of order $4n$ is developed from its first row as shown

$$C = \begin{bmatrix} c_1 & c_2 & \cdots & c_{4n} \\ c_{4n} & c_1 & \cdots & c_{4n-1} \\ \vdots & \vdots & \vdots & \vdots \\ c_2 & c_3 & \cdots & c_1 \end{bmatrix}$$

it is conjectured to exist only for the order 4 Ryser [148].

That is the only existing case, up to permutation of rows or columns, is equivalent to

$$\begin{bmatrix} -1 & 1 & 1 & 1 \\ 1 & -1 & 1 & 1 \\ 1 & 1 & -1 & 1 \\ 1 & 1 & 1 & -1 \end{bmatrix}. \tag{5.1}$$

Since C has the same number of ones in each row and column it is *regular*.

We now assume that a circulant Hadamard matrix, C, that is $C(1,-1)$, exists for order $4n$, $n > 1$ first for $4n$ divisible by 8 and then for $4n \equiv 4 \pmod 8$, and we would like to show these assumptions lead to contradictions. Turyn [189] has shown that no circulant Hadamard matrix exists for $4n \equiv 0 \pmod 8$ but we also prove this.

We note that in the following discussion we will permute rows and columns of C but in no cases will rows or columns be negated.

5.3.2 SBIBD Implications

We know that our matrix C which is of order $4n$ with constant row sum, t, can be written with $4n = t^2$ or $4m^2$ as in Theorem 1.11.

Example 5.1: *Let* $\mathbf{e}$ *be a* $1 \times 4n$ *matrix of all 1s. Then* $\mathbf{e}C = t\mathbf{e}$. *Thus, using* $CC^T = 4nI_{4n}$ *we have*

$$\begin{aligned} \mathbf{e}C(\mathbf{e}C)^\top &= t^2\mathbf{e}\mathbf{e}^\top &&= 4nt^2 \text{ and} \\ \mathbf{e}CC^\top\mathbf{e}^\top &= 4n\mathbf{e}\mathbf{e}^\top &&= 16n^2. \end{aligned}$$

Thus we see that $4n = t^2$ *is a square and may be written as* $4m^2$.

Orthogonality ensures that the row sum is $\pm 2m$. If $4n = 2^a b$, for $a \geq 3$ and b both odd we have that the row length satisfies $4m^2 = 4n = t^2 = 2^a b$, a odd and b odd, which is not possible.

So we have

Theorem 5.1: *There is no circulant Hadamard matrix of order* $4n = 2^a b$, $a \geq 3$ *and* b *both odd.*

However, if $4n = 4m^2$ and the row sum is $= \pm 2m$ we have, we know [166, theorem 10.6], that a regular Hadamard matrix of order $4m^2$ and constant row sum $\pm 2m$ exists if and only if there is an *SBIBD*$(4m^2, 2m^2 \pm m, m^2 \pm m)$. (Both *SBIBD* exist: replace the 0s and 1s of an *SBIBD*$(4m^2, 2m^2 + m, m^2 + m)$ by 1s and 0s and we get the complementary *SBIBD*$(4m^2, 2m^2 - m, m^2 - m)$.)

5.3.3 From ± 1 Matrices to $\pm A, \pm B$ Matrices

The inspiring observation of Hurley et al. [101] which unfortunately does not lead to a proof of the Ryser conjecture as noted by Craigen and Jedwab [34], is that the rows and columns of the circulant Hadamard matrix, C that is $C(1,-1)$, can be reordered by choosing the rows/columns in the order

$$1, 2n+1, 2, 2n+2, 3, 2n+3, \cdots 2n, 4n.$$

This we call **Process** A, B leading to $C' = C(A, B)$.

Hence the blocks are

$$A = \begin{bmatrix} 1 & 1 \\ 1 & 1 \end{bmatrix}, \quad -A = \begin{bmatrix} -1 & -1 \\ -1 & -1 \end{bmatrix}, \quad B = \begin{bmatrix} 1 & -1 \\ -1 & 1 \end{bmatrix}, \quad \text{or} \quad -B = \begin{bmatrix} -1 & 1 \\ 1 & -1 \end{bmatrix}.$$

We write these as $\pm A$ and $\pm B$. Actually $AB = BA = 0$ but that is not used in our work. The inner product of the rows of $\pm A$ is $+2$ and the inner product of the rows of $\pm B$ is -2. The resulting matrix, of order $2n$, written as $C' = C(A, B)$ is circulant in the elements $\pm A$ and $\pm B$.

Example 5.2: *First we note that the following circulant matrix of order 12 is not an Hadamard matrix but given only to illustrate the approach we take:*

$$\begin{bmatrix}
1 & 1 & - & - & - & 1 & 1 & 1 & - & 1 & - & - \\
- & 1 & 1 & - & - & - & 1 & 1 & 1 & - & 1 & - \\
- & - & 1 & 1 & - & - & - & 1 & 1 & 1 & - & 1 \\
1 & - & - & 1 & 1 & - & - & - & 1 & 1 & 1 & - \\
- & 1 & - & - & 1 & 1 & - & - & - & 1 & 1 & 1 \\
1 & - & 1 & - & - & 1 & 1 & - & - & - & 1 & 1 \\
1 & 1 & - & 1 & - & - & 1 & 1 & - & - & - & 1 \\
1 & 1 & 1 & - & 1 & - & - & 1 & 1 & - & - & - \\
- & 1 & 1 & 1 & - & 1 & - & - & 1 & 1 & - & - \\
- & - & 1 & 1 & 1 & - & 1 & - & - & 1 & 1 & - \\
- & - & - & 1 & 1 & 1 & - & 1 & - & - & 1 & 1 \\
1 & - & - & - & 1 & 1 & 1 & - & 1 & - & - & 1
\end{bmatrix}$$

We now rearrange the rows as specified obtaining

$$\begin{bmatrix}
1 & 1 & 1 & 1 & - & - & - & 1 & - & - & 1 & - \\
1 & 1 & 1 & 1 & - & - & 1 & - & - & - & - & 1 \\
- & 1 & 1 & 1 & 1 & 1 & - & - & - & 1 & - & - \\
1 & - & 1 & 1 & 1 & 1 & - & - & 1 & - & - & - \\
- & - & - & 1 & 1 & 1 & 1 & 1 & - & - & - & 1 \\
- & - & 1 & - & 1 & 1 & 1 & 1 & - & - & 1 & - \\
1 & - & - & - & - & 1 & 1 & 1 & 1 & 1 & - & - \\
- & 1 & - & - & 1 & - & 1 & 1 & 1 & 1 & - & - \\
- & - & 1 & - & - & - & - & 1 & 1 & 1 & 1 & 1 \\
- & - & - & 1 & - & - & 1 & - & 1 & 1 & 1 & 1 \\
1 & 1 & - & - & 1 & - & - & - & - & 1 & 1 & 1 \\
1 & 1 & - & - & - & 1 & - & - & 1 & - & 1 & 1
\end{bmatrix}$$

We now replace the 2×2 *submatrices by* $\pm A$ *and* $\pm B$ *and obtain the* 6×6 *matrix:*

$$\begin{bmatrix}
+A & +A & -A & -B & -A & +B \\
-B & +A & +A & -A & -B & -A \\
-A & -B & +A & +A & -A & -B \\
+B & -A & -B & +A & +A & -A \\
-A & +B & -A & -B & +A & +A \\
+A & -A & +B & -A & -B & +A
\end{bmatrix}$$

We note the matrix is now almost circulant but with elements, $\pm A$ and $\pm B$. The process A, B cannot be repeated as the matrix is 6×6. Thus the process leads to

$$\begin{bmatrix} A & -A & -A & A & -B & B \\ -A & A & -A & -B & A & -B \\ -A & -A & A & B & -B & A \\ -B & A & -B & A & -A & -A \\ B & -B & A & -A & A & -A \\ A & B & -B & -A & -A & A \end{bmatrix}.$$

5.3.4 Matrix Specifics

We now switch to the block design notation that $4n = 4m^2$, $m \geq 2$: there are exactly $2m^2 \pm m$ +1s in each row and column of C. First we count the elements and blocks of $C' = C(A, B)$: hence the first two rows of the permuted circulant matrix, $C' = C(A, B)$, can be written, using $\overline{A}$ and $\overline{B}$ for the negatives of A and B, as

$$\underbrace{A\,A\,\cdots\,A}_{x_1}\;\underbrace{\overline{A}\,\overline{A}\,\overline{A}\,\overline{A}\cdots\overline{A}}_{x_2}\;\underbrace{B\,B\,\cdots\,B}_{x_3}\;\underbrace{\overline{B}\,\overline{B}\,\cdots\,\overline{B}}_{x_4}. \tag{5.2}$$

Suppose there are x_1 of the $A\,A\,\cdots\,A$, x_2 of the $\overline{A}\,\overline{A}\,\cdots\,\overline{A}$, x_3 of the $B\,B\,\cdots\,B$, and x_4 of the $\overline{B}\,\overline{B}\,\cdots\,\overline{B}$, where $x_i \geq 0$ for $i = 1, 2, 3$, or 4.

Then counting the order (remembering that $\pm A$ and $\pm B$ are 2×2 matrices) we have

$$2x_1 + 2x_2 + 2x_3 + 2x_4 = 4m^2. \tag{5.3}$$

The inner product of the first two rows is zero so we have

$$2x_1 + 2x_2 - 2x_3 - 2x_4 = 0. \tag{5.4}$$

Hence we have

$$x_1 + x_2 = x_3 + x_4 = m^2. \tag{5.5}$$

5.3.5 Counting Two Ways

So we have $x_1 + x_2 = m^2$ elements $\pm A$ in each row of $C(A, B)$. We also have $x_3 + x_4 = m^2$ elements $\pm B$ in each row. Similarly considering the columns we have that every column contains m^2 elements $\pm A$ and m^2 elements $\pm B$. So we are assuming a $2m^2 \times 2m^2$ matrix, $C(A, B)$, with m^2 elements $\pm A$ and m^2 elements $\pm B$ in each row and column, where $m > 1$.

Permute the columns so we can write the first row formally as

$$\overbrace{A\,A\,\cdots\,A\,\overline{A}\,\overline{A}\,\cdots\,\overline{A}}^{m^2}\;\overbrace{B\,B\,\cdots\,B\,\overline{B}\,\overline{B}\,\cdots\,\overline{B}}^{m^2}.$$

Call this matrix $C(A, B) = [C_1|C_2]$, where C_1 and C_2 are $2m^2 \times m^2$ submatrices.

We now consider the number of $\pm A$ elements in the left hand $2m^2 \times m^2$ submatrix of $C(A, B)$, C_1.

Suppose there are $z_i \geq 0$ rows of C_1, not considering the first, which contain i elements $\pm A$, $i = 0, 1, \ldots, m^2$. Call z_i the **intersection number**. From the intersection numbers the total number of rows in C_1 is

$$z_0 + z_1 + z_2 + z_3 + \cdots + z_{m^2} = 2m^2 - 1. \tag{5.6}$$

We now count the number of $\pm A$ elements in C_1 in two ways: by rows and by columns.

So the total number of elements $\pm A$ in the rows is

$$m^2 + z_1 + 2z_2 + 3z_3 + \cdots + m^2 z_{m^2}. \tag{5.7}$$

The total number of elements $\pm A$ in C_1, counting by columns, is m^4. Hence we have

$$z_1 + 2z_2 + 3z_3 + \cdots + m^2 z_{m^2} = m^4 - m^2. \tag{5.8}$$

From Eqs. (5.6) and (5.8) we see that a possible solution is

$$z_{m^2} = m^2 - 1,\ z_1 = z_2 = \cdots z_{m^2-1} = 0 \text{ and } z_0 = m^2.$$

We have Drs Richard Brent, John Dillon, and Paul Leopardis to thank for the observation that there are further solutions which still need to be analyzed.

Remark 5.1: We now consider an unproven possibility for $C(A, B)$. We again note $C(A, B)$ is circulant in $\pm A$ and $\pm B$. $C(A, B)$ is orthogonal not only on the underlying ± 1 matrix but also formally in terms of a $\pm A, \pm B$ matrix. If we could map $-A$ to -1 and $-B$ to -1 we would have again a ± 1 circulant orthogonal matrix and thus able to repeat process A, B. This is not quite the solution as we have only halved the size of the circulant matrix. However, if we could show how to reduce the size to one quarter we would have an iterative process to keep reducing the size by 4 until we got to $4m^2$, m odd. We would like to be able to show that this is not possible.

5.3.6 For *m* Odd: Orthogonal Design Implications

An orthogonal design or OD, X of order n and type $(s_1, \ldots, s_t)$, s_i positive integers, is an $n \times n$ matrix with entries $\{0, \pm x_1, \ldots, \pm x_t\}$ (the x_i are commuting indeterminates) satisfying

$$XX^{\top} = \left(\sum_{i=1}^{t} s_i x_i^2 \right) I_n,$$

where I_n is the identity matrix of order n. This is denoted as $OD(n; s_1, \ldots, s_t)$.

Such generically orthogonal matrices have played a significant role in the construction of Hadamard matrices (see, e.g. Geramita and Seberry [78] and they have been extensively used in the study of weighing matrices.

When $n \equiv 2 \pmod 4$, $n > 2$, $\sum_{i=1}^{t} s_i \leq n - 1$, see [78, proposition 2.5].

Now, considering, orthogonal designs, we see have assumed a $2m^2 \times 2m^2$ orthogonal matrix, $C(A, B)$, containing m^2 elements $\pm A$ and m^2 elements $\pm B$ per row and column. In Section 5.3.5 we used $AB = 0$, which is stronger than saying we need A and B to be commuting variables. So we need an $OD(2m^2; m^2, m^2)$ with $m > 1$ odd. Geramita and Seberry [78, proposition 2.5] assert this is impossible for odd m given that C_{11} is comprised only of $\pm A$'s and C_{12} is comprised only of $\pm B$'s.

5.3.7 The Case for Order 16

We write H for and Hadamard matrix of order 16 comprising 32 blocks of A's and 32 blocks of B's

$$H = \begin{bmatrix} \overline{A} & \overline{B} & A & B & A & B & A & B \\ \overline{B} & \overline{A} & B & A & B & A & B & A \\ A & B & \overline{A} & \overline{B} & A & B & A & B \\ B & A & \overline{B} & \overline{A} & B & A & B & A \\ A & B & A & B & \overline{A} & \overline{B} & A & B \\ B & A & B & A & \overline{B} & \overline{A} & B & A \\ A & B & A & B & A & B & \overline{A} & \overline{B} \\ B & A & B & A & B & A & \overline{B} & \overline{A} \end{bmatrix} \cong \begin{bmatrix} \overline{A} & A & A & A & \overline{B} & B & B & B \\ A & \overline{A} & A & A & B & \overline{B} & B & B \\ A & A & \overline{A} & A & B & B & \overline{B} & B \\ A & A & A & \overline{A} & B & B & B & \overline{B} \\ \overline{B} & B & B & B & \overline{A} & A & A & A \\ B & \overline{B} & B & B & A & \overline{A} & A & A \\ B & B & \overline{B} & B & A & A & \overline{A} & A \\ B & B & B & \overline{B} & A & A & A & \overline{A} \end{bmatrix}.$$

This is equivalent to a block circulant Hadamard matrix with 4×4 blocks of A's and B's. Replacing the A's and B's by the appropriate ± 1 2×2 matrices. The block circulant Hadamard matrix of order 4 can be written as

$$\begin{bmatrix} A & B \\ B & A \end{bmatrix}$$

which uses only the trivial $SBIBD(4,3,2)$.

There are other possibilities for the first row from choosing the $\overline{B}$ to be in another position. (Question: Does this have any effect on the equivalence class of the matrix of order 16?) This leads us to the proposition.

Conjecture 5.1: There is no circulant Hadamard matrix using 2×2 block structures

$$\begin{bmatrix} 1 & 1 \\ 1 & 1 \end{bmatrix}, \begin{bmatrix} -1 & -1 \\ -1 & -1 \end{bmatrix}, \begin{bmatrix} 1 & -1 \\ -1 & 1 \end{bmatrix} \text{ or } \begin{bmatrix} -1 & 1 \\ 1 & -1 \end{bmatrix}$$

except for order 16.

5.4 Suitable Pairs of NPAF Sequences: Golay Sequences

We recall from Section 1.10.2 that if $X = \{\{a_1, \dots, a_n\}, \{b_1, \dots, b_n\}\}$ are two sequences where $a_i, b_j \in \{1, -1\}$ and $N_X(j) = 0$, N_X the nonperiodic autocorrelation function, for $j = 1, \dots, n-1$, then the sequences in X are called Golay complementary sequences of length n or Golay sequences or Golay pairs, e.g. (writing – for minus 1). We also use $NPAF_X(j) = 0$ for the nonperiodic autocorrelation function.

We would like to use Golay sequences to construct other orthogonal designs, but first, a recap of Lemma 1.23.

Let $X = \{\{a_1, \dots, a_n\}, \{b_1, \dots, b_n\}\}$ be Golay complementary sequences of length n. Suppose k_1 of the a_i are positive and k_2 of the b_i are positive. Then

$$n = (k_1 + k_2 - n)^2 + (k_1 - k_2)^2,$$

and n is even.

5.5 Current Results for Golay Pairs

In the remainder of this chapter Golay pairs and ternary pairs will be assumed to have nonperiodic autocorrelation function, NPAF, zero unless it is specifically stated that periodic autocorrelation function is intended.

Through extended calculations made by hand, Golay demonstrated the existence of two inequivalent pairs at length 10 and a pair at length 26. He also gave rules of composition for forming pairs of lengths $2n$ and $2mn$ from existing pairs of length m and n. His constructions for lengths 2^k give all existing pairs for $0 \leq k \leq 6$.

In his 1974 paper, Turyn [194] gave a construction for forming pairs of length mn from existing pairs of length m and n (see Section 1.10.1).

The first exhaustive search for Golay pairs was conducted at length 26 (Jauregui [105]), taking 75 hours to confirm the single example of inequivalent pairs. In his 1977 master's thesis, Andres [2] showed that a further reduction modulo 2 enables an initial search involving $2^{\frac{n}{2}}$ cases. A further reduction modulo 4 was applied for examples surviving this test. He used one of the equivalences, bringing this to a $2^{\frac{n}{2}-1}$ search, reducing the time taken at $n = 26$ to 1 minute. His work showed nonexistence of pairs at lengths 34, 50, and 58 and produced complete lists of representatives at lengths 8, 10, 16, 20, and 32. Later work by James [104] established the nonexistence of pairs at length 68. Đoković [60] demonstrated how to choose a canonical pair from each class of equivalent pairs. He conducted exhaustive searches at lengths 32 and 40, producing complete lists of canonical pairs.

Table 5.2 Some small weight Golay sequences.

Length	No. of pairs	Equivalence classes	Primitive pairs
1	4	1	1
2	8	1	0
4	32	1	0
8	192	5	0
10	128	2	2
16	1536	36	0
20	1088	25	1
26	64	1	1
32	15,360	336	0
34	0	0	0
40	9728	220	0
50	0	0	0
52	512[a]	12[a]	0[a]
58	0	0	0
64	184,320[a]	3840[a]	0[a]
68	0	0	0
74	0[a]	0[a]	0[a]
80	102,912[a]	?	0[a]
82	0[a]	0[a]	0[a]

Source: Seberry [158, table 1, appendix G], Reproduced with permission of Springer Nature.
[a] Borwein and Ferguson [16]

The work of Borwein and Ferguson [16] outlines improvements which may be made to Andres' algorithm, enabling a $2^{\frac{n}{2}-5}$ search at length 82 with a running time of two weeks. Exhaustive searches have been conducted at all allowable lengths under 100, confirming earlier work and showing the nonexistence of pairs at lengths 74 and 82.

Recent search results by Borwein and Ferguson [16] are summarized in Table 5.2. For all lengths other than 1, 2, 4, and 80, complete lists of canonical pairs were compiled by the search program. The total numbers of pairs agree exactly with those obtained by compositions from the primitive pairs. At length 80, the search was restricted to canonical pairs for which no conjugate is H-regular. The total number of pairs is that determined by composition from the two primitive pairs at length 10 and the single primitive pair of length 20. (The superscript "b" in Table 5.2 indicates work done at Simon Frazer University.)

A Golay pair is said to be **primitive** if it cannot be derived through composition from pairs of shorter lengths. A theory for producing pairs of length 2^n from a primitive pair of length n is developed in [16].

Golay complementary sequences contain no zeros but considerable effort has also been devoted to ternary complementary sequences which still have zero nonperiodic autocorrelation function but contain elements $\{0, \pm 1\}$. Golay complementary sequences of length n are also known as Golay numbers of length n, i.e. n is a Golay number. Craigen and Koukouvinos [36] have made a theory for these ternary sequences and given a table for their existence for some smaller lengths and weights.

5.6 Recent Results for Periodic Golay Pairs

Đoković and Kotsireas [65] show that if $v > 1$ is a periodic Golay number then v is even, it is a sum of 2 squares and satisfies the Arasu–Xiang condition [65, theorem 2, p. 525]. They cite new lengths 34, 50, 58, 74, 82, 122, 136, 202, 226 as those for which periodic sequences exist. Periodic sequences can be used in the construction of T-matrices. The following list is all numbers $2n$ in the range $1, 2, \ldots, 300$ which satisfy the three necessary conditions and for which the question, whether they are periodic Golay numbers, remains open:

$$90, 106, 130, 146, 170, 178, 180, 194, 212, 218, 234, 250, 274, 290, 292, 298.$$

This list may be useful to readers interested in constructing new periodic Golay pairs or finding new periodic Golay numbers.

5.7 More on Four Complementary Sequences

We now discuss other sequences with zero autocorrelation function and expand a little on Section 1.10. We note that using four suitable matrice or equivalently four suitable sequences has proved by far the most useful technique for finding ad hoc order Hadamard matrices, e.g. orders $92, 188, 428, \ldots$.

We now propose to use R.J. Turyn's [188] idea of using structured m-complementary sequences to construct plug-into arrays or full orthogonal designs.

Lemma 5.1: *Consider four $(1, -1)$ sequences $A = \{X, U, Y, V\}$, where*

$$\begin{aligned}
X &= \{x_1 = 1, x_2, x_3, \ldots, x_m, h_m x_m, \ldots, h_3x_3, h_2x_2, h_1x_1 = -1\},\\
U &= \{u_1, u_2, u_3, \ldots, u_m, f_m u_m, \ldots, f_3u_3, f_2u_2, f_1u_1 = 1\},\\
Y &= \{y_1, y_2, \ldots, y_{m-1}, y_m, g_{m-1}y_{m-1}, \ldots, g_3y_3, g_2y_2, g_1y_1\},\\
V &= \{v_1, v_2, \ldots, v_{m-1}, v_m, e_{m-1}v_{m-1}, \ldots, e_3v_3, e_2v_2, e_1v_1\}.
\end{aligned}$$

Then $N_A = 0$ implies that $h_i = f_i$, for $i \geq 2$, and $g_j = e_j$, for $j \geq 1$. Here

$$\begin{aligned}
8m - 2 = {}&\left(\sum_{i=1}^{m}(x_i + x_ih_i)\right)^2 + \left(\sum_{i=1}^{m}(u_i + u_if_i)\right)^2 + \left(y_m + \sum_{i=1}^{m-1}(y_i + y_ig_i)\right)^2\\
&+ \left(v_m + \sum_{i=1}^{m-1}(v_i + v_ie_i)\right)^2.
\end{aligned}$$

Proof: We note that if a, b, x, y, z are all ± 1, $a + b \equiv ab + 1 \pmod 4$, and so $x + xyz \equiv y + z \pmod 4$, Clearly, $N_A(2m - 1) = 0$ gives $-h_1 = f_1 = 1$, and

$$\begin{aligned}
N_A(2m - 2) &= x_1x_2h_2 + x_2h_1x_1 + f_2u_1u_2 + u_2f_1u_1 + g_1y_1^2 + e_1v_1^2\\
&\equiv h_1 + h_2 + f_1 + f_2 + g_1 + e_1 \pmod 4\\
&\equiv h_2f_2 + g_1e_1 + 2 \pmod 4\\
&\equiv 0 \pmod 4.
\end{aligned}$$

This gives $h_2f_2 = g_1e_1$. We proceed by induction to show that $h_if_i = g_{i-1}e_{i-1}$ for all $i \leq m$.

Assume $h_if_i = g_{i-1}e_{i-1}$, i.e. $h_i + f_i + g_{i-1} + e_{i-1} \equiv 0 \pmod 4$ for all $i < k \le m$. Now consider

$$\begin{aligned}
N_A(2m-k) &= (x_1x_kh_k + x_2x_{k-1}h_{k-1} + \cdots + x_{k-1}x_2h_2 + x_kx_1h_1)\\
&\quad +(u_1u_kf_k + u_2u_{k-1}f_{k-1} + \cdots + u_{k-1}u_2f_2 + u_ku_1f_1)\\
&\quad +(y_1y_{k-1}g_{k-1} + y_2y_{k-2}g_{k-2} + \cdots + y_{k-2}y_2g_2 + y_{k-1}y_1g_1)\\
&\quad +(v_1v_{k-1}e_{k-1} + v_2v_{k-2}e_{k-2} + \cdots + v_{k-2}v_2e_2 + v_{k-1}v_1e_1)\\
&\equiv h_1 + \cdots + h_k + f_1 + \cdots + f_k + g_1 + \cdots + g_{k-1}\\
&\quad +e_1 + \cdots + e_{k-1} \pmod 4\\
&\equiv h_kf_k + g_{k-1}e_{k-1} + 2 \pmod 4\\
&\equiv 0 \pmod 4.
\end{aligned}$$

This gives the result for all $i \le m$. Suppose $k = m + j > m$. Then,

$$\begin{aligned}
N_A(2m-k) &= (x_1x_{m-j+1} + \cdots + x_jx_m) + (x_{j+1}h_mx_m + \cdots + x_mh_{j+1}x_{j+1})\\
&\quad +(h_mx_mh_jx_j + \cdots + h_{m-j+1}x_{m-j+1}h_1x_1)\\
&\quad +(u_1u_{m-j+1} + \cdots + u_ju_m) + (u_{j+1}f_mu_m + \cdots + u_mf_{j+1}u_{j+1})\\
&\quad +(f_mu_mf_ju_j + \cdots + f_{m-j+1}u_{m-j+1}f_1u_1)\\
&\quad +(y_1y_{m-j+1} + \cdots + y_jy_m)\\
&\quad +(y_{j+1}g_{m-1}y_{m-1} + \cdots + y_{m-1}g_{j+1}y_{j+1})\\
&\quad +(y_mg_jy_j + \cdots + g_{m-j+1}y_{m-j+1}g_1y_1)\\
&\quad +(v_1v_{m-j+1} + \cdots + v_jm_j)\\
&\quad +(v_{j+1}e_{m-1}v_{m-1} + \cdots + v_{m-1}e_{j+1}v_{j+1})\\
&\quad +(v_me_jv_j + \cdots + e_{m-j+1}v_{m-j+1}e_1v_1)\\
&\equiv h_1f_1 + h_{m-j+1}f_{m-j+1} \pmod 4\\
&\equiv h_{m-j+1}f_{m-j+1} - 1 \pmod 4
\end{aligned}$$

Hence $h_{m-j+1}f_{m-j+1} = 1$. So in general $h_if_i = 1$ and $e_ig_i = 1$.

The last result follows by Lemma 1.22. □

Corollary 5.1: *Consider four $(1, -1)$ sequences $A = \{X, U, Y, V\}$, where*

$$\begin{aligned}
X &= \{x_1 = 1, x_2, x_3, \ldots, x_m, -x_m, \ldots, -x_3, -x_2, -x_1 = -1\},\\
U &= \{u_1 = 1, u_2, u_3, \ldots, u_m, f_mu_m, \ldots, f_3u_3, f_2u_2, f_1u_1 = 1\},\\
Y &= \{y_1, y_2, \ldots, y_{m-1}, y_m, y_{m-1}, \ldots, y_3, y_2, y_1\},\\
V &= \{v_1, v_2, \ldots, v_{m-1}, v_m, e_{m-1}v_{m-1}, \ldots, e_3v_3, e_2v_2, e_1v_1\}.
\end{aligned}$$

Then $N_A = 0$ implies that all e_i are $+1$ and all f_i, $i \ge 2$, are -1. Here $8m - 6$ is the sum of 2 squares.

Similarly we can prove

Corollary 5.2: *Consider four* $(1, -1)$ *sequences* $A = \{X, U, Y, V\}$, *where*

$$X = \{x_1 = 1, x_2, x_3, \ldots, x_m, x_{m+1}, x_m, \ldots, x_3, x_2, x_1 = 1\},$$
$$U = \{u_1 = 1, u_2, u_3, \ldots, u_m, u_{m+1}, f_m u_m \ldots, f_3 u_3, f_2 u_2, -1\},$$
$$Y = \{y_1, y_2, \ldots, y_m, -y_m, \ldots, -y_2, -y_1\},$$
$$V = \{v_1, v_2, \ldots, v_m, e_m v_m, \ldots, e_2 v_2, e_1 v_1\}.$$

which have $N_A = 0$. *Have* $e_i = -1$ *for all i and* $f_i = +1$ *for all i. Here* $8m + 2$ *is the sum of 2 squares.*

Sequences such as those described in the Corollaries 5.1 and 5.2 are Turyn sequences of length ℓ (see Definition 1.62). A sequence $X = \{x_1, \ldots, x_n\}$ is called skew if n is even and $x_i = -x_{n-i+1}$ and symmetric if n is odd and $x_i = x_{n-i+1}$ (see Definition 1.61).

Lemma 5.2: *There exist Turyn sequences of lengths* 2, 4, 6, 8, 3, 5, 7, 13, *and* 15.

Proof: Consider

$$\ell = 2 : X = \{\{1-\},\ \{11\},\ \{1\},\ \{1\}\}$$
$$\ell = 4 : X = \{\{11--\},\ \{11-1\},\ \{111\},\ \{1-1\}\}$$
$$\ell = 6 : X = \{\{111---\},\ \{11-1-1\},\ \{11-11\},\ \{11-11\}\}$$
$$\ell = 8 : X = \{\{11-1-1--\},\ \{1111---1\},\ \{111-111\},\ \{1--1--1\}\}$$
$$\ell = 3 : X = \{\{111\},\ \{11-\},\ \{1-\},\ \{1-\}\}$$
$$\ell = 5 : X = \{\{11-11\},\ \{1111-\},\ \{11--\},\{11--\}\}$$
$$\ell = 7 : X = \{\{111-111\},\ \{11---1-\},\ \{11-1--\},\ \{11-1--\}\}$$
$$\ell = 13 : X = \{\{1111-1-1-1111\},\ \{111--1-1--11-\},$$
$$\{111-11--1---\},\ \{111--1-11---\}\},$$

or

$$X = \{\{111-11-11-111\},\ \{111-1-1-11-\},$$
$$\{11-1-111\},\{1111-1-1----\}\}$$
$$\ell = 15 : X = \{\{11-111-1-111-11\},\ \{111-11---11-11-\},$$
$$\{111--1-11----\},\ \{1----1-1-1111-\}\}.$$

□

Remark 5.2: These sequences were constructed using the Research School of Physical Sciences, Australian National University, DEC-10 System in 1972–1973.

A complete computer search in the case of $\ell = 9, 10, 14$, and 16 gave no solution for any decomposition into squares. The results are listed in Table 5.3.

Edmondson et al. [67] used a combination of mathematics and computer search to show there are no further Turyn sequences of length $n \leq 43$. We note that at this time, year 2017, about length 43 is the upper limit for these types of computer searches.

Conjecture 5.2: The lengths 2, 4, 6, 8, 3, 5, 7, 13, 15 are the only lengths for Turyn sequences.

Four Turyn sequences are used in Lemma 5.3 to construct base sequences.

Table 5.3 Turyn sequences for $4\ell - 6 = x^2 + y^2$.

ℓ	$4\ell - 6 = x^2 + y^2$	Result
6	$18 = 3^2 + 3^2$	Yes
8	$26 = 1^2 + 5^2$	Yes
10	$34 = 3^2 + 5^2$	None
12	$42 \neq x^2 + y^2$	No
14	$50 = 1^2 + 7^2$	None
	$= 5^2 + 5^2$	None
16	$58 = 3^2 + 7^2$	None
18	$66 \neq x^2 + y^2$	No

Source: Seberry and Yamada [166, table 5.2, p. 479], Wiley.

5.8 6-Turyn-Type Sequences

Definition 5.3: 6-Complementary sequences A, B, C, C, D, and D of lengths m, m, m, m, n, and n with elements $0, \pm 1$, and $NPAF = 0$ will be called **6-Turyn-type sequences**.

Kharaghani and Tayfeh-Rezaie [108] in 2004 found six compatible ± 1 complementary sequences with zero NPAF of lengths 36, 36, 36, 36, 35, 35 which can be used in the following theorem to construct Hadamard matrices of order $428 = 4 \times 107$ and $BH(428; 107, 107, 107, 107)$. We quote from Kharaghani and Tayfeh-Rezaie [109, p. 5]:

The following solution was implemented on a cluster of sixteen 2.6 GHz PC's for

$$214 = x^2 + y^2 + z^2 + 2w^2$$

with $x = 0, y = 6, z = 8$, and $w = 5$ and found the following solution after about 12 hours of computation.

$$A = (+++----++-+-+-----++++-++-++++----+-),$$
$$B = (+-+++++--+-+--+--++--++++-++++---++-),$$
$$C = (+-+++++-+--++++-+++-++--+++-+--+---+),$$
$$D = (+++-+-----++--+-+++--+-+-+++-++++-+).$$

Hence:

Theorem 5.2: *Suppose there exist 6-Turyn-type sequences of lengths m, m, m, m, n, n, that is 6 suitable compatible ± 1 complementary sequences with zero NPAF and lengths m, m, m, m, n, n called A, B, C, C, D, D. Then there exist $BH(4(2m+n); 2m+n, 2m+n, 2m+n, 2m+n)$ and an Hadamard matrix of order $4(2m+n)$.*

Proof: Let 0_t be the sequences of t zeros. Write X, Y, where X is the sequence $\{x_1, \dots, x_p\}$ and Y is the sequence $\{y_1, \dots, y_q\}$ for the sequence $\{x_1, x_2, \dots, x_p, y_1, y_2, \dots, y_q\}$ of length $p + q$. Then the required T-sequence for the constructions are

$$\left\{\frac{1}{2}(A+B), 0_{m+n}\right\}, \left\{\frac{1}{2}(A-B), 0_{m+n}\right\}, \left\{\ 0_{2m}, D\right\}, \left\{\ 0_m, C, 0_n\right\}.$$

These may be used in Theorem 3.10. □

A clear description of the proof appears in T. Vis: www.ucdenver/~tvis/coursework/Hadamard.pdf accessed on 2 February 2018.

5.9 Base Sequences

Definition 5.4: Four sequences of elements $+1, -1$ of lengths m, m, n, n which have zero nonperiodic autocorrelation function are called **base sequences**. We denote this by saying $BS(m, n)$ is not empty.

In Table 5.4 base sequences are displayed for lengths $m+1, m+1, m, m$ for $m+1 \in \{2, 3, ..., 30\}$. If X and Y are Golay sequences $\{1, X\}, \{1, -X\}, \{Y\}, \{Y\}$ are base sequences of lengths $m+1, m+1, m, m$. So base sequences exist for all $m = 2^a 10^b 26^c$, a, b, c nonnegative integers. The case for $m = 17$, was found by A. Sproul and J. Seberry, for $m = 23$, by R. Turyn, for $m = 22, 24, 26, 27, 28$, by Koukouvinos et al. [118], for $m = 33, 34, 35$ by Kounias and Sotirakoglou [121] and by Đoković [61, 63] for $m = 37, 38, 39$. These sequences are also discussed in Geramita and Seberry [78, pp. 129–148].

These results and the importance of base sequences for Yang's constructions led to the following conjecture by Đoković [63] called the BSC$(n+1, n)$ or *BSC* conjecture.

Conjecture 5.5 (Đoković/Yang base sequence conjecture): Base sequences $BS(n+1, n)$ exist for all integers n.

The base sequence conjecture (BSC) is true for all $n \leq 39$ [63] and when n is a Golay number. It is worth pointing out that BSC is stronger than the famous Hadamard matrix conjecture. In order to demonstrate the abundance of base sequences, Đoković [61] has attached to $BS(n+1, n)$ a graph Γ_n and computed the Γ_n for $n \leq 35$.

We first note that the following result is due to Kharaghani and Tayfeh-Rezaie [109, p. 5]. It caused a resurgence in interest in sequences as it led, more than a 100 years after Hadamard's work, to a computer proof for the existence of an Hadamard matrix of order 428.

Corollary 5.3: *There are base sequences of lengths* 71, 71, 36, 36 *and therefore T-sequences of length* 107.

Corollary 5.4: *There is a Hadamard matrix of order* 428 *and Baumert–Hall array* $BH(428; 107, 107, 107, 107)$.

Proposition 5.1: Combining the results of this section and Section 3.5, we have Baumert–Hall arrays of order

i) $A = \{1 + 2^a 10^b 26^c, a, b, c$ nonnegative integers$\}$,
ii) $B = \{1, 3, \ldots, 33, 37, 41, 47, 51, 53, 59, 61, 65, 81, 101, 107\}$,
iii) 5β and 9β where $\beta \in A \cup B$,
iv) $2(q+1)\beta$, $q \equiv 3 \pmod 8$ a prime power, $\beta \in A \cup B$.

We shall see that this means

Theorem 5.3: *There are Hadamard matrices of order* $4yh(r+s)w$ *where* y *is a Yang number, and there exist* $WL(h)$, $BS(r, s)$ *and Williamson type matrices of order* w *[62].*

5.10 Yang-Sequences

Base sequences are crucial to Yang's [249–252] constructions for finding longer T-sequences of odd length.

Lemma 5.3: *If there are Turyn sequences of length* $m+1, m+1, m, m$ *there are base sequences of lengths* $2m+2, 2m+2, 2m+1, 2m+1$.

Table 5.4 Base sequences of lengths $m+1, m+1, m, m$.

Length	Sums of Squares	Sequences
$m+1=2$	$2^2+0^2+1^2+1^2$	++,+−,+,+
$m+1=3$	$3^2+1^2+0^2+0^2$	+++,++−,+−,+−
$m+1=4$	$2^2+0^2+3^2+1^2$	++−+,++−−,+++,+−+
$m+1=5$	$3^2+3^2+0^2+0^2$	++−++,++++−,++−−,+−+−
$m+1=5$	$3^2+1^2+2^2+2^2$	++++−,−+++−,++−+,++−+
$m+1=6$	$2^2+0^2+3^2+3^2$	++−+−+,+++−−−,++−++,++−++
$m+1=7$	$3^2+1^2+4^2+0^2$	++−+−++,++−−−+−,−+++++,−−++−+
$m+1=7$	$5^2+1^2+0^2+0^2$	+++−+++,++−−−+−,++−+−−,++−+−−
$m+1=8$	$2^2+0^2+5^2+1^2$	++++−−−+,++−+−+−−,+++−+++,+−−+−−+
$m+1=8$	$4^2+2^2+3^2+1^2$	−+++++−+,+++−−+−+,−++−+++,+−+++−−
$m+1=9$	$5^2+3^2+0^2+0^2$	++++−++−+,−+++−++−+,+++−−−+−,+++−−−+−
$m+1=10$	$4^2+2^2+3^2+3^2$	+++++−−+−+,−++++−−+−+,+++−+++−−,+−+++−++−
$m+1=11$	$5^2+3^2+2^2+2^2$	++−−++++++−,−+−−++++++−,−+++−+−++−,−+++−+−++−
$m+1=11$	$1^2+5^2+0^2+4^2$	++−++−−−−++,−++−−++++++,−−++−+−+−+,−++−+−++++
$m+1=12$	$6^2+0^2+3^2+1^2$	−+++−+−+++++,−++−++−−−++−,−−+++++−++−,−++++−+−+−−
$m+1=12$	$4^2+2^2+5^2+1^2$	++++++−−+−+−,+−−−−−++−+−+,++++−−++−++,+−+−−++−−−+
$m+1=13$	$7^2+1^2+0^2+0^2$	++++−+−+−++++,+++−−+−+−−++−,+++−++−−+−−−,+++−−+−++−−−
$m+1=13$	$5^2+5^2+0^2+0^2$	+−−+−+−−−+−−,++−+−−+−+++++,++−+−−+++−−−,++−+++−−+−−−
$m+1=13$	$3^2+1^2+6^2+2^2$	++++−+−−++−−+,++++−+−−++−−−,+++++−+−+−++,+++−−+−+−−++
$m+1=14$	$6^2+4^2+1^2+1^2$	++++++−−++−+−+,+−−−−−++−−+−+−,
		++++−−−+−−+−−,+−+−−+−−−+++−
$m+1=14$	$2^2+0^2+7^2+1^2$	−−−+++−++++−+−,−−−++−+−−++−++,
		+−+++−+++−+++,+−++−+−−−−+++
$m+1=14$	$6^2+0^2+3^2+3^2$	+++−++−+++++−−,++++−−+−−−−+−+,
		++−−−+++−++−+,+−−++−+−+−+++
$m+1=15$	$7^2+3^2+0^2+0^2$	++−+++−+−+++−++,+++−++−−−++−++−,
		++++−−+−++−−−−,+−−−−+−+−++++−
$m+1=16$	$6^2+4^2+3^2+1^2$	+++++−++−−+++−+−,+−−−−+−−++−−−+−+,
		+−+++++−−+−+−−+,+++−+−++−−−−−++,
$m+1=16$	$6^2+0^2+5^2+1^2$	+−+−−+−−++++++++,+++−+−−+−+−−++−−,
		+−+−+−+++−−++++,++−−−−+++−++−−+
$m+1=17$	$7^2+1^2+4^2+0^2$	++−−−−+−−−+−−+−−−,++++−−+−−−+−−−+−+,
		+−−−−+−+−++−−−−−+,++−+−−−−++−−+−++
$m+1=17$	$5^2+5^2+4^2+0^2$	+−+++−−+−+++−+++−,+++−++−++++−−−+−+,
		+−++−++++−+++−−−,+−−−+++++−+−−+−−
$m+1=17$	$5^2+3^2+4^2+4^2$	+++++−−+−+−+−−+++,+−−++−+−+−++−−−−−,
		+++++−−++−+−++−−,+++++−−++−+−++−−
$m+1=17$	$1^2+1^2+8^2+0^2$	+−−−−++++−+−+−−−+,+++−++−−−+−−−−++−,
		+++++−+++++−−+−+,+−−+−+−++−−+−−++
$m+1=18$	$4^2+2^2+7^2+1^2$	++−−−−+−++−−−+−+−−,+++−−−−+−−++−+−−−+,
		+−+−+−−++−+++++++,++++−−++−−−+−−+−+
$m+1=18$	$4^2+2^2+5^2+5^2$	+−++−+−−+++−−−++++,+−+++−−+−−−−+++++−,
		+++−+++−+++−+−−+−,+++−+++−+++−+−−+−
$m+1=19$	$7^2+3^2+4^2+0^2$	+++−−++−+−++++−++−+,++−+−−+++−+−+−−−−−−,
		++−−+−−++−−−++++++,+−−−−+++−+−+−−++−+

(*Continued*)

Table 5.4 (*Continued*)

Order	Sum of Squares	Sets
$m+1=19$	$3^2 + 1^2 + 8^2 + 0^2$	+−+−−−+−+−++−+−++++,+++++−++−−−−−++−+−−,
		+++++−−−+++−+++−++,+−−+−++−−−++−−++−+
$m+1=19$	$1^2 + 1^2 + 6^2 + 6^2$	++++−−−−+−+−+−−++,+−+−−+−−+++−++++−−−,
		+−−+++−++−+++++−+−,+++++−++−++−−+++−−
$m+1=20$	$2^2 + 0^2 + 7^2 + 5^2$	++−−−−+−−−++−+++++−+,++−+−+−+++−−+−−+−+−−,
		+−++−++−−+++−+−++++,++−−−−−+++−−+−−−−−+
$m+1=21$	$7^2 + 5^2 + 2^2 + 2^2$	++−−++++++−−+++−+−++−,−+−−++++++−−+++−+−++−,
		+−−++++++−+−−−+−+−−+,+−−++++++−+−−−+−+−−+
$m+1=21$	$3^2 + 1^2 + 6^2 + 6^2$	++−−++++++−+−−+−+−−−+,−+−−++++++−+−−+−+−−−+,
		+−−++++++−−++−+−+++−,+−−++++++−−++−+−+++−
$m+1=22$	$6^2 + 0^2 + 7^2 + 1^2$	++−+−+−−−−−−−+−−++−+−−,
		++−−−−+−+−++−−−+−+++−+,
		+−++−++++++−−−++−−+++,
		++−+−++++−−+−−++−−−−+
$m+1=23$	$3^2 + 3^2 + 6^2 + 6^2$	+−+−−+−−−−+−+++++−−++++,
		−++−−++++−++++−−−−−++−+,
		++−+−−+−+++−+++−+−+++−,
		−+−−+++−+++−+++−−+−+++
$m+1=24$	$8^2 + 2^2 + 5^2 + 1^2$	+−−+−−+−+++−−−−−+−−−+−−,
		+−−−+−+−−−−++++++−−+−−−+,
		+++−−−++−−+−−−−+−−−−+−+,
		++−−+−++−+−+++−+−−+−−−+
$m+1=25$	$7^2 + 7^2 + 0^2 + 0^2$	−−−−+++−++−++++++++−+−+−+,
		+−−++++++++−−−−++−+−+++−+,
		−+−+−++−+−−++−−++−+++−−−,
		++−+−−+−++−−−−+++−−++−−+
$m+1=26$	$8^2 + 6^2 + 1^2 + 1^2$	++++++++−−+++−−+−−++−+−+−+,
		+−−−−−−−++−−−++−++−−+−+−+−
		++++++−−−−++−−++−+−−+−+−−,
		+−+−+−−+−++−−++−−−−+++++−
$m+1=27$	$7^2 + 5^2 + 4^2 + 4^2$	++++−−+++−+−−+−+−+−−++−++++
		−+++−−+++−+−−+−+−+−−++−++++
		−−−++−−−+−+++++−+−−++−++++
		−−−++−−−+−+++++−+−−++−++++
$m+1=28$	$4^2 + 2^2 + 3^2 + 9^2$	−++−++−−−++++−−−−+−+−+++−+++
		+++−+−++−+−+−−−+−−−++−−+−+++
		++−+−+++++−−+−++−−+++−−+−−−
		++−++−+++−++−+−−++++++++−−−
$m+1=29$	$3^2 + 1^2 + 2^2 + 10^2$	+−−+++++−+++−−+−+−−+−−+−++−−+
		++−−+−+−−−+−−+++++−−−++++−−+−
		++−+−+++++−−++−−+++−+−−−−−−+
		++++++−+++−−−+−++++−++−+−+−+
$m+1=30$	$8^2 + 6^2 + 3^2 + 3^2$	+++++−++++−+−−++−−++++−+−−−+−+,
		+−−−−+−−−−+−++−−++−−−−+−+++−+−,
		+++−+−−−+−++−−−+−−++++−++++−−,
		+−++++−++++−−+−−−++−+−−−+−++−

Source: Seberry and Yamada [166, table 5.1, p. 474], Wiley.

Proof: Let X, U, Y, V be the Turyn sequences as in the proof of Lemma 5.2.

$$E = \{1, X/Y\}, \quad F = \{-1, X/Y\}, \quad G = \{U/V\}, \quad H = \{U/-V\}$$

are 4-complementary base sequences of lengths $2m+2, 2m+2, 2m+1, 2m+1$, respectively. □

Corollary 5.5: *There are base sequences of lengths $m+1, m+1, m, m$ for m*

a) t, $2t+1$ where there are Turyn sequences of length $t+1$, $t+1$, t, t,
b) 9, 11, 13, 25, 29,
c) for lengths g where there are Golay sequences of length g,
d) 17 (Seberry–Sproul), 23 (Turyn), 22, 24, 26, 27, 28 (Koukouvinos, Kounias, Sotirakoglou) given in Tables 5.4 and 5.5.

Corollary 5.6: *There are base sequences of lengths $m+1, m+1, m, m$ for $m \in \{1, 2, \ldots, 38\} \cup G$, where $G = \{g;\ g = 2^a.10^b.26^c,\ a, b, c$ nonnegative integers\}.*

Thus it becomes important to know for which lengths (and decomposition into squares) T-sequences exist. We note from Lemma 3.2.

Lemma 5.4: *If there are base sequences of length $m+1, m+1, m, m$ there are 4 (disjoint) T-sequences of length $2m+1$.*

Proof: Let X, U, Y, V be the base sequences of lengths $m+1, m+1, m, m$ then

$$\left\{\frac{1}{2}(X+U), 0_m\right\}, \left\{\frac{1}{2}(X-U), 0_m\right\}, \left\{0_{m+1}, \frac{1}{2}(Y+V)\right\}, \left\{0_{m+1}, \frac{1}{2}(Y-V)\right\}$$

are the T-sequences of length $2m+1$ and

$$\{X, Y\}, \{X, -Y\}, \{U, V\}, \{U, V\}$$

are 4 complementary sequences of length $2m+1$. □

Corollary 5.7: *There are T-sequences of lengths t for the following $t < 109$*

$$1, 3, \ldots, 59, 65, 81, 101, 105, 107$$

We pose a further research question in Appendix C, problem C.3.
We now discuss Yang's theorems on T-sequences.

5.10.1 On Yang's Theorems on T-sequences

For a sequence U, we note that the nonperiodic autocorrelation function of the reverse U^* is the same as that of a sequence U: $N_U(j) = N_{U^*}(j)$ from Lemma 1.20.

Now Yang [249–251] found that base sequences can be multiplied by 3,7,13 and $2g+1$, where $g = 2^a 10^b 26^c$, $a, b, c \geq 0$ so we call these integers **Yang numbers**. If y is a Yang number and there are base sequences of lengths m, m, n, n then there are (4-complementary) T-sequences of length $y(m+n)$. This is of most interest when $m+n$ is odd. (Koukouvinos et al. [117] have given a new construction for the Yang number 57.)

Yang numbers currently exist for $y \in \{3, 5, \ldots, 33, 37, 39, 41, 45, 49, 51, 53, 57, 59, 65, 81, \ldots$, and $2g+1 > 81, g \in G\}$ where

$$G = \{g;\ g = 2^a 10^b 26^c,\ a, b, c \text{ nonnegative integers}\}.$$

Base sequences currently exist for $n = m-1$ and $m \in \{1, 2, \ldots, 38\} \cup G$, $m = 71$ and $n = 36$ (see Corollary 5.3).
We reprove and restate Yang's theorems from [252] to illustrate why they work.

Table 5.5 *T*-matrices used.

Order	Sum of Squares	Sets	
31	$3^2+3^2+3^2+2^2$	T_1	$\{1,5,-8,-9,11,-14,24,25,27\}$
		T_2	$\{2,6,10,-12,19,-21,26,-29,30\}$
		T_3	$\{4,7,-16,17,-18,20,22,23,-28\}$
		T_4	$\{3,13,15,-31\}$
39	$6^2+1^2+1^2+1^2$	T_1	$\{17,20,-21,23,24,26,35,38\}$
		T_2	$\{14,15,-16,-18,19,22,25,-34,-36,-37,39\}$
		T_3	$\{-4,-7,-8,10,11,13,28,-29,-31,32,33\}$
		T_4	$\{1,2,-3,-5,6,-9,-12,27,30\}$
43	$4^2+3^2+3^2+3^2$	T_1	$\{1,4,-5,6,7,8,9,-13,-14,15,16,-17,-18,21\}$
		T_2	$\{-2,3,10,11,-12,19,20\}$
		T_3	$\{-22,-23,24,26,29,31,34,36,-39,-41,42\}$
		T_4	$\{-25,-27,28,30,32,33,-35,37,-38,40,43\}$
49	$4^2+4^2+4^2+1^2$	T_1	$\{4,6,-18,19,21,-32,34,44,45,-46\}$
		T_2	$\{-8,9,10,12,14,25,26,28,-36,37,38,-40,-42,-48\}$
		T_3	$\{11,13,22,-23,-24,27,39,-41,47,49\}$
		T_4	$\{-1,2,3,5,7,15,-16,-17,-20,29,-30,-31,33,35,-43\}$
49	$5^2+4^2+2^2+2^2$	T_1	$\{1,-2,-3,5,7,-15,16,17,20,-29,30,31,33,35,-43\}$
		T_2	$\{11,-13,-22,23,24,-27,39,41,47,49\}$
		T_3	$\{-4,6,18,-19,-21,32,34,44,45,-46\}$
		T_4	$\{8,-9,-10,12,14,25,26,28,36,-37,-38,-40,-42,48\}$
55	$5^2+5^2+2^2+1^2$	T_1	$\{1,2,-5,7,8,-9,10,11,-23,-24,27,29,30,-31,32,33,-45,-47,48\}$
		T_2	$\{-14,15,17,-36,37,39,51,52,-53,54,55\}$
		T_3	$\{12,13,-16,18,19,-20,21,22,34,35,-38,-40,-41,42,-43,-44\}$
		T_4	$\{-3,4,6,25,-26,-28,-46,49,50\}$
57	$4^2+4^2+4^2+3^2$	T_1	$\{-24,-25,29,30,-31,32,33,-35,36,37,38,53\}$
		T_2	$\{20,21,22,-23,-26,27,-28,-34,49,50,51,-52,54,55,-56,57\}$
		T_3	$\{5,6,-10,11,-12,13,14,-16,17,18,19,40,-41,42,45,-46,-47,-48\}$
		T_4	$\{1,2,3,-4,-7,8,-9,15,-39,43,44\}$
61	6^2+5^2	T_1	$\{2,7,10,17,18,-26,29,-30,31,-32,35,40,-44,-51,55,61\}$
		T_2	$\{3,4,-8,-11,-12,13,14,15,16,19,22,-25,27,-28,36,-37,-38,41,-42,-47,49,52,56,-57,60\}$
		T_3	$\{-1,5,6,-9,-20,21,-23,-24,-33,-34,39,43,45,46,48,-50,-53,54,-58,59\}$
		T_4	$\{\phi\}$
67	$8^2+1^2+1^2+1^2$	T_1	$\{-1,5,9,13,14,15,18,25,27,29,-31,32,-39,43,50,-67\}$
		T_2	$\{2,-8,-12,16,17,23,-40,41,42,-45,-46,-47,-53,54,-56,65,66\}$
		T_3	$\{-6,7,11,19,20,-21,24,-26,-28,-37,38,44,-49,57,-58,-59,61\}$
		T_4	$\{-3,-4,10,22,30,-33,34,-35,36,48,51,-52,-55,-60,-62,63,64\}$
71	$6^2+5^2+3^2+1^2$	T_1	$\{1,-2,-3,4,5,6,-7,8,9,10,-11,-12,-13,-14,15,16,-17,18,19,-20,21,22,23,24\}$
		T_2	$\{25,26,27,28,-29,30,31,-32,33,34,35,-36,37,-38,39,-40,41,-42,-43,-44,-45,46,47\}$
		T_3	$\{48,49,50,51,-52,-56,57,58,60,-64,65,-66,-71\}$
		T_4	$\{53,54,-55,59,-61,62,-63,-67,-68,69,70\}$

(Continued)

Table 5.5 (*Continued*)

Order	Sum of Squares	Sets	
85	7^2+6^2	T_1	$\{1,2,4,-5,-11,-12,14,-15,21,22,24,-25,31,32,-34,35,-41,-42,-44,45,51,52,-54,55,$ $61,62,64,-65,71,72,-74,75,-81\}$
		T_2	$\{3,-13,23,33,-43,53,63,73,82,83\}$
		T_3	$\{-6,-7,-9,10,16,17,-19,20,-26,-27,-29,30,-36,-37,39,-40,-46,-47,-49,50,$ $56,57,-59,60,66,67,69,-70,76,77,-79,80\}$
		T_4	$\{8,18,28,-38,-48,-58,68,-78,-84,85\}$
87	$7^2+6^2+1^2+1^2$	T_1	$\{-2,-3,5,6,10,11,13,-14,15,16,-17,20,21,24,-25,28,29,-62,-65,66,67,70,-73\}$
		T_2	$\{30,-33,-36,-37,38,41,-47,48,51,52,55,-56,74,75,-78,79,82,83,-86,87\}$
		T_3	$\{1,-4,-7,-8,9,12,18,-19,-22,-23,-26,27,59,-60,61,63,64,68,69,-71,-72\}$
		T_4	$\{-31,-32,34,35,39,40,42,-43,44,-45,46,-49,-50,-53,54,-57,-58,-76,77,80,81,84,-85\}$
91	$5^2+5^2+5^2+4^2$	T_1	$\{-1,-2,3,5,-6,-8,10,11,13,27,28,-29,-31,32,35,38,53,54,-55,-57,58,-60,62,63,65,-79,-82\}$
		T_2	$\{-4,-7,9,12,30,33,34,-36,-37,-39,56,59,61,64,80,-81,-83,84,85\}$
		T_3	$\{17,20,,-22,-25,40,41,-42,-44,45,-48,-51,69,72,74,77,86,88,89,-91\}$
		T_4	$\{-14,-15,16,18,-19,-21,23,24,26,43,46,-47,49,50,52,-66,-67,68,70,$ $-71,73,-75,-76,-78,87,90\}$
93	$6^2+5^2+4^2+4^2$	T_1	$\{2,3,4,5,-6,7,8,-9,-10,11,12,13,-14,15,-16,17,19,21,23,-25,-27,-29,31,-78\}$
		T_2	$\{1,18,-20,-22,24,-26,-28,30,-63,64,-65,66,67,68,-69,-70,71,72,-73,74,75,76,77\}$
		T_3	$\{33,34,35,36,-37,38,39,-40,-41,42,43,44,-45,46,-47,-48,-50,-52,-54,56,58,60,$ $-62,-80,82,84,-86,88,90,-92\}$
		T_4	$\{32,-49,51,53,-55,57,59,-61,79,-81,-83,-85,87,89,91,93\}$

Source: Seberry and Yamada [166, table 5.3, pp. 485–487], Wiley.

5.10.2 Multiplying by $2g+1$, g the Length of a Golay Sequence

Theorem 5.4 ***(Yang [250, p. 378]):*** *Let A, B, C, D be base sequences of lengths m, m, n, n, and $F=(f_k)$ and $G=(g_k)$ be Golay sequences of length s. Then the following Q, R, S, T become 4-complementary sequences (namely, the sum of nonperiodic autocorrelation functions is 0), using X^* to denote the reverse of X.*

$$Q=(Af_s, Cg_1; 0, 0'; Af_{s-1}, Cg_2; 0, 0'; \dots; Af_1, Cg_s; 0, 0'; -B^*, 0')$$
$$R=(Bf_s, Dg_s; 0, 0'; Bf_{s-1}, Dg_{s-1}; 0, 0'; \dots; Bf_1, Dg_1; 0, 0'; A^*, 0')$$
$$S=(0, 0'; Ag_s, -Cf_1; 0, 0'; Ag_{s-1}, -Cf_2; \dots; 0, 0'; Ag_1, -Cf_s; 0, -D^*)$$
$$T=(0, 0'; Bg_1, -Df_1; 0, 0'; Bg_2, -Df_2; \dots; 0, 0'; Bg_s, -Df_s; 0, C^*)$$

where 0 and $0'$ are zero sequences of length m and n.

Furthermore if we define sequences

$$X=\frac{1}{2}(Q+R),\ Y=\frac{1}{2}(Q-R),\ V=\frac{1}{2}(S+T),\ W=\frac{1}{2}(S-T),$$

then these sequences become T-sequences of length $t(2s+1)$, $t=m+n$.

Note: The interesting case for Yang's theorem is for base sequences of lengths m, m, n, n, where $m+n$ is odd for then Yang's theorem produces T-sequences of odd length, for example, $3(m+n)$. It should also be noted that Yang alternates between two notations for base sequences. We use the notation m, m, n, n.

Theorem 5.5 ***(Yang [250]):*** *Suppose E, F, G, H are base sequences of lengths m, m, n, n. Define $A = \frac{1}{2}(E+F)$, $B = \frac{1}{2}(E-F)$, $C = \frac{1}{2}(G+H)$, and $D = \frac{1}{2}(G-F)$ to be suitable (pairwise disjoint) sequences. Then the following sequences are disjoint T-sequences of length $3(m+n)$:*

$$\begin{aligned}
X &= \{A\ C;\quad 0\ 0';\quad B^*\quad 0'\}\\
Y &= \{B\ D;\quad 0\ 0';\quad -A^*\quad 0'\}\\
Z &= \{0\ 0';\quad A\ -C;\quad 0\quad D^*\}\\
W &= \{0\ 0';\quad B\ -D;\quad 0\quad -C^*\}
\end{aligned}$$

and

$$\begin{aligned}
X &= \{B^*\quad 0';\quad A\ C;\quad 0\ 0'\}\\
Y &= \{-A^*\quad 0';\quad B\ D;\quad 0\ 0'\}\\
Z &= \{0\quad D^*;\quad 0\ 0';\quad A\ -C\}\\
W &= \{0\quad -C^*;\quad 0\ 0';\quad B\ -D\}
\end{aligned}$$

where 0 and $0'$ are the zero sequences of lengths m and n, respectively.

5.10.3 Multiplying by 7 and 13

The next two theorems can be used recursively but as the sequences produced are of equal lengths, the next recursive use of the theorems gives sequences of (equal) even length.

Theorem 5.6 ***(Yang [251]):*** *Let E, F, G, H be the base sequences of length m, m, n, n. Let $t = m + n$. Define the suitable sequences $A = \frac{1}{2}(E+F)$, $B = \frac{1}{2}(E-F)$, $C = \frac{1}{2}(G+H)$, and $D = \frac{1}{2}(G-H)$ of lengths m, m, n, and n. Then the following X, Y, Z, W are 4-disjoint T-sequences of length $7t$ (where $\overline{X}$ means negate all the elements of the sequence and $X*$ means reverse all the elements of the sequence):*

$$\begin{aligned}
X &= (\overline{A}, C;\ 0,0';\ A, D;\ 0,0';\ A, C;\ 0,0';\ \overline{B}^*, 0'),\\
Y &= (\overline{B}, D;\ 0,0';\ B, \overline{C};\ 0,0';\ B, D;\ 0,0';\ A^*, 0'),\\
Z &= (0,0';\ A, \overline{C};\ 0,0';\ \overline{B}, \overline{C};\ 0,0';\ A, C;\ 0, \overline{D}^*),\\
W &= (0,0';\ B, \overline{D};\ 0,0';\ A, \overline{D};\ 0,0';\ B, D;\ 0, C^*).
\end{aligned}$$

where 0 and $0'$ are the zero sequences of lengths m and n, respectively.

Theorem 5.7 ***(Yang [251]):*** *Let E, F, G, H be the base sequences of length m, m, n, n. Let $t = m + n$. Define the suitable sequences $A = \frac{1}{2}(E+F)$, $B = \frac{1}{2}(E-F)$, $C = \frac{1}{2}(G+H)$, and $D = \frac{1}{2}(G-H)$ of lengths m, m, n, and n. Then the following X, Y, Z, W are 4-disjoint T-sequences of length $13t$.*

$$\begin{aligned}
Q = (&A, D^*;\ \overline{A}, \overline{C};\ \overline{A}, D^*;\ \overline{A}, C;\ \overline{A}, D^*;\ A, \overline{C};\ 0, C;\\
&0,0';\ 0,0';\ 0,0';\ 0,0';\ 0,0';\ 0,0'),\\
R = (&\overline{B}, C^*;\ B, D;\ B, C^*;\ B, \overline{D};\ B, C^*;\ \overline{B}, D;\ 0, \overline{D};\\
&0,0';\ 0,0';\ 0,0';\ 0,0';\ 0,0';\ 0,0'),\\
S = (&0,0';\ 0,0';\ 0,0';\ 0,0';\ 0,0';\ 0,0';\ \overline{A}, 0';\\
&A, C;\ B^*, \overline{C};\ \overline{A}, \overline{C};\ B^*, \overline{C};\ A, \overline{C};\ B^*, C),\\
T = (&0,0';\ 0,0';\ 0,0';\ 0,0';\ 0,0';\ 0,0';\ B, 0';\\
&\overline{B}, \overline{D};\ A^*, D;\ B, D;\ A^*, D;\ \overline{B}, D;\ A^*, \overline{D}).
\end{aligned}$$

where 0 and $0'$ are the zero sequences of lengths m and n, respectively.

Yang has also shown how to multiply by 11. The sequences obtained are not disjoint and so cannot be used in another iteration but still are vital in that they give complementary sequences of length $11(m+n)$ and hence Hadamard matrices of order $44(m+n)$.

Using the Yang numbers $y = 3, 5, 7, 9, 13, 17, 21, 33, 41, 53, 65, 81$ with base sequences gives T-sequences so

Corollary 5.8: *Yang numbers and base sequences of lengths $m+1, m+1, m, m$ can be used to give T-sequences of lengths $t = y(2m+1)$ for the following $t < 200$*

$$1, 3, \ldots, 41, 45, \ldots, 59, 63, 65, 9, 75, 77, 81, 85, 91, 93, 95, 9, 101, 105, 107, 111, 115, \ldots,$$

$$125, 133, 135, 141, \ldots, 147, 153, 155, 159, 161, 165, 169, 171, 175, 177, 187, 189, 195.$$

The gaps in these sets can sometimes be filled by T-matrices. Thus using Table 5.5 and Corollary 5.7 and noting that T-sequences give T-matrices we have

Lemma 5.5: *T-matrices exist for the following $t < 106$, t odd.*

$$1, 3, \ldots, 71, 75, 77, 81, 85, 87, 91, 93, 95, 99, 101, 105$$

and any t for which a T-sequence exists giving T-sequences for

$$1, 3, \ldots, 69, 75, 77, 81, \ldots, 87, 91, 93, 95, 99, 101, 105, 107, 111, 115, \ldots, 125,$$

$$129, \ldots, 135, 141, \ldots, 147, 153, 155, 159, \ldots, 165, 169, 171, 175, 177, 187, 189, 195.$$

These are given in more detail in Cohen et al.[20], Koukouvinos et al. [115], and Koukouvinos et al. [118]. Further results including multiplication and construction theorems are given in the work of Koukouvinos et al. [116, 117].

5.10.4 Koukouvinos and Kounias Number

We call κ a **Koukouvinos–Kounias number**, or **KK number** [120, 166], if

$$\kappa = g_1 + g_2$$

where g_1 and g_2 are both the lengths of Golay sequences. Then we have

Lemma 5.6: *Let κ be a KK number and y be a Yang number. Then there are T-sequences of length t and $OD(4t; t, t, t, t)$ for $t = y\kappa$.*

This gives T-sequences of lengths

2.101, 2.109, 2.113, 8.127, 2.129, 2.131, 8.151,
8.157, 16.163, 2.173, 4.179, 4.185, 4.193, 2.201.

6

M-structures

6.1 Notations

Table 6.1 gives the notations which are used in this chapter.

6.2 The Strong Kronecker Product

Asymptotic theorems such as Theorem 11.9 change ideas for evaluating construction methods: we consider a method to be more powerful if it lowers the power of 2 for the resultant odd number. Thus Agaian's theorem, which gives Hadamard matrices of order $8mn$ from Hadamard matrices of order $4m$ and $4n$, is more powerful than that of Sylvester which gives a matrix of order $16mn$.

We now see another way to lower the power in a multiplication method. First we introduce some notation.

Let $M = (M_{ij})$ and $N = (N_{gh})$ be orthogonal matrices or t^2 block M-structures of orders tm and tn, respectively, where M_{ij} is of order m $(i,j = 1, 2, \dots, t)$ and N_{gh} is of order n $(g, h = 1, 2, \dots, t)$.

Definition 6.1: The operation $\bigcirc$, the **strong Kronecker product** of two matrices M and N is defined as:

$$M \bigcirc N = \begin{bmatrix} L_{11} & L_{12} & \cdots & L_{1t} \\ L_{21} & L_{22} & \cdots & L_{2t} \\ & & \cdots & \\ L_{t1} & L_{t2} & \cdots & L_{tt} \end{bmatrix}$$

where M_{ij}, N_{ij}, and L_{ij} are of order of m, n, and mn, respectively and

$$L_{ij} = M_{i1} \times N_{1j} + M_{i2} \times N_{2j} + \cdots + M_{it} \times N_{tj},$$

$i,j = 1, 2, \dots, t$. We note that the strong Kronecker product preserves orthogonality but not necessarily with entries in a useful form.

Theorem 6.1: *Let A be an $OD(tm; p_1, \dots, p_u)$ with entries $x_1, \dots, x_u$ and B be an $OD(tn; q_1, \dots, q_s)$ with entries $y_1, \dots, y_s$ then*

$$(A \bigcirc B)(A \bigcirc B)^\top = \left(\sum_{j=1}^{u} p_j x_j^2\right)\left(\sum_{j=1}^{s} q_j y_j^2\right) I_{tmn}.$$

($A \bigcirc B$ is not an orthogonal design but an orthogonal matrix.) If A is a $W(tm, p)$ and B is a weighing matrix $W(tn, q)$ then $A \bigcirc B = C$ satisfies $CC^\top = pqI_{tmn}$.

Hadamard Matrices: Constructions using Number Theory and Algebra, First Edition. Jennifer Seberry and Mieko Yamada.

Table 6.1 Notations used in this chapter.

$M \times N$	The Kronecker product
$M \bigcirc N$	The strong Kronecker product
$W(n,k)$	A weighing matrix of weight k and of order n
$OD(n; s_1, s_2, s, \dots, s_u)$	An orthogonal design of order n and of type $(s_1, s_2, s, \dots, s_u)$
$A * B$	The Hadamard product of matrices A and B

Hereafter, let $H = (H_{ij})$ and $N = (N_{ij})$ of order $4h$ and $4n$, respectively, be 16 block M-structure Hadamard matrices. So

$$H = \begin{bmatrix} H_{11} & H_{12} & H_{13} & H_{14} \\ H_{21} & H_{22} & H_{23} & H_{24} \\ H_{31} & H_{32} & H_{33} & H_{34} \\ H_{41} & H_{42} & H_{43} & H_{44} \end{bmatrix}$$

where

$$\sum_{j=1}^{4} H_{ij}H_{ij}^{\top} = 4hI_h = \sum_{j=1}^{4} H_{ji}H_{ji}^{\top}, \quad \text{for } i = 1,2,3,4, \text{ and}$$

$$\sum_{j=1}^{4} H_{ij}H_{kj}^{\top} = 0 = \sum_{j=1}^{4} H_{ji}H_{jk}^{\top}, \quad \text{for } i \neq k,\ i,k = 1,2,3,4,$$

and similarly for N.

For ease of writing we define

$$X_i = \tfrac{1}{2}\left(H_{i1} + H_{i2}\right) \quad Y_i = \tfrac{1}{2}\left(H_{i1} - H_{i2}\right)$$
$$Z_i = \tfrac{1}{2}\left(H_{i3} + H_{i4}\right) \quad W_i = \tfrac{1}{2}\left(H_{i3} - H_{i4}\right)$$

where $i = 1,2,3,4$. Then both $X_i \pm Y_i$ and $Z_i \pm W_i$ are $(1,-1)$ matrices with $X_i * Y_i = 0$ and $Z_i * W_i = 0$, $*$ the Hadamard product.

Let

$$S = \frac{1}{2}\begin{bmatrix} H_{11}+H_{12} & -H_{11}+H_{12} & H_{13}+H_{14} & -H_{13}+H_{14} \\ H_{21}+H_{22} & -H_{21}+H_{22} & H_{23}+H_{24} & -H_{23}+H_{24} \\ H_{31}+H_{32} & -H_{31}+H_{32} & H_{33}+H_{34} & -H_{33}+H_{34} \\ H_{41}+H_{42} & -H_{41}+H_{42} & H_{43}+H_{43} & -H_{43}+H_{44} \end{bmatrix}.$$

Obviously S is a $(0,1,-1)$ matrix.

Write

$$R = \begin{bmatrix} Y_1 & X_1 & W_1 & Z_1 \\ Y_2 & X_2 & W_2 & Z_2 \\ Y_3 & X_3 & W_3 & Z_3 \\ Y_4 & X_4 & W_4 & Z_4 \end{bmatrix},$$

also a $(0,1,-1)$ matrix.

We note $S \pm R$ is a $(1,-1)$ matrix, $R * S = 0$ and by the previous theorem.

$$SS^{\top} = RR^{\top} = 2hI_{4h}.$$

Lemma 6.1: *If there exists an Hadamard matrix of order* $4h$ *there exists an* $OD(4h; 2h, 2h)$ *and hence a* $W(4h, 2h)$.

Proof: Form S and R as above. Now $H = S + R$. Note $HH^\top = SS^\top + RR^\top + SR^\top + RS^\top = 4hI_{4h}$ and $SS^\top = RR^\top = 2hI_{4h}$. Hence $SR^\top + RS^\top = 0$. Let x and y be commuting variables then $E = xS + yR$ is the required orthogonal design. □

The following idea was first introduced by R. Craigen [25].

Definition 6.2: An **orthogonal pair** or ± 1 matrices, S and R of order n, satisfy

i) $SS^\top + RR^\top = 2nI_n$,
ii) $SR^\top = RS^\top = 0$,

6.3 Reducing the Powers of 2

The basic Kronecker construction for Hadamard matrices introduces many powers of 2. We know from Lemma 1.9 that Hadamard matrices of orders $4m$ and $4n$ have as their Kronecker product an Hadamard matrix of order $16mn$. We now study some theorems which, using the internal structures of Hadamard matrices, allow us to obtain smaller powers of 2 in composition theorems. The first such theorem was given by Agaian:

Theorem 6.2 *(Agaian's theorem [1]):* *If there exist Hadamard matrices of orders $4m$ and $4n$, then there exists an Hadamard matrix of order $8mn$.*

We give a variation and it's proof due to Craigen [25]:

Lemma 6.2: *If there exist Hadamard matrices, H and K, of order $4m$ and $4n$, then there exists an orthogonal pair of order $4mn$, that is, two amicable ± 1 matrices R and S satisfying*

i) $RR^\top + SS^\top = 8mnI_{4mn}$,
ii) $SR^\top = RS^\top = 0$,

and thus an Hadamard matrix of order $8mn$.

Proof: We write

$$H = \begin{bmatrix} H_1 \\ H_2 \\ H_3 \\ H_4 \end{bmatrix} \quad \text{and} \quad K = \begin{bmatrix} K_1 \\ K_2 \\ K_3 \\ K_4 \end{bmatrix}.$$

Each H_i being of size $m \times 4m$ and each K_i being of size $n \times 4n$, $i = 1, 2, 3, 4$. We now take

$$R = \frac{1}{2}(H_1 + H_2)^\top \times K_1 + \frac{1}{2}(H_1 - H_2)^\top \times K_2,$$
$$S = \frac{1}{2}(H_3 + H_4)^\top \times K_3 + \frac{1}{2}(H_3 - H_4)^\top \times K_4.$$

Clearly both R and S are square ± 1 matrices, and

$$RR^\top + SS^\top = \frac{1}{2}(H_1^\top H_1 + H_2^\top H_2 + H_3^\top H_3 + H_4^\top H_4) \times 4nI$$

since

$$H_1^\top H_1 + H_2^\top H_2 + H_3^\top H_3 + H_4^\top H_4 = H^\top H,$$

we have

$$RR^{\top} + SS^{\top} = 8mnI_{4mn}.$$

Since $K_iK_j^{\top} = 0$ for $i \neq j$, R and S are as claimed. Thus

$$\begin{bmatrix} R & S \\ S & R \end{bmatrix}$$

is an Hadamard matrix of order $8mn$. □

Using Craigen [25], who called R and S orthogonal pairs, and Seberry and Xianmo Zhang [167], who used the strong Kronecker product and M-structures, we have

Lemma 6.3: *If there exist Hadamard matrices of order* $4m$ *and* $4n$, *then there exist two disjoint, amicable* $W(4mn, 2mn)$.

Proof: Let R and S be the matrices constructed in Lemma 6.2. Let $X = \frac{1}{2}(R + S)$ and $Y = \frac{1}{2}(R - S)$. We calculate

$$XX^{\top} = YY^{\top} = 2mnI_{4mn}.$$

X and Y are disjoint, $X * Y = 0$ ($*$ the Hadamard product) since R and S are ± 1 matrices. Therefore X and Y are the desired matrices. □

Using Agaian's theorem on four Hadamard matrices of orders $4m$, $4n$, $4p$, and $4q$ gives an Hadamard matrix of order $32mnpq$. However, the following theorem reduces the power of 2 to 16.

Theorem 6.3 ***(Craigen–Seberry–Xianmo Zhang's theorem [37]):*** *If there exist Hadamard matrices of orders* $4m$, $4n$, $4p$, *and* $4q$, *then there exists an Hadamard matrix of order* $16mnpq$.

Proof: By Lemma 6.3, there exist two disjoint $W(4mn, 2mn)$, X and Y and two ± 1 matrices R and S of order $4pq$ satisfying (*i*) and (*ii*) of Lemma 6.2.

Let $H = X \times S + Y \times R$, Then H is a ± 1 matrix and

$$\begin{aligned} HH^{\top} &= XX^{\top} \times SS^{\top} + YY^{\top} \times RR^{\top} = 2mnI_{4mn} \times (SS^{\top} + RR^{\top}) \\ &= 2mnI_{4mn} \times 8pqI_{4pq} = 16mnpqI_{16mnpq}. \end{aligned}$$

Thus H is the required Hadamard matrix. □

Using previous products, the smallest order which can be obtained from using 12 Hadamard matrices is 10: the Craigen–Seberry–Zhang product three times and the Agaian–Sarukhanyan product twice. If the initial 12 Hadamard matrices have exponent equal to 2^{10}, then the resulting construction has exponent equal to 2^9. de Launey (now deceased) believed that such theorems were essential to finally proving the Hadamard conjecture. To this end he proved

Theorem 6.4 ***(de Launey's theorem [44]):*** *Suppose there are 12 Hadamard matrices with the orders* $4a, 4b, 4c, \cdots, 4k, 4\ell$. *Then there is an orthogonal pair of order* $256abcd \cdots k\ell$, *and an Hadamard matrix of order* $512abcd \cdots k\ell$.

Proof: Use the Craigen–Seberry–Zhang product to obtain Hadamard matrices of orders $16abcd$ and $16efgh$. Partition these matrices as in the proof of Lemma 6.2 and write $G_i = H_i^{\top}$. Then the matrices G_i are $16abcd \times abcd$ matrices, and the matrices K_i are $efgh \times 16efgh$ matrices.

Next, use the remaining four Hadamard matrices in Lemma 6.3 to obtain disjoint weighing matrices, $W(4ij, 2ij)$, W and X of order $4ij$, and disjoint weighing matrices, $W(4k\ell, 2k\ell)$, Y, and Z of order $4k\ell$.

Define the matrices A and B by the following equations. and

$$\begin{aligned}2A = {} & W \times Y \times [(G_1 + G_2) \times K_1 + (G_1 - G_2) \times K_2]\\ & + W \times Z \times [(G_3 + G_4) \times K_3 + (G_3 - G_4) \times K_4]\\ & + X \times Y \times [(G_5 + G_6) \times K_5 + (G_5 - G_6) \times K_6]\\ & + X \times Z \times [(G_7 + G_8) \times K_7 + (G_7 - G_8) \times K_8]\end{aligned}$$

and

$$\begin{aligned}2B = {} & W \times Y \times [(G_9 + G_{10}) \times K_9 + (G_9 - G_{10}) \times K_{10}]\\ & + W \times Z \times [(G_{11} + G_{12}) \times K_{11} + (G_{11} - G_{12}) \times K_{12}]\\ & + X \times Y \times [(G_{13} + G_{14}) \times K_{13} + (G_{13} - G_{14}) \times K_{14}]\\ & + X \times Z \times [(G_{15} + G_{16}) \times K_{15} + (G_{15} - G_{16}) \times K_{16}].\end{aligned}$$

It is simple to check that A and B are an orthogonal pair and satisfy Lemma 6.2. □

This process can probably be repeated using appropriate intermediate steps. Related research problems by de Launey's are in Appendix C.

6.4 Multiplication Theorems Using *M*-structures

In this section the reader wishing more details of constructions is referred to Seberry and Yamada [165]. As shown in Section 4.2.1 the power of M-structures comprising wholly circulant or type one blocks, as in the Welch and Ono–Sawade–Yamamoto structures of order 4×5 and 4×9 respectively, permits them to multiply onto the order of T-matrices.

Named after Mieko Yamada and Masahiko Miyamoto these structures have proved to be very powerful in attacking the question "if there is an Hadamard matrix of order $4t$ is there an Hadamard matrix of order $8t + 4$."

M-structures provide another variety of "plug in" matrices which have yet to be fully exploited.

Table A.1 of survey [166] gives the knowledge of Williamson matrices in 1992.

Before continuing, a refresh on definitions; an M-structure is an orthogonal matrix of order $4t$ that can be divided into 16 $t \times t$ blocks M_{ij} and if the orthogonal matrix can be partitioned into 64 $s \times s$ blocks M_{ij} then it is a 64 block M-structure (see Definitions 4.1 and 4.2).

An Hadamard matrix made from (symmetric) Williamson matrices W_1, W_2, W_3, W_4 is an M-structure with

$$\begin{aligned}W_1 &= M_{11} = M_{22} = M_{33} = M_{44},\\ W_2 &= M_{12} = -M_{21} = M_{34} = -M_{43},\\ W_3 &= M_{13} = -M_{31} = -M_{24} = M_{42},\\ W_4 &= M_{14} = -M_{41} = M_{23} = -M_{32}.\end{aligned}$$

An Hadamard matrix made from four circulant (or type 1) matrices A_1, A_2, A_3, A_4 of order n is an M-structure with

$$\begin{aligned}A_1 &= M_{11} = M_{22} = M_{33} = M_{44},\\ A_2 &= M_{12}R = -M_{21}R = RM_{34}^{\top} = -RM_{43}^{\top},\end{aligned}$$

$$A_3 = M_{13}R = -M_{31}R = -RM_{24}^{\mathsf{T}} = RM_{42}^{\mathsf{T}},$$
$$A_4 = M_{14}R = -M_{41}R = RM_{23}^{\mathsf{T}} = -RM_{32}^{\mathsf{T}}$$

where R is the back diagonal identity matrix in Definition 1.10 and Lemma 1.5 makes AR back circulant.

Theorem 6.5: *Suppose there is an M-structure orthogonal matrix of order 4m with each block circulant or type one. Then there is an M-structure orthogonal matrix of order 4mt where t is the order of T-matrices.*

Further

Theorem 6.6: *Let $N = (N_{ij})$, $i,j = 1,2,3,4$ be an Hadamard matrix of order 4n of M-structure. Further let T_{ij}, $i,j = 1,2,3,4$ be 16 $(0,+1,-1)$ type 1 or circulant matrices of order t which satisfy*

$$\begin{aligned}
&(i)\quad T_{ij} * T_{ik} = 0,\ T_{ji} * T_{ki} = 0,\ j \neq k, (* \text{ the Hadamard product})\\
&(ii)\quad \sum_{k=1}^{4} T_{ik} \text{ is a } (1,-1) \text{matrix},\\
&(iii)\quad \sum_{k=1}^{4} T_{ik}T_{ik}^{\mathsf{T}} = tI_t = \sum_{k=1}^{4} T_{ki}T_{ki}^{\mathsf{T}},\\
&(iv)\quad \sum_{k=1}^{4} T_{ik}T_{jk}^{\mathsf{T}} = 0 = \sum_{k=1}^{4} T_{ki}T_{kj}^{\mathsf{T}},\ i \neq j.
\end{aligned} \tag{6.1}$$

Then there is a M-structure Hadamard matrix of order 4nt.

Corollary 6.1: *If there exists an Hadamard matrix of order 4h and an orthogonal design $OD(4u; u_1, u_2, u_3, u_4)$, then an $OD(8hu; 2hu_1, 2hu_2, 2hu_3, 2hu_4)$ exists. In particular the u_i's can be equal.*

This gives the theorem of Agaian and Sarukhanyan (see [1]) as a corollary by setting all variables equal to 1.

We now give as a corollary a result, motivated by (and a little stronger than) that of Agaian and Sarukhanyan (see [1]).

Corollary 6.2: *Suppose there are Williamson or Williamson type matrices of orders u and v. Then there are Williamson type matrices of order 2uv. If the matrices of orders u and v are symmetric, the matrices of order 2uv are also symmetric. If the matrices of orders u and v are circulant and/or type one the matrices of order 2uv are type 1.*

Proof: Suppose A, B, C, D are (symmetric) Williamson or Williamson type matrices of order u then they are pairwise amicable. Define

$$E = \frac{1}{2}(A+B),\quad F = \frac{1}{2}(A-B),\quad G = \frac{1}{2}(C+D),\quad H = \frac{1}{2}(C-D),$$

then E, F, G, H are pairwise amicable (and symmetric) and satisfy

$$EE^{\mathsf{T}} + FF^{\mathsf{T}} + GG^{\mathsf{T}} + HH^{\mathsf{T}} = 2uI_u.$$

Now define

$$T_1 = \begin{bmatrix} E & 0 \\ 0 & E \end{bmatrix},\quad T_2 = \begin{bmatrix} F & 0 \\ 0 & F \end{bmatrix},\quad T_3 = \begin{bmatrix} 0 & G \\ G & 0 \end{bmatrix},\quad \text{and } T_4 = \begin{bmatrix} 0 & H \\ H & 0 \end{bmatrix},$$

so that $T_1T_1^\top + T_2T_2^\top + T_3T_3^\top + T_4T_4^\top = 2uI_{2u}$ and

$$\begin{aligned} T_1 &= T_{11} = T_{22} = T_{33} = T_{44}, \\ T_2 &= T_{12} = -T_{21} = T_{34} = -T_{43}, \\ T_3 &= T_{13} = -T_{31} = -T_{24} = T_{42}, \\ T_4 &= T_{14} = -T_{41} = T_{23} = -T_{32}, \end{aligned}$$

in the theorem. Note T_1, T_2, T_3, T_4 are pairwise amicable. If A, B, C, D were circulant (or type 1) they would be type 1 of order $2u$.

Let X, Y, Z, W be the Williamson or Williamson type (symmetric) matrices of order v. Then X, Y, Z, W are pairwise amicable and

$$XX^\top + YY^\top + ZZ^\top + WW^\top = 4vI_v.$$

Then

$$\begin{aligned} L &= T_1 \times X + T_2 \times Y + T_3 \times Z + T_4 \times W \\ M &= -T_1 \times Y + T_2 \times X + T_3 \times W - T_4 \times Z \\ N &= -T_1 \times Z - T_2 \times W + T_3 \times X + T_4 \times Y \\ P &= -T_1 \times W + T_2 \times Z - T_3 \times Y + T_4 \times X. \end{aligned}$$

are 4 Williamson type (symmetric) matrices of order $2uv$. If the matrices of orders u and v were circulant or type 1 these matrices are type 1. □

6.5 Miyamoto's Theorem and Corollaries via *M*-structures

In this subsection, we reformulate Miyamoto's [137] results so that symmetric Williamson-type matrices can be obtained. The results given here are due to Miyamoto, Seberry, and Yamada.

Lemma 6.4 (*Miyamoto's Lemma Reformulated by Seberry–Yamada [165]*): *Let U_i, V_j, $i, j = 1, 2, 3, 4$ be $(0, +1, -1)$ matrices of order n which satisfy*

i) U_i, U_j, $i \neq j$ are pairwise amicable,
ii) V_i, V_j, $i \neq j$ are pairwise amicable,
iii) $U_i \pm V_i$, $(+1, -1)$ matrices, $i = 1, 2, 3, 4$,
iv) the row sum of U_1 is 1, and the row sum of U_j is zero, $j = 2, 3, 4$,
v) $\sum_{i=1}^{4} U_iU_i^\top = (2n+1)I - 2J$, $\sum_{i=1}^{4} V_iV_i^\top = (2n+1)I$.

Then there are 4 Williamson type matrices of order $2n + 1$. Hence there is a Williamson type Hadamard matrix of order $4(2n + 1)$. If U_i and V_i are symmetric, $i = 1, 2, 3, 4$ then the Williamson-type matrices are symmetric.

Proof: Let S_1, S_2, S_3, S_4 be 4 $(+1, -1)$-matrices of order $2n$ defined by

$$S_j = U_j \times \begin{bmatrix} 1 & 1 \\ 1 & 1 \end{bmatrix} + V_j \times \begin{bmatrix} 1 & -1 \\ -1 & 1 \end{bmatrix}.$$

So the row sum of $S_1 = 2$ and of $S_i = 0$, $i = 2, 3, 4$. Now define

$$X_1 = \begin{bmatrix} 1 & -e_{2n} \\ -e_{2n}^\top & S_1 \end{bmatrix} \quad \text{and} \quad X_i = \begin{bmatrix} 1 & e_{2n} \\ e_{2n}^\top & S_i \end{bmatrix}, \quad i = 2, 3, 4.$$

First note that since $U_i, U_j, i \neq j$ and $V_i, V_j, i \neq j$ are pairwise amicable,

$$\begin{aligned}
S_iS_j^{\mathrm{T}} &= \left(U_i \times \begin{bmatrix} 1 & 1 \\ 1 & 1 \end{bmatrix} + V_i \times \begin{bmatrix} 1 & -1 \\ -1 & 1 \end{bmatrix}\right)\left(U_j^{\mathrm{T}} \times \begin{bmatrix} 1 & 1 \\ 1 & 1 \end{bmatrix} + V_j^{\mathrm{T}} \times \begin{bmatrix} 1 & -1 \\ -1 & 1 \end{bmatrix}\right) \\
&= U_iU_j^{\mathrm{T}} \times \begin{bmatrix} 2 & 2 \\ 2 & 2 \end{bmatrix} + V_iV_j^{\mathrm{T}} \times \begin{bmatrix} 2 & -2 \\ -2 & 2 \end{bmatrix} \\
&= S_jS_i^{\mathrm{T}}.
\end{aligned}$$

(Note this relationship is valid if and only if conditions (i) and (ii) of the theorem are valid.)

$$\begin{aligned}
\sum_{i=1}^{4} S_iS_i^{\mathrm{T}} &= \sum_{i=1}^{4} U_iU_i^{\mathrm{T}} \times \begin{bmatrix} 2 & 2 \\ 2 & 2 \end{bmatrix} + \sum_{i=1}^{4} V_iV_i^{\mathrm{T}} \times \begin{bmatrix} 2 & -2 \\ -2 & 2 \end{bmatrix} \\
&= 2\begin{bmatrix} 2(2n+1)I - 2J & -2J \\ -2J & 2(2n+1)I - 2J \end{bmatrix} \\
&= 4(2n+1)I_{2n} - 4J_{2n}.
\end{aligned}$$

Next we observe

$$X_1X_i^{\mathrm{T}} = \begin{bmatrix} 1-2n & e_{2n} \\ e_{2n}^{\mathrm{T}} & -J + S_1S_i^{\mathrm{T}} \end{bmatrix} = X_iX_1^{\mathrm{T}} \qquad i = 2, 3, 4,$$

and

$$X_iX_j^{\mathrm{T}} = \begin{bmatrix} 1+2n & e_{2n} \\ e_{2n}^{\mathrm{T}} & J + S_iS_j^{\mathrm{T}} \end{bmatrix} = X_jX_i^{\mathrm{T}} \qquad i \neq j,\ i,j = 2, 3, 4.$$

Further

$$\begin{aligned}
\sum_{i=1}^{4} X_iX_i^{\mathrm{T}} &= \begin{bmatrix} 1+2n & -3e_{2n} \\ -3e_{2n}^{\mathrm{T}} & J + S_1S_1^{\mathrm{T}} \end{bmatrix} + \sum_{i=2}^{4} \begin{bmatrix} 1+2n & e_{2n} \\ e_{2n}^{\mathrm{T}} & J + S_iS_i^{\mathrm{T}} \end{bmatrix} \\
&= \begin{bmatrix} 4(2n+1) & 0 \\ 0 & 4J + 4(2n+1)I - 4J \end{bmatrix}.
\end{aligned}$$

Thus, we have shown that X_1, X_2, X_3, X_4 are 4 Williamson type matrices of order $2n+1$. Hence there is a Williamson type Hadamard matrix of order $4(2n+1)$. □

Many powerful corollaries which give many new results exist by suitable choices in the theorem. For example

Corollary 6.3: *Let $q \equiv 1 \pmod 4$ be a prime power. Then there are symmetric Williamson type matrices of order $q+2$ whenever $\frac{1}{2}(q+1)$ is a prime power or $\frac{1}{2}(q+3)$ is the order of a symmetric conference matrix. Also there exists a Williamson Hadamard matrix of order $4(q+2)$.*

Corollary 6.4: *Let $q \equiv 1 \pmod 4$ be a prime power. Then*

i) if there are Williamson type matrices of order $(q-1)/4$ or an Hadamard matrix of order $\frac{1}{2}(q-1)$, there exist Williamson type matrices of order q;
ii) if there exist symmetric conference matrices of order $\frac{1}{2}(q-1)$ or a symmetric Hadamard matrix of order $\frac{1}{2}(q-1)$, then there exist symmetric Williamson type matrices of order q.

Hence there exists a Williamson Hadamard matrix of order $4q$.

Corollary 6.5: *Let $q \equiv 1 \pmod 4$ be a prime power or $q+1$ be the order of a symmetric conference matrix. Let $2q-1$ be a prime power. Then there exist symmetric Williamson type matrices of order $2q+1$ and a Williamson Hadamard matrix of order $4(2q+1)$.*

Note that this last corollary is a modified version of Miyamoto's Corollary 5 (original unpublished manuscript).

Theorem 6.7 (Miyamoto's theorem [137] reformulated by Seberry and Yamada [166]): *Let U_{ij}, V_{ij}, $i,j = 1,2,3,4$ be $(0,+1,-1)$ matrices of order n that satisfy*

i) U_{ki}, U_{kj} *are pairwise amicable,* $k = 1,2,3,4, i \neq j$,
ii) V_{ki}, V_{kj} *are pairwise amicable,* $k = 1,2,3,4, i \neq j$,
iii) $U_{ki} \pm V_{ki}$, $(+1,-1)$ *matrices,* $i,k = 1,2,3,4$,
iv) the row sum of U_{ii} *is 1, and the row sum of* U_{ij} *is zero,* $i \neq j$, $i,j = 1,2,3,4$,
v) $\sum_{i=1}^{4} U_{ji}U_{ji}^{\top} = (2n+1)I - 2J$, $\sum_{i=1}^{4} V_{ji}V_{ji}^{\top} = (2n+1)I$, $j = 1,2,3,4$,
vi) $\sum_{i=1}^{4} U_{ji}U_{ki}^{\top} = 0$, $\sum_{i=1}^{4} V_{ji}V_{ki}^{\top} = 0, j \neq k, j,k = 1,2,3,4$.

If conditions (i)–(v) hold, there are 4 Williamson-type matrices of order $2n+1$ and thus a Williamson type Hadamard matrix of order $4(2n+1)$. Furthermore if the matrices U_{ki} and V_{ki} are symmetric for all $i,j = 1,2,3,4$, the Williamson matrices obtained of order $2n+1$ are also symmetric.

If conditions (iii)–(vi) hold, there is an M-structure Hadamard matrix of order $4(2n+1)$.

Proof: We prove the first assertion. Let S_{ij}, $i,j = 1,2,3,4$, be 16 $(+1,-1)$-matrices of order $2n$ defined by

$$S_{ij} = U_{ij} \times \begin{bmatrix} 1 & 1 \\ 1 & 1 \end{bmatrix} + V_{ij} \times \begin{bmatrix} 1 & -1 \\ -1 & 1 \end{bmatrix}.$$

So the row sum of $S_{ii} = 2$ and of $S_{ij} = 0$, $i \neq j$, $i,j = 1,2,3,4$. Now define

$$X_{11} = \begin{bmatrix} -1 & -e \\ -e^{\top} & S_{11} \end{bmatrix} \quad X_{12} = \begin{bmatrix} 1 & e \\ e^{\top} & S_{12} \end{bmatrix} \quad X_{13} = \begin{bmatrix} 1 & e \\ e^{\top} & S_{13} \end{bmatrix} \quad X_{14} = \begin{bmatrix} -1 & e \\ e^{\top} & S_{14} \end{bmatrix}$$

$$X_{21} = \begin{bmatrix} 1 & e \\ e^{\top} & S_{21} \end{bmatrix} \quad X_{22} = \begin{bmatrix} -1 & -e \\ -e^{\top} & S_{22} \end{bmatrix} \quad X_{23} = \begin{bmatrix} 1 & e \\ e^{\top} & S_{23} \end{bmatrix} \quad X_{24} = \begin{bmatrix} -1 & e \\ e^{\top} & S_{24} \end{bmatrix}$$

$$X_{31} = \begin{bmatrix} 1 & e \\ e^{\top} & S_{31} \end{bmatrix} \quad X_{32} = \begin{bmatrix} 1 & e \\ e^{\top} & S_{32} \end{bmatrix} \quad X_{33} = \begin{bmatrix} -1 & -e \\ -e^{\top} & S_{33} \end{bmatrix} \quad X_{34} = \begin{bmatrix} -1 & e \\ e^{\top} & S_{34} \end{bmatrix}$$

$$X_{41} = \begin{bmatrix} -1 & e \\ e^{\top} & -S_{41} \end{bmatrix} \quad X_{42} = \begin{bmatrix} 1 & e \\ e^{\top} & -S_{42} \end{bmatrix} \quad X_{43} = \begin{bmatrix} -1 & e \\ e^{\top} & -S_{43} \end{bmatrix} \quad X_{44} = \begin{bmatrix} -1 & -e \\ -e^{\top} & -S_{44} \end{bmatrix}$$

Thus $X_{41}, X_{42}, X_{43}, X_{44}$ are 4 Williamson type matrices of order $2n+1$ and thus a Williamson type Hadamard matrix of order $4(2n+1)$ exists.

Assume that the conditions (*iii*)–(*vi*) hold. The assumption (*iv*) assures

$$\sum_{i=1}^{4} S_{ki}S_{ji}^{\top} = 0, j \neq k,$$

then

$$\sum_{i=1}^{4} X_{ji}X_{ki}^{\top} = 0.$$

Since $\sum_{i=1}^{4} X_{ji}X_{ji}^{\top} = 4(2n+1)I_{2n+1}, j = 1,2,3,4$, there is an M-structure Hadamard matrix of order $4(2n+1)$. □

Note that if we write our M-structure from the theorem as

$$\begin{bmatrix} -1 & 1 & 1 & -1 & -e & e & e & e \\ 1 & -1 & 1 & -1 & e & -e & e & e \\ 1 & 1 & -1 & -1 & e & e & -e & e \\ 1 & 1 & 1 & 1 & -e & -e & -e & e \\ -e^\top & e^\top & e^\top & e^\top & S_{11} & S_{12} & S_{13} & S_{14} \\ e^\top & -e^\top & e^\top & e^\top & S_{21} & S_{22} & S_{23} & S_{24} \\ e^\top & e^\top & -e^\top & e^\top & S_{31} & S_{32} & S_{33} & S_{34} \\ -e^\top & -e^\top & -e^\top & e^\top & S_{41} & S_{42} & S_{43} & S_{44} \end{bmatrix}$$

and we can see Yamada's matrix with trimming [236] or the J. Wallis–Whiteman [216] matrix with a border embodied in the construction.

In [159] it is shown, using the propus variation of the Williamson array, where $B = C$, that parts (*i*) – –(*iii*) of Miyamoto's theorem can be used to find

Corollary 6.6: *Let* $q \equiv 1 \pmod 4$ *be a prime power and* $\frac{1}{2}(q+1)$ *be a prime power or the order of the core of a symmetric conference matrix (this happens for* $q = 89$*). Then there exist symmetric Williamson type matrices of order* $2q+1$ *and a symmetric propus-type Hadamard matrix of order* $4(2q+1)$.

Proof: Let R be the back-diagonal identity matrix and Q be the Paley core. If $p \equiv 3 \pmod 4$ is a prime power, we set $U_1 = I, U_2 = U_3 = QR, U_4 = 0$ of order p, and if $p \equiv 1 \pmod 4$ is a prime power, we set $U_1 = I, U_2 = U_3 = Q, U_4 = 0$ of order p. Hence $\sum_{k=1}^{4} U_k U_k^\top = (2q+1)I - 2J$.

From Turyn's result (the matrix in Eq. (3.14)) we set, for $p \equiv 1 \pmod 4$, $V_1 = P, V_2 = V_3 = I$, and $V_4 = S$, and for $p \equiv 3 \pmod 4$, $V_1 = P, V_2 = V_3 = R$, and $V_4 = S$, so $\sum_{k=1}^{4} V_k V_k^\top = (2q+1)I$. □

This gives new symmetric propus-type Hadamard matrices.

Corollary 6.7: *Suppose there exists a symmetric conference matrix of order* $q+1 = 4t+2$ *and an Hadamard matrix of order* $4t = q-1$*. Then there is an Hadamard matrix with M-structure of order* $4(4t+1) = 4q$*. Further, if the Hadamard matrix is symmetric, the Hadamard matrix of order* $4q$ *is of the form*

$$\begin{bmatrix} X & Y \\ -Y & X \end{bmatrix},$$

where X, Y *are amicable and symmetric.*

In a similar fashion, we consider the following lemma so symmetric 8-Williamson type matrices can be obtained.

Lemma 6.5 *(Seberry and Yamada [165])*: *Let* $U_i, V_j, i, j = 1, \ldots, 8$ *be* $(0, +1, -1)$ *matrices of order* n *which satisfy*

i) $U_i, U_j, i \neq j$ *are pairwise amicable,*
ii) $V_i, V_j, i \neq j$ *are pairwise amicable,*
iii) $U_i \pm V_i$, $(+1, -1)$ *matrices,* $i = 1, \ldots, 8$,
iv) the row (column) sums of U_1 *and* U_2 *are both* 1, *and the row sum of* U_i, $i = 3, \ldots, 8$ *is zero,*
v) $\sum_{i=1}^{8} U_i U_i^\top = 2(2n+1)I - 4J$, $\sum_{i=1}^{8} V_i V_i^\top = 2(2n+1)I$.

Then there are 8-Williamson type matrices of order $2n+1$. Furthermore, if the U_i and V_i are symmetric, $i = 1, \dots, 8$, then the 8-Williamson-type matrices are symmetric. Hence there is a M-structure Hadamard matrix of order $8(2n+1)$.

Proof: Let $S_1, \dots, S_8$ be 8 $(+1,-1)$-matrices of order $2n$ defined by

$$S_j = U_j \times \begin{bmatrix} 1 & 1 \\ 1 & 1 \end{bmatrix} + V_j \times \begin{bmatrix} 1 & -1 \\ -1 & 1 \end{bmatrix}.$$

So the row sums of S_1 and S_2 are both 2 and of $S_i = 0$, $i = 3, \dots, 8$. Now define

$$X_j = \begin{bmatrix} 1 & -e_{2n} \\ -e_{2n}^{\mathrm{T}} & S_j \end{bmatrix},\ j = 1, 2 \text{ and } X_i = \begin{bmatrix} 1 & e_{2n} \\ e_{2n}^{\mathrm{T}} & S_i \end{bmatrix},\ i = 3, \dots, 8.$$

Thus we have that $X_1, \dots, X_8$ are 8-Williamson-type matrices of order $2n+1$.

Hence there is a M-structure Hadamard matrix of order $8(2n+1)$ obtained by replacing the variables of an orthogonal design $OD(8; 1, 1, 1, 1, 1, 1, 1, 1)$ by the 8-Williamson-type matrices. □

Some very powerful corollaries are:

Corollary 6.8 *(Seberry and Yamada [165])*: *Let $q+1$ be the order of amicable Hadamard matrices $I+S$ and P. Suppose there exist 4 Williamson-type matrices of order q. Then there exist Williamson-type matrices of order $2q+1$. Furthermore, there exists a 64 block M-structure Hadamard matrix of order $8(2q+1)$.*

Corollary 6.9: *Let q be a prime power and $(q-1)/2$ be the order of (symmetric) 4 Williamson-type matrices. Then there exist (symmetric) 8-Williamson-type matrices of order q and a 64 block M-structure Hadamard matrix of order $8q$.*

Corollary 6.10: *Let $q \equiv 1 \pmod 4$ be a prime power or $q+1$ be the order of a symmetric conference matrix. Suppose there exist (symmetric) 4 Williamson-type matrices of order q. Then there exist (symmetric) 8-Williamson-type matrices of order $2q+1$ and a 64 block M-structure Hadamard matrix of order $8(2q+1)$.*

Proof: Form the Paley core Q. Thus we choose

$$U_1 = I + Q,\quad U_2 = I - Q,\quad U_3 = U_4 = Q,\quad U_5 = U_6 = U_7 = U_8 = 0$$
$$\text{and } V_1 = V_2 = 0,\quad V_3 = V_4 = I,\quad V_{i+4} = W_i,$$

$i = 1, 2, 3, 4$, where W_i are (symmetric) Williamson-type matrices. Then

$$\sum_{i=1}^{8} U_i U_i^{\mathrm{T}} = 2(2q+1)I - 4J,\quad \sum_{i=1}^{8} V_i V_i^{\mathrm{T}} = 2(2q+1)I.$$

These U_i and V_i are then used in Lemma 6.5 to obtain the (symmetric) 8-Williamson-type matrices. □

This corollary gives 8-Williamson-type matrices for many new orders but it does not give new Hadamard matrices for these orders.

Using Tuyrn's [165] construction for Williamson matrices of order 9^t gives another construction for orders 49^t and $8\{2 \cdot 9^t + 1\}$

Also we have the following theorem:

Theorem 6.8 *(Seberry and Yamada [165])*: *Let U_{ij}, V_{ij}, $i,j = 1, \dots, 8$ be $(0,+1,-1)$ matrices of order n which satisfy*

i) U_{ki}, U_{kj} *are pairwise amicable,* $k = 1, \dots, 8$, $i \neq j$,
ii) V_{ki}, V_{kj} *are pairwise amicable,* $k = 1, \dots, 8$, $i \neq j$,
iii) $U_{ki} \pm V_{ki}$, $(+1, -1)$ *matrices,* $i, k = 1, \dots, 8$,
iv) *the row(column) sum of* U_{ab} *is* 1 *for* $(a, b)\epsilon\{(i, i), (i, i+1), (i+1, i)\}$, $i = 1, 3, 5, 7$, *the row(column) sum of* U_{aa} *is* -1 *for* $(a, a) = 2, 4, 6, 8$ *and otherwise, and the row(column) sum of* U_{ij}, $i \neq j$ *is zero,*
v) $\sum_{i=1}^{8} U_{ji}U_{ji}^{\top} = 2(2n+1)I - 4J$, $\sum_{i=1}^{8} V_{ji}V_{ji}^{\top} = 2(2n+1)I$, $j = 1, \dots, 8$,
vi) $\sum_{i=1}^{8} U_{ji}U_{ki}^{\top} = 0$, $\sum_{i=1}^{8} V_{ji}V_{ki}^{\top} = 0$, $j \neq k$, $j, k = 1, \dots, 8$.

If (i)–(v) hold, there are 8-Williamson-type matrices of order $2n+1$ *and thus a M-structure Hadamard matrix of order* $8(2n+1)$. *Further, if* U_{7i}, V_{7i} *are symmetric,* $1 \leq i \leq 8$, *then the 8-Williamson-type matrices are symmetric.*

If (iii)–(vi) hold, there is a 64 block M-structure Hadamard matrix of order $8(2n+1)$.

Proof: Let S_{ij} be 64 $(+1, -1)$-matrices of order $2n$ defined by

$$S_{ij} = U_{ij} \times \begin{bmatrix} 1 & 1 \\ 1 & 1 \end{bmatrix} + V_{ij} \times \begin{bmatrix} 1 & -1 \\ -1 & 1 \end{bmatrix}.$$

So the row(column) sum of S_{ii}, $S_{i,i+1}$, $S_{i+1,i}$ $i = 1, 3, 5, 7$ is 2, the row(column) sum of S_{ii} is -2 for (i, i), $i = 2, 4, 6, 8$ and otherwise, the row(column) sum of $S_{ij} = 0$, $i \neq j$. Now define

$$X_{11} = \begin{bmatrix} -1 & -e \\ -e^{\top} & S_{11} \end{bmatrix}, X_{12} = \begin{bmatrix} -1 & -e \\ -e^{\top} & S_{12} \end{bmatrix}, X_{13} = \begin{bmatrix} 1 & e \\ e^{\top} & S_{13} \end{bmatrix}, \quad X_{14} = \begin{bmatrix} 1 & e \\ e^{\top} & S_{14} \end{bmatrix},$$

$$X_{15} = \begin{bmatrix} 1 & e \\ e^{\top} & S_{15} \end{bmatrix}, \quad X_{16} = \begin{bmatrix} 1 & e \\ e^{T} & S_{16} \end{bmatrix}, \quad X_{17} = \begin{bmatrix} -1 & e \\ e^{\top} & S_{17} \end{bmatrix}, \quad X_{18} = \begin{bmatrix} -1 & e \\ e^{\top} & S_{18} \end{bmatrix},$$

$$X_{21} = \begin{bmatrix} -1 & -e \\ -e^{\top} & S_{21} \end{bmatrix}, X_{22} = \begin{bmatrix} 1 & e \\ e^{\top} & S_{22} \end{bmatrix}, \quad X_{23} = \begin{bmatrix} 1 & e \\ e^{\top} & S_{23} \end{bmatrix}, \quad X_{24} = \begin{bmatrix} -1 & -e \\ -e^{\top} & S_{24} \end{bmatrix},$$

$$X_{25} = \begin{bmatrix} 1 & e \\ e^{\top} & S_{25} \end{bmatrix}, \quad X_{26} = \begin{bmatrix} -1 & -e \\ -e^{\top} & S_{26} \end{bmatrix}, X_{27} = \begin{bmatrix} -1 & e \\ e^{\top} & S_{27} \end{bmatrix}, \quad X_{28} = \begin{bmatrix} 1 & -e \\ -e^{\top} & S_{28} \end{bmatrix},$$

$$X_{31} = \begin{bmatrix} 1 & e \\ e^{\top} & S_{31} \end{bmatrix}, \quad X_{32} = \begin{bmatrix} 1 & e \\ e^{\top} & S_{32} \end{bmatrix}, \quad X_{33} = \begin{bmatrix} -1 & -e \\ -e^{\top} & S_{33} \end{bmatrix}, X_{34} = \begin{bmatrix} -1 & -e \\ -e^{\top} & S_{34} \end{bmatrix},$$

$$X_{35} = \begin{bmatrix} 1 & e \\ e^{\top} & S_{35} \end{bmatrix}, \quad X_{36} = \begin{bmatrix} 1 & e \\ e^{\top} & S_{36} \end{bmatrix}, \quad X_{37} = \begin{bmatrix} -1 & e \\ e^{\top} & S_{37} \end{bmatrix}, \quad X_{38} = \begin{bmatrix} -1 & e \\ e^{\top} & S_{38} \end{bmatrix},$$

$$X_{41} = \begin{bmatrix} 1 & e \\ e^{\top} & S_{41} \end{bmatrix}, \quad X_{42} = \begin{bmatrix} -1 & -e \\ -e^{\top} & S_{42} \end{bmatrix}, X_{43} = \begin{bmatrix} -1 & -e \\ -e^{\top} & !S_{43} \end{bmatrix}, X_{44} = \begin{bmatrix} 1 & e \\ e^{\top} & S_{44} \end{bmatrix},$$

$$X_{45} = \begin{bmatrix} 1 & e \\ e^{\top} & S_{45} \end{bmatrix}, \quad X_{46} = \begin{bmatrix} -1 & -e \\ -e^{\top} & S_{46} \end{bmatrix}, X_{47} = \begin{bmatrix} -1 & e \\ e^{\top} & S_{47} \end{bmatrix}, \quad X_{48} = \begin{bmatrix} 1 & -e \\ -e^{\top} & S_{48} \end{bmatrix},$$

$$X_{51} = \begin{bmatrix} 1 & e \\ e^{\top} & S_{51} \end{bmatrix}, \quad X_{52} = \begin{bmatrix} 1 & e \\ e^{\top} & S_{52} \end{bmatrix}, \quad X_{53} = \begin{bmatrix} 1 & e \\ e^{\top} & S_{53} \end{bmatrix}, \quad X_{54} = \begin{bmatrix} 1 & e \\ e^{\top} & S_{54} \end{bmatrix},$$

$$X_{55} = \begin{bmatrix} -1 & -e \\ -e^{\top} & S_{55} \end{bmatrix}, X_{56} = \begin{bmatrix} -1 & -e \\ -e^{\top} & S_{56} \end{bmatrix}, X_{57} = \begin{bmatrix} -1 & e \\ e^{\top} & S_{57} \end{bmatrix}, X_{58} = \begin{bmatrix} -1 & e \\ e^{\top} & S_{58} \end{bmatrix},$$

$$X_{61} = \begin{bmatrix} 1 & e \\ e^{\top} & S_{61} \end{bmatrix}, X_{62} = \begin{bmatrix} -1 & -e \\ -e^{\top} & S_{62} \end{bmatrix}, X_{63} = \begin{bmatrix} 1 & e \\ e^{\top} & S_{63} \end{bmatrix}, X_{64} = \begin{bmatrix} -1 & -e \\ -e^{\top} & S_{64} \end{bmatrix},$$

$$X_{65} = \begin{bmatrix} -1 & -e \\ -e^{\top} & S_{65} \end{bmatrix}, X_{66} = \begin{bmatrix} 1 & e \\ e^{\top} & S_{66} \end{bmatrix}, X_{67} = \begin{bmatrix} -1 & e \\ e^{\top} & S_{67} \end{bmatrix}, X_{68} = \begin{bmatrix} 1 & -e \\ -e^{\top} & S_{68} \end{bmatrix},$$

$$X_{71} = \begin{bmatrix} 1 & -e \\ -e^{\top} & S_{71} \end{bmatrix}, X_{72} = \begin{bmatrix} 1 & -e \\ -e^{\top} & S_{72} \end{bmatrix}, X_{73} = \begin{bmatrix} 1 & -e \\ -e^{\top} & S_{73} \end{bmatrix}, X_{74} = \begin{bmatrix} -1 & -e \\ -e^{\top} & S_{74} \end{bmatrix},$$

$$X_{75} = \begin{bmatrix} 1 & -e \\ -e^{\top} & S_{75} \end{bmatrix}, X_{76} = \begin{bmatrix} 1 & -e \\ -e^{\top} & S_{76} \end{bmatrix}, X_{77} = \begin{bmatrix} 1 & e \\ e^{\top} & S_{77} \end{bmatrix}, X_{78} = \begin{bmatrix} 1 & e \\ e^{\top} & S_{78} \end{bmatrix},$$

$$X_{81} = \begin{bmatrix} 1 & -e \\ -e^{\top} & S_{81} \end{bmatrix}, X_{82} = \begin{bmatrix} -1 & e \\ e^{\top} & S_{82} \end{bmatrix}, X_{83} = \begin{bmatrix} 1 & -e \\ -e^{\top} & S_{83} \end{bmatrix}, X_{84} = \begin{bmatrix} 1 & e \\ e^{\top} & S_{84} \end{bmatrix},$$

$$X_{85} = \begin{bmatrix} 1 & -e \\ -e^{\top} & S_{85} \end{bmatrix}, X_{86} = \begin{bmatrix} -1 & e \\ e^{\top} & S_{86} \end{bmatrix}, X_{87} = \begin{bmatrix} 1 & e \\ e^{\top} & S_{87} \end{bmatrix}, X_{88} = \begin{bmatrix} -1 & -e \\ -e^{\top} & S_{88} \end{bmatrix}.$$

Then provided conditions (*i*)–(*v*) hold and $S_{7i}^{\top} = S_{7i}$, $i = 1, \ldots, 8$ are symmetric, X_{7i}, $i = 1, \ldots, 8$ are symmetric 8-Williamson-type matrices. Otherwise X_{7i}, $i = 1, \ldots, 8$ are 8-Williamson-type matrices. This can be verified by straightforward checking. Hence there is a M-structure Hadamard matrix of order $8(2n + 1)$.

Table 6.2 *M*-structure Hadamard matrix of order 8(2*n* + 1).

−1	−1	1	1	1	1	-1	−1	$-e$	$-e$	e	e	e	e	e	e
−1	1	1	1	−1	1	−1	1	$-e$	e	e	$-e$	e	$-e$	e	$-e$
1	1	-1	−1	1	1	-1	−1	e	e	$-e$	$-e$	e	e	e	e
1	−1	-1	1	1	−1	-1	1	e	$-e$	$-e$	e	e	$-e$	e	$-e$
1	1	1	1	−1	−1	−1	−1	e	e	e	e	$-e$	$-e$	e	e
1	−1	1	−1	−1	1	−1	1	e	$-e$	e	$-e$	$-e$	e	e	$-e$
1	1	1	1	1	1	1	1	$-e$	$-e$	$-e$	$-e$	$-e$	-e	e	e
1	−1	1	−1	1	−1	1	−1	$-e$	e	$-e$	$-e$	e	e	$-e$	e
$-e^{\top}$	$-e^{\top}$	$e^{\top}$	$e^{\top}$	$e^{\top}$	$e^{\top}$	$e^{\top}$	$e^{\top}$	S_{11}	S_{12}	S_{13}	S_{14}	S_{15}	S_{16}	S_{17}	S_{18}
$-e^{\top}$	$e^{\top}$	$e^{\top}$	$-e^{\top}$	$e^{\top}$	$-e^{\top}$	$e^{\top}$	$-e^{\top}$	S_{21}	S_{22}	S_{23}	S_{24}	S_{25}	S_{26}	S_{27}	S_{28}
$e^{\top}$	$e^{\top}$	$-e^{\top}$	$-e^{\top}$	$e^{\top}$	$e^{\top}$	$e^{\top}$	$e^{\top}$	S_{31}	S_{32}	S_{33}	S_{34}	S_{35}	S_{36}	S_{37}	S_{38}
$e^{\top}$	$-e^{\top}$	$-e^{\top}$	$e^{\top}$	$e^{\top}$	$-e^{\top}$	$e^{\top}$	$-e^{\top}$	S_{41}	S_{42}	S_{43}	S_{44}	S_{45}	S_{46}	S_{47}	S_{48}
$e^{\top}$	$e^{\top}$	$e^{\top}$	$e^{\top}$	$-e^{\top}$	$-e^{\top}$	$e^{\top}$	$e^{\top}$	S_{51}	S_{52}	S_{53}	S_{54}	S_{55}	S_{56}	S_{57}	S_{58}
$e^{\top}$	$-e^{\top}$	$e^{\top}$	$-e^{\top}$	$-e^{\top}$	$e^{\top}$	$e^{\top}$	$-e^{\top}$	S_{61}	S_{62}	S_{63}	S_{64}	S_{65}	S_{66}	S_{67}	S_{68}
$-e^{\top}$	$-e^{\top}$	$-e^{\top}$	$-e^{\top}$	$-e^{\top}$	$-e^{\top}$	$e^{\top}$	$e^{\top}$	S_{71}	S_{72}	S_{73}	S_{74}	S_{75}	S_{76}	S_{77}	S_{78}
$-e^{\top}$	$e^{\top}$	$-e^{\top}$	$e^{\top}$	$-e^{\top}$	$e^{\top}$	$e^{\top}$	$-e^{\top}$	S_{81}	S_{82}	S_{83}	S_{84}	S_{85}	S_{86}	S_{87}	S_{88}

Source: Adapted from Seberry and Yamada [166, p. 510], Wiley

If conditions (*iii*)–(*vi*) hold then straightforward verification shows the 64 block M-structure X_{ij} is an Hadamard matrix of order $8(2n+1)$. □

Corollary 6.11: *Let q be an odd prime power and suppose there exist Williamson-type matrices of order $\frac{1}{2}(q-1)$. Then there exists an M-structure Hadamard matrix of order $8q$.*

Corollary 6.12: *Let $q = 2m + 1 \equiv 9 \pmod{16}$ be a prime power. Suppose there are Williamson-type matrices of order q. Then there is a M-structure Hadamard matrix of order $8(2q+1)$.*

The analogous Yamada–J. Wallis–Whiteman structure to Theorem 6.8 is given in Table 6.2.

We pose a further research problem in Appendix C concerning trimming and bordering of M-structures.

7

Menon Hadamard Difference Sets and Regular Hadamard Matrices

7.1 Notations

Table 7.1 gives the notations which are used in this chapter.

7.2 Menon Hadamard Difference Sets and Exponent Bound

A difference set over an abelian group with parameters $v = 4u^2$, $k = 2u^2 - u$, and $\lambda = u^2 - u$ is called a **Menon Hadamard difference set**. Alternative names, Hadamard difference sets, Menon difference sets are used.

In 1965 Turyn [189] introduced the concept of self-conjugate and gave the exponent bound of the Sylow p-subgroup S of an abelian group G containing a difference set. Namely, if a prime divisor p of v is self-conjugate (mod $\exp(G/U)$), then $\exp S \leq \frac{|U|}{p^a}|S|$ where U is a subgroup of G such that $U \cap S = \{1_G\}$ and $p^{2a}|n$. This bound is called Turyn's exponent bound.

In 1963, Yamamoto [244] introduced the concept of the decomposition field of a difference set and showed that the decomposition field of the difference set over an abelian group cannot be real. Then he showed a necessary condition for existence is given by Diophantine equations and the property of self-conjugate for a prime divisor p of n is equivalent to that the decomposition field of p is real [244].

In 1998, Schmidt [151] developed a new exponent bound for any group (not necessary abelian) containing a difference set without self-conjugacy condition. Namely, he gave the exponent bound of G/U where G is a group and U is a normal subgroup such that G/U is cyclic. See more details in [15].

Assume that the order of G is a power of 2, namely G has a Menon Hadamard difference set with $v = 4 \cdot 2^{2t}$, $k = 2 \cdot 2^{2t} - 2^t$, and $\lambda = 2^{2t} - 2^t$. Kraemer [122] and Davis [39] proved Turyn's bound is a necessary and sufficient condition (for more details , see [41]).

Theorem 7.1 ***(Kraemer [122], Davis [39]):*** *There exists an abelian $(2^{2t+2}, 2^{2t+1} - 2^t, 2^{2t} - 2^t)$ Menon Hadamard difference set if and only if exp$(G) \leq 2^{t+2}$.*

For $u = 3$, Turyn showed the existence of a $(36, 15, 6)$ difference set [189]. For the case where $u > 3$ is a prime, we have the following theorem.

Theorem 7.2 ***(McFarland [135]):*** *If $u > 3$ is a prime, then there does not exist a Menon Hadamard difference set.*

This reduces us to the case where u is a composite integer.

Hadamard Matrices: Constructions using Number Theory and Algebra, First Edition. Jennifer Seberry and Mieko Yamada.

Table 7.1 Notations used in this chapter.

$\boldsymbol{F}_q$, $GF(q)$	A finite field with q elements for a prime power q
$(4u^2, 2u^2 \pm u, u^2 \pm u)$ difference set	Menon Hadamard difference set
$\boldsymbol{Z}$	The rational integer ring
$\boldsymbol{Z}[G]$	A group ring of a group G over $\boldsymbol{Z}$
$\boldsymbol{Z}_n = \boldsymbol{Z}/n\boldsymbol{Z}$	A residue class ring modulo n
ζ_m	A primitive mth root of unity
$\widehat{K^+}$	The group of additive characters of a finite field K
T_K	The absolute trace from a finite field K to a prime subfield of K
S_i	eth cyclotomic class, $0 \leq i \leq e-1$
$X_i = \sum_{x \in S_i} \zeta_p^{T_K(x)}$	The Gaussian period for eth cyclotomic class S_i, $0 \leq i \leq e-1$
Z_n	A cyclic group of order n
J	The matrix of all entries 1
$V_k(q)$	k-Dimensional vector space over $GF(q)$
$PG(k-1, q)$	The projective space of dimension $k-1$ over $GF(q)$
$\mathcal{A}_n$	$\boldsymbol{Z}/2^n\boldsymbol{Z}$.
$\mathcal{R}_n = GR(2^n, s)$	A Galois ring of characteristic 2^n and of an extension degree s
$\mathcal{R}_n^\times$	The unit group of $\mathcal{R}_n$
$\mathfrak{p}_n = 2\mathcal{R}_n$	A unique maximal ideal of $\mathcal{R}_n$
$\mathcal{E}$	The set of principal units
$\alpha^F = \alpha_0^2 + 2\alpha_1^2 + \cdots + 2^{n-1}\alpha_{n-1}^2$	A Frobenius automorphism of $\mathcal{R}_n$
$\mathcal{T}_n = \{0, 1, \xi, \ldots, \xi^{2^s-2}\}$	Teichmüller system.
$T_n(\alpha) = \alpha + \alpha^F + \cdots + \alpha^{F^{s-1}}$	The relative trace from $GR(2^n, s)$ to $\mathcal{A}_n$
$\tau_{n-\ell} : \mathcal{R}_n \to \mathcal{R}_{n-\ell}$	A ring homomorphism from $\mathcal{R}_n$ to $\mathcal{R}_{n-\ell}$
λ_β	An additive character of $\mathcal{R}_n$
$\alpha * \beta$	An operation of $GR(2^n, s)$
$G(\chi, \lambda_\beta)$	The Gauss sum over a Galois ring associated with a multiplicative character χ and an additive character λ_β

7.3 Menon Hadamard Difference Sets and Regular Hadamard Matrices

The existence of a $(4u^2, 2u^2 \pm u, u^2 \pm u)$ Menon Hadamard difference set is equivalent to the existence of a regular Hadamard matrix of order $4u^2$.

Lemma 7.1: *Let H be a $(1, -1)$ matrix if order $n > 2$ and $N = \frac{1}{2}(H + J)$ where J is the matrix with all entries* 1. *Then H is a regular Hadamard matrix of order $n = 4u^2$ if and only if N is the $(1, 0)$ incidence matrix of a $(4u^2, 2u^2 \pm u, u^2 \pm u)$ Menon Hadamard difference set.*

For $u = 3^t$, a power of 3, Mukhopadhyay [139] and Turyn [195] showed that there exists Hadamard matrices of order $4u^2$. Turyn used the matrix representation of $Z_2 \times Z_2 \times \boldsymbol{Z}_3^{2t}$ and $Z_4 \times \boldsymbol{Z}_3^{2t}$.

Let $Z_2 \times Z_2 = \{1, a, b, ab\}$ and G be an abelian group of order u^2. For the symmetric elements A, B, C, and D in $\boldsymbol{Z}[G]$ with ± 1 coefficients, we denote the matrix representation of A, B, C, and D by X, Y, Z, and W, respectively. Then the matrix representation of $A + aB + bC + abD \in \boldsymbol{Z}[Z_2 \times Z_2 \times G]$ is

$$\begin{bmatrix} X & Y & Z & W \\ Y & X & W & Z \\ Z & W & X & Y \\ W & Z & Y & X \end{bmatrix}.$$

Lemma 7.2: *Let G be an abelian group of order u^2. For the symmetric elements A, B, C, and D in $\mathbf{Z}[G]$ with ± 1 coefficients, the matrix representation of $A + aB + bC + abD \in \mathbf{Z}[Z_2 \times Z_2 \times G]$ is an Hadamard matrix of order $4u^2$ if and only if $\tilde{A} + a\tilde{B} + b\tilde{C} + ab\tilde{D}$ is a $(4u^2, 2u^2 - u, u^2 - u)$ Menon Hadamard difference set where $\tilde{A}, \tilde{B}, \tilde{C}$, and $\tilde{D}$ are obtained from A, B, C, and D respectively by replacing the coefficient -1 by 0.*

We have the following theorem under the above conditions.

Theorem 7.3 (*Turyn's composition theorem [195]*): *Let G_i be an abelian group of order u^2 for $i = 1, 2$ and $Z_2 \times Z_2 = \{1, a, b, ab\}$. Assume that symmetric elements $A_i, B_i, C_i, D_i \in \mathbf{Z}[G_i]$ have ± 1 coefficients and matrix representation of $A_i + aB_i + bC_i + abD_i$ is a Hadamard matrix of order $4u_i^2$ for $i = 1, 2$. The subsets of $\mathbf{Z}[G_1 \times G_2]$ are defined by*

$$U_1 = \frac{1}{2}\left\{(A_1 + B_1)A_2 + (A_1 - B_1)B_2\right\},$$
$$U_2 = \frac{1}{2}\left\{(A_1 + B_1)C_2 + (A_1 - B_1)D_2\right\},$$
$$U_3 = \frac{1}{2}\left\{(C_1 + D_1)A_2 + (C_1 - D_1)B_2\right\},$$
$$U_4 = \frac{1}{2}\left\{(C_1 + D_1)C_2 + (C_1 - D_1)D_2\right\}.$$

Then U_i has ± 1 coefficient and the matrix representation of $U_1 + aU_2 + bU_3 + abU_4 \in \mathbf{Z}[Z_2 \times Z_2 \times G_1 \times G_2]$ is an Hadamard matrix of order $4u_1^2u_2^2$.

We have an Hadamard matrix with the same order $4u_1^2u_2^2$ by replacing $Z_2 \times Z_2$ by Z_4.

Turyn proved the existence of a family of Hadamard matrices of order $4 \cdot 3^{2t}$ in $Z_2 \times Z_2 \times \mathbf{Z}_3^{2t}$ and $Z_4 \times \mathbf{Z}_3^{2t}$ from $(36, 15, 6)$ difference set.

We obtain the next theorem from the theorems in the following sections.

Theorem 7.4: *There exists a $(4u^2, 2u^2 - u, u^2 - u)$ Menon Hadamard difference set where $u = 2^{e_1}3^{e_2}m^2, e_1, e_2 \in \{0, 1\}$, and m is a positive integer.*

7.4 The Constructions from Cyclotomy

Xia [226] constructed a family of Menon Hadamard difference sets when u is an odd prime power. Later Xiang–Chen [232] gave a character theoretic proof based on cyclotomy.

Let q be an odd prime power. Put $q - 1 = ef$. Let $S_i, 0 \le i \le e - 1$ be eth cyclotomic classes of $K = GF(q)$. Denote the absolute trace function by T_K and a primitive pth root of unity by ζ_p. Then we recall the Gaussian periods for eth cyclotomic class.

$$X_i = \sum_{x \in S_i} \zeta_p^{T_K(x)}, \quad i = 0, 1, \ldots, e - 1$$

from Chapter 2.

The following theorem determines cyclotomic periods explicitly for these special cases and is the basis for the proof of Xia's construction.

Theorem 7.5 (Myerson [141]): *Assume j is the smallest positive integer such that $p^j \equiv -1 \pmod{e}$. Let $q = p^a, a = 2jr$. Then the Gaussian periods are given by the following:*

(i) *If $r, p, \frac{p^j+1}{e}$ are all odd, then*
$$X_{e/2} = \frac{(e-1)p^{jr} - 1}{e} \quad \text{and} \quad X_j = \frac{-1 - p^{jr}}{e}, \quad j \neq \frac{e}{2}.$$

(ii) *In all other cases,*
$$X_0 = \frac{-1 - (-1)^r (e-1)p^{jr}}{e} \quad \text{and} \quad X_j = \frac{(-1)^r p^{jr} - 1}{e} \quad j \neq 0.$$

Assume $p \equiv -1 \pmod 4$. Let $q = p^{2s}$ where s is a positive even integer. Applying Theorem 7.5 to the case for $e = 4, j = 1$, and $r = s$, we have

$$X_0 = \frac{1}{4}(p^s - 1) - p^s \text{ and } X_j = \frac{1}{4}(p^s - 1),\ j \neq 0. \tag{7.1}$$

Denote the group of additive characters of K by $\widehat{K^+}$. Let g be a primitive element of K and $B_i = \{\chi_{g^k} \in \widehat{K^+} \mid k \equiv i \pmod 4\}$, $i = 0, 1, 2, 3$. Let C_i, $i = 0, 1, 2, 3$ be 4th cyclotomic classes and S_i, $0 \le i \le p^s$ be $(p^s + 1)$st cyclotomic classes.

We see $K = GF(q)$ is a $2s$-dimensional vector space V over $GF(p)$. For $\beta \neq 0$, we define the quadratic form as $Q(x) = T_K(\beta x^{1+p^s})$, $x \in K$. The corresponding bilinear form is $B(x, y) = T_K(\beta x y^{p^s} + \beta x^{p^s} y)$ for all $x, y \in K$. We let $Rad\ V = \{y \in V | B(x,) = 0 \text{ or all } x \in V\}$.

Lemma 7.3 (Xiang and Chen [232]): *If $\beta^{p^s-1} \neq -1$, then $Rad\ V = \{0\}$ and if $\beta^{p^s-1} = -1$, then $Rad\ V = V$.*

Proof: Since $T_K(x) = T_K(x^p)$, for $y \neq 0 \in V$, we have,
$$B(x,y) = T_K\left(\beta x y^{p^s} + \beta x^{p^s} y\right) = T_K\left(\beta^{p^s} x^{p^s} y + \beta x^{p^s} y\right) = T_K\left(\left(\beta^{p^s} y + \beta y\right) x^{p^s}\right) = 0$$
for all $x \in V$. Hence if $\beta^{p^s} + \beta = 0$, then $Rad\ V = V$ and if $\beta^{p^s} + \beta \neq 0$, then $Rad\ V = \{0\}$. □

We put $L_0 = S_1 \bigcup S_5 \bigcup \cdots \bigcup S_{1+4\left(\frac{1}{4}(p^s-1)-1\right)}$. For the nontrivial character $\chi_\beta \in \widehat{K^+}$, χ_β is either nontrivial in every class S_{4k+1}, $0 \le k \le \frac{1}{4}(p^s - 1) - 1$, or trivial on exactly one class S_{4k+1} for some k for Lemma 7.3.

Lemma 7.4 (Xiang and Chen [232]): *We have*
$$\chi_\beta(C_0) = \begin{cases} \frac{1}{4}(p^s - 1) - p^s & \text{if } \chi_\beta \in B_0, \\ \frac{1}{4}(p^s - 1) & \text{if } \chi_\beta \notin B_0, \end{cases}$$
and
$$\chi_\beta(L_0) = \begin{cases} -\frac{1}{4}(p^s - 1) & \text{if } \chi_\beta \in N, \\ p^s - \frac{1}{4}(p^s - 1) & \text{if } \chi_\beta \in T, \end{cases}$$
where
$$T = \left\{ \chi_\beta \in \widehat{K^+} \mid \chi_\beta \text{ is trivial in exactly one class } S_{4k+1} \text{ for some } k \right\}$$

and

$$N = \left\{ \chi_\beta \in \widehat{K^+} \mid \chi_\beta \text{ is nontrivial in every class } S_{4k+1}, 0 \le k \le \frac{1}{4}(p^s - 1) - 1 \right\}.$$

Proof: From the Eq. (7.1), we find $\chi_\beta(C_0)$. Write $e = p^s + 1$. Since $p^s \equiv -1 \pmod{e}$, we have from Theorem 7.5,

$$X_0 = p^s - 1 \quad \text{and} \quad X_j = -1, \quad j \neq 0.$$

If $\chi_\beta \in N$, then $\chi_\beta(L_0) = -\frac{1}{4}(p^s - 1)$ and if $\chi_\beta \in T$, then $\chi_\beta(L_0) = -\left(\frac{1}{4}(p^s-1) - 1\right) + p^s - 1 = p^s - \frac{1}{4}(p^s - 1)$. □

Assume t is a positive odd integer. Let $L_1 = S_{t+1} \bigcup S_{t+5} \bigcup \cdots \bigcup S_{t+1+4(\frac{p^s-1}{4}-1)}$. We define

$$D_0 = C_0 \bigcup L_0,\ D_2 = C_2 \bigcup L_0,\ D_1 = C_t \bigcup L_1,\ D_3 = C_{t+2} \bigcup L_1.$$

We see that $D_2 = g^{p^s+1} D_0, D_1 = g^t D_0, D_3 = g^{p^s+1} D_1$. Let

$$D_i^* = \left\{ \chi_\beta \in \widehat{K^+} \backslash \{\chi^0\} \mid \chi_\beta(D_i) \neq 0 \right\},\ i = 0, 1, 2, 3.$$

Lemma 7.5 *(Xiang and Chen [232]):*

(i) $|D_i| = \frac{1}{2}(q - \sqrt{q})$.

(ii) $\chi_\beta(D_i) = \pm p^s$ *or* 0 *where* χ_β *is a nontrivial character.*

(iii) Let $T' = \left\{ \chi_\beta \in \widehat{K^+} | \chi_\beta(L_1) = p^s - \frac{1}{4}(p^s - 1) \right\}$,

and $N' = \left\{ \chi_\beta \in \widehat{K^+} | \chi_\beta(L_1) = -\frac{1}{4}(p^s - 1) \right\}$.

$$D_0^* = (T \cap B_2) \cup (N \cap B_0),$$
$$D_1^* = (T' \cap B_t) \cup (N' \cap B_{t+2}),$$
$$D_2^* = (T \cap B_0) \cup (N \cap B_2),$$
$$D_3^* = (T' \cap B_{t+2}) \cup (N' \cap B_t),$$

and

$$|D_i^*| = \frac{1}{4}(p^{2s} - 1),\ i = 0, 1, 2, 3.$$

Proof: (i) For $i = 0, 2$, $|D_i| = |C_i| + |L_0| = \frac{1}{4}(p^s - 1) + \frac{p^{2s}-1}{p^s+1}\frac{1}{4}(p^s - 1) = \frac{1}{2}(p^{2s} - p^s) = \frac{1}{2}(q - \sqrt{q})$. We obtain $|D_1| = |D_3| = \frac{1}{2}(q - \sqrt{q})$ similarly.

(ii) From Lemma 7.4, $\chi_\beta(D_0) = \chi_\beta(C_0) + \chi_\beta(L_0) = 0$ or $\pm p^s$. Since

$$\chi_\beta(C_2) = \begin{cases} \frac{1}{4}(p^s - 1) - p^s & \text{if } \chi_\beta \in B_2, \\ \frac{1}{4}(p^s - 1) & \text{if } \chi_\beta \notin B_2, \end{cases}$$

$\chi_\beta(D_2) = \pm p^s$ or 0. Similarly we have

$$\chi_\beta(L_1) = \begin{cases} -\frac{1}{4}(p^s - 1) & \text{if } \chi_\beta \in N_1, \\ p^s - \frac{1}{4}(p^s - 1) & \text{if } \chi_\beta \in T_1, \end{cases}$$

where

$$T_1 = \left\{ \chi_\beta \in \widehat{K^+} \mid \chi_\beta \text{ is trivial in exactly one class } S_{4k+1+t} \text{ for some } k \right\}$$

and

$$N_1 = \left\{ \chi_\beta \in \widehat{K^+} \mid \chi_\beta \text{ is nontrivial in every class } S_{4k+1+t}, 0 \le k \le \frac{1}{4}(p^s - 1) - 1 \right\}.$$

Thus we have $\chi_\beta(D_i) = \pm p^s$ or 0 for $i = 1, 3$.

(*iii*) Assume that χ_{g^i} is trivial on S_{4k+1}. Then

$$T_K(g^{(p^s+1)t+4k+1+i}) = 0, \quad \text{for all } t \in \{0, 1, \dots, p^s - 2\}.$$

We put $\beta = g^{4k+1+i}$ and assume that $\beta^{p^s-1} = -1$. Then the above equality holds if and only if

$$i + 4k + 1 = \frac{1}{2}\left(p^s + 1\right) j, \quad 1 \le j \le 2\left(p^s - 1\right), j : \text{odd}$$

from Lemma 7.3. We see

$$i = \frac{1}{2}\left(p^s + 1\right) j - (4k + 1) \equiv 0, 2 \pmod 4,$$

that is, $\chi_{g^i} \in B_0$ or B_2. Hence we obtain $D_0^* = (T \cap B_2) \cup (N \cap B_0)$ from Lemma 7.4. Since $(T \cap B_0) \cup (N \cap B_0) = B_0$ and $|B_0 \cap T| = |B_2 \cap T| = \frac{1}{2}|T|$,

$$|D_0^*| = |B_2 \cap T| + |B_0 \cap N| = |B_2 \cap T| + |B_0| - |B_0 \cap T| = |B_0| = \frac{1}{4}\left(p^{2s} - 1\right).$$

Similarly, we have $D_2^* = (T \cap B_0) \cup (N \cap B_2)$ and $|D_2^*| = \frac{1}{4}\left(p^{2s} - 1\right)$.
Assume that χ_{g^i} is trivial on S_{4k+1+t}. Then we obtain

$$i = \frac{1}{2}\left(p^s + 1\right) j - (4k + 1 + t) \equiv 1, 3 \pmod 4$$

by replacing $4k + 1$ by $4k + 1 + t$ in the above proof. Assume $t \equiv 1 \pmod 4$.

$$\chi_{g^i}(C_1) = \begin{cases} \frac{1}{4}(p^s - 1) & \text{if } \chi_{g^i} \in B_1, \\ \frac{1}{4}(p^s - 1) - p^s & \text{if } \chi_{g^i} \in B_3. \end{cases}$$

Hence we have $D_1^* = (T' \cap B_1) \cup (N' \cap B_3)$, $D_3^* = (T' \cap B_3) \cup (N' \cap B_1)$ and $|D_1^*| = |D_3^*| = \frac{1}{4}\left(p^{2s} - 1\right)$. We obtain similar results for the case $t \equiv 3 \pmod 4$. With both cases, the assertion follows. □

Let $E = \{1, a, b, ab\}$ be the Klein 4 group (an elementary 2 group) and $C = \{1, c, c^2, c^3\}$ be a cyclic group of order 4.

Theorem 7.6 ***(Xia [226], Xiang and Chen [232]):*** *Assume $p \equiv 3 \pmod 4$. $D = D_0 + aD_1 + bD_2 + ab\overline{D}_3$ is a $(4p^{2s}, 2p^{2s} - p^s, p^{2s} - p^s)$ Menon Hadamard difference set over $E \times K$ where $\overline{D}_3 = K \backslash D_3$. Further D is reversible, $D^{-1} = D$. $D' = D_0 + cD_1 + c^2D_2 + c^3\overline{D}_3$ is a $(4p^{2s}, 2p^{2s} - p^s, p^{2s} - p^s)$ Menon Hadamard difference set over $C \times K$.*

Proof: Let $\psi = \phi \otimes \chi_\beta$ be a character of $E \times K$ where ϕ is a character of E and χ_β is a character of K.

$$\psi(D) = \chi_\beta(D_0) + \phi(a)\chi_\beta(D_1) + \phi(b)\chi_\beta(D_2) + \phi(ab)\chi_\beta(\overline{D_3}).$$

1) If ψ is the trivial character, then $\psi(D) = \frac{3}{2}(q - \sqrt{q}) + (q - \frac{1}{2}(q - \sqrt{q})) = 2p^{2s} - p^s$.

2) If ϕ is a nontrivial character and χ_β is a trivial character, then

$$\psi(D) = |D_0| + \phi(a)|D_1| + \phi(b)|D_2| + \phi(ab)(|K| - |D_3|) = \phi(ab)p^s.$$

3) Assume ϕ and χ_β are nontrivial. Then D_0^*, D_1^*, D_2^*, and D_3^* are a partition of $K^*\backslash\{\chi^0\}$. Hence

$$\psi(D) = \pm p^s$$

from Lemma 7.5. Furthermore $\psi(D) = \overline{\psi(D)}$ holds.

We obtain D' is a $(4p^{2s}, 2p^{2s} - p^s, p^{2s} - p^s)$ Menon Hadamard difference set over $C \times K$ similarly. □

7.5 The Constructions Using Projective Sets

Wilson and Xiang [225] extended Xia's construction of Menon Hadamard difference sets based on projective geometry.

Let p be an odd prime and $q = p^t$. Let $V_k(q)$ be a k-dimensional vector space over $GF(q)$. Let $PG(k-1, q) = (V_k(q) - \{0\})/GF(q)^\times$ denote the projective space of dimension $k-1$ over $GF(q)$. A **projective** (n, k, h_1, h_2) **set** $\mathcal{O} = \{\langle y_1\rangle, \langle y_2\rangle, \ldots, \langle y_n\rangle\}$ is a proper non-empty subset of n points of the projective space $PG(k-1, q)$ with the property that every hyperplane in $PG(k-1, q)$ meets $\mathcal{O}$ in either h_1 points or h_2 points.

Let $\Omega = \{x \in V_k(q) \mid \langle x\rangle \in \mathcal{O}\}$ be the set of $V_k(Q)$ corresponding to a projective set $\mathcal{O} = \{\langle y_1\rangle, \langle y_2\rangle, \ldots, \langle y_n\rangle\}$.

Lemma 7.6: *$\mathcal{O}$ is a projective (n, k, h_1, h_2) set if and only if $\chi(\Omega) = qh_1 - n$ or $qh_2 - n$ for every nontrivial additive character χ of $GF(q^k)$.*

We consider the projective 3 space $\Sigma_3 = PG(3, q)$ over $GF(q)$. The corresponding vector space is $W = \boldsymbol{F}_q^4$. A **spread** in Σ_3 is a set of $q^2 + 1$ projective lines which are disjoint each other and a partition of the points of Σ_3.

Example 7.1: *The following $q^2 + 1$ projective lines*

$$L_\gamma = \{(\beta, \gamma\beta) \in GF(q^2) \times GF(q^2) \mid \beta \in GF(q^2)^\times\},$$
$$L_\infty = \{(0, \beta) \in GF(q^2) \times GF(q^2) \mid \beta \in GF(q^2)^\times\}$$

form a spread of Σ_3, which is called a regular spread.

A projective $\left(\frac{1}{4} \cdot \frac{q^4-1}{q-1}, 4, \frac{1}{4}(q-1)^2, \frac{1}{4}(q+1)^2\right)$ set in Σ_3 is called **Type Q**.
We denote the set $\{x \in W \mid \langle x\rangle \in X\}$ of W by $\tilde{X}$ for the set X of Σ_3.

Theorem 7.7 (Wilson and Xiang [225], Chen [18]): *Let p be an odd prime and $q = p^t$. Let $S = \{L_1, L_2, \ldots, L_{q^2+1}\}$ be a spread of Σ_3. If there exist two subsets C_0 and C_1 of Type Q in Σ_3 such that*

$$|C_0 \cap L_j| = \frac{1}{2}(q+1) \text{ for } 1 \leq j \leq \frac{1}{2}\left(q^2+1\right)$$

and

$$|C_1 \cap L_j| = \frac{1}{2}(q+1) \text{ for } \frac{1}{2}\left(q^2+1\right) < j \leq q^2+1,$$

then there exists a $\left(4q^4, 2q^4 - q^2, q^4 - q^2\right)$ Menon Hadamard difference set in $E \times \boldsymbol{Z}_p^{4t}$ where E is a group of order 4.

Proof: Put $s = \frac{1}{2}(q^2+1)$. Let χ be a nontrivial additive character of W. Let W^* be the set of additive characters of W. The sets $C_2 = \{L_1 \cup L_2 \cup \cdots \cup L_s\} \backslash C_0$ and $C_3 = \{L_{s+1} \cup L_{s+2} \cup \cdots \cup L_{2s}\} \backslash C_1$ are also the subsets of Type Q. From Lemma 7.6,

$$\chi(\tilde{C}_i) = \frac{1}{4}(q^2-1) - q^2 \text{ or } \frac{1}{4}(q^2-1), \quad i = 0, 1, 2, 3$$

for a nontrivial character χ. We define

$$U = \left\{\chi \in W^* \mid \chi(\tilde{C}_0) = \frac{1}{4}(q^2-1) - q^2\right\},$$
$$V = \left\{\chi \in W^* \mid \chi(\tilde{C}_0) = \frac{1}{4}(q^2-1)\right\}.$$

Let $\mathcal{L}_1 = \left\{x \in W \mid \langle x\rangle \in \bigcup_{j=1}^{s} L_j\right\}$. Since $S = \{L_1, L_2, \dots, L_{q^2+1}\}$ is a spread,

$$\chi(\mathcal{L}_1) = \begin{cases} -\frac{1}{2}(q^2+1) & \text{if } \chi \in N_1, \\ \frac{1}{2}(q^2-1) & \text{if } \chi \in T_1 \end{cases} \tag{7.2}$$

where $N_1 = \{\chi \in W^*\backslash\{\chi^0\} \mid \chi$ is nontrivial on every L_j, $1 \le j \le s\}$ and $T_1 = \{\chi \in W^*\backslash\{\chi^0\} \mid \chi$ is trivial on L_j, for some j, $1 \le j \le s\}$.

We let $L_j^{\perp} = \{\chi \in W^* \mid \chi \text{ is trivial on } L_j\}$. Then $|L_j^{\perp}| = q^2$. For $1 \le j \le s$, we define $\alpha_j = |L_j^{\perp}\backslash\{\chi^0\} \cap U|$ and $\beta_j = |L_j^{\perp}\backslash\{\chi^0\} \cap V|$. Then $\alpha_j + \beta_j = q^2 - 1$. For every $\chi \in L_j^{\perp}$, we have

$$\chi(\tilde{C}_0) = \chi(\tilde{C}_0\backslash(\tilde{C}_0 \cap L_j)) + |\tilde{C}_0 \cap L_j|.$$

Therefore

$$\sum_{\chi\in L_j^{\perp}} \chi(\tilde{C}_0) = \sum_{w\in\tilde{C}_0\backslash(\tilde{C}_0\cap L_j)} \sum_{\chi\in L_j^{\perp}} \chi(w) + q^2|\tilde{C}_0 \cap L_j| = q^2|\tilde{C}_0 \cap L_j|.$$

On the other hand

$$\sum_{\chi\in L_j^{\perp}} \chi(\tilde{C}_0) = \frac{1}{4}(q^4-1) + \alpha_j\left(\frac{1}{4}(q^2-1) - q^2\right) + \beta_j\frac{1}{4}(q^2-1).$$

Comparing two equalities and since $|\tilde{C}_0 \cap L_j| = \frac{1}{2}(q^2-1)$ for every j, $1 \le j \le s$, we have $\beta_j = q^2 - 1$, then $\alpha_j = 0$ and $T_1 \cap U = \emptyset$. So we have

$$\chi(\tilde{C}_2) = \chi(\mathcal{L}_1) - \chi(\tilde{C}_0) = \begin{cases} \frac{1}{4}(q^2-1) - q^2 & \text{if } \chi \in N_1 \cap V, \\ \frac{1}{4}(q^2-1) & \text{if } \chi \in (T_1 \cap V) \cup (N_1 \cap U). \end{cases}$$

Similarly to the above, we define

$$X = \left\{\chi \in W^* \mid \chi(\tilde{C}_1) = \frac{1}{4}(q^2-1) - q^2\right\},$$
$$Y = \left\{\chi \in W^* \mid \chi(\tilde{C}_1) = \frac{1}{4}(q^2-1)\right\},$$
$$N_2 = \left\{\chi \in W^*\backslash\{\chi^0\} \mid \chi \text{ is nontrivial on every } L_j, s+1 \le j \le 2s\right\},$$
$$T_2 = \left\{\chi \in W^*\backslash\{\chi^0\} \mid \chi \text{ is trivial on } L_j, \text{ for some } j, s+1 \le j \le 2s\right\}.$$

We see that $T_2 = N_1$ and $N_2 = T_1$. Furthermore we define

$$\mathcal{L}_2 = \left\{x \in W \mid \langle x\rangle \in \bigcup_{j=s+1}^{2s} L_j\right\},$$

then

$$\chi(\mathcal{L}_2) = \begin{cases} -\frac{1}{2}\left(q^2+1\right) & \text{if } \chi \in N_2, \\ \frac{1}{2}\left(q^2-1\right) & \text{if } \chi \in T_2. \end{cases}$$

We obtain $T_2 \cap X = \emptyset$ and

$$\chi(\tilde{C}_3) = \chi(\mathcal{L}_2) - \chi(\tilde{C}_1) = \begin{cases} \frac{1}{4}\left(q^2-1\right) - q^2 & \text{if } \chi \in N_2 \cap Y, \\ \frac{1}{4}\left(q^2-1\right) & \text{if } \chi \in (T_2 \cap Y) \cup (N_2 \cap X) \end{cases}$$

in a similar way.

Assume that A is the union of any $\frac{1}{4}(q^2-1)$ lines from $\{L_{s+1}, \dots, L_{2s}\}$ and B is the union of any $\frac{1}{4}(q^2-1)$ lines from $\{L_1, \dots, L_s\}$. For corresponding sets $\tilde{A}$ and $\tilde{B}$ of A and B, we define

$$D_0 = \tilde{C}_0 \bigcup \tilde{A},\ D_1 = \tilde{C}_1 \bigcup \tilde{B},\ D_2 = \tilde{C}_2 \bigcup \tilde{A},\ D_3 = \tilde{C}_3 \bigcup \tilde{B}. \tag{7.3}$$

Note that $\tilde{A}$ and $\tilde{B}$ do not contain 0.

1) Assume $\chi \in T_1 = N_2$, that is, χ is trivial on L_j for some j, $1 \le j \le s$.
   We have $\chi(D_0) = \chi(\tilde{C}_0) + \chi(\tilde{A}) = \frac{1}{4}(q^2-1) - \frac{1}{4}(q^2-1) = 0$ and $\chi(D_2) = 0$.

$$\chi(\tilde{B}) = \begin{cases} q^2 - \frac{1}{4}\left(q^2-1\right) & \text{if } L_j \in B, \\ -\frac{1}{4}\left(q^2-1\right) & \text{if } L_j \notin B. \end{cases} \tag{7.4}$$

   Hence

$$\chi(D_1) = \begin{cases} 0 & \text{if } \chi \in Y, L_j \notin B \text{ or } \chi \in X, L_j \in B, \\ q^2 & \text{if } \chi \in Y, L_j \in B, \\ -q^2 & \text{ir } \chi \in X, L_j \notin B. \end{cases}$$

   and

$$\chi(D_3) = \begin{cases} 0 & \text{if } \chi \in Y, L_j \in B \text{ or } \chi \in X, L_j \notin B, \\ -q^2 & \text{if } \chi \in Y, L_j \notin B, \\ q^2 & \text{if } \chi \in X, L_j \in B. \end{cases}$$

   This shows exactly one of $\chi(D_1)$ and $\chi(D_3)$ is $\pm q^2$ and the other is 0.
2) If $\chi \in T_2 = N_1$, we obtain $\chi(D_1) = \chi(D_3) = 0$ and exactly one of $\chi(D_0), \chi(D_2)$ is $\pm q^2$ and the other is 0.
   Let $E = \{a_0, a_1, a_2, a_3\}$ be an abelian group of order 4. We define

$$D = a_0(W \backslash D_0) \bigcup a_1 D_1 \bigcup a_2 D_2 \bigcup a_3 D_3. \tag{7.5}$$

   Then

$$|D| = \frac{1}{2}\left(3(q^4 - q^2)\right) + \left(q^4 - \frac{1}{2}(q^4 - q^2)\right) = 2q^4 - q^2$$

   since

$$|\tilde{C}_i| = \frac{1}{2}(q+1)\frac{1}{2}(q^2+1)(q-1) = \frac{1}{4}(q^4-1),\ i = 0, 1, 2, 3$$

   and

$$|\tilde{A}| = |\tilde{B}| = \frac{1}{4}(q^2-1)(q^2-1) = \frac{1}{4}\left((q^2-1)^2\right).$$

   Let ϕ be a character of E.

3) If χ is a nontrivial character, then

$$|\phi \otimes \chi(D)| = |\phi(a_i)\chi(D_i)| = q^2 \quad \text{for some } i \in \{0,1,2,3\}.$$

4) If χ is trivial and ϕ is nontrivial, then

$$\begin{aligned}\phi \otimes \chi(D) &= \phi(a_0)\left(q^4 - \frac{1}{2}\left(q^4 - q^2\right)\right) + \frac{1}{2}\left(q^4 - q^2\right)\left(\phi(a_1) + \phi(a_2) + \phi(a_3)\right) \\ &= \phi(a_0)q^2.\end{aligned}$$

Thus $|\phi \otimes \chi(D)| = |\phi(a_0)q^2| = q^2$. From Lemma 1.16, D is a $(4q^4, 2q^4 - q^2, q^4 - q^2)$ Menon Hadamard difference set. □

Example 7.2 (Wilson and Xiang [225]): *We let $\Sigma_3 = PG(3,3)$ and g be a primitive element of $W = GF(3^4)$. The points of Σ_3 are represented by $\langle 1\rangle, \langle g\rangle, \dots, \langle g^{39}\rangle$. Let*

$$L_i = \left\{\langle g^i\rangle, \langle g^{10+i}\rangle, \langle g^{20+i}\rangle, \langle g^{30+i}\rangle\right\}, \quad 0 \le i \le 9,$$
$$\mathcal{L}_1 = \left\{x \in W | x \in L_0 \cup L_2 \cup L_4 \cup L_6 \cup L_8\right\},$$

and

$$\mathcal{L}_2 = \left\{x \in W | x \in L_1 \cup L_3 \cup L_5 \cup L_7 \cup L_9\right\}.$$

$S = \left\{L_0, L_1, \dots, L_9\right\}$ *is a spread of Σ_3. Let*

$$C_0 = \left\{\langle 1\rangle, \langle g^4\rangle, \dots, \langle g^{4i}\rangle, \dots, \langle g^{36}\rangle\right\}$$

and

$$C_1 = \left\{\langle g\rangle, \langle g^5\rangle, \dots, \langle g^{4i+1}\rangle, \dots, \langle g^{37}\rangle\right\}.$$

We see that $|C_0 \cap L_{2i}| = \frac{1}{2}(q+1) = 2, 0 \le i \le 4$ and $|C_1 \cap L_{2i+1}| = 2, 0 \le i \le 4$.
Let χ_β be a additive character of W, $\beta \in W$. Then

$$\chi_1(\tilde{L_5}) = \sum_{j=0}^{3} \sum_{\alpha \in \langle g^{10j+5}\rangle} \chi_1(\alpha) = 8 \quad and \quad \chi_1(\tilde{L_i}) = -1, i \ne 5.$$

Hence $\chi_1 \in N_1$. Thus $\chi_1(\mathcal{L}_1) = -5$ and $\chi_1(\mathcal{L}_2) = 4$. It is easy to evaluate $\chi_\beta(\mathcal{L}_1)$ and $\chi_\beta(\mathcal{L}_2)$.
We have

$$\chi_1(\tilde{C}_0) = -7 \text{ and } \chi_\beta(\tilde{C}_0) = 2 \text{ for } \beta = g, g^2, g^3,$$
$$\chi_{g^3}(\tilde{C}_1) = -7 \text{ and } \chi_\beta(\tilde{C}_1) = 2 \text{ for } \beta = 1, g, g^2.$$

Hence C_0 and C_1 are a projective $(10,4,1,4)$ set, that is, type Q. We see that $C_2 = (L_0 \cup L_2 \cup L_4 \cup L_6 \cup L_8)\backslash C_0$ and $C_3 = (L_1 \cup L_3 \cup L_5 \cup L_7 \cup L_9)\backslash C_1$ are also of a type Q.
For example, we let $A = \left\{L_1, L_3\right\}$ and $B = \left\{L_2, L_4\right\}$. Since $\chi_1 \in N_1 = T_2$ and $L_5 \notin B$, we have $\chi_1(A) = -2$ and $\chi_1(B) = -2$. On the other hand, $\chi_{g^4}(A) = 7$ and $\chi_{g^4}(B) = -2$ from $\chi_{g^4} \in T_2$.
Let D_0, D_1, D_2, and D_3 be defined as in (7.3) and D be as in (7.5). Then

$$|D_i| = 36 = \frac{1}{2}(q^4 - q^2), \quad i = 0,1,2,3 \text{ and}$$
$$|D| = 153 = 2q^4 - q^2,$$

since $|C_i| = 20 = \frac{1}{4}(q^4 - 1)$ and $|A| = 16 = \frac{1}{4}(q^2 - 1)^2$.

Thus we have

$$\chi_1(D_0) = \chi_1(\tilde{C}_0) + \chi_1(\tilde{A}) = -7 - 2 = -9,$$
$$\chi_1(D_1) = \chi_1(\tilde{C}_1) + \chi_1(\tilde{B}) = 2 - 2 = 0,$$
$$\chi_1(D_0) = \chi_1(\tilde{C}_2) + \chi_1(\tilde{A}) = 2 - 2 = 0,$$
$$\chi_1(D_1) = \chi_1(\tilde{C}_3) + \chi_1(\tilde{B}) = 2 - 2 = 0.$$

We see the conditions (ii) and (iv) are satisfied. Let ϕ be a character of E. For a nontrivial character χ_β, $\left|\phi(a_i)\chi_\beta(D_i)\right| = 3^2$ for some $i \in 0, 1, 2, 3$. For a trivial character χ^0 and a nontrivial character ϕ, $\phi \otimes \chi^0(D) = 45\phi(a_0) + 36(\phi(a_1) + \phi(a_2) + \phi(a_3)) = 3^2\phi(a_0)$. Consequently the subset D is a $(4 \cdot 3^4, 2 \cdot 3^4 - 3^2, 3^4 - 3^2)$ Menon Hadamard difference set.

Remark 7.1: If $E = \mathbf{Z}_2 \times \mathbf{Z}_2$, then the Hadamard difference set in Theorem 7.7 is reversible.

7.5.1 Graphical Hadamard Matrices

A symmetric Hadamard matrix is said to be **graphical** if it has a constant diagonal. If an Hadamard matrix H is graphical with the diagonal δ, then $\frac{1}{2}(J - \delta H)$ is the adjacency matrix of a graph with n vertices.

Definition 7.1: Let H be a regular graphical Hadamard matrix with row sum ℓ and the diagonal δ. We say H is **of positive type** or **type** $+1$ if $\ell\delta > 0$ and of **negative type** or **type** -1 if $\ell\delta < 0$.

Lemma 7.7 (*Haemers and Xiang [90]*): *Let t be a positive integer and D be a reversible $(4t^2, 2t^2 + t, t^2 + t)$ Menon Hadamard difference set in a group G of order $4t^2$ such that $1_G \notin D$. Then there exists a regular graphical Hadamard matrix of negative type of order $4t^2$.*

Proof: Let A be the $(0, 1)$ incidence matrix of D. Since D is reversible and $1_G \notin D$, A is symmetric and has 0 on the diagonal. If we let $H = J - 2A$, then H is also symmetric and 1 on the diagonal. The row sum of H is $-2t < 0$. Since $A^2 = t^2I + (t^2 + t)J$, $HH^\top = 4t^2I$. Hence H is a regular graphical Hadamard matrix of negative type. □

Example 7.3: *The matrix*

$$H_4 = \begin{bmatrix} 1 & - & - & - \\ - & 1 & - & - \\ - & - & 1 & - \\ - & - & - & 1 \end{bmatrix}$$

is a regular graphical Hadamard matrix of negative type with the diagonal 1 and row sum -2.
Let $G = \langle c \rangle$ be the cyclic group generated by c. The set $D = \{c, c^2, c^3\}$ is a reversible $(4, 3, 2)$ Menon Hadamard difference set. The $(0, 1)$ incidence matrix of D is

$$A = \begin{bmatrix} 0 & 1 & 1 & 1 \\ 1 & 0 & 1 & 1 \\ 1 & 1 & 0 & 1 \\ 1 & 1 & 1 & 0 \end{bmatrix}$$

and we see $A = \frac{1}{2}(J - H_4)$.

Theorem 7.8 (*Haemers and Xiang [90]*): *Let m be a positive odd integer. Then there exists a regular graphical Hadamard matrix of order $4m^4$.*

Proof: From Theorem 7.7, Remark 7.1, and Theorem 7.3, there exists a reversible $(4m^4, 2m^4 - m^2, m^4 - m^2)$ Menon Hadamard difference set in $\mathbf{Z}_2 \times \mathbf{Z}_2 \times G$ where G is an abelian group of order m^4. Let $E = \mathbf{Z}_2 \times \mathbf{Z}_2 = \{1, a, b, ab\}$ and $D_i' = W \backslash D_i, i = 0, 1, 2, 3$. We write $D = D_0' \cup aD_1 \cup bD_2 \cup abD_3$, then D is a reversible Menon Hadamard difference set. We see $1_G \in D_0'$ and $1_G \notin D_i$, $i = 1, 2, 3$. Let $\overline{D} = (E \times G) \backslash D$ be the complement of D. If we put $D' = ab\overline{D} = abD_0 \cup bD_1' \cup aD_2' \cup D_3'$, then D' is a reversible $(4m^4, 2m^4 + m^2, m^4 + m^2)$ Menon Hadamard difference set and $1_{E \times G} \notin D'$. From Lemma 7.7, there exists a regular graphical Hadamard matrix of order $4m^4$ of negative type. We can obtain a regular graphical Hadamard matrix of positive type by Seidel switching from this Hadamard matrix of negative type (see [90]). □

7.6 The Construction Based on Galois Rings

7.6.1 Galois Rings

In 1994, Sloane et al. showed that the known nonlinear codes are the image of linear codes in the Galois ring $GR(2^2, 4)$ of characteristic 4 under the Gray map. It was breakthrough and gave an entirely new approach to coding theory. The underlying algebraic structure in the research of combinatorics extends from finite fields to Galois rings.

We give the definition of Galois rings of characteristic 2^n. The results and facts are almost applicable to Galois rings of any prime characteristic.

Denote $\mathbf{Z}/2^n\mathbf{Z}$ by $\mathcal{A}_n$. Let $\varphi(x) \in \mathcal{A}_n[x]$ be a primitive basic irreducible polynomial of degree s and denote the root of $\varphi(x)$ by ξ. Then $\mathcal{A}_n[x]/\varphi(x)$ is a Galois extension of $\mathcal{A}_n$ and is called a **Galois ring** of characteristic 2^n and of an extension degree s, denoted by $GR(2^n, s)$. The extension ring of $\mathcal{A}_n$ obtained by adjoining ξ is isomorphic to $\mathcal{A}_n[x]/\varphi(x)$. For easy reference, we put $\mathcal{R}_n = GR(2^n, s)$.

$\mathcal{R}_n$ is a local ring and has a unique maximal ideal $\mathfrak{p}_n = 2\mathcal{R}_n$. Every ideal of $\mathcal{R}_n$ is $\mathfrak{p}_n^\ell = 2^\ell \mathcal{R}_n, 1 \le \ell \le n-1$ and $\mathcal{R}_n$ has the structure $\mathfrak{p} \supset \mathfrak{p}^2 \supset \cdots \supset \mathfrak{p}^{n-1}$. The residue class field $\mathcal{R}_n/\mathfrak{p}_n$ is isomorphic to a finite field $GF(2^s)$. We take the Teichmüller system $\mathcal{T}_n = \{0, 1, \xi, \dots, \xi^{2^s-2}\}$ as a set of complete representatives of $\mathcal{R}_n/\mathfrak{p}_n$.

Any element α of $\mathcal{R}_n$ is uniquely represented as

$$\alpha = \alpha_0 + 2\alpha_1 + \cdots + 2^{n-1}\alpha_{n-1}, \quad \alpha_i \in \mathcal{T}_n \quad (0 \le i \le n-1)$$

and also uniquely represented as

$$\alpha = \gamma_0 + \gamma_1\xi + \cdots + \gamma_{s-1}\xi^{s-1}, \quad \gamma_i \in \mathcal{A}_n \quad (0 \le i \le s-1).$$

We denote the set of principal units $1 + 2a, a \in \mathcal{R}_{n-1}$ by $\mathcal{E} = \{1 + 2a | a \in \mathcal{R}_{n-1}\}$. The unit group $\mathcal{R}_n^\times$ is a direct product of a cyclic group $\langle \xi \rangle$ and $\mathcal{E}$. An arbitrary element α of $\mathcal{R}_n^\times$ is uniquely represented as

$$\alpha = \xi^t e = \xi^t(1 + 2a), \ a \in \mathcal{R}_{n-1}, \ e \in \mathcal{E}.$$

7.6.2 Additive Characters of Galois Rings

For any element $\alpha = \alpha_0 + 2\alpha_1 + \cdots + 2^{n-1}\alpha_{n-1}$ of $\mathcal{R}_n$, $\alpha_i \in \mathcal{T}_n$ $(0 \le i \le n-1)$, we define a ring automorphism $F : \mathcal{R}_n \longrightarrow \mathcal{R}_n$ as

$$\alpha^F = \alpha_0^2 + 2\alpha_1^2 + \cdots + 2^{n-1}\alpha_{n-1}^2,$$

which is called a **Frobenius automorphism of a Galois ring**. The relative trace T_n from $\mathcal{R}_n$ to $\mathcal{A}_n$ is defined by

$$T_n(\alpha) = \alpha + \alpha^F + \cdots + \alpha^{F^{s-1}}.$$

Let $\tilde{\varphi}(x)$ be a primitive basic irreducible polynomial of $\mathcal{R}_n$ and $\tilde{\xi}$ be a root of $\tilde{\varphi}(x)$. Put $\varphi(x) \equiv \tilde{\varphi}(x) \pmod{2^{n-\ell}}$. Then $\varphi(x)$ is a primitive basic irreducible polynomial of $\mathcal{R}_{n-\ell}$. Let ξ be a root of $\varphi(x)$. We define the homomorphism $\tau_{n-\ell} : \mathcal{R}_n \to \mathcal{R}_{n-\ell}$ as

$$\tau_{n-\ell}\left(\sum_{i=0}^{s-1} \tilde{\gamma}_i \tilde{\xi}^i\right) = \sum_{i=0}^{s-1} \gamma_i \xi^i$$

where $\gamma_i \equiv \tilde{\gamma}_i \pmod{2^{n-\ell}}$, $\tilde{\gamma}_i \in \mathcal{A}_n$, and $\gamma_i \in \mathcal{A}_{n-\ell}$. The following commutative relation holds,

$$\tau_{n-\ell} T_n = T_{n-\ell} \tau_{n-\ell}.$$

We will, in particular, denote $\tau_1 : \mathcal{R}_n \to GF(2^s)$ by μ. Let tr be the relative trace from $GF(2^s)$ to $\mathcal{A}_1 = GF(2)$ (the definition is given in Chapter 2). Then we have the commutability, $\mu \cdot T_n = tr \cdot \mu$.

Lemma 7.8 *(Yamada [240, 242])*: *The additive characters of $\mathcal{R}_n$ are given by*

$$\lambda_\beta(\alpha) = \zeta_{2^n}^{T_n(\beta\alpha)}$$

where $\beta \in \mathcal{R}_n$ and ζ_{2^n} is a primitive 2^n-th root of unity.

We let $\lambda_1(\mathcal{R}_n^\times) = \sum_{\alpha \in \mathcal{R}_n^\times} \lambda_1(\alpha)$ and $\lambda_1(\mathfrak{p}_n) = \sum_{\alpha \in \mathfrak{p}_n} \lambda_1(\alpha)$.

Lemma 7.9: *For a nontrivial additive character λ_1, we have*

$$\lambda_1(\mathcal{R}_n^\times) = 0 \quad and \quad \lambda_1(\mathfrak{p}_n) = 0.$$

7.6.3 A New Operation

We introduce a new operation $*$ in $\mathcal{R}_n$, $n \geq 2$. For elements α and β, we let

$$\alpha * \beta = \alpha + \beta + 2\alpha\beta.$$

We shall write $\alpha^{*m} = \alpha * \alpha * \cdots * \alpha$, ($m$ factors α) for the sake of convenience. Theorem XVI.9 in McDonald's book [134] is helpful to prove the following theorem.

Theorem 7.9: *Assume that $n \geq 2$. Let $g_1 = 1, g_2, \ldots, g_s$ be a free $\mathcal{A}_n$-basis. Let $\mu : \mathcal{R}_n \to GF(2^s)$ be the map defined by $\mu(\alpha) \equiv \alpha \pmod 2$ and b be an element of $\mathcal{R}_n$ such that $x^2 + x = \mu(b)$ has no solution in $GF(2^s)$. Then the elements of the ring $\mathcal{R}_n$ will form an abelian group with respect to this new operation and $\mathcal{R}_n$ is a product of cyclic groups as follows,*

$$\mathcal{R}_n = \langle -1 \rangle * \langle 2b \rangle * \prod_{j=2}^{s} \langle g_j \rangle$$

where $|\langle -1 \rangle| = 2$, $|\langle 2b \rangle| = 2^{n-1}$, and $|\langle g_j \rangle| = 2^n$, $2 \leq j \leq s$.

7.6.4 Gauss Sums Over $GR(2^{n+1}, s)$

Let $\tilde{\chi}$ be a multiplicative character of $\mathcal{R}_{n+1}^\times$. If the order of $\tilde{\chi}$ is a power of 2, 2^m, then $\tilde{\chi}(\xi) = 1$, since $|\langle \xi \rangle| = 2^s - 1$ and $gcd(2^m, 2^s - 1) = 1$. For $\xi^t(1 + 2\alpha)$, $\xi^u(1 + 2\beta) \in \mathcal{R}_{n+1}^\times$, we have

$$\tilde{\chi}(\xi^t(1 + 2\alpha) \cdot \xi^u(1 + 2\beta)) = \tilde{\chi}((1 + 2\alpha)(1 + 2\beta)) = \tilde{\chi}(1 + 2(\alpha * \beta)).$$

Thus a multiplicative character $\tilde{\chi}$ can be regarded as a multiplicative character χ of the group $\mathcal{R}_n$ with respect to the new operation. We extend $\tilde{\chi}$ as the character of $\mathcal{R}_{n+1}$ by defining $\tilde{\chi}(\alpha) = 0$ for any element $\alpha \in \mathfrak{p}_{n+1}$. Denote the trivial character by $\tilde{\chi}^0$. For a multiplicative character $\tilde{\chi}$ of $\mathcal{R}_{n+1}$ and an additive character λ_β of $\mathcal{R}_{n+1}$, we define the Gauss sum over $\mathcal{R}_{n+1}$

$$G(\tilde{\chi}, \lambda_\beta) = \sum_{\alpha \in \mathcal{R}_{n+1}} \tilde{\chi}(\alpha)\lambda_\beta(\alpha).$$

Lemma 7.10: *For $\beta = 2^h(1 + 2\beta_0)\xi^u \in \mathcal{R}_{n+1}$, we have*

$$G(\tilde{\chi}, \lambda_\beta) = \tilde{\chi}^{-1}(\frac{\beta}{2^h})G(\tilde{\chi}, \lambda_{2^h}).$$

7.6.5 Menon Hadamard Difference Sets Over $GR(2^{n+1}, s)$

We assume that $n \geq 3$ is odd and s is odd. We define the subsets $A \subset \mathcal{R}_n$, $A'_\ell \subset \mathcal{R}_{n-\ell}$ with $1 \leq \ell \leq \frac{n-3}{2}$ and $n \geq 5$, and $B \subset \mathcal{R}_{\frac{n+1}{2}}$ as follows.

$$A = \bigcup_{m=0}^{2^{n-1}-1} \langle -1 \rangle * \prod_{j=2}^{s-1} \langle g_j \rangle * \langle 2b \rangle * g_s^{*m},$$

$$A'_\ell = \bigcup_{m=0}^{2^{n-2\ell-1}-1} \langle -1 \rangle * \prod_{j=2}^{s-1} \langle g_j \rangle * \langle 2b \rangle * \langle g_s^{*2^{n-2\ell}} \rangle * g_s^{*m},$$

$$B = \langle -1 \rangle * \prod_{j=2}^{s-1} \langle g_j \rangle * \langle g_s^{*2} \rangle * \langle 2b \rangle.$$

For convenience sake, we use same symbols g_j $(j = 2, \dots, s)$ and $2b$, though the basis g_j $(j = 2, \dots, s)$ and $2b$ are contained in different rings $\mathcal{R}_n$, R_{n-l}, or $\mathcal{R}_{\frac{n+1}{2}}$.

We define the subset $D_{\mathcal{R}^\times_{n+1}} \subset \mathcal{R}^\times_{n+1}$, $D_{\mathfrak{p}^\ell_{n+1}} \subset \mathfrak{p}^\ell_{n+1} - \mathfrak{p}^{\ell+1}_{n+1}$, from A, A'_l, with $1 \leq \ell \leq \frac{n-3}{2}$ and $n \geq 5$, and B as follows,

$$D_{\mathcal{R}^\times_{n+1}} = \left\{(1 + 2\alpha)\xi^t | \alpha \in A, t = 0, 1, \dots, 2^s - 2\right\},$$
$$D_{\mathfrak{p}^\ell_{n+1}} = \left\{2^\ell(1 + 2\alpha)\xi^t | \alpha \in A'_l, t = 0, 1, \dots, 2^s - 2\right\},\ 1 \leq \ell \leq \frac{1}{2}(n-3), n \geq 5,$$
$$D_{\mathfrak{p}^{(n-1)/2}_{n+1}} = \left\{2^{\frac{1}{2}(n-1)}(1 + 2\alpha)\xi^t | \alpha \in B, t = 0, 1, \dots, 2^s - 2\right\}$$

where ξ is the root of a primitive basic irreducible polynomial of degree s. Furthermore we put

$$D_{n+1} = D_{\mathcal{R}^\times_{n+1}} \bigcup_{\ell=1}^{\frac{1}{2}(n-3)} D_{\mathfrak{p}^\ell_{n+1}} \bigcup D_{\mathfrak{p}^{(n-1)/2}_{n+1}}.$$

For the case $n = 3$, we let

$$D_4 = D_{\mathcal{R}^\times_4} \bigcup D_{\mathfrak{p}_4}.$$

Theorem 7.10 *(Yamada [242])*: *For every even integer $n + 1 \geq 4$ and every odd degree $s \geq 3$, there exists a difference set D_{n+1} with parameters*

$$v = 2^{(n+1)s}, k = 2^{(n+1)s/2-1}(2^{(n+1)s/2} - 1), \lambda = 2^{(n+1)s/2-1}(2^{(n+1)s/2-1} - 1)$$

over the Galois ring $GR(2^{n+1}, s)$. This difference set D_{n+1} is embedded in the ideal part of a difference set D_{n+3} over $GR(2^{n+3}, s)$. This means that there exists an infinite family of difference sets with this embedding system over Galois rings.

For the case that the degree s is even, we define the subsets as follows:

$$A = \bigcup_{m=0}^{2^{n-2}-1} \langle -1 \rangle * \prod_{j=2}^{s} \langle g_j \rangle * (2b)^{*m},$$

$$A'_l = \bigcup_{m=0}^{2^{n-2l-2}-1} \langle -1 \rangle * \prod_{j=2}^{s} \langle g_j \rangle * \langle 2b^{*2^{(n-1-2l)}} \rangle * (2b)^{*m},$$

$$B = \prod_{j=2}^{s-1} \langle g_j \rangle * \langle -1 \rangle * \langle g_s^{*2} \rangle * \langle 2b \rangle.$$

We have the similar theorem from $D_{\mathcal{R}^{\times}_{n+1}}$, $D_{\mathfrak{p}^{\ell}_{n+1}}$, $1 \le \ell \le \frac{1}{2}(n-3)$, $n \ge 5$, and $D_{\mathfrak{p}^{(n-1)/2}_{n+1}}$ (see [243] for further discussion).

7.6.6 Menon Hadamard Difference Sets Over $GR(2^2, s)$

The structure of a Galois ring of characteristic 4 is simpler than that of a Galois ring of $2^n \ge 2^3$.

We may regard β for the principal unit $1 + 2\beta \in \mathcal{E}$ is the element of $K = GF(2^s)$. For $1 + 2\alpha, 1 + 2\beta \in \mathcal{E}$,

$$(1 + 2\alpha)(1 + 2\beta) = 1 + 2(\alpha + \beta), \quad \alpha, \beta \in K.$$

Hence the unit group $\mathcal{E}$ is an elementary abelian group of order 2^s isomorphic with the additive group of K. In other word, the introduced operation $*$ turns out to be the addition of a finite field when the characteristic 2^2.

Let S_a be a coset of $\mathcal{E}$ in $\mathcal{R}_2$ containing $1 + 2a$,

$$S_a = \left\{ \xi^a(1 + 2a), m = 0, \ldots, 2^s - 2 \right\}.$$

Theorem 7.11 ***(Yamamoto and Yamada [248]):*** *Let A be a subset of $K = GF(2^s)$ with $|A| = 2^{s-1}$. Then $D = \cup_{a \in A} S_a$ is a $\left(2^{2s}, 2^{s-1}(2^s - 1), 2^{s-1}(2^{s-1} - 1)\right)$ Menon Hadamard difference set if and only if the equation*

$$\sum_{a \in A+b} z_a = 2^{s-1} e_b, \quad e_b \in \{\pm 1, \pm i\}$$

is satisfied for all $b \in K$ where $i = \sqrt{-1}$ and

$$z_a = \sum_{m=0}^{2^s-1} i^{T_2(1+2a)\xi^m}.$$

Corollary 7.1 ***(Yamamoto and Yamada [248]):*** *If A is a subgroup of K, then D is a $\left(2^{2s}, 2^{s-1}(2^s - 1), 2^{s-1}(2^{s-1} - 1)\right)$ Menon Hadamard difference set.*

Example 7.4 ***(Yamamoto and Yamada [248]):*** *Let $s = 3$ and $\varphi(x) = x^3 + 2x^2 + x + 3$. Denote the root of $\varphi(x)$ by ξ and the primitive element of $K = GF(2^3)$ by δ. The symbol xyz in the table stands for the element $\xi^\ell = x + y\xi + z\xi^2$.*

ℓ	xyz	$T_2(\xi^\ell)$
0	100	3
1	010	2
2	001	2
3	132	1
4	233	2
5	331	1
6	121	1

Let $A = \{0, 1, \delta, \delta^3\}$. *Then we have*

$$z_0 = \sum_{m=0}^{6} i^{T_2(\xi^m)} = -3 + 2i,$$

$$z_1 = \sum_{m=0}^{6} i^{T_2(3\xi^m)} = -3 - 2i,$$

$$z_\delta = \sum_{m=0}^{6} i^{T_2((1+2\delta)\xi^m)} = \sum_{m=0}^{6} i^{T_2(\xi^m)}(-1)^{tr(\delta^{m+1})} = 1 - 2i,$$

$$z_{\delta^3} = \sum_{m=0}^{6} i^{T_2((1+2\delta^3)\xi^m)} = \sum_{m=0}^{6} i^{T_2(\xi^m)}(-1)^{tr(\delta^{m+3})} = 1 + 2i.$$

Thus we have $\sum_{a\in A} z_a = -4$. *For* $b = \delta^2$, $A + \delta^2 = \{\delta^2, \delta^4, \delta^5, \delta^6\}$. *We know that* $\sum_{a\in A+\delta^2} z_a = 4$. *By verifying* $\sum_{a\in A+b} z_a$ *for every element b of K,* $D = \cup_{a\in A} S_a$ *is a* $(2^6, 2^2(2^3-1), 2^2(2^2-1))$ *Menon Hadamard difference set.*

8

Paley Hadamard Difference Sets and Paley Type Partial Difference Sets

8.1 Notations

Table 8.1 gives the notations which are used in this chapter.

8.2 Paley Core Matrices and Gauss Sums

We recognize a Paley core matrix Q as a representation matrix of a element $\sum_{\alpha \in GF(1)} \psi(\alpha)\alpha$ of a group ring $\boldsymbol{Z}[GF(q)]$ where ψ is a quadratic character of $GF(q)$. The equation $QQ^{\top} = nI - J$ follows from the norm of the Gauss sum.

Let q be a prime power and $\boldsymbol{Z}$ be a rational integer ring. Put $F = GF(q)$. We consider the group ring $\boldsymbol{Z}[F]$ of the additive group F^+ over $\boldsymbol{Z}$. For an element $\mathcal{D} = \sum_{\alpha \in D} \alpha$ of $\boldsymbol{Z}[F]$, we set

$$\mathcal{D}^{-1} = \sum_{\alpha \in D} (-\alpha).$$

Let ψ be a quadratic character of F. For $x, y \in F$ we write

$$Q = (\psi(x - y))$$

and, as in Eq. (1.5) define

$$M = \begin{bmatrix} 0 & \boldsymbol{e}^{\top} \\ \psi(-1)\boldsymbol{e} & Q \end{bmatrix}$$

where $\boldsymbol{e}$ is the vector with all entries 1.

Though the following lemma was proved in Chapter 1, we give an another proof based on number theory.

Lemma 8.1: *The diagonal elements of Q are all 0 and the other elements are in $\{1, -1\}$. The matrix Q satisfies*

i) $Q^{\top} = \begin{cases} Q & \text{if } q \equiv 1 \pmod 4, \\ -Q & \text{if } q \equiv 3 \pmod 4, \end{cases}$

ii) $Q\boldsymbol{e} = \boldsymbol{0}$, $QJ = 0$,

iii) $QQ^{\top} = qI - J$, *where J is the matrix with all entries* 1.

Hadamard Matrices: Constructions using Number Theory and Algebra, First Edition. Jennifer Seberry and Mieko Yamada.

Table 8.1 Notations used in this chapter.

$\boldsymbol{F}_q$, $GF(q)$	A finite field with q elements for a prime power q
$F^{\times}$	The multiplicative group of a finite field F
ϕ^0	The trivial character of a finite field
I	The identity matrix
X^{T}	The transpose of the matrix X
$\boldsymbol{Z}$	The rational integer ring
Q	A Paley core matrix
T_α	The regular representation matrix of $\alpha \in GF(q)$
$T_{F/K}$	The relative trace from F to K where K is a finite field and F is an extension of K
$N_{F/K}$	The relative norm from F to K where K is a finite field and F is an extension of K
$g(\chi, \eta)$	The Gauss sum associated with a character χ and an additive character η
PDS	A partial difference set
$(v, \frac{1}{2}(v-1), \frac{1}{4}(v-3)$ difference set	Paley Hadamard difference set
$(v, \frac{1}{2}(v-1), \frac{1}{4}(v-5), \frac{1}{4}(v-1))$ *PDS*	A Paley type partial difference set
$\Sigma_3 = PG(3, q)$	The project space of dimension 3 over $GF(q)$
$V_k(q)$	k-Dimensional vector space over $GF(q)$
$\gcd(a, b)$	The greatest common divisor of integers a and b

Proof: For a map from an element γ to $\gamma + \alpha$, we denote the regular representation matrix by T_α, where γ and α are elements of F. Then the matrix Q can be written as

$$Q = \sum_{\alpha \in F} \psi(\alpha) T_\alpha$$

and Q is the right regular representation matrix of the element

$$\gamma = \sum_{\alpha \in F} \psi(\alpha)\alpha \in \boldsymbol{Z}[F].$$

We note that the representation matrix of

$$\gamma^{-1} = \sum_{\alpha \in F} \psi(\alpha)(-\alpha)$$

is Q^{T}.

We get the relations $Q^{\mathsf{T}} = Q$ or $Q^{\mathsf{T}} = -Q$ from $\psi(-1) = (-1)^{\frac{1}{2}(q-1)}$. The relation (*ii*) holds from the relation in Theorem 1.7.

$$\begin{aligned} \gamma\gamma^{-1} &= \sum_{\alpha \in F} \psi(\alpha)\alpha \cdot \sum_{\beta \in F} \psi(\beta)(-\beta) \\ &= \sum_{\alpha \in F} \sum_{\beta \in F} \psi(\alpha)\psi(\beta)(\alpha - \beta) \\ &= \sum_{\gamma \in F} \sum_{\alpha \in F} \psi(\alpha)\psi(\alpha - \gamma)\gamma \end{aligned}$$

from Lemma 1.11,

$$= \sum_{\gamma \neq 0} (-1)\gamma + (q-1)0. \tag{8.1}$$

From $T_0 = I$, we obtain the relation (*iii*). □

Let η be an additive character of F. Then $\eta(\gamma) = \sum_{\alpha\in F} \psi(\alpha)\eta(\alpha)$ is the Gauss sum $g(\psi, \eta)$ associated with ψ and η. The Eq. (8.1) is true if $\eta(\gamma)\eta(\gamma^{-1}) = q$ for a nontrivial character η or 0 for the trivial character. However, the norm relation of the Gauss sum $\eta(\gamma) = g(\psi, \eta)$ was given as

$$g(\psi,\eta)\overline{g(\psi,\eta)} = \begin{cases} q & \text{if } \eta \neq \eta^0, \\ 0 & \text{if } \eta = \eta^0, \end{cases}$$

from Theorems 1.7 and 2.1. Hence Eq. (8.1) is true.

Example 8.1:

1) *Let* $q = 5$.

F	**0**	**1**	**2**	**3**	**4**
$\psi(\alpha)$	0	1	–	–	1

$\gamma = \sum_{\alpha\in F} \psi(\alpha)\alpha = 0\cdot\mathbf{0} + 1\cdot\mathbf{1} + (-1)\cdot\mathbf{2} + (-1)\cdot\mathbf{3} + 1\cdot\mathbf{4}$.

$$T_0 = \begin{bmatrix} 1&0&0&0&0 \\ 0&1&0&0&0 \\ 0&0&1&0&0 \\ 0&0&0&1&0 \\ 0&0&0&0&1 \end{bmatrix}, \quad T_1 = \begin{bmatrix} 0&1&0&0&0 \\ 0&0&1&0&0 \\ 0&0&0&1&0 \\ 0&0&0&0&1 \\ 1&0&0&0&0 \end{bmatrix},$$

$$T_2 = \begin{bmatrix} 0&0&1&0&0 \\ 0&0&0&1&0 \\ 0&0&0&0&1 \\ 1&0&0&0&0 \\ 0&1&0&0&0 \end{bmatrix}, \quad T_3 = \begin{bmatrix} 0&0&0&1&0 \\ 0&0&0&0&1 \\ 1&0&0&0&0 \\ 0&1&0&0&0 \\ 0&0&1&0&0 \end{bmatrix} = T_2^{\mathsf{T}},$$

$$T_4 = \begin{bmatrix} 0&0&0&0&1 \\ 1&0&0&0&0 \\ 0&1&0&0&0 \\ 0&0&1&0&0 \\ 0&0&0&1&0 \end{bmatrix} = T_1^{\mathsf{T}}.$$

The matrix representation of γ *is*

$$Q = \sum_{\alpha\in F} \psi(\alpha)T_\alpha = \begin{bmatrix} 0&1&-&-&1 \\ 1&0&1&-&- \\ -&1&0&1&- \\ -&-&1&0&1 \\ 1&-&-&1&0 \end{bmatrix} = Q^{\mathsf{T}}.$$

2) *Let* $q = 7$.

F	**0**	**1**	**2**	**3**	**4**	**5**	**6**
$\psi(\alpha)$	0	1	1	–	1	–	–

$$\gamma = \sum_{\alpha\in F} \psi(\alpha)\alpha$$
$$= 0\cdot\mathbf{0} + 1\cdot\mathbf{1} + 1\cdot\mathbf{2} + (-1)\cdot\mathbf{3} + 1\cdot\mathbf{4} + (-1)\cdot\mathbf{5} + (-1)\cdot\mathbf{6}$$

The matrix representation of γ is

$$Q = \sum_{\alpha \in F} \psi(\alpha) T_\alpha = \begin{bmatrix} 0 & 1 & 1 & - & 1 & - & - \\ - & 0 & 1 & 1 & - & 1 & - \\ - & - & 0 & 1 & 1 & - & 1 \\ 1 & - & - & 0 & 1 & 1 & - \\ - & 1 & - & - & 0 & 1 & 1 \\ 1 & - & 1 & - & - & 0 & 1 \\ 1 & 1 & - & 1 & - & - & 0 \end{bmatrix} = -Q^{\mathrm{T}}$$

and $M + I$ is an Hadamard matrix of order 8.

Paley core matrix Q can be written in the following form:

Theorem 8.1: *Let $F = GF(q^2)$ be a quadratic extension of $K = GF(q)$. Let ξ be a generator of $F^\times$ and put $\theta = \xi^{(q+1)/2}$. The Paley matrix Q is written as*

$$Q = \left(\psi(N_{F/K}\alpha)\psi\left(T_{F/K}\theta\beta\alpha^{-1}\right)\right)_{\alpha,\beta \in F^\times/K^\times}.$$

Proof: See Yamamoto [245]. □

8.3 Paley Hadamard Difference Sets

Let D_q be a set of quadratic residues of $F = GF(q)$ and ψ be a quadratic character of F. Denote the trivial character by ψ^0. Let η be an additive character of F. We see $|D_q| = \frac{1}{2}(q-1)$. The corresponding element of D_q in $\boldsymbol{Z}[F]$ is

$$D_q = \sum_{d \in D_q} = \frac{1}{2}\left\{\sum_{\alpha \in F}(\psi(\alpha) + \psi^0(\alpha))\alpha\right\}.$$

For a nontrivial additive character η,

$$\eta(D_q) = \frac{1}{2}\left\{g(\psi,\eta) + \sum_{\alpha \in F}\eta(\alpha)\right\}.$$

Assume $q = p^s \equiv 3 \pmod 4$. From Theorem 2.2,

$$\eta(D_q) = \frac{1}{2}\left\{i^s\sqrt{q} - 1\right\}$$

where i is a primitive fourth root of unity and s is odd. Since $|\eta(D_q)| = \frac{1}{2}\sqrt{q+1}$, D_q is a $\left(q, \frac{1}{2}(q-1), \frac{1}{4}(q-3)\right)$ difference set from Lemma 1.16.

Definition 8.1: When $v \equiv 3 \pmod 4$ be a positive odd integer, we call a difference set with parameters $\left(v, \frac{1}{2}(v-1), \frac{1}{4}(v-3)\right)$ in a group G of order v a **Paley Hadamard difference set**. When v is a prime power the difference set D_q is called **Paley difference set**.

We show some constructions of Paley Hadamard difference set for the case that v is not prime power.

8.3.1 Stanton–Sprott Difference Sets

Theorem 8.2 ***(Stanton–Sprott difference sets [178]):*** *Let q and $q+2$ be twin prime powers. Denote the quadratic characters in $GF(q)$ and $GF(q+2)$ by the χ and ψ, respectively. Then*

$$D = \left\{(x,y) \mid x \in GF(q)^{\times}, y \in GF(q+2)^{\times}, \chi(x) = \psi(y)\right\} \bigcup \{(x,0) \mid x \in GF(q)\}$$

is a $\left(q(q+2), \frac{1}{2}(q(q+2)-1), \frac{1}{4}(q(q+2)-3)\right)$ Paley Hadamard difference set in $GF(q) \oplus GF(q+2)$.

Proof: We let $F_1 = GF(q)$ and $F_2 = GF(q+2)$. We see

$$|D| = \frac{1}{2}(q-1)\frac{1}{2}(q+1) + \frac{1}{2}(q-1)\frac{1}{2}(q+1) + q = \frac{1}{2}\left(q(q+2)-1\right).$$

Let $\mathcal{D}$ be an element of the group ring $\boldsymbol{Z}(F_1 \oplus F_2)$. From Lemma 1.16, we prove $|\lambda(\mathcal{D})|^2 = \frac{1}{4}(q+1)^2$ where λ is a nontrivial additive character of $F_1 \oplus F_2$. Let χ^0 and ψ^0 be the trivial characters of $F_1^{\times}$ and $F_2^{\times}$ respectively, Then

$$\mathcal{D} = \frac{1}{2}\sum_{y\in F_2^{\times}}\sum_{x\in F_1^{\times}} \left(\chi(x)\psi(y) + \chi^0(x)\psi^0(y)\right)(x,y) + \sum_{x\in F_1} \chi^0(x)(x,0).$$

We put $\lambda = \lambda_1\lambda_2$ where λ_1 and λ_2 are additive nontrivial characters of F_1 and F_2, respectively. Hence

$$\begin{aligned}
\lambda(\mathcal{D}) &= \frac{1}{2}\sum_{y\in F_2^{\times}}\sum_{x\in F_1^{\times}} \left(\chi(x)\psi(y)\lambda_1(x)\lambda_2(y)\right. \\
&\quad \left. +\chi^0(x)\psi^0(y)\lambda_1(x)\lambda_2(y)\right) + \sum_{x\in F_1} \lambda_1(x) \\
&= \frac{1}{2}\left\{\sum_{x\in F_1^{\times}} \chi(x)\lambda_1(x) \sum_{y\in F_2^{\times}} \psi(y)\lambda_2(y) + 1\right\} \\
&= \frac{1}{2}\left\{g(\chi,\lambda_1)g(\psi,\lambda_2) + 1\right\}.
\end{aligned}$$

Since one of q and $q+2$ is congruent 3 (mod 4),

$$\overline{g(\chi,\lambda_1)g(\psi,\lambda_2)} = \chi(-1)g(\chi,\lambda_1)\psi(-1)g(\psi,\lambda_2) = -g(\chi,\lambda_1)g(\psi,\lambda_2).$$

Thus we have

$$\begin{aligned}
|\lambda(\mathcal{D})|^2 &= \lambda(\mathcal{D})\overline{\lambda(\mathcal{D})} \\
&= \frac{1}{2}\left\{g(\chi,\lambda_1)g(\psi,\lambda_2)+1\right\}\frac{1}{2}\left\{\overline{g(\chi,\lambda_1)g(\psi,\lambda_2)}+1\right\} \\
&= \frac{1}{4}\left\{g(\chi,\lambda_1)\overline{g(\chi,\lambda_1)} + g(\psi,\lambda_2)\overline{g(\psi,\lambda_2)} + 1\right\} \\
&= \frac{1}{4}(q+1)^2.
\end{aligned}$$

□

We call this difference set a **Stanton–Sprott difference set** or a **twin prime powers difference set.**

Example 8.2: *Let $q = 3$. We notice that $q+2 = 5$ is also a prime. We denote the quadratic characters in $GF(3)$ and $GF(5)$ by χ and ψ, respectively. Then we the set*

$$D = \{(x,y) \mid x \in GF(3)^{\times}, y \in GF(5)^{\times}, \chi(x) = \psi(y)\} \bigcup \{(x,0) \mid x \in GF(3)\}$$
$$= \{(1,1),(1,4),(2,2),(2,3),(0,0),(1,0),(2,0)\}$$

is a $(15,7,3)$ *Paley Hadamard difference set in* $GF(3) \oplus GF(5)$.

8.3.2 Paley Hadamard Difference Sets Obtained from Relative Gauss Sums

Theorem 8.3: *Let q be a prime power and $r \geq 3$ be an integer. Let $F = GF(q^r)$ and $K = GF(q)$. Denote the relative trace from F to K by $T_{F/K}$. Then*

$$D = \left\{ n \left(\bmod \frac{q^r - 1}{q - 1} \right) \mid T_{F/K}(\xi^n) = 0 \right\}$$

is a difference set with the parameters

$$v = \frac{q^r - 1}{q - 1}, \quad k = \frac{q^{r-1} - 1}{q - 1}, \quad \lambda = \frac{q^{r-2} - 1}{q - 1}$$

where ξ is a generator of F.

Proof: We define the polynomial

$$f(x) = \sum_{m \in D} x^m \quad (\bmod x^{\frac{q^r-1}{q-1}} - 1).$$

Then, we prove the theorem by showing that

$$f(x)f(x^{-1}) = q^{r-2} + \frac{q^{r-2} - 1}{q - 1} \sum_{m \in Z_v} x^m$$

where $Z_v = Z/\frac{q^r-1}{q-1}Z$. Let $\mathcal{L}, \mathcal{L}_0, \mathcal{L}_1$ be as in Theorem 2.17, such that $\mathcal{L}$ is a system of representatives of the quotient group $F^{\times}/K^{\times}$,

$$\mathcal{L} = \mathcal{L}_0 + \mathcal{L}_1,\ \mathcal{L}_0 = \{\beta \mid T_{F/K}\beta = 0\},\ \mathcal{L}_1 = \{\beta \mid T_{F/K}\beta = 1\}.$$

Let ζ be a $\frac{q^r-1}{q-1}$-st root of unity. We have $f(\zeta) = \sum_{m \in D} \zeta^m = \sum_{\beta \in \mathcal{L}_0} \chi(\beta)$ where χ is a character of order $\frac{q^r-1}{q-1}$ in F. Thus it is sufficient to verify

$$\sum_{\beta \in \mathcal{L}_0} \chi(\beta) \sum_{\beta \in \mathcal{L}_0} \overline{\chi(\beta)} = q^{r-2} + \frac{q^{r-2} - 1}{q - 1} \sum_{m \in Z_v} \zeta^m$$

for every ζ. We notice that χ restricted to K is the trivial character. From the proof of Theorem 2.17,

$$\sum_{\beta \in \mathcal{L}_0} \chi(\beta) = \begin{cases} \frac{1}{q} g_F(\chi_F), & \text{if } \chi \neq \chi^0. \\ \frac{q^{r-1}-1}{q-1}, & \text{if } \chi = \chi^0. \end{cases}$$

Hence we obtain

$$\sum_{\beta \in \mathcal{L}_0} \chi(\beta) \sum_{\beta \in \mathcal{L}_0} \overline{\chi(\beta)} = \begin{cases} \frac{1}{q^2} g_F(\chi_F)\overline{g_F(\chi_F)} = q^{r-2}, & \text{if } \chi \neq \chi^0 \text{ or } \zeta \neq 1. \\ \left(\frac{q^{r-1}-1}{q-1}\right)^2, & \text{if } \chi = \chi^0 \text{ or } \zeta = 1. \end{cases}$$

□

Corollary 8.1: *In the case $q = 2$, a $(2^r - 1, 2^{r-1} - 1, 2^{r-2} - 1)$ difference set is Paley Hadamard difference set for every $r \geq 3$.*

Example 8.3: *$q = 2$ and $r = 4$.*

n	0	1	2	3	4	5	6	7	8	9	10	11	12	13	14
$T_{F/K}(\alpha^n)$	0	0	0	1	0	0	1	1	0	1	0	1	1	1	1

From Corollary 8.1,

$$D = \{0, 1, 2, 4, 5, 8, 10\}.$$

is a $(15, 7, 3)$ Paley Hadamard difference set.

We see that a Paley Hadamard difference set is not necessarily skew. We observe that the Paley Hadamard difference set in Example 8.2 is not skew. Assume that a $(v, \frac{v-1}{2}, \frac{v-3}{4})$ Paley Hadamard difference set D in a group G is skew, namely G is a disjoint union of D, D^{-1}, and $\{0\}$. Denote the $(0, 1)$ incidence matrix of D by A. If we let $A' = 2A - J$, then A' is a $(1, -1)$ matrix and satisfies $A'J = -J$ and $A'A'^{\top} = (v + 1)I - J$. Then

$$H = \begin{bmatrix} 1 & \boldsymbol{e}^{\top} \\ -\boldsymbol{e} & A' \end{bmatrix}$$

is a skew Hadamard matrix of order $v + 1$.

Remark 8.1: Singer [169] showed that points and hyperplanes of $PG(n, q)$ form a SBIBD. From Theorem 1.22, it yields a difference set with the same parameters in Theorem 8.3, which is called a **Singer difference set**.

8.3.3 Gordon–Mills–Welch Extension

Gordon–Mills–Welch showed the recursive construction of circulant difference sets based on Singer difference sets [86]. Yamamoto modified their proof [246].

Theorem 8.4 *(Gordon–Mills–Welch extension [86, 246])*: *Let q be a prime power and $m \geq 3, n \geq 3$ be integers. Let*

$$K = GF(q), \quad F = GF\left(q^m\right), \text{and } L = GF\left(q^{mn}\right)$$

Denote a primitive element of L by ξ. Let D be a

$$\left(\frac{q^m - 1}{q - 1}, \frac{q^{m-1} - 1}{q - 1}, \frac{q^{m-2} - 1}{q - 1}\right) \text{ circulant difference set.}$$

Then we define the subset

$$\Delta = \left(\Delta_{L/F}^{(0)} + \frac{q^{mn} - 1}{q^m - 1}\boldsymbol{Z}_{\frac{q^m-1}{q-1}}\right) \cup \left(\Delta_{L/F}^{(1)} + \frac{q^{mn} - 1}{q^m - 1}D\right) \left(\bmod \frac{q^{mn} - 1}{q - 1}\right)$$

where

$$\Delta_{L/F}^{(0)} = \left\{\ell \mid T_{L/F}(\xi^{\ell}) = 0\right\} \text{ and } \Delta_{L/F}^{(1)} = \left\{\ell \mid T_{L/F}(\xi^{\ell}) = 1\right\}.$$

Then Δ is a $\left(\frac{q^{mn}-1}{q-1}, \frac{q^{mn-1}-1}{q-1}, \frac{q^{mn-2}-1}{q-1}\right)$ circulant difference set.

Singer difference set has the same parameters in the above theorem. However, the subset D is not necessary a Singer difference set. If $\left(s, \frac{q^m-1}{q-1}\right) = 1$, then sD is a circulant difference set as well. We construct a family of circulant difference sets recursively starting from a circulant difference set with the same parameters as a Singer difference set. The resultant difference sets are called **Gordon–Mills–Welch difference sets** and the construction is called **Gordon–Mills–Welch extension**.

Example 8.4 *(Yamamoto [246])*: *Let $q = 2$, $m = 3$, and $n = 2$. Let $K = GF(2)$, $F = GF(2^3)$, and $L = GF(2^6)$. The polynomials $x^3 + x^2 + 1$ and $x^6 + x + 1$ are primitive polynomials F/K and L/K, respectively.*

n	0	1	2	3	4	5	6
$T_{F/K}(\alpha^n)$	1	1	1	0	1	0	0

$D = \{n \pmod 7 | T_{F/K}(\xi^n) = 0\} = \{3, 5, 6\}$. $\Delta^{(0)}_{L/F} = \{0\}$.
$\Delta^{(1)}_{L/F} = \{11, 21, 22, 25, 37, 42, 44, 50\}$. *We take $s = 3$ and $3D = \{1, 2, 4\}$.*

$$\begin{aligned}\Delta_1 &= \left(\Delta^{(0)}_{L/F} + 9\boldsymbol{Z}_7\right) \cup \left(\Delta^{(1)}_{L/F} + 9\{3, 5, 6\}\right)\\ &= \{0, 1, 2, 3, 4, 6, 7, 8, 9, 12, 13, 14, 16, 18, 19, 24,\\ &\quad 26, 27, 28, 32, 33, 35, 36, 38, 41, 45, 48, 49, 52, 54, 56\}\\ \Delta_2 &= \left(\Delta^{(0)}_{L/F} + 9\boldsymbol{Z}_7\right) \cup \left(\Delta^{(1)}_{L/F} + 9\{1, 2, 4\}\right)\\ &= \{0, 5, 9, 10, 15, 17, 18, 20, 23, 27, 29, 30, 31, 34, 36,\\ &\quad 39, 40, 43, 45, 46, 47, 51, 53, 54, 55, 57, 58, 59, 60, 61, 62\}\end{aligned}$$

These are $(63, 31, 15)$ *circulant difference sets.*

8.4 Paley Type Partial Difference Set

An important family of PDS is a Paley type in an elementary abelian group $F = GF(q)$ as follows.

Theorem 8.5: *Let $q \equiv 1 \pmod 4$ be an odd prime power. Let D_q be the set of nonzero squares of $F = GF(q)$. Then D_q is a $\left(q, \frac{1}{2}(q-1), \frac{1}{4}(q-5), \frac{1}{4}(q-1)\right)$ PDS.*

Proof: From Theorem 1.21, if we put $\lambda = \frac{1}{4}(q-5)$ and $\mu = \frac{1}{4}(q-1)$, then

$$\frac{1}{2}\left\{\lambda - \mu \pm \sqrt{(\lambda-\mu)^2 + 4(k-\mu)}\right\} = \frac{1}{2}(-1 \pm \sqrt{q}).$$

Denote a quadratic character of $GF(q)$ by ψ. We define the element $\mathcal{D}_q = \sum_{\alpha \in D_q} \alpha$ in $\boldsymbol{Z}[F]$. For a nontrivial additive character η, it is sufficient to prove that

$$\eta(\mathcal{D}_q) = \frac{1}{2}\left(-1 \pm \sqrt{q}\right).$$

In fact,

$$\eta(\mathcal{D}_q) = \frac{1}{2}\left(g(\psi^0, \eta) + g(\psi, \eta)\right\} = \frac{1}{2}\left(\pm\sqrt{q} - 1\right)$$

from Theorem 2.2. Hence we know D_q is a $\left(q, \frac{1}{2}(q-1), \frac{1}{4}(q-5), \frac{1}{4}(q-1)\right)$ *PDS.* □

In general, let $v \equiv 1 \pmod 4$ be a positive odd integer. We call PDS with parameters $\left(v, \frac{1}{2}(v-1), \frac{1}{4}(v-5), \frac{1}{4}(v-1)\right)$ a **Paley type partial difference set**.

Ma has studied PDS as a generalization of difference sets, though the earlier results were written in terms of strongly regular graphs and related topics. Ma [130] raised several research questions. These are given in Appendix C.

Davis [40] and Leung and Ma [125] constructed Paley type PDS in p-groups that are not elementary abelian groups. Polhill gave the construction of Paley type PDSs in groups with the orders which are not a prime power and we will show his construction in the next section. It is Polhill's construction that can be used to develop solutions to the further research questions posed by Ma given in Appendix C.5.

We need the following lemma on a necessary and sufficient condition of the existence of Paley type PDSs in an abelian group G.

Lemma 8.2: *The subset* $\boldsymbol{D}$ *is a Paley type partial difference set in* G *of order* v *with* $1_G \notin D$ *if and only if*

$$\chi(D) = \frac{1}{2}\left(-1 \pm \sqrt{v}\right) \tag{8.2}$$

for all nontrivial character χ *of* G *and* $|\boldsymbol{D}| = \frac{1}{2}(v-1)$.

Denote the $(0,1)$ incidence matrix of a Paley type partial difference set D in G by A. From Theorem 1.20, $A^2 = \frac{1}{4}(v-1)J - A + \frac{1}{4}(v-1)I$. We let $N' = 2A - J + I$ and

$$K = \begin{bmatrix} 0 & \boldsymbol{e}^{\top} \\ \boldsymbol{e} & N' \end{bmatrix}.$$

Then K is a symmetric matrix with the diagonal 0 and the other elements ± 1. $N'^2 = vI - J$, $N'\boldsymbol{e} = 0$, and $K^2 = vI$. The matrix

$$H = \begin{bmatrix} K+I & K-I \\ K-I & -K-I \end{bmatrix}$$

is an Hadamard matrix of order $2(v+1)$.

8.5 The Construction of Paley Type PDS from a Covering Extended Building Set

Polhill [147] constructed the Paley-type PDS in the group whose order is not a prime power by using a covering extended building set.

We give the definition of a covering extended building set which was first introduced by Davis and Jedwab [42].

Definition 8.2: Let G be an abelian group and m be a positive real number. A subset B of G is called a **building block with modulus** m if

$$|\chi(\mathsf{B})| \in \{0, m\}$$

is satisfied for every nontrivial character χ.

Let U be a subgroup of G and $\mathcal{B}$ be the set of h building blocks with modulus m. Assume that $h-1$ blocks of $\mathcal{B}$ have cardinality a and the last one has cardinality $a+m$. The collection $\mathcal{B}$ is called an $(a, m, h, +)$ **extended building set on** G **with respect to** U if the following conditions are satisfied:

i) If χ is trivial on U and nontrivial on G, then exactly one building block B has $\chi(\mathsf{B}) \neq 0$.
ii) If χ is nontrivial on U, then no building block B in $\mathcal{B}$ has $\chi(\mathsf{B}) \neq 0$.

In the case $U = \{1\}$, the extended building set is called **covering**.

An $(a, m, h, -)$ extended building set on G with respect to U is defined similarly replacing $a + m$ by $a - m$.

Notation 8.3: For the convenience, we abbreviate an extended building set as an EBS.

Example 8.5 (*Beth et al.* [15]): *Let* $G = \langle x, y, z \mid x^4 = y^4 = z^2 = 1 \rangle \cong \mathbf{Z}_4^2 \times \mathbf{Z}_2$. *We define three sets*

$$B_1 = \{1, x, xz, x^2z, x^2yz, xy, xy^3z, y^3\},$$
$$B_2 = \{1, x^3, x^3y^2z, x^2y^2z, yz, xy^3z, xy^3, x^2y\},$$
$$B_3 = \{1, x^2, y^2, x^2y^2\}.$$

Then these sets are a covering $(8, 4, 3, -)$ *extended building set on G. For the character such that* $\chi(x) = -1$, $\chi(y) = -i$, *and* $\chi(z) = -1$, *we have* $\chi(B_2) = -4i$, $\chi(B_1) = 0$ *and* $\chi(B_3) = 0$ *where* $i = \sqrt{-1}$. *For the character such that* $\chi(x) = \chi(y) = \chi(z) = -1$, *then* $\chi(B_1) = \chi(B_2) = 0$ *and* $\chi(B_3) = 4$. *We can verify (i) for the other characters.*

Based on the above facts, Polhill [147] introduced an extended building set with amicable Paley type PDS and the composition criteria.

We call a set (D_0, D_1, D_2, D_3, P) in a group G of order n^2 an **extended building set with amicable Paley type PDS**, if the set satisfies the following:

i) The sets $G \backslash D_0, D_1, D_2, D_3$ are a $(\frac{1}{2}(n^2 - n), n, 4, +)$ covering extended building set.
ii) G has a Paley type PDS P.
iii) For a nontrivial character χ, if $\chi(D_0) = \pm n$ (or $\chi(D_2) = \pm n$), then $\chi(P) = \frac{1}{2}(\pm n - 1)$.
iv) For a nontrivial character χ, if $\chi(D_1) = \pm n$ (or $\chi(D_3) = \pm n$), then $\chi(P) = \frac{1}{2}(\mp n - 1)$.

Thus we see the subsets D_0, D_1, D_2, and D_3 given in by (7.3) and $P = \tilde{A} \cup \tilde{B}'$ form a covering extended building set with amicable Paley type PDS.

Furthermore if (D_0, D_1, D_2, D_3, P) has the following additional properties, then we say it satisfies the **composition criteria**:

v) $P = \{(D_0 \cap D_2) \cup ((G \backslash D_1) \cap (G \backslash D_3))\} - \{0\}$ such that $D_0 \cap D_2$ and $(G \backslash D_1) \cap (G \backslash D_3)$ are disjoint;
vi) $G - P = (D_1 \cap D_3) \cup ((G \backslash D_0) \cap (G \backslash D_2))$ such that $D_1 \cap D_3$ and $(G \backslash D_0) \cap (G \backslash D_2)$ are disjoint.

Example 8.6 (*Polhill* [147]): $G = \mathbf{Z}_3^2$ *contains four subgroups of order* 3, *that is*

$$H_0 = \{(0,0), (1,0), (2,0)\},\ H_1 = \{(0,0), (1,1), (2,2)\},$$
$$H_2 = \{(0,0), (0,1), (0,2)\},\ H_3 = \{(0,0), (1,2), (2,1)\}.$$

Since $\mathbf{Z}_3^2$ *is the additive group of* $F = GF(3^2)$, *we can write the subgroups by using a primitive element* θ *of* F *as*

$$H_0 = \{0, 1, \theta^4\},\ H_1 = \{0, \theta^3, \theta^7\},\ H_2 = \{0, \theta, \theta^5\},\ H_3 = \{0, \theta^2, \theta^6\}.$$

Denote the additive character of F *for* $\beta \in F$ *by* χ_β. *We verify*

$$\chi_1(H_0) = \chi_1(H_1) = \chi_1(H_2) = 0 \text{ and } \chi_1(H_3) = 3.$$

We verify that exactly one subgroup H_i *has* $\chi_\beta(H_i) = 3$ *and the other subgroups have* $\chi_\beta(H_i) = 0$ *for every additive character* χ_β. *Then* $(G \backslash H_0, H_1, H_2, H_3)$ *is a* $(3, 3, 4, +)$ *covering extended building set.*

We let $H_i^* = H_i \backslash \{(0,0)\}$ *for* $i = 0, 1, 2, 3$. *We define*

$$P = H_0^* \cup H_2^* = \{(1,0), (2,0), (0,1), (0,2)\}.$$

Then we verify

$$P = \{(H_0 \cap H_2) \cup ((G \backslash H_1) \cap (G \backslash H_3))\} - \{(0,0)\}.$$

Thus we have

$$\chi_\beta(P) = \chi_\beta(H_0^*) + \chi_\beta(H_2^*) = \begin{cases} -2 & \text{if } \chi_\beta(H_0) = \chi_\beta(H_2) = 0, \\ 1 & \text{if } \chi_\beta(H_0) = 3 \text{ or } \chi_\beta(H_2) = 3, \end{cases}$$

that is , the conditions (iii) and (iv) hold. Since $\chi_\beta(P)$ satisfies the Eq. (8.2), $P = H_0^* \cup H_2^*$ is a $(9, 4, 1, 2)$ Paley type partial difference set. Hence $(H_0, H_1, H_2, H_3, H_0^* \cup H_2^*)$ is a covering extended building set with amicable Paley type PDS $H_0^* \cup H_2^*$ satisfying the composition criteria (v) and (vi).

Lemma 8.3 ***(Polhill [147]):*** *For an odd prime p and a positive integer t, there exists a building set with amicable Paley type PDS in $\boldsymbol{F}_q^4$ that satisfies the composition criteria.*

Proof: Let $q = p^t$ be an odd prime power and $\Sigma_3 = PG(3, q)$. The corresponding vector space is $W = \boldsymbol{F}_q^4$.

Let

$$\Omega = \left\{x \in V_k(q) \mid \langle x \rangle \in \mathcal{O}\right\}$$

be the set of k-dimensional vector space $V_k(Q)$ corresponding to a (n, k, h_1, h_2) projective set $\mathcal{O} = \{\langle y_1 \rangle, \langle y_2 \rangle, \dots, \langle y_n \rangle\}$. The subsets C_0 and C_1 are of Type Q in Σ_3 and $\left\{L_1, L_2, \dots, L_{q^2+1}\right\}$ is a spread of Σ_3 as in Theorem 7.7.

We denote the set $\{x \in W \mid \langle x \rangle \in X\}$ of W by $\tilde{X}$ for the set X of Σ_3.

Put $s = \frac{1}{2}(q^2 + 1)$. Let A be the union of any $\frac{1}{4}(q^2 - 1)$ lines from $\{L_{s+1}, \dots, L_{2s}\}$, B be the union of any $\frac{1}{4}(q^2 - 1)$ lines from $\{L_1, \dots, L_s\}$, C_0, and C_1 be as in Theorem 7.7.

We consider the subsets D_0, D_1, D_2, and D_3 given by (7.3), that is

$$D_0 = \tilde{C}_0 \bigcup \tilde{A}, \quad D_1 = \tilde{C}_1 \bigcup \tilde{B}, \quad D_2 = \tilde{C}_2 \bigcup \tilde{A}, \quad D_3 = \tilde{C}_3 \bigcup \tilde{B}$$

where

$$C_2 = \left\{L_1 \cup L_2 \cup \cdots \cup L_s\right\} \backslash C_0 \text{ and } C_3 = \left\{L_{s+1} \cup L_{s+2} \cup \cdots \cup L_{2s}\right\} \backslash C_1.$$

Notice that $\tilde{A}$ and $\tilde{B}$ do not contain 0. We notice C_2 and C_3 are also the subsets of Type Q.

Let $T_1 = \left\{\chi \in W^* \backslash \{\chi^0\} \mid \chi \text{ is trivial on } L_j \text{ for some } j, 1 \le j \le s\right\}$ and $N_1 = \left\{\chi \in W^* \backslash \{\chi^0\} \mid \chi \text{ is nontrivial on every } L_j, 1 \le j \le s\right\}$.

We see

$$|D_i| = \frac{1}{2}\left(q^4 - q^2\right) \quad \text{for } i = 0, 1, 2, 3$$

and

$$|W \backslash D_0| = q^4 - \frac{1}{2}\left(q^4 - q^2\right) = \frac{1}{2}\left(q^4 - q^2\right) + q^2$$

For a nontrivial character χ, we know that $\chi(D_i) = \pm q^2$ for exactly one subset D_i and $\chi(D_i) = 0$ for all other subsets D_i. Thus $W \backslash D_0, D_1, D_2$, and D_3 form a $(\frac{1}{2}(q^4 - q^2), q^2, 4, +)$ covering extended building set.

We put

$$\tilde{B}' = (\tilde{C}_0 \cup \tilde{C}_2) \backslash \tilde{B} \text{ and } \tilde{A}' = (\tilde{C}_1 \cup \tilde{C}_3) \backslash \tilde{A}.$$

A' and B' is a union of $\frac{1}{4}(q^2 + 3)$ lines in Σ_3. Let χ be a nontrivial character. Assume that $\chi \in T_1$ and χ is trivial on L_j, for some j, $1 \le j \le s$. Then either $L_j \in B$ or $L_j \in B'$. If $\chi \in N_1$, then χ is nontrivial on every L_j in B and B'. From the Eqs. (7.2) and (7.4), we have

$$\chi\left(\tilde{B}'\right) = \begin{cases} q^2 - \frac{1}{4}\left(q^2 + 3\right) & \text{if } L_j \in B', \\ -\frac{1}{4}\left(q^2 + 3\right) & \text{if } L_j \notin B'. \end{cases}$$

We let $P = \tilde{A} \cup \tilde{B}'$. If a nontrivial character χ is trivial on L_j, $1 \le j \le s$, $\chi(\tilde{A}) = -\frac{1}{4}(q^2 - 1)$.

Then

$$\chi(P) = \begin{cases} q^2 - \frac{1}{2}\left(q^2+1\right) & \text{if } L_j \in B', \\ -\frac{1}{2}\left(q^2+1\right) & \text{if } L_j \notin B'. \end{cases}$$

If χ is trivial on some $L_j, s+1 \le j \le 2s$, we have

$$\chi(P) = \begin{cases} q^2 - \frac{1}{2}\left(q^2+1\right) & \text{if } L_j \in A, \\ -\frac{1}{2}\left(q^2+1\right) & \text{if } L_j \notin A \end{cases}$$

similarly. Namely, $\chi(P) = \frac{1}{2}(-1 \pm q^2)$.

From Theorem 1.21, we know P is a $\left(q^4, \frac{1}{2}(q^4-1), \frac{1}{4}(q^4-5), \frac{1}{4}(q^4-1)\right)$ Paley type PDS and $0 \notin P$.

Furthermore, we see that if $\chi(D_0) = \pm q^2$ (or $\chi(D_2) = \pm q^2$), then $\chi(P) = \frac{1}{2}(\pm q^2 - 1)$ and if $\chi(D_1) = \pm q^2$ (or $\chi(D_3) = \pm q^2$), then $\chi(P) = \frac{1}{2}(\mp q^2 - 1)$ from the proof of Theorem 7.7.

From the Eq. (7.3) we let,

$$D_0 = \tilde{C}_0 \bigcup \tilde{A}, \quad D_1 = \tilde{C}_1 \bigcup \tilde{B}, \quad D_2 = \tilde{C}_2 \bigcup \tilde{A}, \quad D_3 = \tilde{C}_3 \bigcup \tilde{B}$$

such that $\tilde{C}_0 \cap \tilde{C}_2 = \emptyset$ and $\tilde{C}_1 \cap \tilde{C}_3 = \emptyset$. Then we have

$$D_0 \cap D_2 = \left(\tilde{C}_0 \bigcup \tilde{A}\right) \cap \left(\tilde{C}_2 \bigcup \tilde{A}\right) = \tilde{A}$$

and

$$\begin{aligned} (\boldsymbol{F}_q^4 \backslash (\tilde{C}_1 \bigcup \tilde{B})) \cap (\boldsymbol{F}_q^4 \backslash (\tilde{C}_3 \bigcup \tilde{B})) &= \boldsymbol{F}_q^4 - (\tilde{C}_1 \cup \tilde{C}_3 \cup \tilde{B}) \\ &= (\boldsymbol{F}_q^4 - (\tilde{C}_1 \cup \tilde{C}_3)) \cap (\boldsymbol{F}_q^4 - \tilde{B}) \\ &= (\tilde{C}_0 \cup \tilde{C}_2 \cup \{0\}) \cap (\tilde{B}' \cup \{0\}) \\ &= \tilde{B}' \cup \{0\}. \end{aligned}$$

Hence

$$(D_0 \cap D_2) \cup ((\boldsymbol{F}_q^4 \backslash D_1) \cap (\boldsymbol{F}_q^4 \backslash D_3)) - \{0\} = \tilde{A} \cup \tilde{B}'.$$

We verify the condition (*vi*) is satisfied similarly. □

Example 8.7: *Let $D_i, i = 0, 1, 2, 3, A$ and B as in Example 7.2. Then $B' = \{L_0, L_6, L_8\}$, $\chi_1(\tilde{B}') = -3$ and $\chi_{g^5}(\tilde{B}') = 6$. We put $P = \tilde{A} \cup \tilde{B}'$. Then $\chi_1(D_0) = 9$ and $\chi_1(D_1) = \chi_1(D_2) = \chi_1(D_3) = 0$ and $\chi_1(P) = 4 = \frac{1}{2}(3^2 - 1)$. Further, we have $\chi_{g^5}(D_3) = -9$ and $\chi_{g^5}(D_0) = \chi_{g^5}(D_1) = \chi_{g^5}(D_2) = 0$ and $\chi_{g^5}(P) = -5 = \frac{1}{2}(-3^2 - 1)$.*

Polhill gave the recursive construction of covering EBS with amicable PDS. The following lemma is helpful to prove the theorem.

Lemma 8.4: *Let (D_0, D_1, D_2, D_3, P) be a covering EBS in G with amicable Paley PDS P that satisfies the composition criteria. We let $P = (D_2 \cap D_3) \cup ((G\backslash D_0) \cap (G\backslash D_1)) - \{0\}$.*

Then

i) $D_0 \cap D_1 \subset D_2 \cup D_3$ *and* $(G\backslash D_2) \cap (G\backslash D_3) \subset (G\backslash D_0) \cup (G\backslash D_1)$.
$D_2 \cap D_3 \subset D_0 \cup D_1$ *and* $(G\backslash D_0) \cap (G\backslash D_1) \subset (G\backslash D_2) \cup (G\backslash D_3)$.

ii) *P is a union of disjoint sets* $D_0 \cap P$, $D_1 \cap P$, $(G\backslash D_2) \cap P$, *and* $(G\backslash D_3) \cap P$. *Similarly* $P^c = G\backslash P - \{0\}$ *is a union of disjoint sets* $D_2 \cap P^c$, $D_3 \cap P^c$, $(G\backslash D_0) \cap P^c$, *and* $(G\backslash D_1) \cap P^c$

Proof: The assertion (i) follows from that $(D_0 \cap D_1)$ and $((G\backslash D_2) \cap (G\backslash D_3))$ are disjoint, and $(D_2 \cap D_3)$ and $((G\backslash D_0) \cap (G\backslash D_1))$ are disjoint.

From (i), we have

$$\begin{aligned}&(D_0 \cap P) \cup (D_1 \cap P) \cup ((G\backslash D_2) \cap P) \cup ((G\backslash D_3) \cap P)\\ &\quad= (D_0 \cup D_1 \cup (G\backslash D_2) \cup (G\backslash D_3)) \cap P\\ &\quad\supset (D_2 \cap D_3 \cup (G\backslash D_0) \cap (G\backslash D_1)) \cap P = P,\end{aligned}$$

and

$$\begin{aligned}&(D_2 \cap P^c) \cup (D_3 \cap P^c) \cup ((G\backslash D_0) \cap P^c) \cup ((G\backslash D_1) \cap P^c)\\ &\quad= (D_2 \cup D_3 \cup (G\backslash D_0) \cup (G\backslash D_1)) \cap P^c = P^c.\end{aligned}$$

Assume that $x \in (D_0 \cap P) \cap ((G\backslash D_2) \cap P)$. Since $x \in G\backslash D_2, x \in P^c$. It contradicts to the assumption $x \in P$. It is easy to prove that the other subsets are disjoint. □

Based on Lemmas 8.3 and 8.4, we have the following theorem.

Theorem 8.6 *(Polhill [147])*: *Let G_1 be an abelian group of order n^2 and G_2 be an abelian group of order m^2. Let (D_0, D_1, D_2, D_3, P) be a $(\frac{1}{2}(n^2 - n), n^2, 4, +)$ covering EBS with amicable Paley type PDS P in G_1.*

Let (E_0, E_1, E_2, E_3, R) be a $(\frac{1}{2}(m^2 - m), m^2, 4, +)$ covering EBS with amicable Paley type PDS R in G_2 that satisfies the composition criteria.

Further suppose any subset of E_0, E_1, E_2, E_3 does not contain 0.

Then there exists a $(\frac{1}{2}((nm)^2 - nm), (nm)^2, 4, +)$ covering EBS with amicable Paley type PDS in $G_1 \times G_2$.

Proof: Let χ be a character of G_1 and ψ be a character of G_2. Assume that

$$\chi(P) = \begin{cases} \frac{1}{2}(\mp n - 1) & \text{if } \chi(D_0) = \pm n \text{ or } \chi(D_1) = \pm n,\\ \frac{1}{2}(\pm n - 1) & \text{if } \chi(D_2) = \pm n \text{ or } \chi(D_3) = \pm n,\end{cases}$$

and

$$\psi(R) = \begin{cases} \frac{1}{2}(\mp m - 1) & \text{if } \psi(E_0) = \pm m \text{ or } \psi(E_1) = \pm m,\\ \frac{1}{2}(\pm m - 1) & \text{if } \psi(E_2) = \pm m \text{ or } \chi(E_3) = \pm m.\end{cases}$$

Let $R = (E_2 \cap E_3) \cup ((G_2\backslash E_0) \cap (G_2\backslash E_1)) - \{0\}$. Then $G_2\backslash R = (E_0 \cap E_1) \cup ((G_2\backslash E_2) \cap (G_2\backslash E_3))$ and put $R^c = (G_2\backslash R) - \{0\}$.

We define the subset Y of $G_1 \times G_2$ by

$$\begin{aligned}Y = &(D_0 \times (E_3 \cap R^c)) \cup ((G_1\backslash D_0)) \times ((G_2\backslash E_3 \cap R)) \cup (D_1 \times (E_2 \cap R^c))\\ &\cup ((G_1\backslash D_1) \times ((G_2\backslash E_2) \cap R)) \cup ((G_1\backslash D_2) \times (E_0 \cap R))\\ &\cup (D_2 \times ((G_2\backslash E_0) \cap R^c)) \cup ((G_1\backslash D_3) \times (E_1 \cap R))\\ &\cup (D_3 \times ((G_2\backslash E_1) \cap R^c)) \cup (P \times \{0\}).\end{aligned}$$

We see the subsets of Y are pairwise disjoint since (E_0, E_1, E_2, E_3, R) satisfy the composition criteria. We define the sets of $G_1 \times G_2$ by

$$\begin{aligned}
F_0 &= (D_1 \times ((G_2 \backslash E_0) \cap (G_2 \backslash E_1))) \cup ((G_1 \backslash D_1) \times (E_0 \cap E_1)) \\
&\quad \cup (D_3 \times ((G_2 \backslash E_0) \cap E_1)) \cup ((G_1 \backslash D_3) \times (E_0 \cap (G_2 \backslash E_1))), \\
F_1 &= (D_0 \times ((G_2 \backslash E_0) \cap (G_2 \backslash E_1))) \cup ((G_1 \backslash D_0) \times (E_0 \cap E_1)) \\
&\quad \cup (D_2 \times (E_0 \cap (G_2 \backslash E_1))) \cup ((G_1 \backslash D_2) \times ((G_2 \backslash E_0) \cap E_1)), \\
F_2 &= (D_1 \times (E_2 \cap (G_2 \backslash E_3))) \cup ((G_1 \backslash D_1) \times (E_3 \cap (G_2 \backslash E_2))) \\
&\quad \cup (D_3 \times ((G_2 \backslash E_2) \cap (G_2 \backslash E_3))) \cup ((G_1 \backslash D_3) \times (E_2 \cap E_3)), \\
F_3 &= (D_0 \times ((G_2 \backslash E_2) \cap E_3)) \cup ((G_1 \backslash D_0) \times (E_2 \cap (G_2 \backslash E_3))) \\
&\quad \cup (D_2 \times ((G_2 \backslash E_2) \cap (G_2 \backslash E_3))) \cup ((G_1 \backslash D_2) \times (E_2 \cap E_3)).
\end{aligned}$$

It is sufficient to prove that (F_0, F_1, F_2, F_3, Y) satisfies the following conditions:

i) $|Y| = \frac{1}{2}\left((nm)^2 - 1\right)$, $|F_i| = \frac{1}{2}\left((nm)^2 - nm\right)$, $i = 0, 1, 2, 3$,
ii) For a nontrivial character ϕ of $G_1 \times G_2$, $\phi(F_i) = \pm nm$ for some i and $\phi(F_j) = 0$ for $j \neq i$.
iii) $\phi(Y) = \frac{1}{2}(-1 \pm nm)$.
iv)

$$\phi(Y) = \begin{cases} \frac{1}{2}(\mp nm - 1) & \text{if } \phi(F_0) = \pm nm \text{ or } \phi(F_1) = \pm nm, \\ \frac{1}{2}(\pm nm - 1) & \text{if } \phi(F_2) = \pm nm \text{ or } \phi(F_3) = \pm nm. \end{cases}$$

From Lemma 8.4, we have

$$\begin{aligned}
|Y| &= |D_0| \left\{ |E_2 \cap R^c| + |E_3 \cap R^c| + \left|(G_2 \backslash E_0) \cap R^c\right| + \left|(G_2 \backslash E_1) \cap R^c\right| \right\} \\
&\quad + |G_1 \backslash D_0| \left\{ |E_0 \cap R| + |E_1 \cap R| + \left|(G_2 \backslash E_2) \cap R\right| + \left|(G_2 \backslash E_3) \cap R\right| \right\} \\
&\quad + |P| \\
&= \frac{1}{2}(n^2 - n)\,|R^c| + \frac{1}{2}(n^2 + n)\,|R| + \frac{1}{2}(n^2 - 1) = \frac{1}{2}\left((nm)^2 - 1\right).
\end{aligned}$$

$$\begin{aligned}
|F_0| &= |D_0| \left\{ \left|(G_2 \backslash E_0) \cap (G_2 \backslash E_1)\right| + \left|(G_2 \backslash E_0) \cap E_1\right| \right\} \\
&\quad + |G_1 \backslash D_0| \left\{ |E_0 \cap E_1| + \left|E_0 \cap (G_2 \backslash E_1)\right| \right\} \\
&= \frac{1}{2}(n^2) \left\{ \left|(G_2 \backslash E_0) \cap (G_2 \backslash E_1)\right| + \left|(G_2 \backslash E_0) \cap E_1\right| + |E_0 \cap E_1| \right. \\
&\quad \left. + \left|E_0 \cap (G_2 \backslash E_1)\right| \right\} \\
&\quad + \frac{1}{2}n \left\{ |E_0 \cap E_1| + \left|E_0 \cap (G_2 \backslash E_1)\right| \right. \\
&\quad \left. - \left|(G_2 \backslash E_0) \cap (G_2 \backslash E_1)\right| - \left|(G_2 \backslash E_0) \cap E_1\right| \right\} \\
&= \frac{1}{2}(n^2)\,|G_2| + \frac{1}{2}n \left\{ |E_0| - |G_2 \backslash E_0| \right\} \\
&= \frac{1}{2}\left((nm)^2 - nm\right).
\end{aligned}$$

The other sets F_1, F_2, and F_3 have the same cardinality as F_0.

We write $\phi = \chi \cdot \psi$ and assume that ϕ is nontrivial.

Case 1
We assume that χ is trivial on G_1 and ψ is nontrivial on G_2.

$$\begin{aligned}
\phi(Y) = |D_0| &\left\{ \psi(E_2 \cap R^c) + \psi(E_3 \cap R^c) + \psi((G_2 \backslash E_0) \cap R^c) \right. \\
&\left. + \psi((G_2 \backslash E_1) \cap R^c) \right\} \\
&+ |G_1 \backslash D_0| \left\{ \psi(E_0 \cap R) + \psi(E_1 \cap R) + \psi((G_2 \backslash E_2) \cap R) \right. \\
&\left. + \psi((G_2 \backslash E_3) \cap R) \right\} + |P| . \\
= &\frac{1}{2}(n^2 - n)\, \psi(R^c) + \frac{1}{2}(n^2 + n)\, \psi(R) + \frac{1}{2}(n^2 - 1) \\
\phi(F_0) = |D_1| &\left\{ \psi((G_2 \backslash E_0) \cap (G_2 \backslash E_1)) + \psi((G_2 \backslash E_0) \cap E_1) \right\} \\
&+ |G_1 \backslash D_1| \left\{ \psi(E_0 \cap E_1) + \psi(E_0 \cap (G_2 \backslash E_1)) \right\} \\
= &\frac{1}{2}(n^2 - n)\psi(G_2 \backslash E_0) + \frac{1}{2}(n^2 + n)\psi(E_0).
\end{aligned}$$

Similarly we have

$$\begin{aligned}
\phi(F_1) &= \frac{1}{2}(n^2 - n)\psi(G_2 \backslash E_1) + \frac{1}{2}(n^2 + n)\psi(E_1), \\
\phi(F_2) &= \frac{1}{2}(n^2 - n)\psi(G_2 \backslash E_3) + \frac{1}{2}(n^2 + n)\psi(E_3), \\
\phi(F_3) &= \frac{1}{2}(n^2 - n)\psi(G_2 \backslash E_2) + \frac{1}{2}(n^2 + n)\psi(E_2).
\end{aligned}$$

We see that $\phi(F_i), i = 0, 1, 2, 3$ satisfy the condition (ii).
If $\psi(E_0) = \pm m$ and $\psi(E_i) = 0$ for $i \neq 0$, then

$$\begin{aligned}
\psi(R) &= \frac{\mp m - 1}{2}, \\
\phi(F_0) &= \frac{1}{2}(n^2 - n)(\psi(G_2) - \psi(E_0)) + \frac{1}{2}(n^2 + n)\, \psi(E_0) = \pm nm,
\end{aligned}$$

and

$$\phi(F_i) = 0 \quad \text{for } i \neq 0.$$

Thus

$$\begin{aligned}
\phi(Y) &= \frac{1}{2}(n^2 - n) \cdot \frac{1}{2}(\pm m - 1) + \frac{1}{2}(n^2 + n) \cdot \frac{1}{2}(\mp m - 1) + \frac{1}{2}(n^2 - 1) \\
&= \frac{1}{2}(\mp nm - 1).
\end{aligned}$$

We verify that the remaining cases also satisfy the condition (iv).

Case 2

We assume that χ is nontrivial on G_1 and ψ is trivial on G_2.

$$\begin{aligned}\phi(Y) = \chi(D_0)&\left\{ |E_3 \cap R^c| - \left|(G_2 \backslash E_3) \cap R\right| \right\}\\ +\chi(D_1)&\left\{ |E_2 \cap R^c| - \left|(G_2 \backslash E_2) \cap R\right| \right\}\\ +\chi(D_2)&\left\{ \left|(G_2 \backslash E_0) \cap R^c\right| - |E_0 \cap R| \right\}\\ +\chi(D_3)&\left\{ \left|(G_2 \backslash E_1) \cap R^c\right| - |E_1 \cap R| \right\} + \chi(P).\end{aligned}$$

Since the subsets E_0, E_1, E_2, and E_3 do not contain 0,

$$\begin{aligned}&|E_3 \cap R^c| - \left|(G_2 \backslash E_3) \cap R\right| = |E_3| - |E_3 \cap R| - (|R| - |E_3 \cap R|) = \frac{1}{2}(-m+1),\\ &\left|(G_2 \backslash E_0) \cap R^c\right| - |E_0 \cap R| = |R^c| - |E_0| + |E_0 \cap R| - |E_0 \cap R| = \frac{1}{2}(m-1).\end{aligned}$$

We have the similar results for the remaining cases. Thus

$$\begin{aligned}\phi(Y) = \chi(D_0)&\frac{1}{2}(-m+1) + \chi(D_1)\frac{1}{2}(-m+1) + \chi(D_2)\frac{1}{2}(m-1)\\ &+\chi(D_3)\frac{1}{2}(m-1) + \chi(P).\end{aligned}$$

Now

$$\begin{aligned}\phi(F_0) &= \chi(D_1)\left\{ \left|(G_2 \backslash E_0) \cap (G_2 \backslash E_1)\right| - |E_0 \cap E_1| \right\}\\ &\quad + \chi(D_3)\left\{ \left|(G_2 \backslash E_0) \cap E_1\right| - \left|E_0 \cap (G_2 \backslash E_1)\right| \right\}\\ &= \chi(D_1)\left\{ |G_2| - |E_0 \cup E_1| - |E_0 \cap E_1| \right\} + \chi(D_3)\left\{ |E_1| - |E_0| \right\}\\ &= \chi(D_1)\left\{ |G_2| - (|E_0| + |E_1|) \right\}\\ &= m\chi(D_1).\end{aligned}$$

Similarly we have

$$\phi(F_1) = m\chi(D_0), \quad \phi(F_2) = m\chi(D_3), \quad \phi(F_3) = m\chi(D_2).$$

Then $\phi(F_i), i = 0, 1, 2, 3$ satisfy the condition *(ii)*.

We assume that $\phi(F_0) = \pm nm$. It implies $\chi(D_1) \pm n$ and $\chi(P) = \frac{1}{2}(\mp n - 1)$. Then

$$\phi(Y) = \pm n\frac{1}{2}(-m+1) + \frac{1}{2}(\mp n - 1) = \frac{1}{2}(\mp nm - 1).$$

We can verify the remaining cases, then $\phi(F_i), i = 0, 1, 2, 3$ satisfy the condition *(iv)*.

Case 3

We assume that both χ and ψ are nontrivial.

$$\begin{aligned}\phi(Y) &= \chi(D_0)\left\{\psi(E_3\cap R^c) - \psi((G_2\backslash E_3)\cap R)\right\}\\ &\quad + \chi(D_1)\left\{\psi(E_2\cap R^c) - \psi((G_2\backslash E_2)\cap R)\right\}\\ &\quad + \chi(D_2)\left\{\psi((G_2\backslash E_0)\cap R^c) - \psi(E_0\cap R)\right\}\\ &\quad + \chi(D_3)\left\{\psi((G_2\backslash E_1)\cap R^c) - \psi(E_1\cap R)\right\} + \chi(P)\\ &= \chi(D_0)\left\{\psi(E_3) - \psi(R)\right\} + \chi(D_1)\left\{\psi(E_2) - \psi(R)\right\}\\ &\quad + \chi(D_2)\left\{\psi(R^c) - \psi(E_0)\right\} + \chi(D_3)\left\{\psi(R^c) - \psi(E_1)\right\} + \chi(P)\end{aligned}$$

and

$$\begin{aligned}\phi(F_0) &= \chi(D_1)\left\{\psi((G_2\backslash E_0)\cap (G_2\backslash E_1)) - \psi(E_0\cap E_1)\right\}\\ &\quad + \chi(D_3)\left\{\psi((G_2\backslash E_0)\cap E_1) - \psi(E_0\cap (G_2\backslash E_1))\right\}\\ &= -\chi(D_1)\left(\psi(E_0)+\psi(E_1)\right) + \chi(D_3)\left(\psi(E_1)-\psi(E_0)\right).\end{aligned}$$

Similarly we have

$$\begin{aligned}\phi(F_1) &= -\chi(D_0)\left(\psi(E_0)+\psi(E_1)\right) + \chi(D_2)\left(\psi(E_0)-\psi(E_1)\right),\\ \phi(F_2) &= \chi(D_1)\left(\psi(E_2)-\psi(E_3)\right) - \chi(D_3)\left(\psi(E_2)+\psi(E_3)\right),\\ \phi(F_3) &= \chi(D_0)\left(\psi(E_3)-\psi(E_2)\right) - \chi(D_2)\left(\psi(E_2)+\psi(E_3)\right).\end{aligned}$$

Assume that $\chi(D_0)\neq 0$ and $\psi(E_1)\neq 0$. Then $\phi(F_0)=\phi(F_2)=\phi(F_3)=0$ and $\phi(F_1) = -\chi(D_0)\psi(E_1) = -(\pm n)(\pm m)$. Thus we know that the condition *(ii)* is satisfied. Furthermore we see that

$$\phi(Y) = \begin{cases} \frac{1}{2}((\pm n)(\pm m) - 1), & \text{if } \phi(F_2) = (\pm n)(\pm m) \text{ or } \phi(F_3) = (\pm n)(\pm m),\\ \frac{1}{2}(-(\pm n)(\pm m) - 1), & \text{if } \phi(F_0) = (\pm n)(\pm m) \text{ or } \phi(F_1) = (\pm n)(\pm m).\end{cases}$$

This completes the proof. □

From Lemma 8.3, Theorem 8.6 and Example 8.6, we obtain the following theorem.

Theorem 8.7 *(Polhill [146, 147])*: *Let $n > 1$ be a positive odd integer and $n = p_1^{t_1} p_2^{t_2} \dots , p_\ell^{t_\ell}$ be the prime factorization. There exists a Paley type partial difference set in a group $G = \mathbf{Z}_3^{2t} \times \mathbf{Z}_{p_1}^{4t_1} \times \dots \times \mathbf{Z}_{p_\ell}^{4t_\ell}$ of order n^4 and $9n^4$ where $t = 0, 1$ and $t_i \in \mathbf{Z}_{>0}, 1 \le i \le \ell$.*

8.6 Constructing Paley Hadamard Difference Sets

Weng and Hu [218] constructed Paley Hadamard difference sets from Paley type PDS and skew Hadamard difference sets. Their theorem is a generalization of Stanton and Sprott's work [178]. We will discuss skew Hadamard difference sets in Chapter 10.

Theorem 8.8 (*Weng and Hu [218]*): *Assume that there exists a $(v_1, \frac{1}{2}(v_1-1), \frac{1}{4}(v_1-5), \frac{1}{4}(v_1-1))$ Paley type partial difference set P with $0 \notin P$ in the group G_1 of order v_1 and there exists a $(v_2, \frac{1}{2}(v_2-1), \frac{1}{4}(v_2-3))$ skew Hadamard difference set S in the group G_2 of order v_2. We put $v = v_1v_2$. Then we define the subset D of $G_1 \times G_2$ by*

$$D = (P \times S) \bigcup \left(G_1 - P - \left\{0_{G_1}\right\}\right) \times \left(G_2 - S - \left\{0_{G_2}\right\}\right).$$

i) If $v_2 = v_1 - 2$, then $(\{0\} \times G_2) \cup D$ is a $(v, \frac{1}{2}(v-1), \frac{1}{4}(v-3))$ Paley Hadamard difference set in the group $G_1 \times G_2$.
ii) If $v_2 = v_1 + 2$, then $(G_1 \times \{0\}) \cup D$ is a $\left(v, \frac{1}{2}(v-1), \frac{1}{4}(v-3)\right)$ Paley Hadamard difference set in the group $G_1 \times G_2$.

Proof: It is easy to verify that $\left|(\{0\} \times G_2) \cup D\right| = \left|(G_1 \times \{0\}) \cup D\right| = \frac{1}{2}(v_1v_2 - 1)$. Let χ be a character of G_1 and ψ be a character of G_2. For a nontrivial character χ, $\chi(P) = \frac{1}{2}(-1 \pm \sqrt{v_1})$ and for a nontrivial character ψ, $\psi(S) = \frac{1}{2}(-1 \pm \sqrt{-v_2})$. We write a character of $G_1 \times G_2$ as $\phi = \chi \cdot \psi$.

From Lemma 1.16, it is sufficient to prove

$$\begin{aligned}\left|(\phi(\{0\} \times G_2) \cup D)\right| &= \left|\phi\left((G_1 \times \{0\}) \cup D\right)\right| \\ &= \frac{1}{2}\left(\sqrt{v_1v_2+1}\right) = \begin{cases}\frac{1}{2}(v_1-1) & \text{when } v_2 = v_1 - 2,\\ \frac{1}{2}(v_1+1) & \text{when } v_2 = v_1 + 2\end{cases}\end{aligned}$$

for a nontrivial character ϕ of $G_1 \times G_2$.

Assume that both characters χ and ψ are nontrivial. Then

$$\begin{aligned}\phi(D) &= \chi(P)\psi(S) + (-\chi(P) - 1)(-\psi(S) - 1) \\ &= \frac{1}{2}\{(2\chi(P)+1)(2\psi(S)+1)+1\} = \frac{1}{2}\left(\pm\sqrt{-v_1v_2}+1\right).\end{aligned}$$

We know that $\phi((\{0\} \times G_2) \cup D) = \phi((G_1 \times \{0\}) \cup D) = \phi(D)$. Assume that χ is trivial and ψ is nontrivial. Then

$$\begin{aligned}\phi(D) &= |P|\psi(S) + \left(v_1 - \frac{1}{2}(v_1-1) - 1\right)(-\psi(S) - 1) \\ &= \frac{1}{2}(v_1-1)\psi(S) + \frac{1}{2}(v_1-1)(-\psi(S)-1) = -\frac{1}{2}(v_1-1).\end{aligned}$$

When $v_2 = v_1 - 2$, $\phi((\{0\} \times G_2) \cup D) = -\frac{1}{2}(v_1 - 1)$. When $v_2 = v_1 + 2$, $\phi((G_1 \times \{0\}) \cup D) = \frac{1}{2}(v_1 + 1)$.

Assume that χ is nontrivial and ψ is trivial. Then

$$\phi(D) = \chi(P)|S| + (-\chi(P) - 1)\frac{1}{2}(v_2 - 1) = -\frac{1}{2}(v_2 - 1).$$

When $v_2 = v_1 - 2$, $\phi((\{0\} \times G_2) \cup D) = \frac{1}{2}(v_1 - 1)$. When $v_2 = v_1 + 2$, $\phi((G_1 \times \{0\}) \cup D) = -\frac{1}{2}(v_1 + 1)$. □

We know that R in Theorem 8.6 does not contain 0. Hence we have the following corollary.

Corollary 8.2 (*Polhill [147]*): *Let n be an odd integer such that $9^tn^4 \pm 2$ is a prime power for $t = 0, 1$. Then there exists a Paley Hadamard difference set in a group of order $9^tn^4(9^tn^4 \pm 2)$.*

Example 8.8 (*Polhill [146]*):

1) *From Theorem 8.7, there exists a Paley type PDS in a group $\mathbf{Z}_3^2 \times \mathbf{Z}_5^4$. Since $v - 2 = 3^2 \cdot 5^4 - 2 = 5623 \equiv 3 \pmod 4$ is a prime, there exists a skew Hadamard matrix in $GF(5623)$. Hence the group $\mathbf{Z}_3^2 \times \mathbf{Z}_5^4 \times \mathbf{Z}_{5623}$ contains a Paley Hadamard difference set.*
2) *Since there exists a Paley type PDS in a group $\mathbf{Z}_3^2 \times \mathbf{Z}_5^8$ and $v - 2 = 3^2 \cdot 5^8 - 2 = 3515625 - 2 = 3515623 \equiv 3 \pmod 4$ is a prime, we have a Paley Hadamard difference set in a group $\mathbf{Z}_3^2 \times \mathbf{Z}_5^8 \times \mathbf{Z}_{3515623}$.*

9

Skew Hadamard, Amicable, and Symmetric Matrices

9.1 Notations

Table 9.1 gives the notations which are used in this chapter.

9.2 Introduction

In the paper, Geramita et al. [76], the following remarkable pairs of matrices are given:

$$X = \begin{bmatrix} x_1 & x_2 \\ x_2 & -x_1 \end{bmatrix}; \qquad Y = \begin{bmatrix} y_1 & y_2 \\ -y_2 & y_1 \end{bmatrix}. \tag{9.1}$$

$$X = \begin{bmatrix} x_1 & x_2 & x_3 & x_3 \\ -x_2 & x_1 & x_3 & -x_3 \\ x_3 & x_3 & -x_1 & -x_2 \\ x_3 & -x_3 & x_2 & -x_1 \end{bmatrix}; \qquad Y = \begin{bmatrix} y_1 & y_2 & y_3 & y_3 \\ y_2 & -y_1 & y_3 & -y_3 \\ -y_3 & -y_3 & y_2 & -y_1 \\ -y_3 & y_3 & -y_1 & y_2 \end{bmatrix}. \tag{9.2}$$

They have the property that $XY^\top = YX^\top$.

Definition 9.1: Let X be an $OD(n; u_1, \dots, u_s)$ on variables $\{x_1, x_2, \dots, x_s\}$ and Y an $OD(n; v_1, \dots, v_t)$ on variables $\{y_1, y_2, \dots, y_t\}$. X and Y are called **amicable orthogonal designs** if X and Y have the property that $XY^\top = YX^\top$. We write this as $AOD(n : (u_1, \dots, u_s); (v_1, \dots, v_t))$. This property is called **amicability**.

Amicable orthogonal designs have flowered from the concept of amicable Hadamard matrices. See Section 1.7 for more details. So we start our study of skew Hadamard matrices by investigating amicable Hadamard matrices.

9.3 Skew Hadamard Matrices

Skew Hadamard matrices have played a significant part in the search for Hadamard matrices (see Williamson [224], Goethals and Seidel [81], Spence [172], and J. Wallis [197–199]). Szekeres [181] realized that skew Hadamard matrices were of interest in themselves and, in fact, equivalent to doubly regular tournaments.

Once attention was focused on skew Hadamard matrices, J. Wallis [200] realized that in order to form skew Hadamard matrices of order mn from ones of orders m and n, a notion of "amicability" could be decisive. This was the origin of the idea of amicable Hadamard matrices.

Hadamard Matrices: Constructions using Number Theory and Algebra, First Edition. Jennifer Seberry and Mieko Yamada.

Table 9.1 Notations used in this chapter.

I	The identity matrix
J	The matrix with every entry 1
$\mathbf{e}$	The vector with all entries 1
$X^{\top}$	The transpose of the matrix X
$M \times N$	The Kronecker product of matrices M and N
$M * N$	The Hadamard product of matrices M and N
$OD(n; s_1, s_2, s, \dots, s_u)$	An orthogonal design of order n and of type $(s_1, s_2, s, \dots, s_u)$
$AOD(n : (u_1, \dots, u_s); (v_1, \dots, v_t))$	Amicable orthogonal designs
$n - \{v; k_1, k_2, \dots, k_n; \lambda\}$	Supplementary difference sets
$n - \{v; k; \lambda\}$	Supplementary difference sets with $k = k_1 = \cdots = k_n$
$4 - \left\{v; k_1, k_2, k_3, k_4; \sum_{i=1}^{4} k_i - v\right\}$	Hadamard supplementary difference sets
(i,j)	A cyclotomic number
C_i	A cyclotomic class
R	The back diagonal identity matrix

9.3.1 Summary of Skew Hadamard Orders

Skew Hadamard matrices are known for the following orders (the reader should consult [204, p. 451], [153], and Geramita and Seberry [78]).

SI	$2^t \Pi k_i$	t, k_i, all nonnegative positive integers $k_i - 1 \equiv 3 \pmod 4$ a prime power [144]
SII	$(p-1)^u + 1$	p the order of a skew Hadamard matrix, $u > 0$ an odd integer [191]
SIII	$2(q+1)$	$q \equiv 5 \pmod 8$ a prime power [181] or $2q+1$ a prime power and $q \equiv 1 \pmod 4$ a prime
SIV	$2(q+1)$	$q = p^t$ is a prime power with $p \equiv 5 \pmod 8$ and $t \equiv 2 \pmod 4$ [182, 219]
SV	$4m$	$m \in$ {odd integers between 3 and 31 inclusive} [99, 183]
		$m \in \{37, 43, 67, 113, 127, 157, 163, 181, 213, 241, 631\}$ [54, 56, 64]
SVI	$mn(n-1)$	n the order of amicable orthogonal designs of types $((1, n-1); (n))$ and nm the order of an orthogonal design of type $(1, m, mn-m-1)$ [153]
SVII	$4(q+1)$	$q \equiv 9 \pmod{16}$ a prime power [216]
SVIII	$(\lvert t \rvert + 1)(q+1)$	$q = s^2 + 4t^2 \equiv 5 \pmod 8$ a prime power and $\lvert t \rvert + 1$ the order of a skew Hadamard matrix [209]
SIX	$4(q^2+q+1)$	q a prime power and $q^2+q+1 \equiv 3, 5,$ or $7 \pmod 8$ a prime power or $2(q^2+q+1)+1$ a prime power [175]
SX	$2^t q$	$q = s^2 + 4r^2 \equiv 5 \pmod 8$ a prime power and an orthogonal design $OD(2^t; 1, a, b, c, c + \lvert r \rvert)$ exists where $1 + a + b + 2c + \lvert r \rvert = 2^t$ and $a(q+1) + b(q-4) = 2^t$ [153]
SXI	$h^k + 1$	h the order of an pair of amicable cores [157]
SXII	hm	h the order of a skew Hadamard matrix, m the order of amicable Hadamard matrices [213]

Spence [176] has found a new construction for *SIV* and Whiteman [220] a new construction for *SI* when $k_i - 1 \equiv 3 \pmod 8$. These are of considerable interest because of the structure involved and have use in the construction of orthogonal designs.

9.4 Constructions for Skew Hadamard Matrices

We now turn our attention to skew Hadamard matrices in general.

In 1970 Jennifer Wallis [202] used a computer to obtain skew Hadamard matrices. She used a variation of Williamson matrices, called good matrices, where $(A - I)^{\top} = -(A - I)$, but B , C, and D remain symmetric by using the Seberry–Williamson variation of the Williamson array, where R is the back diagonal identity matrix.

$$\begin{bmatrix} A & BR & CR & DR \\ -BR & A & DR & -CR \\ -CR & -DR & A & BR \\ -DR & CR & -BR & A \end{bmatrix}.$$

The following first rows generate the required matrices: the results for 21, 25 were found by Hunt, for 27, 29, 31 by Szekeres and the remainder by (Seberry) Wallis. These establish class SV.

3

1	−1	1
1	−1	−1
1	−1	−1
1	1	1

5

1	−1	−1	1	1
1	−1	−1	−1	−1
1	−1	−1	−1	−1
1	−1	1	1	−1

7

1	−1	−1	−1	1	1	1
1	−1	−1	−1	−1	−1	−1
1	−1	−1	1	1	−1	−1
1	−1	1	−1	−1	1	−1

9

1	−1	−1	−1	1	−1	1	1	1
1	−1	−1	−1	1	1	−1	−1	−1
1	1	−1	1	−1	−1	1	−1	1
1	1	1	−1	1	1	−1	1	1

11

1	−1	−1	−1	−1	1	−1	1	1	1	1
1	−1	−1	−1	−1	1	1	−1	−1	−1	−1
1	1	−1	1	−1	1	1	−1	1	−1	1
1	1	−1	1	1	−1	−1	1	1	−1	1

13

1	−1	−1	−1	−1	1	−1	1	−1	1	1	1	1
1	−1	1	1	1	−1	1	1	−1	1	1	1	−1
1	1	−1	−1	1	−1	1	1	−1	1	−1	−1	1
1	1	1	1	−1	−1	1	1	−1	−1	1	1	1

15

1	−1	−1	−1	−1	−1	1	−1	1	−1	1	1	1	1	1
1	−1	1	−1	−1	1	1	−1	−1	1	1	−1	−1	1	−1
1	1	1	−1	−1	1	−1	1	1	−1	1	−1	−1	1	1
1	1	−1	1	1	1	1	−1	−1	1	1	1	1	−1	1

17

1	−1	−1	−1	−1	−1	1	−1	−1	1	1	−1	1	1	1	1	1
1	1	−1	−1	−1	1	−1	−1	1	1	−1	−1	1	−1	−1	−1	1
1	1	−1	−1	−1	1	−1	1	−1	−1	1	−1	1	−1	−1	−1	1
1	−1	−1	1	−1	1	−1	−1	−1	−1	−1	−1	1	−1	1	−1	−1

19

1	−1	1	−1	−1	−1	−1	−1	1	1	−1	−1	1	1	1	1	1	−1	1
1	1	−1	1	1	1	1	1	−1	−1	−1	−1	1	1	1	1	1	−1	1
1	−1	1	−1	−1	−1	−1	1	1	−1	−1	1	1	−1	−1	−1	−1	1	−1
1	−1	−1	1	−1	1	−1	1	1	−1	−1	1	1	−1	1	−1	1	−1	−1

21

1	−1	−1	−1	−1	−1	1	−1	−1	1	−1	1	−1	1	1	−1	1	1	1	1	1
1	1	1	−1	−1	1	−1	1	1	1	−1	−1	1	1	1	−1	1	−1	−1	1	1
1	1	−1	1	−1	−1	−1	−1	−1	1	−1	−1	1	−1	−1	−1	−1	−1	1	−1	1
1	−1	1	−1	−1	1	1	−1	−1	−1	1	1	−1	−1	−1	1	1	−1	−1	1	−1

23

1 −1 −1 −1 −1 −1 −1 −1 1 1 −1 1 −1 1 −1 −1 1 1 1 1 1 1 1
1 1 −1 −1 1 −1 −1 1 1 1 1 −1 −1 1 1 1 1 −1 −1 1 −1 −1 1
1 1 −1 −1 −1 1 −1 1 −1 1 −1 1 1 −1 1 −1 1 −1 1 −1 −1 −1 1
1 −1 −1 −1 −1 1 −1 −1 1 −1 −1 1 1 −1 −1 1 −1 −1 1 −1 −1 −1 −1

25

1 1 −1 −1 −1 −1 −1 1 −1 1 −1 −1 −1 1 1 1 −1 1 −1 1 1 1 1 1 −1
1 1 −1 1 1 1 1 −1 1 −1 1 1 −1 −1 1 1 −1 1 −1 1 1 1 1 −1 1
1 −1 −1 −1 1 −1 −1 1 1 1 1 −1 −1 −1 −1 1 1 1 1 −1 −1 1 −1 −1 −1
1 −1 −1 1 −1 1 1 1 −1 −1 1 −1 −1 −1 −1 1 −1 −1 1 1 1 −1 1 −1 −1

27

1 1 −1 −1 −1 1 −1 1 −1 −1 1 1 −1 1 −1 1 −1 −1 1 1 −1 1 −1 1 1 1 −1
1 1 1 1 −1 −1 −1 1 1 −1 1 1 −1 1 1 −1 1 1 −1 1 1 −1 −1 −1 1 1 1
1 1 1 1 −1 1 1 1 −1 1 1 −1 −1 −1 −1 −1 −1 1 1 −1 1 1 1 −1 1 1 1
1 1 −1 −1 −1 1 1 1 1 −1 1 −1 1 −1 −1 1 −1 1 −1 1 1 1 1 −1 −1 −1 1

29

1 1 −1 −1 1 −1 −1 −1 −1 1 −1 1 −1 1 1 −1 −1 1 −1 1 −1 1 1 1 1 −1 1 1 −1
1 1 −1 −1 −1 1 1 1 1 −1 1 1 −1 −1 1 1 −1 −1 1 1 −1 1 1 1 1 −1 −1 −1 1
1 1 −1 1 −1 −1 −1 −1 1 1 −1 1 −1 −1 1 1 −1 −1 1 −1 1 1 −1 −1 −1 −1 1 −1 1
1 1 1 1 1 1 1 −1 −1 −1 1 −1 1 −1 1 1 −1 1 −1 1 −1 −1 −1 1 1 1 1 1 1

31

1 1 −1 −1 1 1 −1 −1 −1 −1 1 −1 −1 −1 1 −1 1 −1 1 1 1 −1 1 1 1 1 −1 −1 1 1 −1
1 −1 1 1 −1 1 −1 −1 −1 1 −1 1 1 1 −1 −1 −1 −1 1 1 1 −1 1 −1 −1 −1 1 −1 1 1 −1
1 1 1 −1 −1 −1 1 −1 −1 1 −1 −1 1 1 1 −1 −1 1 1 1 −1 −1 1 −1 −1 1 −1 −1 −1 1 1
1 −1 1 1 1 1 1 1 1 1 −1 −1 1 −1 1 −1 −1 1 −1 1 −1 −1 1 1 1 1 1 1 1 1 −1

Comments on their construction and further references can be found in Awyzio and Seberry [3].

9.4.1 The Goethals–Seidel Type

Goethals and Seidel modified the Williamson matrix so that the matrix entries did not have to be circulant and symmetric. Their matrix which was originally given to construct a skew Hadamard matrix of order 36 [81], has been valuable in constructing many new Hadamard matrices. The Goethals and Seidel theorem is initially discussed in Theorem 3.5.

Recently Đoković [54, 56] and Đoković, et al. [64] have carried out computer searches for circulant matrices which can be used in the Goethals–Seidel array and found matrices to give skew Hadamard matrices of order $4n$, $n = 37, 43, 67, 113, 127, 157, 163, 181, 213, 241$ and 631.

The following two pairs of four sets are $4 - (37; 18, 18, 16, 13; 28)$ and $4 - (37; 18, 15, 15, 15; 26)$ supplementary difference sets, respectively, found by Đoković [56], which may be used to construct circulant $(1, -1)$ matrices which give, using the Goethals-Seidel array, skew Hadamard matrices of order $4.37 = 148$:

$1, 3, 4, 10, 14, 17, 18, 21, 22, 24, 25, 26, 28, 29, 30, 31, 32, 35$
$1, 6, 8, 9, 10, 11, 12, 14, 16, 17, 22, 23, 26, 27, 29, 31, 35, 36$
$0, 5, 6, 7, 8, 11, 13, 18, 19, 23, 24, 27, 32, 33, 34, 36$
$0, 2, 5, 11, 13, 15, 17, 19, 20, 22, 27, 35, 36$

1, 7, 9, 10, 12, 14, 16, 17, 18, 22, 24, 26, 29, 31, 32, 33, 34, 35
1, 5, 6, 7, 8, 10, 13, 18, 19, 23, 24, 26, 32, 33, 34
2, 5, 11, 13, 14, 15, 18, 19, 20, 24, 27, 29, 31, 32, 36
2, 5, 6, 8, 9, 12, 13, 14, 15, 16, 19, 20, 23, 29, 31

The following four sets, also found by Đoković [56], give $4-(43;21,21,21,15;35)$ supplementary difference sets and may be used similarly, to form a skew Hadamard matrix of order $4.43=172$:

2, 3, 5, 7, 8, 12, 18, 19, 20, 22, 26, 27, 28, 29, 30, 32, 33, 34, 36, 39, 42 (twice)
1, 3, 4, 5, 6, 10, 11, 12, 16, 19, 20, 21, 23, 24, 31, 33, 35, 36, 38, 40, 41
0, 6, 7, 10, 18, 23, 24, 26, 28, 29, 30, 31, 34, 38, 40.

9.4.2 An Adaption of Wallis–Whiteman Array

We note the following adaptation of the Goethals–Seidel matrix which does not require the matrix entries to be circulant at all.

Theorem 9.1 ***(J. Wallis and Whiteman [216]):*** *Suppose X, Y, and W are type 1 incidence matrices and Z is a type 2 incidence matrix of $4-\{v;k_1,k_2,k_3,k_4;\sum_{i=1}^{4}k_i-v\}$ supplementary difference sets, then if*

$$A=2X-J,\qquad B=2Y-J,\qquad C=2Z-J,\qquad D=2W-J,$$

$$H=\begin{bmatrix} A & B & C & D\\ -B^{\top} & A^{\top} & -D & C\\ -C & D^{\top} & A & -B^{\top}\\ -D^{\top} & -C & B & A^{\top}\end{bmatrix} \tag{9.3}$$

is an Hadamard matrix of order $4v$.

Further, if A is skew-type ($C^{\top}=C$ as Z is of type 2) then H is skew Hadamard.

This matrix can be used when the sets are from any finite abelian group. We now show how (9.3) may be further modified to obtain useful results.

Theorem 9.2 ***(J. Wallis and Whiteman[216]):*** *Suppose X, Y, and W are type 1 incidence matrices and Z is a type 2 incidence matrix of $4-\{2m+1;m;2(m-1)\}$ supplementary difference sets. Then, if*

$$A=2X-J\qquad B=2Y-J\qquad C=2Z-J\qquad D=2W-J$$

and e is the $1\times(2m+1)$ matrix with every entry 1,

$$H=\begin{bmatrix} -1 & -1 & -1 & -1 & e & e & e & e\\ 1 & -1 & 1 & -1 & -e & e & -e & e\\ 1 & -1 & -1 & 1 & -e & e & e & -e\\ 1 & 1 & -1 & -1 & -e & -e & e & e\\ e^{\top} & e^{\top} & e^{\top} & e^{\top} & A & B & C & D\\ -e^{\top} & e^{\top} & -e^{\top} & e^{\top} & -B^{\top} & A^{\top} & -D & C\\ -e^{\top} & e^{\top} & e^{\top} & -e^{\top} & -C & D^{\top} & A & -B^{\top}\\ -e^{\top} & -e^{\top} & e^{\top} & e^{\top} & -D^{\top} & -C & B & A^{\top}\end{bmatrix}$$

is an Hadamard matrix of order $8(m+1)$. Further if A is skew-type H is skew Hadamard.

Delsarte et al. [50] show important result that if there exists a $W(n,n-1)$ for $n\equiv 0 \pmod 4$ there exists a skew symmetric $W(n,n-1)$. This is used in the next theorem which uses orthogonal designs and is due to Seberry.

The results for skew Hadamard matrices are far less complete than for Hadamard matrices. Most of Table 9.2 is from [166] but updated for 213 and 631 in [64].

Theorem 9.3 (*Seberry [153]*): *Let $q \equiv 5 \pmod 8$ be a prime power and $p = \frac{1}{2}(q+1)$ be a prime. Then there is a skew Hadamard matrix of order $2^t p$ where $t \geq [2\log_2(p-2)] + 1$.*

In the Table 9.2 the lowest power of 2 for which a skew Hadamard matrix is known is indicated. For example, the entry 193 3 means a skew Hadamard matrix of order 2^3.193 is known, the entry 59. Means a skew Hadamard matrix of order 2^t.59 is not yet known for any t.

Table 9.2 Orders for which skew Hadamard matrices exist.

q	*t*	*q*	*t*	*q*	*t*	*q*	*t*	*q*	*t*
1		59	.	117		175		233	4
3		61		119	4	177	.	235	3
5		63		121	3	179	8	237	
7		65	4	123		181		239	4
9		67		125		183		241	
11		69	3	127		185		243	
13		71		129	3	187		245	4
15		73		131		189		247	6
17		75		133	3	191	.	249	4
19		77		135		193	3	251	6
21		79		137		195		253	4
23		81	3	139		197		255	
25		83		141		199		257	4
27		85		143		201	3	259	5
29		87		145	5	203		261	3
31		89	4	147		205	3	263	
33		91		149	4	207		265	4
35		93	3	151	5	209	4	267	4
37		95		153	3	211		269	8
39	3	97	9	155		213		271	
41		99		157		215		273	
43		101	10	159		217	4	275	4
45		103	3	161		219	4	277	5
47	4	105		163		221		279	
49	4	107	.	165		223	3	281	
51		109	9	167	4	225	4	283	.
53		111		169	5	227		285	3
55		113		171		229	3	287	4
57		115		173		231		289	3

(Continued)

Table 9.2 *(Continued)*

q	*t*	*q*	*t*	*q*	*t*	*q*	*t*	*q*	*t*
291		371		451		531		571	
293		373	7	453		533		573	3
295	5	375		455	4	535		575	4
297		377	6	457	.	537	5	577	.
299	4	379		459	3	539	4	579	5
301	3	381		461	17	541	3	581	4
303	3	383		463	7	543	5	583	3
305	4	385	3	465	3	545		585	
307		387		467		547		587	
309	3	389	15	469	3	549	3	589	5
311	.	391	4	471		551		591	
313		393		473	5	553	3	593	
315		395		475	4	555		595	3
317	6	397	5	477		557	.	597	4
319	3	399		479	.	559		599	.
321		401	.	481	3	561		601	5
323		403	5	483		563		603	
325	5	405		485	4	565	3	605	4
327		407		487	5	567		607	
329	6	409	3	489	3	569	4	609	3
331	3	411		491	.	571	3	611	6
333		413	4	493	3	573	3	613	3
335	7	415		495		575	4	615	
337	18	417		497		577	.	617	
339		419	4	499		579	5	619	
341	4	421		501		581	4	621	3
343	6	423	4	503		583	3	623	4
345	4	425		505	.	585		625	3
347	.	427		507		587		627	4
349	3	429	3	509	.	589	5	629	.
351		431		511		591		631	
353	4	433	3	513	4	553	3	633	
355		435	4	515	5	555		635	
357		437		517	6	557	.	637	4
359	4	439		519	4	559		639	
361	3	441	3	521		561		641	6
363		443	6	523	7	563		643	.
365		445	3	525		565	3	645	
367		447		527	4	567		647	.
369	4	449	.	529	3	569	4	649	7

(Continued)

Table 9.2 (*Continued*)

q	*t*	*q*	*t*	*q*	*t*	*q*	*t*	*q*	*t*
651		721	5	791		861	4	931	
653	.	723	3	793	3	863	4	933	.
655	4	725	6	795	3	865	4	935	
657	5	727		797		867		937	5
659	.	729	4	799		869	4	939	
661	.	731	5	801		871		941	6
663		733	.	803	9	873		943	4
665		735		805	4	875		945	
667	6	737	7	807		877	.	947	6
669	3	739	.	809	.	879		949	3
671		741		811		881	6	951	
673	7	743		813		883	.	953	.
675		745	6	815		885	4	955	3
677		747		817	5	887		957	4
679	3	749	.	819		889	5	959	4
681	4	751	3	821	6	891	3	961	3
683		753		823	3	893		963	
685		755		825		895		965	4
687		757		827		897	5	967	
689	4	759	4	829	.	899	6	969	4
691		761	.	831		901	3	971	6
693	4	763	11	833		903	4	973	4
695	4	765	4	835		905	4	975	
697	4	767		837		907	5	977	
699	3	769	3	839	.	909	4	979	5
701		771		841	8	911		981	
703	3	773	.	843		913	4	983	
705		775		845	6	915		985	3
707	4	777	4	847		917	4	987	
709	.	779	4	849	3	919	3	989	4
711		781	3	851	.	921		991	3
713		783		853	3	923		993	
715		785	7	855		925	3	995	4
717	4	787	5	857	4	927	4	997	.
719	4	789	3	859	3	929	.	999	

Source: Updated from Seberry and Yamada [166, table 7.1, pp. 499–501], Wiley.

9.5 Szekeres Difference Sets

To prove Case SIII, a new concept is introduced due to Szekeres [181] which arose while he was considering tournaments. Szekeres used supplementary difference sets with one symmetry condition ($a \in M \Rightarrow -a \notin M$) to construct skew Hadamard matrices. He pointed out to Seidel that there was no skew Hadamard matrix of order

36 known (at that time). This, in turn, led to Goethals and Seidel publishing their array, which was to prove so significant and useful.

Definition 9.2: Let G be an additive abelian group of order $2m+1$. Then two subsets, M and N, of G, which satisfy

i) M and N are m-sets,
ii) $a \in M \Rightarrow -a \notin M$,
iii) for each $d \in G$, $d \neq 0$, the equations $d = a_1 - a_2$, $d = b_1 - b_2$ have together $m-1$ distinct solution vectors for $a_1, a_2 \in M$, $b_1, b_2 \in N$,

will be called **Szekeres difference sets**. Alternatively, $2-\{2m+1;\ m;\ m-1\}$ supplementary difference sets, M and $N \subset G$, are called Szekeres difference sets, if $a \in M \Rightarrow -a \notin M$.

The following shows such sets exist.

Theorem 9.4 ***(Szekeres [183]):*** *If $q = 4m+3$ is a prime power and $G = \mathbf{Z}/(2m+1)\mathbf{Z}$ is the cyclic group of order $2m+1$, then there exist Szekeres difference sets M and N in G with $a \in M \Rightarrow -a \notin M$ and $b \in N \Rightarrow -b \in N$.*

Proof: Let x be a primitive root of $GF(q)$ and $Q = \{x^{2b} : b = 0, \dots, 2m\}$ the set of quadratic residues in $GF(q)$. Define M and N by the rules

$$a \in M \iff x^{2a} - 1 \in Q, \tag{9.4}$$

$$b \in N \iff x^{2b} + 1 \in Q. \tag{9.5}$$

Since

$$-1 = x^{2m+1} \notin Q,$$

$$x^{2a} - 1 \in Q \Rightarrow x^{-2a} - 1 = -x^{-2a}(x^{2a} - 1) \notin Q,$$

$$x^{2b} + 1 \in Q \Rightarrow x^{-2b} + 1 = x^{-2b}(x^{2b} + 1) \in Q,$$

so that $a \in M \Rightarrow -a \notin M$, $b \in N \Rightarrow -b \in N$, that is, condition (ii) of Definition 9.2 is satisfied. Condition (i) is obtained from the cyclotomic number $(0,0) = \frac{q-3}{4} = m$. Also, writing N' for the complement of N, gives

$$b' \in N' \quad \text{if} \quad -(x^{2b'} + 1) \in Q. \tag{9.6}$$

We note $|N'| = m+1$ and $b' \in N' \Rightarrow -b' \in N'$. Suppose

$$d = \alpha - a \neq 0, \qquad a, \alpha \in M, \tag{9.7}$$

where

$$x^{2a} = 1 + x^{2(i-d)}, \tag{9.8}$$

$$x^{2\alpha} = 1 + x^{2j}, \tag{9.9}$$

by (9.4) for suitable $i, j \in G$. Then

$$x^{2\alpha} = x^{2(a+d)} = x^{2d} + x^{2i},$$

by (9.7) and (9.8). Hence

$$x^{2d} - 1 = x^{2j} - x^{2i}, \tag{9.10}$$

where $x^{2j}+1 \in Q$ by (9.9). Similarly, if

$$d = b' - \beta' \neq 0, \qquad b', \beta' \in N', \tag{9.11}$$

where

$$-x^{2\beta'} = 1 + x^{2(i-d)}, \tag{9.12}$$
$$-x^{2b'} = 1 + x^{2j}, \tag{9.13}$$

for some $i, j \in G$, producing

$$-x^{2b'} = -x^{2(d+\beta')} = x^{2d} + x^{2i}.$$

Hence again

$$x^{2d} - 1 = x^{2j} - x^{2i},$$

with $-(x^{2j}+1) \in Q$ by (9.13).

Conversely for every solution, $i, j \in G$, of Eq.(9.10), we can determine uniquely $\alpha \in M$ or $b' \in N'$ from (9.9) or (9.13) depending on whether $1 + x^{2j} = x^{2d} + x^{2i}$ is in Q or not. Hence a or β' from (9.7) (9.11) so that (9.8) or (9.12) is also satisfied. Thus the total number of solutions of (9.7) and (9.11) is equal to the number of solutions of (9.10) which is $m = \frac{q-3}{4}$ from cyclotomic numbers (0, 0) and (1, 1). Thus the sets M and N' are $2-\{2m+1; m, m+1; m\}$ *sds*. It follows that M and $N = G - N'$ are Szekeres difference sets. □

Example 9.1: *Consider $q = 23$ which has primitive root 5 and quadratic residues $Q = \{1, 2, 3, 4, 6, 8, 9, 12, 13, 16, 18\}$. Hence*

$$M = \{1, 2, 5, 7, 8\} \text{ and } N = \{0, 1, 3, 8, 10\}$$

are Szekeres difference sets.

9.5.1 The Construction by Cyclotomic Numbers

We let $q = ef + 1$ be a prime power. For $e = 2, 4$, and 8, we determined cyclotomic numbers in Chapter 1 and discussed the relation between cyclotomic numbers and Jacobi sums in Chapter 2.

In this section, we construct an infinite family of Szekeres difference sets by cyclotomic numbers of order 8. An infinite family of skew Hadamard matrices follows from these Szekeres difference sets.

Theorem 9.5 (Whiteman [219]): *Let $q = p^t$ be a prime power such that $p \equiv 5 \pmod 8$ and $t \equiv 2 \pmod 4$. Let $C_i, 0 \le i \le 7$, be cyclotomic classes of order 8 in $GF(q)$. We define the subsets*

$$\mathcal{A} = C_0 \cup C_1 \cup C_2 \cup C_3 \quad \text{and} \quad \mathcal{B} = C_0 \cup C_1 \cup C_6 \cup C_7.$$

Then $\mathcal{A}$ and $\mathcal{B}$ are Szekeres difference sets, that is $2-\left\{q; \frac{1}{2}(q-1); \frac{1}{2}(q-3)\right\}$ sds.

Proof: We see $|\mathcal{A}| = |\mathcal{B}| = \frac{1}{2}(q-1)$. Let

$$\lambda_{i,j}^k = \left|\left\{(x,y) | x - y = d, x \in C_i, y \in C_j, d \in C_k\right\}\right|.$$

Let (i,j), $0 \le i, j \le 7$, be cyclotomic numbers of order 8.

The equation $x - y = d$ is equivalent to the equation $xd^{-1} - yd^{-1} = 1$. Then $\lambda_{i,j}^k = (i-k, j-k)$. When we let $N_k^{\mathcal{A}} = \left|\left\{(x,y) | x - y = d, x, y \in \mathcal{A}, d \in C_k\right\}\right|$,

$$N_k^{\mathcal{A}} = \sum_{j=0}^{3} \sum_{i=0}^{3} (i-k, j-k).$$

Similarly we define $N_k^{\mathcal{B}}$ and

$$N_k^{\mathcal{B}} = \sum_{j=0,1,6,7} \sum_{i=0,1,6,7} (i-k, j-k).$$

We will show $N_k^{\mathcal{A}} + N_k^{\mathcal{B}} = \frac{1}{2}(q-3)$ for any $k, 0 \le k \le 7$.

We put $q = 8f + 1$. Since $p \equiv 5 \pmod 8$ and $t \equiv 2 \pmod 4$, f is an odd integer. We have $-1 = \xi^{\frac{1}{2}(q-1)} = \xi^{4f} \in C_4$ where ξ is a primitive element of $GF(q)$. It follows that if $a \in \mathcal{A}$, then $-a \notin \mathcal{A}$. Condition (ii) in Definition 9.2 is satisfied.

From (i) of Lemma 1.25 and $h = \frac{1}{2}e = 4$, we have $(i,j) = (j+4, i+4)$. It follows $N_k^{\mathcal{A}} = N_{k+4}^{\mathcal{A}}$. Furthermore

$$\begin{aligned} N_{k+2}^{\mathcal{A}} &= \sum_{j=0}^{3} \sum_{i=0}^{3} (i-(k+2), j-(k+2)) \\ &= \sum_{j=0}^{1} \sum_{i=0}^{1} \{(i+6-k, j+6-k) + (i+6-k, j-k) + (i-k, j+6-k) \\ &\quad + (i-k, j-k)\} \\ &= \sum_{j=0,1,6,7} \sum_{i=0,1,6,7} (i-k, j-k) \\ &= N_k^{\mathcal{B}}. \end{aligned}$$

Thus we have

$$\begin{aligned} N_0^{\mathcal{A}} + N_0^{\mathcal{B}} &= N_2^{\mathcal{A}} + N_2^{\mathcal{B}} = N_4^{\mathcal{A}} + N_4^{\mathcal{B}} = N_6^{\mathcal{A}} + N_6^{\mathcal{B}} \\ N_1^{\mathcal{A}} + N_1^{\mathcal{B}} &= N_3^{\mathcal{A}} + N_3^{\mathcal{B}} = N_5^{\mathcal{A}} + N_5^{\mathcal{B}} = N_7^{\mathcal{A}} + N_7^{\mathcal{B}}. \end{aligned}$$

Hence it suffices to evaluate $N_0^{\mathcal{A}} + N_0^{\mathcal{B}}$ and $N_1^{\mathcal{A}} + N_1^{\mathcal{B}}$.

When we put $g = \xi^{\frac{q-1}{p-1}}$, g is a primitive element of $GF(p)$. Since $p \equiv 5 \pmod 8$, 2 is a quadratic nonresidue in $GF(p)$. Then if we let $2 = g^r$, then r is odd. $2^{r\frac{q-1}{p-1}}$ is not a fourth power in $GF(q)$, since $r\frac{q-1}{p-1} = r(p^{t-1} + p^{t-2} + \cdots + 1) \equiv rt \equiv 2 \pmod 4$.

From (1.27), we have,

$$\begin{aligned} N_0^{\mathcal{A}} &= A+B+C+D+I+J+K+L+N+O+N+M+J+O+O+I, \\ N_1^{\mathcal{A}} &= I+H+M+K+J+A+I+N+K+B+J+O+L+C+K+N, \\ N_0^{\mathcal{B}} &= A+I+N+J+B+J+M+K+G+L+N+M+H+M+O+I, \\ N_1^{\mathcal{B}} &= I+H+O+O+J+A+I+N+L+F+J+K+M+G+O+N. \end{aligned}$$

From Lemma 1.28,

$$N_0^{\mathcal{A}} + N_0^{\mathcal{B}} = N_1^{\mathcal{A}} + N_1^{\mathcal{B}} = \frac{1}{64}(32q - 96) = \frac{1}{2}(q-3).$$

□

We have an infinite family of skew Hadamard matrices by the following theorem.

Theorem 9.6: *Let G be a finite abelian group of order $2m+1$. Assume that there exist Szekeres difference sets. Then there exists a skew Hadamard matrix of order $4(m+1)$.*

Proof: Let $\mathcal{A}$ and $\mathcal{B}$ be Szekeres difference sets such that $a \in \mathcal{A} \to -a \notin \mathcal{A}$. Let X be a type 1 $(1,-1)$ incidence matrix of $\mathcal{A}$ and Y be a type 2 $(1,-1)$ incidence matrix of $\mathcal{B}$. Then $X\boldsymbol{e} = Y\boldsymbol{e} = -\boldsymbol{e}$ where $\boldsymbol{e}$ is the all one row vector of length $2m+1$. We see

$$X = -I + U, \quad U^\top = -U, \quad X^\top = -I - U, \quad Y^\top = Y$$

and

$$XX^\top + YY^\top = 4(m+1)I - 2J$$

from Eq. (1.10). Hence the matrix

$$\begin{bmatrix} 1 & 1 & \boldsymbol{e} & \boldsymbol{e} \\ -1 & 1 & \boldsymbol{e} & -\boldsymbol{e} \\ -\boldsymbol{e}^\top & -\boldsymbol{e}^\top & -X & -Y \\ -\boldsymbol{e}^\top & \boldsymbol{e}^\top & Y & -X \end{bmatrix}$$

is a skew Hadamard matrix of order $4(m+1)$. □

Corollary 9.1: *Let $q = p^t$ be a prime power such that $p \equiv 5 \pmod 8$ and $t \equiv 2 \pmod 4$. There exists an infinite family of skew Hadamard matrices of order $2(q+1)$.*

9.6 Amicable Hadamard Matrices

Szekeres and Whiteman (see Wallis [211, p. 32]) have independently shown that there exist Szekeres difference sets of size $\frac{(p^t-1)}{2}$ when $p \equiv 5 \pmod 8$ is a prime power and $t \equiv 2 \pmod 4$. But in this case both sets M, N satisfy the condition, $x \in M, N \longrightarrow -x \notin M, N$, and as yet these sets have not been used to construct amicable Hadamard matrices. Nevertheless, the next theorem and corollary indicate the way these Szekeres difference sets may, in some cases, be used:

Theorem 9.7: *Suppose there exist $(1,-1)$ matrices A, B, C, D of order n satisfying:*

$$C = I + U, \quad U^\top = -U, \quad A^\top = A, \quad B^\top = B, \quad D^\top = D,$$

$$AA^\top + BB^\top = CC^\top + DD^\top = 2(n+1)I - 2J,$$

and with $e = [1, \dots, 1]$ a $1 \times n$ matrix

$$eA^\top = eB^\top = eC^\top = eD^\top = e, \quad AB^\top = BA^\top, \quad \text{and} \quad CD^\top = DC^\top.$$

Then if

$$X = \begin{bmatrix} 1 & 1 & e & e \\ 1 & -1 & -e & e \\ e^\top & -e^\top & A & -B \\ e^\top & e^\top & -B & -A \end{bmatrix}, \qquad Y = \begin{bmatrix} 1 & 1 & e & e \\ -1 & 1 & e & -e \\ -e^\top & -e^\top & C & D \\ -e^\top & e^\top & -D & C \end{bmatrix},$$

X is a symmetric Hadamard matrix and Y is a skew Hadamard matrix both of order $2(n+1)$. Further, if

$$AC^\top - BD^\top \quad \text{and} \quad BC^\top + AD^\top$$

are symmetric, X and Y are amicable Hadamard matrices of order $2(n+1)$.

The next result illustrates how the conditions of the theorem can be satisfied;

Corollary 9.2: *Let G be an additive abelian group of order $2m+1$. Suppose there exist Szekeres difference sets, M and N, in G such that $x \in N \Rightarrow -x \in N$.*

Further suppose there exist $2 - \{2m+1;\ m+1;\ m+1\}$ *supplementary difference sets P and S in G such that* $x \in X \Rightarrow -x \in X$ *for* $X \in \{P, S\}$. *Then there exist amicable Hadamard matrices of order* $4(m+1)$.

Proof: Form the type 1 $(1, -1)$ incidence matrix C of M. Form the type 2 $(1, -1)$ incidence matrices, D, A, B of N, P, S, respectively. Now use the properties of type 1 and type 2 matrices in the theorem.

Since M and N are Szekeres difference sets, and P and S are $2 - \{2m+1;\ m+1; m+1\}$ *sds*, we have $MM^\top + NN^\top = PP^\top + SS^\top = 4(m+1)I - 2J$. The matrices M, N, P, and S satisfy the conditions of Theorem 9.7. □

In these theorems, circulant and back circulant can be used to replace type 1 and type 2 incidence matrices, respectively, when the orders are prime.

We now wish to show that sets satisfying the conditions of Corollary 9.2 exist for some orders $n \equiv 1 \pmod 4$.

To find M and N, we use the result of Szekeres in Theorem 9.4. Then we have:

Corollary 9.3: *There exist amicable Hadamard matrices of order* $4(m+1)$ *whenever* $2m+1 \equiv 1 \pmod 4$ *is a prime power and* $4m+3$ *is a prime power.*

Proof: With $q = 4m+3$, we form Szekeres difference sets M and N with m elements from Theorem 9.4. Put $n = 2m+1 \equiv 1 \pmod 4$ and choose $Q = \{x^{2b} : b = 0, 1, \dots, m-1\}$ and $S = xQ$, where x is a primitive element of $GF(n)$. Then Q and S are $2 - \{2m+1; m; m-1\}$ *sds*. Further $y \in Q \Rightarrow -y \in Q$, and $y \in S \Rightarrow -y \in S$ since $-1 = x^{\frac{n-1}{2}} \in Q$. So the complements of Q and S satisfying the corollary exist. □

This justifies Case AIII.

Theorem 9.8: *Let* $q \equiv 5 \pmod 8$ *be a prime power and* $q = s^2 + 4t^2$ *be its proper representation with* $s \equiv 1 \pmod 4$. *Suppose there are amicable orthogonal designs of types* $(1, 2r-1)$ *and* (r, r) *AOD*$(2r:\ (1, 2r-1); (r, r))$, $2r = |t| + 1$. *Then there exist amicable Hadamard matrices of order* $(|t|+1)(q+1)$.

Proof: Using the theory of cyclotomy, we can show that for the q of the enunciation,

$$C_0 \text{ and } C_1 \text{ and } |t| \text{ copies of } C_0 \text{ and } C_2$$

are $(|t|+1) - \left\{q; \frac{1}{2}(q-1);\ (|t|+1)\frac{1}{4}(q-3)\right\}$ *sds*, with the property that

$$x \in C_0 \text{ and } C_1 \Rightarrow -x \notin C_0 \text{ and } C_1$$

and

$$y \in C_0 \text{ and } C_2 \Rightarrow -y \in C_0 \text{ and } C_2,$$

where C_i, $i = 0, 1, 2, 3$ are cyclotomic classes of order 4.

Also $\frac{1}{2}(|t|+1)$ copies of each of C_0 and C_2 and C_1 and C_3 are

$$(|t|+1) - \left\{q; \frac{1}{2}(q-1); (|t|+1)\frac{1}{4}(q-3)\right\} \textit{sds}.$$

with the property that

$$y \in C_0 \text{ and } C_2 \Rightarrow -y \in C_0 \text{ and } C_2,$$

$$z \in C_1 \text{ and } C_3 \Rightarrow -z \in C_1 \text{ and } C_3.$$

Let A be the type $1(1, -1)$ incidence matrix of C_0 and C_1, and B and C be the type $2(1, -1)$ incidence matrices of C_0 and C_2 and C_1 and C_3, respectively.

Then

$$AJ = BJ = CJ = -J, \quad (A+I)^{\mathrm{T}} = -(A+I), \quad B^{\mathrm{T}} = B, \quad C^{\mathrm{T}} = C,$$

$$\begin{aligned} AA^{\mathrm{T}} + |t|BB^{\mathrm{T}} &= \frac{(|t|+1)}{2}(BB^{\mathrm{T}} + CC^{\mathrm{T}}) \\ &= (|t|+1)(q+1)I - (|t|+1)J. \end{aligned}$$

Let $P = x_0U + x_1V$ and $Q = x_3X + x_4Y$ be the $AOD(2r: \ (1, 2r-1); (r, r))$. Further, let e be the $1 \times q$ matrix of all ones. Clearly, we may assume that $U = I$, $V^{\mathrm{T}} = -V$, $X^{\mathrm{T}} = X$, $Y^{\mathrm{T}} = Y$, for if not, we pre-multiply P and Q by the same matrix W until U, V, X, Y do have the required properties. Now

$$E = \left[\begin{array}{c|c} U+V & (U+V)\times e \\ \hline (-U+V)\times e^{\mathrm{T}} & U\times(-A) + V\times B \end{array}\right]$$

and

$$F = \left[\begin{array}{c|c} X+Y & (X+Y)\times e \\ \hline (X+Y)\times e^{\mathrm{T}} & X\times B + Y\times C \end{array}\right]$$

are the required amicable Hadamard matrices. □

We note that $AOD((1, 2r-1); (r, r))$ certainly exist where $2r$ is a power of 2 (see Corollary 9.6). Hence we have:

Corollary 9.4: *Let $q \equiv 5 \pmod 8$ be a prime power and $q = s^2 + 4t^2$ be its proper representation with $s \equiv 1$ (mod 4). Further, suppose $|t| = 2^r - 1$ for some r. Then there exist amicable Hadamard matrices of order $2^r(q+1)$.*

In particular, this leads to two results published elsewhere (Wallis [201], [207]) which now become corollaries:

Corollary 9.5: *Let $q \equiv 5 \pmod 8$ be a prime power, and suppose $q = s^2 + 4$ or $q = s^2 + 36$ with $s \equiv 1 \pmod 4$. Then there exist amicable Hadamard matrices of orders $2(s^2+5)$ or $4(s^2+37)$, respectively.*

Theorem 9.8 and Corollary 9.4 justify Case AIV. The case AV is covered by the paper on the existence of amicable cores [157]:

We know that amicable Hadamard matrices exist for the following orders

AI	2^t	t a nonnegative integer
AII	p^r+1	p^r (prime power)$\equiv 3 \pmod 4$
AIII	$2(q+1)$	$2q+1$ is a prime power, q (prime)$\equiv 1 \pmod 4$
AIV	$(\lvert t\rvert+1)(q+1)$	q (prime power)$\equiv 5 \pmod 8 = s^2+4t^2$, $s \equiv 1$ (mod 4), and $\lvert t\rvert+1$ is the order of amicable orthogonal designs of type $(1, \lvert t\rvert); (\frac{1}{2}(\lvert t\rvert+1))$.
	$2^r(q+1)$	q (prime power)$\equiv 5 \pmod 8 = s^2+4(2^r-1)^2$, $s \equiv 1 \pmod 4$, r some integer
AV	p^r+1	p the order of amicable cores and r any odd integer [157]
AVI	S	where S is a product of the above orders

AVI and AI will be proved first. AVI is proved in Theorem 1.16.

Many families of skew Hadamard matrices are found as part of a family of amicable orthogonal designs.

9.7 Amicable Cores

To make our next result clear we first define what we mean by different types of cores.

Definition 9.3 (Amicable Hadamard Cores): Let $A = I + S$ and B be **amicable Hadamard matrices** of order $n + 1$ which can be written in the form

$$A = \begin{bmatrix} 1 & e \\ -e^{\top} & I + W \end{bmatrix}$$

and

$$B = \begin{bmatrix} -1 & e \\ e^{\top} & R + V \end{bmatrix},$$

where e is a $1 \times n$ vector of all +1s. Let R be the back diagonal identity matrix. W and V satisfy

$$\begin{aligned} &V^{\top} = V,\ VW^{\top} = WV,\ W^{\top} = -W,\ RW^{\top} = WR,\ WJ = VJ = 0, \\ &VV^{\top} = WW^{\top} = nI_n - J_n. \end{aligned} \tag{9.14}$$

Then $I + W$, $R + V$ will be said to be **amicable cores of amicable Hadamard matrices**. We call W and V **amicable Hadamard cores** when the properties of Eq. (9.14) are satisfied. We call W the **skew symmetric core** and V the **symmetric (partner) core**.

Now $C = I + W$ or $C = R + V$ are said to be **amicable cores of the Hadamard matrix**, and

$$CC^{\top} = (n + 1)I_n - J_n,\ CJ = JC = J.$$

Example 9.2 ***(Amicable Hadamard matrices and their cores):*** *The following two Hadamard matrices are amicable Hadamard matrices.*

$$A = \begin{bmatrix} 1 & 1 & 1 & 1 \\ -1 & 1 & 1 & -1 \\ -1 & -1 & 1 & 1 \\ -1 & 1 & -1 & 1 \end{bmatrix} \quad \text{and} \quad B = \begin{bmatrix} -1 & 1 & 1 & 1 \\ 1 & -1 & 1 & 1 \\ 1 & 1 & 1 & -1 \\ 1 & 1 & -1 & 1 \end{bmatrix}$$

with the following two matrices

$$I + W = \begin{bmatrix} 1 & 1 & -1 \\ -1 & 1 & 1 \\ 1 & -1 & 1 \end{bmatrix} \quad \text{and} \quad R + V = \begin{bmatrix} -1 & 1 & 1 \\ 1 & 1 & -1 \\ 1 & -1 & 1 \end{bmatrix}$$

as amicable cores of the amicable Hadamard matrices. Note they satisfy all the properties of Eq. (9.14).

We recall that Paley showed that for $p \equiv 3 \pmod 4$ be a prime power. The following theorem quoted from Geramita and Seberry [78, theorem 5.52] extends this

Theorem 9.9 ***(Amicable Paley cores):*** *Let $p \equiv 3 \pmod 4$ be a prime power. Then there exist amicable Hadamard matrices of order $p + 1$.*

Example 9.3 ***(Amicable Paley cores):*** *First we note that Paley cores apply to prime power orders only.*

To illustrate we use Paley's original construction [144, 214] to form a skew symmetric matrix, Q, with zero diagonal and other entries ± 1 satisfying $QJ = 0$, $QQ^{\top} = pI - J$, $Q^{\top} = -Q$. This matrix Q is called the Paley core.

We now write W for Q to be consistent with the remainder of this paper. Let R be the back diagonal identity matrix of order p, when p is prime, and the type 2 (see [214] for definitions) equivalent when p is a prime power. Write $V = WR$. Then W and V are amicable Hadamard cores. We also (loosely) call them amicable Paley cores. From the proof of Lemma [214, lemma 2.2] V is symmetric.

Writing W and V for the amicable Paley cores we see that we can write the amicable Hadamard matrices as

$$M_{p+1} = \begin{bmatrix} 1 & e \\ -e^{\top} & I+W \end{bmatrix}$$

and

$$N_{p+1} = \begin{bmatrix} -1 & e \\ e^{\top} & R+V \end{bmatrix}.$$

where e is a $1 \times p$ vector of all $+1$s.

$I + W$ and $R + V$ are amicable cores of amicable Hadamard matrices. W and V are amicable Hadamard cores.

9.8 Construction for Amicable Hadamard Matrices of Order 2^t

Seberry and Yamada [166, p.535] give amicable Hadamard matrices which were known in 1992.

Unfortunately the very first family of amicable Hadamard designs (and indeed amicable orthogonal designs) which were foreshadowed in Geramita et al. [77], has never been explicitly cited. We use Lemma [214, lemma 2.2] which has this family as a corollary.

Corollary 9.6: *Let t be a positive integer. Then there exist $AOD(2^t; 1, 2^t - 1; 2^t)$ or amicable Hadamard matrices for every 2^t.*

Proof: We write this proof in some detail to ensure we have the symmetric and skew symmetric cores for Theorem 9.10. We note the amicable Hadamard matrices of order 2:

$$M_2 = \begin{bmatrix} 1 & 1 \\ -1 & 1 \end{bmatrix} \quad \text{and} \quad N_2 = \begin{bmatrix} 1 & 1 \\ 1 & -1 \end{bmatrix}.$$

From Example 1.17 once gives us the 4×4 matrices

$$M_4 = \begin{bmatrix} 1 & 1 & 1 & 1 \\ -1 & 1 & 1 & -1 \\ -1 & -1 & 1 & 1 \\ -1 & 1 & -1 & 1 \end{bmatrix} \quad \text{and} \quad N_4 = \begin{bmatrix} 1 & 1 & 1 & 1 \\ 1 & -1 & 1 & -1 \\ 1 & 1 & -1 & -1 \\ 1 & -1 & -1 & 1 \end{bmatrix}.$$

Here all the properties of Eq. (9.14) hold.

We note that M_4 is Hadamard of skew-type, a skew Hadamard matrix and N_4 is symmetric. This example is not yet in the form we need for Theorem 9.10 but multiplying column 1 and then rows 2–4 of N_4 by -1, to give N_4', which does not alter the symmetry of amicable Hadamard property of M_4 and N_4' gives the desired form.

Then iterative use of Theorem 1.16 gives Hadamard matrices of order $q = 2^t$ where M_{2^t} is skew-type and $N_{2^t} = (n_{ij})$ is symmetric and the elements

$$(2, 2^t),\ (3, 2^t - 1),\ \ldots, (2^t, 2)$$

of N_{2^t} are all -1. This corresponds to R for N'_{2^t}. That is, using induction, we can write the amicable Hadamard matrices $M_{2^t} = I + S$ and N'_{2^t} as

$$M_{2^t} = \begin{bmatrix} 1 & e \\ -e^{\top} & I+W \end{bmatrix}$$

and

$$N'_{2^t} = \begin{bmatrix} -1 & e \\ e^{\top} & R+V \end{bmatrix},$$

where e is a $1 \times 2^t - 1$ vector of all $+1$s and W and V are amicable Hadamard cores. R is the back diagonal identity matrix of order $2^t - 1$.

We note M_{2^t} and N'_{2^t} satisfy all the properties of Eq. (9.14). □

9.9 Construction of Amicable Hadamard Matrices Using Cores

In early papers Belevitch [10, 11] and Goldberg [85] showed that the core of a skew Hadamard matrix, of order $n+1$, could be used to generate a core of a skew Hadamard matrix of order n^3+1. Seberry Wallis [203] realized that this construction could be extended to orders n^5+1 and n^7+1. These were further generalized by Turyn [191] using G-strings to orders n^r+1, where $r > 0$ is an odd integer. We now give the results in considerable detail to try to make the constructions as clear as possible.

Theorem 9.10 *(Belevitch–Goldberg theorem):* *Suppose W is a skew-symmetric core of size $n \equiv 3 \pmod 4$ then*

$$I \times J \times W + W \times I \times J + J \times W \times I + W \times W \times W$$

is a core of order n^3.

Remark 9.1: If $I+W$ and $R+V$ are amicable Hadamard cores then the symmetric companion of the above skew symmetric core is

$$R \times J \times V + V \times R \times J + J \times V \times R + V \times V \times V$$

We now use part of corollary 3.12 of [214] which shows that is W is a skew symmetric (symmetric) core of size $n \equiv 3 \pmod 4$ then there exists a skew symmetric (symmetric) core of size n^r for all odd $r > 1$.

Example 9.4: *The skew symmetric core of order n^5 from a skew symmetric core of order n is the sum of*

$$I \times J \times I \times J \times W; \text{ and } I \times J \times W \times W \times W;$$

plus

$$W \times W \times W \times W \times W,$$

plus their circulants

$$W \times I \times J \times I \times J; \; W \times I \times J \times W \times W;$$
$$J \times W \times I \times J \times I; \; W \times W \times I \times J \times W;$$
$$I \times J \times W \times I \times J; \; W \times W \times W \times I \times J;$$
$$J \times I \times J \times W \times I\;; \; J \times W \times W \times W \times I.$$

The symmetric core will have the same form with I replaced by R and W replaced by V. So it becomes the sum of

$$R \times J \times R \times J \times V; \text{ and } R \times J \times V \times V \times V; V \times V \times V \times V \times V,$$

plus their circulants

$$V \times R \times J \times R \times J; V \times R \times J \times V \times V; J \times V \times R \times J \times R; V \times V \times R \times J \times V;$$

$$R \times J \times V \times R \times J; V \times V \times V \times R \times J; J \times R \times J \times V \times R; J \times V \times V \times V \times R.$$

These amicable cores, that is the skew symmetric and the symmetric cores, are amicable term by term.

We now use this to construct amicable Hadamard matrices of order $n^r + 1$ from amicable Hadamard matrix cores of order $n + 1$. This is illustrated by the Belevitch–Goldberg construction Theorem 9.10 for $n = 3$ and by Example 9.4 for $n = 5$.

We define G-string and symmetric G-string which are a generalization of Theorem 9.10 and Example 9.4.

Definition 9.4: Let $I + W$ and $R + V$ be amicable cores of amicable Hadamard matrices of order n. Assume r is odd. A ***G*-string of length** r is a Kronecker product of r matrices from $\{I, J, W\}$ with the following conditions,

i) I is always followed by J and J is preceded by I,
ii) the last matrix is considered to be followed by a first matrix, that is, the circulant of a G-string is also a G-string.

By replacing W by V and I by R, we define a **symmetric *G*-string**.

Lemma 9.1 *(Turyn [191])*: *Let G_i be a G-string and $W^\top = \epsilon W$, $\epsilon = -1$.*

i) $G_i^\top = \epsilon G_i$.
ii) $G_i G_j^\top = 0$ *if* $G_i \neq G_j$.
iii) $G_i * G_j = 0$ *if* $G_i \neq G_j$ *where* $*$ *is a Hadamard product.*

Proof: (*i*) follows from the fact that each G-string has an odd number of W's and $W^\top = \epsilon W$. We obtain (*ii*) and (*iii*) from $JW = WJ = 0$ and $W \star I = 0$. □

Lemma 9.1 is true for a symmetric core V and $\epsilon = 1$. Now we let W_r is the sum of all G-strings with the length r.

Lemma 9.2 *(Turyn [191])*: *W_r has ± 1 entries except 0 on the diagonal.*

Proof: From Lemma 9.1, we know all entries of W_r are ± 1 or 0. The diagonal of W_r is 0 as all G-strings have 0 diagonal. We assume that the (i,j), $i \neq j$ entry of G-string is not zero.

Write $i = (i_1, \dots, i_r)$ and $j = (j_1, \dots, j_r)$ where (i_ℓ, j_ℓ) is a entry of ℓth matrix of G-string.

If $i_\ell \neq j_\ell$ for all ℓ, then the G-string should be $W \times \cdots \times W$. We assume there exists $i_\ell = j_\ell$ for some ℓ. Since the circulant of a G-string is also a G-string, we may assume $i_1 = j_1$. The first matrix should be I and the second matrix is J.

If $i_3 = j_3$, then the third matrix is I and if $i_3 \neq j_3$, then the third matrix is W. In this way, we obtain the G-string uniquely by the giving nonzero entry (i,j). It means the entries of W_r except the diagonal are ± 1. □

Lemma 9.3 *(Turyn [191])*: $W_r W_r^\top = n^r I_r - J_r$.

Proof: From Lemma 9.1, we let $W_r W_r^\top = \sum G_i G_i^\top$. Since

$$
\begin{aligned}
WW^\top &= nI - J \\
(W \times W)(W^\top \times W^\top) &= n^2 I \times I - nI \times J - nJ \times I + nJ \times J \\
(I \times J)(I \times J) &= I \times nJ,
\end{aligned}
$$

the matrices in $W_r W_r^\top$ are I and J. Let P be a Kronecker product of r matrices from $\{I, J\}$. We assume that P has b $I \times J$'s and c J's not preceded by I. Then there are $r - c - 2b$ I's not followed by J. For example, $P = I \times J \times J \times I \times J$ has $b = 2, c = 1$, and $r - c - 2b = 0$.

Assume that j $I \times J$'s occur in G-string G_ℓ. Then the coefficient of a string P in $G_\ell G_\ell^\top$ is

$$n^{b-j}(-1)^{b-j} n^j (-1)^c n^{r-c-2b}.$$

Hence the coefficient of P in $W_r W_r^\top$ is

$$(-1)^c n^{r-c-2b} \sum_{j=0}^{b} n^{b-j}(-1)^{b-j} n^j = 0$$

if $b > 0$.

If $P = I_r = I \times \cdots \times I$, then $b = 0, c = 0, r - c - 2b = r$, and the coefficient is n^r. If $P = J_r = J \times \cdots \times J$, then $b = 0, c = r, r - c - 2b = 0$, and the coefficient is $(-1)^r = -1$. Consequently $W_r W_r = n^r I_r - J_r$. □

Let V_r be the sum of all symmetric G-strings with the length r. We obtain similar lemmas by replacing W_r by V_r.

Theorem 9.11: *Assume n be an odd integer. Let $I + W$ and $R + V$ be amicable cores of order n of amicable Hadamard matrices. Let W_r be the sum of all G-strings with the length r and V_r be the sum of all symmetric G-strings with the length r. Then*

$$I + W_r \quad \textit{and} \quad R + V_r$$

be amicable cores of order n^r.

Proof: We verify Eq. (9.14). Since $I + W$ and $R + V$ are amicable cores, we have

$$V_r^\top = V_r, \quad W_r J = V_r J = 0, \quad RW_r^\top = W_r R, \text{ and } V_r W_r^\top = W_r V_r.$$

Together with Lemma 9.3 and $V_r V_r^\top = n^r I_r - J_r$, the assertion follows. □

Corollary 9.7: *Suppose there exist amicable Hadamard matrices of order $n + 1$ with amicable cores of order n. Then there exist amicable Hadamard matrices of order $n^r + 1$, for all odd $r \geq 1$.*

9.10 Symmetric Hadamard Matrices

It has taken many years to find an equivalent of the Williamson [223] and Goethals and Seidel [82] arrays to construct symmetric Hadamard matrices. The first effort, which parallels the Williamson construction, is for propus Hadamard matrices [160]. Propus is a construction method for symmetric orthogonal ± 1 matrices, using four symmetric circulant matrices A, $B = C$, and D of order t, see [160], where

$$AA^\top + 2BB^\top + DD^\top = 4tI_t, \tag{9.15}$$

based on the array

$$\begin{bmatrix} A & B & B & D \\ B & D & -A & -B \\ B & -A & -D & B \\ D & -B & B & -A \end{bmatrix}.$$

The related theorem, for reference, is the Propus Symmetric Hadamard Construction [160] given in Theorem 1.17.

Building on the Balonin–Seberry theorem using luchshie matrices (see Theorem 1.18 and Definition 1.30), we have a more relaxed variation of the propus array, called the Balonin–Seberry array, which allows the construction of symmetric Hadamard matrices.

9.10.1 Symmetric Hadamard Matrices Via Computer Construction

In Seberry and Balonin [160] we show that for

i) $q \equiv 1 \pmod 4$, a prime power, propus matrices exist for order $t = \frac{1}{2}(q+1)$, and thus symmetric Hadamard matrices of order $2(q+1)$;
ii) $q \equiv 1 \pmod 4$, a prime power, and $\frac{1}{2}(q+1)$ a prime power or the order of the core of a symmetric conference matrix (this happens for $q = 89$), the required symmetric propus-type Hadamard matrices of order $4(2q+1)$ exist;
iii) $t \equiv 3 \pmod 4$, a prime, such that D-optimal designs, constructed using two circulant matrices of order t, one of which must be circulant and symmetric, the propus-type symmetric Hadamard matrices of order $4t$ exist.

9.10.2 Luchshie Matrices Known Results

The results given in this section were found by N. A. Balonin, M. B. Sergeev, and their students Dmytri Karbovskiy and Yuri Balonin, some are still are to be published.

Table 9.3 gives the numbers of inequivalent solutions for the existence of symmetric Hadamard matrices constructed via the Propus and Balonin–Seberry array for values symmetric Hadamard matrices of order $4t$, where $t \in \{1, \dots, 103\}$. In Table 9.3, $L(x, y)$ means there are x inequivalent luchshie matrices for this order and y means there are y inequivalent propus matrices for this order and Williamson matrices are denoted by "W."

Table 9.3 Existence of luchshie matrices.

n	*W*	*L*	*n*	*W*	*L*	*n*	*W*	*L*
4	1	$L(1,1)$	76	6	$L(3,1)$	148	4	$L(5,1)$
12	1	$L(1,1)$	84	7	$L(2,1)$	156	1	$L(2,0)$
20	1	$L(1,1)$	92	1	$L(2,0)$	164	1	$L(2,1)$
28	2	$L(2,1)$	100	10	$L(3,1)$	172	2	$L(5,0)$
36	3	$L(1,1)$	108	6	$L(4,1)$	180	1	$L(5,1)$
44	1	$L(1,0)$	116	1	$L(1,0)$	188	0	$L(4,0)$
52	4	$L(3,1)$	124	2	$L(4,1)$	196	1	$L(4,1)$
60	4	$L(2,1)$	132	5	$L(4,0)$	204	2	$L(2,1)$
68	4	$L(2,0)$	140	0	$L(2,0)$	212	0	$L(3,0)$

Good and suitable matrices exist for orders 35, 43, and 59 but do not exist for Williamson matrices and luchshie matrices exist for orders 17, 23, 29, 33, 35, 39, 43, 47, 103 whereas propus matrices do not exist for these orders.

We note that of these orders 23, 29, 33, 35, 39, 43, 47 also needed extra work after Williamson [223] to find solutions. In addition, D. Đoković and colleagues and N. A. Balonin, M. B. Sergeev, and their students Dmytri Karbovskiy and Yuri Balonin have found luchshie matrices for orders $t \in S = \{51, 53, 67, 71, 73, 77, 79, 83, 103, 109, 113, 151, 157, 307, \dots, \}$ [4–6, 52]. These are denoted by ℓ in Table 9.4.

Table 9.4 gives the existence and a construction method for 147 of the 200 possible symmetric Hadamard matrices of order ≤200. This has been verified for two infinite families [160] (see (i) and (ii) above) and for values symmetric Hadamard matrices of order $4t$, where $t \in S$ (S given above).

Table 9.4 Orders of known symmetric Hadamard matrices.

q	t	How	q	t	How	q	t	How	q	t	How
1		$a1$	59		·	117		$a1$	175		$c1$
3		$a1$	61		$c1$	119		·	177		$c1$
5		$a1$	63		$a1$	121		$c1$	179		·
7		$c1$	65		·	123		$a1$	181		$c1$
9		$c1$	67		ℓ	125		$a1$	183		·
11		$a1$	69		$c1$	127		·	185		$a1$
13		$c1$	71		$a1$	129		$c1$	187		$c1$
15		$a1$	73		ℓ	131		$a1$	189		·
17		$a1$	75		$c1$	133		·	191		·
19		$c1$	77		$a1$	135		$c1$	193		·
21		$a1$	79		$c1$	137		$a1$	195		$c1$
23		ℓ	81		·	139		$c1$	197		$a1$
25		$c1$	83		$a1$	141		$a1$	199		$c1$
27		$a1$	85		$c1$	143		$a1$	201		$c1$
29		$a2$	87		$a1$	145		$c1$	203		$a1$
31		$c1$	89		$a2$	147		$a1$	205		$c1$
33		$a1$	91		$c1$	149		·	207		$a1$
35		$a1$	93		·	151		ℓ	209		·
37		$c1$	95		$a1$	153		·	211		$c1$
39		ℓ	97		$c1$	155		$a1$	213		·
41		$a1$	99		$c1$	157		$c1$	215		$a1$
43		ℓ	101		·	159		$c1$	217		$c1$
45		$a1$	103		ℓ	161		$a1$	219		·
47		ℓ	105		$a1$	163		(ii)	221		$a1$
49		$c1$	107		(ii)	165		$a1$	223	3	$a1$
51		$c1$	109		ℓ	167		·	225		$c1$
53		$a1$	111		$a1$	169		$c1$	227		$a1$
55		$c1$	113		$a2$	171		$a1$	229		$c1$
57		$a1$	115		$c1$	173		$a1$	231		$c1$

(Continued)

Table 9.4 *(Continued)*

q	*t*	How	*q*	*t*	How	*q*	*t*	How	*q*	*t*	How
233		*a*2	275		(*ii*)	317		·	359	4	*a*1
235		·	277		·	319		·	361		·
237		*a*1	279		*c*1	321		*a*1	363		*a*1
239	4	*a*1	281		*a*1	323		*a*1	365		*a*1
241		·	283		·	325		·	367		*c*1
243		*a*1	285		*c*1	327		*a*1	369		·
245		·	287		·	329		·	371		*a*1
247		·	289		*c*1	331		*c*1	373		·
249		·	291		*a*1	333		·	375		*a*1
251		·	293		*a*1	335		·	377		·
253		·	295		·	337		*c*1	379		*c*1
255		*a*1	297		*a*1	339		*c*1	381		*a*1
257		·	299		·	341		·	383		*a*1
259		·	301		*c*1	343		·	385		*c*1
261		*c*1	303		·	345		·	387		*c*1
263		*a*1	305		·	347		(*ii*)	389		·
265		*c*1	307		*c*1	349		·	391		·
267		·	309		*c*1	351		*c*1	393		*a*1
269		·	311		·	353		·	395		*a*1
271		*c*1	313		*c*1	355		*c*1	397		(*i*)
273		*a*1	315		*a*1	357		*a*1	399		*c*1

Source: Constructed from Seberry and Yamada original survey [166], Wiley.

We use the notation "*ai*" to denote existence because there are amicable Hadamard matrices of four times that order; "*ci*" denotes existence because of the existence of conference matrices of twice that order [166]; and "ℓ" denotes existence because luchshie matrices exist for that order having been found by Balonin, Sergeev, Đoković, and their students [4–6]. If no symmetric Hadamard matrix is yet known the entry is given by "·."

The table gives the odd part *q* of an order, the smallest power of 2, *t*, for which the symmetric Hadamard matrix is known and a construction method. If no number is given the power of 2 will be 2. For example a symmetric Hadamard matrix of order $2^2 \times 237$ is known but only a symmetric Hadamard matrix of order $2^4 \times 239$.

10

Skew Hadamard Difference Sets

10.1 Notations

Table 10.1 gives the notations which are used in this chapter.

10.2 Skew Hadamard Difference Sets

Let q be a prime power and $q \equiv 3 \pmod 4$. A Paley type I core matrix associates with a $(q, \frac{1}{2}(q-1), \frac{1}{4}(q-3))$ difference set consisting of nonzero squares of $GF(q)$. This Paley Hadamard difference set is a skew difference set and was the only known example of a skew Hadamard difference set for many years.

There were two conjectures on skew Hadamard difference sets in abelian groups.

i) If an abelian group G contains a skew Hadamard difference set, then G is necessarily an elementary abelian group.
ii) Up to equivalence, the Paley difference sets are the only skew Hadamard difference sets in an abelian group.

The first conjecture is still open. The second conjecture was solved negatively. The second and third examples of a family of skew Hadamard difference sets by Ding and Yuan are inequivalent to the Paley difference set [53]. Muzychuk [140] constructed many inequivalent skew Hadamard difference sets over an elementary abelian group of order q^3.

10.3 The Construction by Planar Functions Over a Finite Field

10.3.1 Planar Functions and Dickson Polynomials

Let $q = p^t$ be a power of a prime. We define a permutation polynomial of $K = \boldsymbol{F}_q$.

Definition 10.1: If a polynomial $f : K \to K$ is a bijection, then a polynomial f is called a **permutation polynomial**.

Example 10.1 ***(Lidl and Niederreiter [127, p. 352]):*** *The following polynomials are permutation polynomials.*

i) $f(x) = x$ *for any* q.
ii) $f(x) = x^2$ *if* $q \equiv 0 \pmod 2$.
iii) $f(x) = x^3 - ax$ *(a not a square) if* $q \equiv 0 \pmod 3$.

We define a planar function.

Hadamard Matrices: Constructions using Number Theory and Algebra, First Edition. Jennifer Seberry and Mieko Yamada.

Table 10.1 Notations used this chapter.

ζ_m	A primitive mth root of unity
$\mathbf{Z}[F]$	A group ring of a finite field F over $\mathbf{Z}$
$(\mathbf{Z}/n\mathbf{Z})^\times$	Multiplicative group of integers modulo n
$\mathbf{Q}(\zeta_m)$	A mth cyclotomic field, a number field by adjoining a Primitive mth root of unity to $\mathbf{Q}$, the field of numbers
$D_n(x, u)$	Dickson polynomial
$g_K(\chi)$	The Gauss sum over a finite field K associated with a multiplicative character χ
φ	Euler's function
$\langle p \rangle$	A cyclic group generated by p
$\mathcal{P}$	A prime ideal in $\mathbf{Q}(\zeta_m)$ lying above p
$\langle t \rangle$	A fractional part of a real number t
$\widetilde{\vartheta}_{F/K}(\chi_F)$	The normalized relative Gauss sum
$ord_m p$	The order of p modulo m
$gcd(a, b)$	The greatest common divisor of integers a and b
$J(\chi_1, \chi_2)$	The Jacobi sum for multiplicative characters χ_1 and χ_2

Definition 10.2: The polynomial $f : K \to K$ is a **planar function** if $f(x + a) - f(x)$ is a permutation polynomial for all nonzero element $a \in K$.

Example 10.2 *(Ding and Yuan [53])*: *Let p be an odd prime. The known planar functions are as follows:*

i) $f(x) = x^2$.
ii) $f(x) = x^{p^k+1}$ *where* $\dfrac{t}{gcd(t, k)}$ *is odd (Dembowski and Ostrom [51]).*
iii) $f(x) = x^{(3^k+1)/2}$ *where* $p = 3$*, k is odd and* $gcd(t, k) = 1$ *(Coulter and Matthews [24]).*
iv) $f(x) = x^{10} + x^6 - x^2$ *where* $p = 3$*, and* $t = 2$ *or t is odd (Coulter and Matthews [24]).*

Let t be odd. For $u \in K^\times$, the Dickson polynomial $D_n(x, u)$ of the first kind is defined by

$$D_n(x, u) = \sum_{j=0}^{\lfloor \frac{n}{2} \rfloor} \frac{n}{n-j} \binom{n-j}{j} (-u)^j x^{n-2j}.$$

Theorem 10.1: *The Dickson polynomial* $D_n(x, u), u \in K^\times$ *is a permutation polynomial if and only if* $gcd(n, q^2 - 1) = 1$.

Proof: See Lidl and Niederreiter [127]. □

Let $n = 5$ and $p = 3$. Then we have $D_5(x, -u) = x^5 - ux^3 - u^2x$. For any $u \in \mathbf{F}_{3^t}$, we define

$$g_u(x) = D_5(x^2, -u) = x^{10} - ux^6 - u^2x^2.$$

We let $K = \mathbf{F}_{3^t}$ throughout this section.

Lemma 10.1 *(Ding and Yuan [53])*: *Assume t is odd. For* $u \in K^\times$, $g_u(x)$ *is a planar function from K to K.*

Proof: It is sufficient to prove that $g_u(x + a) - g_u(x) = ax^9 + ua^3x^3 + (a^9 + u^2a)x + g_u(a)$ is a permutation polynomial for all nonzero elements $a \in K$. If $L_{a,u}(x) = x^9 + ua^2x^3 + (a^8 + u^2)x \in K[x]$ is a permutation polynomial, that is, $L_{a,u}(b) \neq 0$ for all $b \neq 0$, then $g_u(x+a) - g_u(x)$ is a permutation polynomial. We assume $L_{a,u}(b) = 0$ for some $b \neq 0$. Then we have $(a^4 + b^4)^2 + (a^2b^2 - u)^2 = 0$. If $a^4 + b^4 = 0$, then $-1 = \left(\frac{a^2}{b^2}\right)^2$ and if $a^4 + b^4 \neq 0$, then $-1 = \left(\frac{a^2b^2-u}{a^4+b^4}\right)^2$. The both cases contradict to the fact that -1 is a quadratic nonresidue of K. Therefore $g_u(x)$ is a planar function. □

Lemma 10.2 (Ding and Yuan [53]): *Assume t is odd. Let $u \in K$, $x, y \in K$. Then $g_u(x) + g_u(y) = 0$ if and only if $(x, y) = (0, 0)$.*

Proof: Put $s = x + y$ and $r = x - y$. Then $x^2 + y^2 = -(s^2 + r^2)$ and $x^4 + y^4 = -(s^4 + r^4)$. We have

$$\begin{aligned} g_u(x) + g_u(y) &= x^{10} - ux^6 - u^2x^2 + y^{10} - uy^6 - u^2y^2 \\ &= (x^2 + y^2)\left\{(x^4 + y^4)^2 - x^2y^2(x^2 - y^2)^2 - u(x^2 + y^2)^2 - u^2\right\} \\ &= -(s^2 + r^2)\left\{(s^4 + r^4)^2 - (s^2 - r^2)^2s^2r^2 - u(s^2 + r^2)^2 - u^2\right\} \\ &= -(s^2 + r^2) \\ &\quad \times \left\{s^8 + r^8 - s^6r^2 - s^2r^6 + s^4r^4 - us^4 - ur^4 + us^2r^2 - u^2\right\} \end{aligned}$$

Let $\beta \in \boldsymbol{F}_{3^{2t}}$ such that $\beta^2 = -1$. Then $\beta \notin K$. Thus

$$\begin{aligned} g_u(x) + g_u(y) = &-(s^2 + r^2)\left\{(s^4 + r^4 + s^2r^2 + u) + \beta(s^2r^2 + u)\right\} \\ &\times \left\{(s^4 + r^4 + s^2r^2 + u) - \beta(s^2r^2 + u)\right\}. \end{aligned}$$

Assume that $(s^4 + r^4 + s^2r^2 + u) + \beta(s^2r^2 + u) = 0$. If $s^2r^2 + u \neq 0$, then

$$\beta = -(s^4 + r^4 + s^2r^2 + u)(s^2r^2 + u)^{-1} \in K,$$

a contradiction. Hence $s^2r^2 + u = 0$ and consequently $s^4 + r^4 = 0$. If $s \neq 0$ or $r \neq 0$, then $-1 = \left(\frac{r^2}{s^2}\right)^2$, then it contradicts that -1 is a quadratic nonresidue of K. It follows that $s = r = 0$. If $(s^4 + r^4 + s^2r^2 + u) \pm \beta(s^2r^2 + u) \neq 0$, then $s^2 + r^2 = 0$, that is $s = r = 0$. By definition, $(s, r) = (0, 0)$ if and only if $(x, y) = (0, 0)$. Thus we have the assertion. □

We now obtain a family of skew Hadamard difference sets.

Theorem 10.2 (Ding and Yuan [53]): *Let $u \in K = \boldsymbol{F}_{3^t}$. Assume t is odd. The subset $g_u(K)\backslash\{0\}$ of K is a $(3^t, \frac{3^t-1}{2}, \frac{3^t-3}{4})$ skew Hadamard difference set.*

Proof: Since $g_u(x) = g_u(-x) = D_5\left(x^2, -u\right)$ for $u \in K$ and $g_u(0) = 0$, $|g_u(K)\backslash\{0\}| = \frac{1}{2}\left(3^t - 1\right)$.

Assume that $b \neq 0$. From the definition of a planar function, $g_u(x+a) - g_u(x) = b$ has $3^t - 1$ solutions $(x + a, x)$ for $a \neq 0$.

Since the solutions of $g_u(y) - g_u(x) = b$ contain $(y, 0)$ and $(0, x)$, and $g_u(x) = g_u(-x)$, the number of solutions of $g_u(y) - g_u(x) = b$, $b \neq 0, x \neq 0, y \neq 0$ is $\frac{1}{4}(3^t - 3)$.

From Lemma 10.2, exactly one of the equations $g_u(x) = b$ or $g_u(x) = -b$ has a solution. It follows the subset $g_u(K)\backslash\{0\}$ is skew. □

Example 10.3 (Ding and Yuan [53]): *Let $t = 3, u = -1 \in K = \boldsymbol{F}_{3^3}$. $g_{-1}(x) = x^{10} + x^6 - x^2$. Let γ be a primitive element of $\boldsymbol{F}_{3^3}$.*

$$g_{-1}(K)\backslash\{0\} = \left\{\gamma^{18}, \gamma^{23}, \gamma^2, \gamma^3, \gamma^{25}, \gamma^{17}, \gamma^7, \gamma^{21}, \gamma^6, \gamma, \gamma^{11}, \gamma^9, 1\right\}$$

is a $(3^3, 13, 6)$ skew Hadamard difference set.

10.4 The Construction by Using Index 2 Gauss Sums

Feng, Momihara, and Xiang constructed skew Hadamard difference sets by using index 2 Gauss sums and Momihara constructed them by using normalized relative Gauss sums (we will discuss in the next section). These construction methods are new and interesting. Though methods need the knowledge of number theory, Chapter 2 will help the reader.

10.4.1 Index 2 Gauss Sums

Let $q = p^f$ and N be an order of a multiplicative character χ of $K^\times = \mathbf{F}_q^\times$. Assume that $f = ord_N(p) = \frac{\varphi(N)}{2}$ where φ is the Euler's function.

If the subgroup $\langle p\rangle$ generated by p has index 2 in $(\mathbf{Z}/N\mathbf{Z})^\times$ and $-1 \notin \langle p\rangle$, it is called the **index 2 case**. In this case, it can be shown that N has at most two odd prime divisors. Gauss sums in the index 2 case were evaluated by [253].

Theorem 10.3 ***(Yang and Xia [253] and Feng et al. [71]):*** *Let $p_1 > 3$ be a prime such that $p_1 \equiv 3 \pmod 4$ and m be a positive integer. Put $N = 2p_1^m$. Let p be a prime such that $\gcd(p, N) = 1$. Assume that $\left[(\mathbf{Z}/N\mathbf{Z})^\times : \langle p\rangle\right] = 2$. Let $f = \frac{\varphi(N)}{2}$ and $q = p^f$. Let χ be a multiplicative character of K of order N. For $0 \le t \le m-1$,*

$$g(\chi^{p_1^t}) = \begin{cases} (-1)^{\frac{p-1}{2}(m-1)} p^{\frac{f-1}{2}-hp_1^t} \sqrt{p^*} \left(\dfrac{b + c\sqrt{-p_1}}{2}\right)^{2p_1^t} & \text{if } p_1 \equiv 3 \pmod 8, \\ (-1)^{\frac{p-1}{2}m} p^{\frac{f-1}{2}} \sqrt{p^*} & \text{if } p_1 \equiv 7 \pmod 8, \end{cases}$$

$$g(\chi^{p_1^m}) = (-1)^{\frac{p-1}{2}\frac{f-1}{2}} p^{\frac{f-1}{2}} \sqrt{p^*}$$

where $p^* = (-1)^{\frac{p-1}{2}} p$, h is the class number of $\mathbf{Q}(\sqrt{-p_1})$, and b, c are integers satisfying

$$(i)\ b^2 + p_1c^2 = 4p^h, \quad (ii)\ bp^{\frac{f-h}{2}} \equiv -2 \pmod{p_1}.$$

Example 10.4 ***(Yang and Xia [253]):*** *Let χ be a character of order N.*

i) Let $p_1 = 7$, $N = 14$, and $f = \dfrac{\varphi(14)}{2} = 3$. Take $p = 11$, then

$$\left[(\mathbf{Z}/14\mathbf{Z})^\times : \langle 11\rangle\right] = 2 \text{ and } -1 \notin < 11 >.$$

Then the Gauss sum of order 14 over $\mathbf{F}_{11^3}$, $g(\chi) = -11\sqrt{-11}$.

ii) Let $p_1 = 11$, $N = 22$, $f = \dfrac{\varphi(22)}{2} = 5$. Take $p = 3$, then

$$\left[(\mathbf{Z}/22\mathbf{Z})^\times : \langle 3\rangle\right] = 2 \text{ and } -1 \notin < 3 >.$$

The class number of $\mathbf{Q}(\sqrt{-11})$ is 1.

Since $b^2 + 11c^2 = 4 \cdot 3$ and $b \cdot 3^2 \equiv -2 \pmod{11}$, we have $b = 1, c = \pm 1$. From Theorem 10.3, $g(\chi)$ of order 22 over $\mathbf{F}_{3^5}$ and $g(\overline{\chi})$ are given by

$$\{g(\chi), g(\overline{\chi})\} = \left\{3\sqrt{-3}\,\frac{1}{2}\left(-5 \pm \sqrt{-11}\right)\right\}.$$

10.4.2 The Case that $p_1 \equiv 7 \pmod 8$

Under the conditions of Theorem 10.3, we will construct a family of skew Hadamard difference sets by index 2 Gauss sums for the case that $p_1 \equiv 7 \pmod 8$. Let $F = \boldsymbol{F}_{q^s}$ be an extension of $K = \boldsymbol{F}_q$ with an extension degree s and γ be a primitive element of F where $q = p^f$. Let $N = 2p_1^m$. Let C_0 be the set of Nth power of residues of F and $C_i = \gamma^i C_0, 0 \le i \le N-1$ be Nth cyclotomic classes.

Theorem 10.4 ***(Feng and Xiang [72]):*** *Assume s is odd. We let the conditions of the index 2 cases defined as above. Let I be a subset of* $\boldsymbol{Z}/N\boldsymbol{Z}$ *such that* $\{i \pmod{p_1^m} | i \in I\} = \boldsymbol{Z}/p_1^m\boldsymbol{Z}$. *We define the set*

$$D = \bigcup_{i \in I} C_i.$$

D is a skew Hadamard difference set of F if $p \equiv 3 \pmod 4$ *and D is a Paley type partial difference set of F if* $p \equiv 1 \pmod 4$.

Proof: From $p_1 \equiv 7 \pmod 8$, we have $f = \frac{1}{2}\varphi(N) \equiv 3 \cdot 3^{m-1} \pmod 4$ and f is odd. Hence $q = p^f \equiv 1$ or $3 \pmod 4$ if and only if $p \equiv 1$ or $3 \pmod 4$, respectively. If $p \equiv 3 \pmod 4$, then $-1 \in C_{p_1^m}$ and $-D \cap D = \emptyset$, since s is odd. If $p \equiv 1 \pmod 4$, then $-1 \in C_0$ and $D = -D$. Further we have

$$|D| = p_1^m \frac{q^s - 1}{2p_1^m} = \frac{1}{2}(q^s - 1).$$

Let ϕ be a multiplicative character of $\boldsymbol{F}_q$ of order N. From Theorem 10.3, we have

$$g_K(\phi^{p_1^t}) = g(\phi^{p_1^t}) = (-1)^{\frac{p-1}{2}m} p^{\frac{f-1}{2}} \sqrt{p *}$$

for $0 \le t \le m-1$. From Theorems 2.1 and 10.3, and since

$$\phi^{p_1^t}(-1) = (-1)^{\frac{1}{2}(q-1)p_1^t} = (-1)^{\frac{1}{2}(q-1)} = (-1)^{\frac{1}{2}(p-1)},$$

we have

$$g\left(\phi^{-p_1^t}\right) = \phi^{p_1^t}(-1)\overline{g(\phi^{p_1^t})} = (-1)^{\frac{p-1}{2}}(-1)^{\frac{p-1}{2}m} p^{\frac{f-1}{2}} \overline{\sqrt{p^*}} = g\left(\phi^{p_1^t}\right).$$

By the index 2 assumption $-1 \notin \langle p \rangle$, we see $\langle p \rangle \cap -\langle p \rangle = \emptyset$ and by the assumption $[(\boldsymbol{Z}/N\boldsymbol{Z})^\times : \langle p \rangle] = 2$, we have $\langle p \rangle \cup -\langle p \rangle = (\boldsymbol{Z}/N\boldsymbol{Z})^\times$.

Let $S = \{i \mid 1 \le i \le N-1, i : \text{odd}\}$. Since $p_1 \in S$ and $p_1 \notin (\boldsymbol{Z}/N\boldsymbol{Z})^\times$, all odd integer $j \in S$ are congruent to $\pm p^{\ell} p_1^t$ modulo N.

From Theorem 2.1, $g(\phi^{\pm p^{\ell} p_1^t}) = g(\phi^{p_1^t})$. It follows that

$$g(\phi^{-j}) = (-1)^{\frac{p-1}{2}m} p^{\frac{f-1}{2}} \sqrt{p^*}$$

for all odd integer $j, 1 \le j \le N-1$.

Let χ be a multiplicative character of $F = \boldsymbol{F}_{q^s}^\times$ of order N. Since $N | (q-1)$, χ is a lift of some character of $\boldsymbol{F}_q^\times$ from Theorem 2.3. We let χ^j is a lift of ϕ^j. By using Davenport–Hasse lifting formula, we have

$$\begin{aligned} g_F(\chi^j) = (-1)^{s-1}\left(g_K(\phi^j)\right)^s &= \left((-1)^{\frac{p-1}{2}m} p^{\frac{f-1}{2}} \sqrt{p^*}\right)^s \\ &= (-1)^{\frac{p-1}{2}m} p^{\frac{s(f-1)}{2}} (\sqrt{p^*})^s. \end{aligned} \tag{10.1}$$

Let γ be a primitive element of F. We identify the subset D with the element $\sum_{\alpha \in D} \alpha$ of $\boldsymbol{Z}[F]$. For an additive nontrivial character ψ_a of F, $\psi_a(D) = \psi_1(\gamma^a D)$. Then

$$\psi_a(D) = \sum_{i \in I} \psi_1(\gamma^a C_i) = \frac{1}{N} \sum_{i \in I} \sum_{x \in F^\times} \psi_1\left(\gamma^{a+i} x^N\right).$$

On the other hand,

$$\begin{aligned}\psi_1\left(\gamma^{a+i}x^N\right) &= \frac{1}{q^s-1}\sum_{y\in F^\times}\psi_1(y)\sum_{\theta\in\widehat{F^\times}}\theta\left(\gamma^{a+i}x^N\right)\theta\left(y^{-1}\right)\\ &= \frac{1}{q^s-1}\sum_{\theta\in\widehat{F^\times}}\theta\left(\gamma^{a+i}x^N\right)\sum_{y\in F^\times}\theta^{-1}(y)\psi_1(y)\\ &= \frac{1}{q^s-1}\sum_{\theta\in\widehat{F^\times}}\theta\left(\gamma^{a+i}x^N\right)g_F\left(\theta^{-1}\right).\end{aligned}$$

Then we have

$$\begin{aligned}\psi_a(D) &= \frac{1}{N}\sum_{i\in I}\sum_{x\in F^\times}\frac{1}{q^s-1}\sum_{\theta\in\widehat{F^\times}}\theta\left(\gamma^{a+i}x^N\right)g_F\left(\theta^{-1}\right)\\ &= \frac{1}{N(q^s-1)}\sum_{i\in I}\sum_{\theta\in\widehat{F^\times}}\theta\left(\gamma^{a+i}\right)g_F\left(\theta^{-1}\right)\sum_{x\in F^\times}\theta\left(x^N\right).\end{aligned}$$

Since $x^N \in C_0$, if $\theta = \chi^\ell$ for some $\ell, 0 \le \ell \le N-1$, then θ is trivial on C_0 and $\sum_{x\in F^\times}\theta(x^N) = q^s - 1$. Otherwise the sum is equal to 0. Thus

$$\psi_a(D) = \frac{1}{N}\sum_{i\in I}\chi^\ell\left(\gamma^{a+i}\right)\sum_{\ell=0}^{N-1}g_F\left(\chi^\ell\right).$$

If $\ell = 0$, then $\psi_a(D) = -p_1^m$. If $\ell \equiv 0 \pmod 2, \ell \ne 0$, then $\sum_{i\in I}\chi^\ell(\gamma^i) = 0$, from $(\chi^\ell)^{p_1^m} = \chi_0$ and $I \pmod{p_1^m} \cong \boldsymbol{Z}/p_1^m\boldsymbol{Z}$.

Hence

$$\psi_a(D) = \frac{1}{N}\left\{-p_1^m + \sum_{\substack{\ell:\text{odd}\\ 1\le\ell\le N-1}} g_F\left(\chi^{-\ell}\right)\sum_{i\in I}\chi^\ell\left(\gamma^{a+i}\right)\right\}.$$

Letting $\ell = 1 + 2u, 0 \le u \le p_1^m - 1$ and using (10.1),

$$\psi_a(D) = \frac{1}{N}\left\{-p_1^m + (-1)^{\frac{p-1}{2}m}p^{s(f-1)/2}\left(\sqrt{p^*}\right)^s\sum_{i\in I}\zeta_N^{a+i}\sum_{u=0}^{p_1^m-1}\zeta_{p_1^m}^{u(a+i)}\right\}.$$

For each $a, 0 \le a \le N-1$, there is a unique $i \in I$ such that $p_1^m|(a+i)$. Write $a + i = z_a p_1^m$. Then

$$\begin{aligned}\psi_a(D) &= \frac{1}{N}\left\{-p_1^m + (-1)^{\frac{p-1}{2}m}p^{s(f-1)/2}\left(\sqrt{p^*}\right)^s\zeta_N^{z_a p_1^m}p_1^m\right\}\\ &= \frac{1}{2}\left\{-1 + (-1)^{\frac{p-1}{2}m+z_a}\sqrt{(-1)^{\frac{p-1}{2}}p^{sf}}\right\}.\end{aligned}$$

If $p \equiv 3 \pmod 4$, then

$$\psi_a(D) = \frac{1}{2}\left\{-1 + (-1)^{m+z_a}\sqrt{-p^{sf}}\right\},$$

and if $p \equiv 1 \pmod 4$, then

$$\psi_a(D) = \frac{1}{2}\left\{-1 + (-1)^{z_a}\sqrt{p^{sf}}\right\}.$$

From Lemmas 1.16 and 8.2, we have the theorem. □

Example 10.5 (Feng and Xiang [72]): *Let $p_1 = 7$, $N = 14$, $p = 11$, and $f = ord_{14}(11) = \varphi(14)/2 = 3$. Let C_i, $0 \leq i \leq 13$ be the cyclotomic classes of order 14. If we take $I = \{0, 1, 2, 3, 4, 5, 6\}$, then*

$$D = C_0 \cup C_1 \cup C_2 \cup C_3 \cup C_4 \cup C_5 \cup C_6$$

is a $\left(11^3, \frac{1}{2}(11^3 - 1), \frac{1}{4}(11^3 - 3)\right)$ skew Hadamard difference set of $\boldsymbol{F}_{11^3}$.
If we take $I = \{0, 1, 3, 4, 5, 6, 9\} \pmod 7 = \boldsymbol{Z}/7\boldsymbol{Z}$, then

$$D = C_0 \cup C_1 \cup C_3 \cup C_4 \cup C_5 \cup C_6 \cup C_9$$

is also a skew Hadamard difference set of $\boldsymbol{F}_{11^3}$.

10.4.3 The Case that $p_1 \equiv 3 \pmod 8$

We let $q = p^f$ and $F = \boldsymbol{F}_q$. We assume the following conditions in addition to the index 2 cases, $[(\boldsymbol{Z}/N\boldsymbol{Z})^\times : \langle p \rangle] = 2$, $-1 \notin \langle p \rangle$:

i) $p_1 \equiv 3 \pmod 8$, $p_1 \neq 3$,
ii) $N = 2p_1^m$,
iii) $1 + p_1 = 4p^h$ where h is the class number of $\boldsymbol{Q}(\sqrt{-p_1})$,
iv) p is a prime such that $f = ord_N(p) = \frac{\varphi(p_1^m)}{2}$.

In this section, we will construct a family of skew Hadamard difference sets for the case $p_1 \equiv 3 \pmod 8$, $p_1 \neq 3$ under the above conditions.

The condition (*iii*) with the condition $b^2 + p_1c^2 = 4p^h$ in Theorem 10.3 leads to $b, c \in \{1, -1\}$. Let $\wp$ be a prime ideal lying above p in $\boldsymbol{Q}(\zeta_{q-1})$. Since $\boldsymbol{Q}(\zeta_{q-1})$ contains a quadratic field $\boldsymbol{Q}(\sqrt{-p_1})$, $1 + p_1$ splits in $\boldsymbol{Q}(\zeta_{q-1})$, $1 + p_1 = (1 + \sqrt{-p_1})(1 - \sqrt{-p_1})$. From $1 + p_1 \in \wp$, $1 + \sqrt{-p_1} \in \wp$ or $1 - \sqrt{-p_1} \in \wp$. We choose a prime ideal $\wp$ lying above p such that $1 + \sqrt{-p_1} \in \wp$. Therefore we have $bc = 1$.

Let $C_i = \gamma^i \left\langle \gamma^N \right\rangle$, $0 \leq i \leq N - 1$ be the Nth cyclotomic classes where γ is a primitive element of F. We put

$$J = \langle p \rangle \cup 2\langle p \rangle \cup \{0\} \pmod{2p_1}.$$

Notice that 2 is a quadratic nonresidue in $\boldsymbol{F}_{p_1}$. Since $[(\boldsymbol{Z}/N\boldsymbol{Z})^\times : \langle p \rangle] = 2$, $\langle p \rangle \pmod{p_1}$ has index 2 in $(\boldsymbol{Z}/p_1\boldsymbol{Z})^\times$, that is, $\langle p \rangle \pmod{p_1}$ is the subgroup consisting of quadratic residues in $\boldsymbol{F}_{p_1}$. Hence

$$J \pmod{p_1} = \langle p \rangle \cup 2\langle p \rangle \cup \{0\} \pmod{p_1} = \boldsymbol{Z}/p_1\boldsymbol{Z}. \tag{10.2}$$

Furthermore, $\boldsymbol{Z}/2p_1\boldsymbol{Z}$ is a union of six disjoint sets,

$$J \cup -\langle p \rangle \cup -2\langle p \rangle \cup \{p_1\} = \boldsymbol{Z}/2p_1\boldsymbol{Z}. \tag{10.3}$$

Hence we have $|\langle p \rangle \pmod{2p_1}| = \frac{1}{2}(p_1 - 1)$ and $|J| = p_1$.

We define the subset

$$D = \bigcup_{i=0}^{p_1^{m-1}-1} \bigcup_{j \in J} C_{2i + p_1^{m-1}j}$$

of F. If $p \equiv 1 \pmod 4$, then $D = -D$ and if $p \equiv 3 \pmod 4$, $D \cap -D = \emptyset$.

Lemma 10.3 (Feng et al. [71]): *If a ranges over $\boldsymbol{Z}/N\boldsymbol{Z}$, $\psi_a(D)$ has exactly two distinct values.*

Proof: Set

$$A = (-1)^{\frac{p-1}{2}(m-1)} p^{\frac{f-1}{2} - h} \sqrt{p^*} \text{ and } B = (-1)^{\frac{p-1}{2}\frac{f-1}{2}} p^{\frac{f-1}{2}} \sqrt{p^*}.$$

Since $p_1 \equiv 3 \pmod 8$, we have

$$f - 1 = \frac{1}{2}\varphi(N) \equiv 3^{m-1} - 1 \equiv (-1)^{m-1} - 1 \pmod 4.$$

Thus

$$\frac{1}{2}(f-1) \equiv \begin{cases} 0 \pmod 2, & \text{if } m \equiv 1 \pmod 2 \\ 1 \pmod 2, & \text{if } m \equiv 0 \pmod 2. \end{cases}$$

Hence we have $(-1)^{\frac{1}{2}(f-1)} = (-1)^{m-1}$ and $p^h A = B$. Let $\chi = \chi_{2p_1^m}$ be a multiplicative character of order N over F. We denote a character of order β by χ_β. In the similar discussion in the proof of Theorem 10.4, we have

$$\boldsymbol{\psi_a} = N\psi_a(D) = \sum_{\ell=0}^{N-1} \sum_{i=0}^{p_1^{m-1}-1} \sum_{j \in J} \chi^{-\ell}\left(\gamma^{a+2i+p_1^{m-1}j}\right) g_F\left(\chi^{\ell}\right)$$

for $a \in F^\times$. We distinguish $\ell \pmod{2p_1^m}$, $\ell \neq 0$. If $gcd(\ell, 2p_1^m) > 1$, then $\ell = 2\ell'$, $\ell' \not\equiv 0 \pmod{p_1}$ or $\ell = p_1\ell'$, $\ell' \not\equiv 0 \pmod{p_1^{m-1}}$ or $\ell = p_1^m$. If $gcd(\ell, 2p_1^m) = 1$, then $\ell \in \langle p \rangle$ or $\ell \in -\langle p \rangle$ from the index 2 conditions.

We set

$$T_a(l) = \sum_{i=0}^{p_1^{m-1}-1} \sum_{j \in J} \chi^{-\ell}\left(\gamma^{a+2i+p_1^{m-1}j}\right) g_F\left(\chi^{\ell}\right).$$

Thus we have

$$\boldsymbol{\psi_a} = \sum_{\ell=0}^{N-1} T_a(l).$$

Case 1

$$\ell = 0.\ T_a(0) = g_F(\chi^0) \sum_{i=0}^{p_1^{m-1}-1} \sum_{j \in J} \chi^0\left(\gamma^{a+2i+p_1^{m-1}j}\right) = -p_1^{m-1}|J| = -p_1^m.$$

Case 2

$\ell = 2\ell'$, $\ell' \not\equiv 0 \pmod{p_1}$. From the Eq. (10.2), $\sum_{j \in J} \chi_{p_1}^{-\ell'}(\gamma^j) = 0$, then

$$\sum_{j \in J} \chi_{2p_1^m}^{-2\ell'}\left(\gamma^{a+2i+p_1^{m-1}j}\right) = \chi_{2p_1^m}^{-2\ell'}\left(\gamma^{a+2i}\right) \sum_{j \in J} \chi_{2p_1^m}^{-2p_1^{m-1}\ell'}\left(\gamma^j\right) = 0$$

and $T_a(2\ell') = 0$.

Case 3

$\ell = p_1\ell'$, $\ell' \not\equiv 0 \pmod{p_1^{m-1}}$. From $\sum_{i=0}^{p_1^{m-1}-1} \chi_{p_1^{m-1}}^{-\ell'}(\gamma^i) = 0$,

$$\begin{aligned} &\sum_{i=0}^{p_1^{m-1}-1} \sum_{j \in J} \chi_{2p_1^m}^{-p_1\ell'}\left(\gamma^{a+2i+p_1^{m-1}j}\right) \\ &= \chi_{2p_1^m}^{-p_1\ell'} \chi\left(\gamma^a\right) \sum_{i=0}^{p_1^{m-1}-1} \chi_{2p_1^m}^{-p_1\ell'} \chi\left(\gamma^{2i}\right) \sum_{j \in J} \chi_{2p_1^m}^{-p_1\ell'}\left(\gamma^{p_1^{m-1}j}\right) = 0, \end{aligned}$$

and $T_a(p_1\ell') = 0$.

Case 4

$\ell = p_1^m$. From $\sum_{j\in J} \chi_2 \left(\gamma^{p_1^{m-1}j}\right) = \sum_{j\in \mathbf{Z}/p_1\mathbf{Z}} (-1)^j = 1$,

$$\sum_{i=0}^{p_1^{m-1}-1} \sum_{j\in J} \chi_{2p_1^m}^{-p_1^m} \left(\gamma^{a+2i+p_1^{m-1}j}\right)$$
$$= \chi_2\left(\gamma^a\right) \sum_{i=0}^{p_1^{m-1}-1} \sum_{j\in J} \chi_2 \left(\gamma^{p_1^{m-1}j}\right) = (-1)^a p_1^{m-1}. \tag{10.4}$$

Then we have $T_a(p_1^m) = B(-1)^a p_1^{m-1}$ from $B = g_F(\chi_{2p_1^m}^{p_1^m})$ from Theorem 10.3.

Case 5

$\ell \in \langle p\rangle$ or $\ell \in -\langle p\rangle$. If $\ell \in \langle p\rangle$, then from Theorem 2.1 $g_F(\chi_{2p_1^m}^{\ell}) = g_F(\chi_{2p_1^m})$. From Theorem 10.3 and $bc = 1, b, c \in \{1, -1\}$, we obtain

$$g_F\left(\chi_{2p_1^m}^{\ell}\right) = g_F(\chi_{2p_1^m}) = A\left(\frac{1}{2}\left(b + c\sqrt{-p_1}\right)\right)^2 = \frac{1}{4}\left(1 - p_1 + 2\sqrt{-p_1}\right)A.$$

If $\ell \in -\langle p\rangle$, then

$$g_F(\chi_{2p_1^m}^{\ell}) = \chi_{2p_1^m}(-1)\overline{g_F(\chi_{2p_1^m}^{-\ell})} = \chi_{2p_1^m}(-1)\overline{g_F(\chi_{2p_1^m})} = \frac{1}{4}\left(1 - p_1 - 2\sqrt{-p_1}\right)A$$

by $\chi_{2p_1^m}(-1)\overline{\sqrt{p^*}} = \sqrt{p^*}$.

We put

$$S_{\langle p\rangle} = \sum_{\ell\in\langle p\rangle} \sum_{i=0}^{p_1^{m-1}-1} \sum_{j\in J} \chi_{2p_1^m}^{-\ell} \left(\gamma^{a+2i+p_1^{m-1}j}\right).$$

Then

$$\sum_{\ell\in\langle p\rangle} T_a(\ell) = \frac{1}{4}\left(1 - p_1 + 2\sqrt{-p_1}\right) A S_{\langle p\rangle}$$

and

$$\sum_{\ell\in-\langle p\rangle} T_a(\ell) = \frac{1}{4}\left(1 - p_1 - 2\sqrt{-p_1}\right) A \overline{S_{\langle p\rangle}}.$$

Any element $\ell \in \langle p\rangle, 0 \le \ell \le 2p_1^m - 1$, can be written as

$$\ell = x + 2p_1 y,\ x \in \langle p\rangle \pmod{2p_1},\ y \in \{0, 1, \dots, p_1^{m-1} - 1\}.$$

Then we have

$$S_{\langle p\rangle} = \sum_{\ell\in\langle p\rangle} \sum_{i=0}^{p_1^{m-1}-1} \sum_{j\in J} \chi_{2p_1^m}^{-\ell} \left(\gamma^{a+2i+p_1^{m-1}j}\right)$$
$$= \sum_{\substack{x\in\langle p\rangle \\ (\mathrm{mod}\ 2p_1)}} \sum_{j\in J} \sum_{i=0}^{p_1^{m-1}-1} \chi_{2p_1^m}^{-x} \left(\gamma^{a+2i+p_1^{m-1}j}\right) \sum_{y=0}^{p_1^{m-1}-1} \chi_{2p_1^m}^{-2p_1 y} \left(\gamma^{a+2i}\right).$$

There exists a unique $i \in \{0, 1, \dots, p_1^{m-1} - 1\}$ such that $a + 2i = p_1^{m-1}z_a, z_a \in \mathbf{Z}/2p_1\mathbf{Z}$. Then

$$\begin{aligned} S_{\langle p\rangle} &= p_1^{m-1} \sum_{\substack{x\in\langle p\rangle \\ (\text{mod } 2p_1)}} \sum_{j\in J} \chi_{2p_1^m}^{-x}\left(\gamma^{p_1^{m-1}z_a + p_1^{m-1}j}\right) \\ &= p_1^{m-1} \sum_{\substack{x\in\langle p\rangle \\ (\text{mod } 2p_1)}} \chi_{2p_1}^{-x}(\gamma^{z_a}) \sum_{j\in J} \chi_{2p_1}^{-j}(\gamma) \end{aligned}$$

since $jx \in J$. On the other hand, since $\langle p\rangle \pmod{p_1}$ is the set Q of quadratic residues of $\mathbf{Z}/p_1\mathbf{Z}$, for $i \neq 0, p_1$,

$$\begin{aligned} X_i &= \sum_{\substack{x\in\langle p\rangle \\ (\text{mod } 2p_1)}} \chi_{2p_1}^{-x}(\gamma^i) = \sum_{\substack{x\in\langle p\rangle \\ (\text{mod } 2p_1)}} \chi_2(\gamma^i)\, \chi_{p_1}^{-x}(\gamma^i) = (-1)^i \sum_{x\in Q} \zeta_{p_1}^{-xi} \\ &= (-1)^i \frac{1}{2} \left\{ \sum_{x\in \mathbf{F}_{p_1}^\times} (1+\phi(x))\psi(-ix) \right\} = (-1)^i \frac{1}{2}\left\{-1 + g(\phi, \psi_{-i})\right\} \\ &= (-1)^i \frac{1}{2}\left\{-1 + \phi(i)\sqrt{-p_1}\right\} \end{aligned}$$

by Theorems 2.1 and 2.2 where ϕ is a quadratic character and ψ is an additive character of $\mathbf{F}_{p_1}$. For $i = 0$, $X_0 = \frac{1}{2}(p_1 - 1)$ and for $i = p_1$, $X_{p_1} = -\frac{1}{2}(p_1 - 1)$.

From the definition of J, by letting $i = 1, x = j$ and $i = 2, x = j$ in X_i, we have

$$\begin{aligned} \sum_{j\in J} \chi_{2p_1}^{-j}(\gamma) &= \sum_{\substack{j\in\langle p\rangle \\ (\text{mod } 2p_1)}} \chi_{2p_1}^{-j}(\gamma) + \sum_{\substack{j\in 2\langle p\rangle \\ (\text{mod } 2p_1)}} \chi_{2p_1}^{-j}(\gamma) + \chi_{2p_1}^{0}(\gamma) \\ &= (-1)\frac{1}{2}\left(-1 + \sqrt{-p_1}\right) + \frac{1}{2}\left(-1 - \sqrt{-p_1}\right) + 1 = 1 - \sqrt{-p_1} \end{aligned}$$

It follows that $S_{\langle p\rangle} = p_1^{m-1}(1 - \sqrt{-p_1})X_{z_a}$. Hence

$$\sum_{\ell\in\langle p\rangle} T_a(\ell) = \frac{p_1^{m-1}A}{4}\left(1 - p_1 + 2\sqrt{-p_1}\right)\left(1 - \sqrt{-p_1}\right)X_{z_a}$$

Consequently we obtain

$$\begin{aligned} \boldsymbol{\psi_a} &= -p_1^m + B(-1)^a p_1^{m-1} + \frac{p_1^{m-1}A}{4}\left(1 - p_1 + 2\sqrt{-p_1}\right)\left(1 - \sqrt{-p_1}\right)X_{z_a} \\ &+ \frac{p_1^{m-1}A}{4}\left(1 - p_1 - 2\sqrt{-p_1}\right)\left(1 + \sqrt{-p_1}\right)\overline{X_{z_a}}. \end{aligned}$$

We distinguish $\boldsymbol{\psi_a}$ by the value of $z_a \in \mathbf{Z}/2p_1\mathbf{Z}$ from (10.3).

i) For $z_a = 0$, we have $a \equiv 0 \pmod 2$ and

$$\begin{aligned} \boldsymbol{\psi_a} &= -p_1^m + Bp_1^{m-1} + \frac{1}{4}p_1^{m-1}A(1 - p_1 + 2\sqrt{-p_1})(1 - \sqrt{-p_1})\frac{p_1 - 1}{2} \\ &+ \frac{1}{4}p_1^{m-1}A(1 - p_1 - 2\sqrt{-p_1})(1 + \sqrt{-p_1})\frac{p_1 - 1}{2} \\ &= -p_1^m + Bp_1^{m-1} + \frac{1}{4}p_1^{m-1}A(p_1^2 - 1) \\ &= -p_1^m + \frac{1}{4}p_1^{m-1}(A(p_1^2 - 1) + 4B) \end{aligned}$$

since $p^h A = B$ and $4p^h = p_1 + 1$,

$$= -p_1^m + \frac{1}{4}p^m A(p_1 + 1)$$
$$= -p_1^m + p_1^m p^h A.$$

ii) For $z_a = p_1$, we have $a \equiv 1 \pmod 2$ and $\boldsymbol{\psi_a} = -p_1^m - p_1^m p^h A$.
iii) For $z_a \in \langle p \rangle$, we have $a \equiv 1 \pmod 2$ and $\boldsymbol{\psi_a} = -p_1^m + p_1^m p^h A$.
iv) For $z_a \in -\langle p \rangle$, we have $a \equiv 1 \pmod 2$ and $\boldsymbol{\psi_a} = -p_1^m - p_1^m p^h A$.
v) For $z_a \in 2\langle p \rangle$, we have $a \equiv 0 \pmod 2$ and $\boldsymbol{\psi_a} = -p_1^m + p_1^m p^h A$.
vi) For $z_a \in -2\langle p \rangle$, we have $a \equiv 0 \pmod 2$ and $\boldsymbol{\psi_a} = -p_1^m - p_1^m p^h A$.

It follows $\psi_a(D) = \psi(\gamma^a D)$ has exactly two values when a ranges over $\boldsymbol{Z}/N\boldsymbol{Z}$. □

Theorem 10.5 ***(Feng et al. [71]):*** *We define the subset*

$$D = \bigcup_{i=0}^{p_1^{m-1}-1} \bigcup_{j \in J} C_{2i+p_1^{m-1}j}$$

of F. The set D is a skew Hadamard difference set if $p \equiv 3 \pmod 4$ *or D is a Paley type partial difference set if* $p \equiv 1 \pmod 4$.

Proof: From Lemma 10.3 and $1 + p_1 = 4p^h$, we have

$$\psi_a(D) = \frac{1}{N}\boldsymbol{\psi_a} = \frac{1}{2}\left\{-1 \pm (-1)^{\frac{p-1}{2}(m-1)} p^{\frac{f-1}{2}} \sqrt{p^*}\right\}$$
$$= \frac{1}{2}\left\{-1 \pm (-1)^{\frac{p-1}{2}(m-1)} \sqrt{(-1)^{\frac{p-1}{2}} p^f}\right\}.$$

Then the assertion follows from Lemmas 1.16 and 8.2. □

Example 10.6 ***(Feng et al. [71]):*** *Table 10.2 shows all possible skew Hadamard difference sets and Paley type partial difference sets for* $p_1 \le 10^6$ *except when* $p_1 = 11$ *and* $m = 1$ *[71, p. 431].*

Table 10.2 Hadamard difference sets and Paley PDS for $p_1 \le 10^6$.

p	N	h	b	v
5	$2{\cdot}19^m$	1	1	$5^{9\cdot19^{m-1}}$
17	$2{\cdot}67^m$	1	1	$17^{33\cdot67^{m-1}}$
3	$2{\cdot}107^m$	3	1	$3^{53\cdot107^{m-1}}$
41	$2{\cdot}163^m$	1	1	$41^{81\cdot163^{m-1}}$
5	$2{\cdot}499^m$	3	1	$5^{249\cdot499^{m-1}}$

Source: Feng et al. [71, table 3, p. 431], Reproduced with permission of Springer Nature.

10.5 The Construction by Using Normalized Relative Gauss Sums

10.5.1 More on Ideal Factorization of the Gauss Sum

Let $m = 2^t p_1^{e_1} p_2^{e_2} \cdots p_\ell^{e_\ell}$, $e_i \geq 1$ for $i = 1, \ldots, \ell$ and put $m' = mp_1$. Let p be a prime number and $gcd(p, m) = 1$. Let f and f' be the smallest integers such that

$$p^f \equiv 1 \pmod{m}, \quad p^{f'} \equiv 1 \pmod{m'}.$$

Put $F = GF(p^{f'})$, $K = GF(p^f)$ and $\boldsymbol{Z}_{m'}^\times = (\boldsymbol{Z}/m'\boldsymbol{Z})^\times$ and $\boldsymbol{Z}_m^\times = (\boldsymbol{Z}/m\boldsymbol{Z})^\times$. Assume that the group $\langle p \rangle$ is of index g in both $\boldsymbol{Z}_{m'}^\times$ and $\boldsymbol{Z}_m^\times$. Then we have $\varphi(m') = f'g$ and $\varphi(m) = fg$. Since $\varphi(m') = \varphi(m)p_1$, we have $f' = fp_1$.

Put $\zeta_m = e^{2\pi i/m}$. We give the prime factorization of the Gauss sum of order m in $\boldsymbol{Q}(\zeta_m)$ by Theorem 2.22.

We put $q' = p^{f'}$ and $q = p^f$. For the multiplicative character χ_m of K of order m, we have

$$g\left(\chi_m\right) \sim \mathcal{P}^{\sum_{c \in \boldsymbol{Z}_m^\times/\langle p \rangle} \sum_{i=0}^{f-1} \left\langle \frac{cp^i}{m} \right\rangle \sigma_c^{-1}} \tag{10.5}$$

where $\mathcal{P}$ is a prime ideal in $\boldsymbol{Q}(\zeta_m)$ lying above p. For the multiplicative character $\chi_{m'}$ of F of order m',

$$g\left(\chi_{m'}\right) \sim \mathcal{P}'^{\sum_{c \in \boldsymbol{Z}_{m'}^\times/\langle p \rangle} \sum_{i=0}^{f'-1} \left\langle \frac{cp^i}{m'} \right\rangle \sigma_c^{-1}} \tag{10.6}$$

where $\mathcal{P}'$ is a prime ideal in $\boldsymbol{Q}(\zeta_{m'})$ lying above $\mathcal{P}$.

The prime ideal $\mathcal{P}$ does not split and ramify in $\boldsymbol{Q}(\zeta_{m'})$, it remains inert, that is, $\mathcal{P} = \mathcal{P}'$. Hence the prime ideal factorization of $g(\chi_m)$ in $\boldsymbol{Q}(\zeta_{m'})$ is given by

$$g\left(\chi_m\right) \sim \mathcal{P}'^{\sum_{c \in \boldsymbol{Z}_{m'}^\times/\langle p \rangle} \sum_{i=0}^{f-1} \left\langle \frac{cp^i}{m} \right\rangle \sigma_c^{-1}}. \tag{10.7}$$

Notice that the character χ_m of K is an induced character of $\chi_{m'}$ of F.

10.5.2 Determination of Normalized Relative Gauss Sums

We need the following lemma to determine the values of normalized relative Gauss sums under the conditions in Section 10.5.1.

Lemma 10.4: *For a fixed element $a \in \boldsymbol{Z}_{m'}^\times / \langle p \rangle$, we define the subsets A and B,*

$$A = \left\{ap^j \mid 0 \leq j \leq f' - 1\right\},$$

$$B = \left\{ bm + cp^\ell \mid 0 \leq b \leq \frac{f'}{f} - 1 = p_1 - 1, \right.$$

$$\left. c \in \boldsymbol{Z}_m^\times / \langle p \rangle,\, c \equiv a \pmod{m},\, 0 \leq \ell \leq f - 1 \right\}.$$

Then $A = B$.

Proof: If we put $j = fj' + \ell$, $0 \leq j \leq f' - 1$, $0 \leq \ell \leq f - 1$ then we can write $ap^j = bm + cp^\ell$ from $ap^j = ap^{fj' + \ell} \equiv cp^\ell \pmod{m}$ where $c \equiv a \pmod{m}$. Hence $A \subset B$. The elements of B are all distinct. Then $|B| = p_1 f = f' = |A|$. It follows $A = B$. □

From Lemma 10.4, we have

$$\sum_{j=0}^{f'-1} \left\langle \frac{ap^j}{m'} \right\rangle = \sum_{b=0}^{p_1-1} \sum_{\ell=0}^{f-1} \left\langle \frac{bm + cp^\ell}{m'} \right\rangle = \sum_{b=0}^{p_1-1} \sum_{l=0}^{f-1} \left\langle \frac{b}{p_1} \right\rangle + \sum_{b=0}^{p_1-1} \sum_{\ell=0}^{f-1} \left\langle \frac{cp^\ell}{m'} \right\rangle$$

$$= \frac{1}{2}(f' - f) + \sum_{\ell=0}^{f-1} \left\langle \frac{cp^\ell}{m} \right\rangle.$$

Theorem 10.6 ***(Momihara [138]):*** *Under the conditions of Section 10.4.1, a normalized relative Gauss sum*

$$\tilde{\vartheta}(\chi_{m'}, \chi_m) = \frac{g_F(\chi_{m'})}{p^{\frac{f'-f}{2}} g_K(\chi_m)}$$

is a $2m'$-th root of unity if m' is odd or m'-th root of unity if m' is even.

Proof: Let $T' = \mathbf{Z}_{m'}^{\times} / \langle p \rangle$. From Theorem 2.22, Lemma 10.4 and Eqs. (10.6), (10.7),

$$\begin{aligned} g_F(\chi_{m'}) &\sim \mathfrak{p}'^{\sum_{c' \in T'} \sum_{i=0}^{f'-1} \left\langle \frac{c'p^i}{m'} \right\rangle \sigma_{c'}^{-1}} \\ &\sim \mathfrak{p}'^{\sum_{c' \in T'} \frac{f'-f}{2} \sigma_{c'}^{-1} + \sum_{c' \in T'} \sum_{i=0}^{f-1} \left\langle \frac{c'p^i}{m} \right\rangle \sigma_{c'}^{-1}} \\ &\sim p^{\frac{f'-f}{2}} g_K(\chi_m). \end{aligned}$$

Hence $\tilde{\vartheta}(\chi_{m'}, \chi_m)$ is a unit of $\mathbf{Q}(\zeta_{m'})$. The all conjugates of $\tilde{\vartheta}(\chi_{m'}, \chi_m)$ has absolute value 1. It follows $\tilde{\vartheta}(\chi_{m'}, \chi_m)$ is $2m'$-th root of unity if m' is odd or m'-th root of unity if m' is even. □

Lemma 10.5: *Let $d = 2\gcd(m', p-1)$ if m' is odd or $d = \gcd(m', p-1)$ if m' is even. Then*

$$\tilde{\vartheta}(\chi_{m'}, \chi_m)^d = 1.$$

Proof: For the automorphism $\sigma_{p,1}$ defined by $\sigma_{p,1}(\zeta_{m'}) = \zeta_{m'}^p$, $\sigma_{p,1}(\zeta_p) = \zeta_p$, $\sigma_{p,1}(\tilde{\vartheta}(\chi_{m'}, \chi_m)) = \tilde{\vartheta}(\chi_{m'}, \chi_m)$. If m' is odd, then

$$\tilde{\vartheta}(\chi_{m'}, \chi_m)^{2\gcd(m',\, p-1)} = 1,$$

since $\tilde{\vartheta}(\chi_{m'}, \chi_m)$ is a $2m'$-th root of unity from Theorem 10.6. If m' is even, then

$$\tilde{\vartheta}(\chi_{m'}, \chi_m)^{\gcd(m',\, p-1)} = 1,$$

from Theorem 10.6. □

Theorem 10.7: *Assume that $2 \parallel m'$, $2 \parallel m$, and $\gcd(\frac{m'}{2}, p-1) = \gcd(\frac{m}{2}, p-1) = 1$. Then*

$$\tilde{\vartheta}(\chi_{m'}, \chi_m) = (-1)^{\frac{(p-1)(f'-f)}{4}}.$$

Proof: From the assumption, $m_0' = \frac{m'}{2}$, $m_0 = \frac{m}{2}$ are both odd. We denote a character of order m_0' of F by $\chi_{m_0'}$ and a character of order m_0 of K by χ_{m_0}. In particular, χ_2' and χ_2 are quadratic characters of F and K, respectively. Then the characters $\chi_{m'}$ and χ_m can be written as

$$\chi_{m'} = \chi_{m_0'} \cdot \chi_2', \quad \chi_m = \chi_{m_0} \cdot \chi_2.$$

From Lemma 10.5,

$$\frac{g_F(\chi_{m_0'})}{g_K(\chi_{m_0})} = \pm p^{\frac{f'-f}{2}}.$$

The ratio $g_F(\chi_{m_0'})/g_K(\chi_{m_0})$ is fixed by the automorphism $\sigma_{2,1}$

$$\left(\frac{g_F(\chi_{m_0'})}{g_K(\chi_{m_0})}\right)^{\sigma_{2,1}} = \frac{g_F(\chi_{m_0'}^2)}{g_K(\chi_{m_0}^2)} = \pm p^{\frac{f'-f}{2}} = \frac{g_F(\chi_{m_0'})}{g_K(\chi_{m_0})}.$$

Therefore, the Davenport–Hasse product formula (Theorem 2.5) yields

$$\frac{g_F(\chi^2_{m'_0})}{g_K(\chi^2_{m_0})} = \frac{g_F(\chi_{m'_0})}{g_K(\chi_{m_0})} \cdot \frac{g_F(\chi_{m'_0}\chi'_2)}{g_K(\chi_{m_0}\chi_2)} \cdot \frac{g_K(\chi_2)}{g_F(\chi'_2)}$$

from $\chi_{m'_0}(4) = \chi_{m_0}(4) = 1$. Together with the above, we have

$$\frac{g_F(\chi_{m'})}{g_K(\chi_m)} = \frac{g_F(\chi'_2)}{g_K(\chi_2)} = (-1)^{\frac{(p-1)(f'-f)}{4}} p^{\frac{f'-f}{2}}$$

from Theorem 2.2. This implies the theorem. □

Theorem 10.8: *Assume m' is odd. If $gcd(m', p-1) = 1$, then $\tilde{\vartheta}(\chi_{m'}, \chi_m) = 1$.*

Proof: From Lemma 10.5, $\tilde{\vartheta}(\chi_{m'}, \chi_m) = \pm 1$. We assume $\tilde{\vartheta}(\chi_{m'}, \chi_m) = -1$, that is, $g_F(\chi_{m'})/g_K(\chi_m) = -p^{\frac{f'-f}{2}}$. In the proof of Theorem 10.7, we have

$$\frac{g_F(\chi_{m'}\chi'_2)}{g_K(\chi_m\chi_2)} = \frac{g_F(\chi'_2)}{g_K(\chi_2)}.$$

It follows

$$\frac{J(\chi_{m'}, \chi'_2)}{J(\chi_m, \chi_2)} = \frac{g_F(\chi_{m'})}{g_K(\chi_m)} = -p^{\frac{f'-f}{2}}$$

that is

$$J(\chi_{m'}, \chi'_2) = -p^{\frac{f'-f}{2}} J(\chi_m, \chi_2). \tag{10.8}$$

From Corollary 2.1,

$$J(\chi_{m'}, \chi'_2) \equiv -p^{f'} \pmod{2(1-\zeta_{m'})}, \quad J(\chi_m, \chi_2) \equiv -p^f \pmod{2(1-\zeta_m)}.$$

Since $p^{f'} \equiv 1 \pmod{m'}$, $p^{f'} \equiv 1 \pmod{2(1-\zeta_{m'})}$, and $J(\chi_{m'}, \chi'_2) \equiv -1 \pmod{2(1-\zeta_{m'})}$. A primitive mth root of unity ζ_m is a power of $\zeta_{m'}$. Then $p^f \equiv 1 \pmod{2(1-\zeta_{m'})}$ from $p^f \equiv 1 \pmod{2(1-\zeta_m)}$. Hence

$$J(\chi_m, \chi_2) \equiv -1 \pmod{2(1-\zeta_{m'})}.$$

By substituting these into Eq. (10.8), we have

$$-1 \equiv -p^{f\frac{p_1-1}{2}}(-1) \equiv 1 \pmod{2(1-\zeta_{m'})}.$$

This is a contradiction. Hence $\tilde{\vartheta}(\chi_{m'}, \chi_m) = 1$. □

10.5.3 A Family of Skew Hadamard Difference Sets

Let p_1 be an odd prime number and p be a prime such that $gcd(p_1^e, p-1) = 1$. Assume that $\langle p \rangle$ is of index g in $(\boldsymbol{Z}/2p_1^e\boldsymbol{Z})^\times$ for every integer $e \geq 1$. When we let $m = 2p_1^e$, $ord_m p = \varphi(m)/g$.

We assume $m = 2p_1^e$ and p satisfies the above conditions.

Put $h = 2p_1, m = 2p_1^e, e \geq 2$, and $\varphi(h) = d \cdot g$, $\varphi(m) = f \cdot g$. Let γ_0 and γ be primitive elements of $\boldsymbol{F}_{p^d}$ and $\boldsymbol{F}_{p^f}$, respectively.

Let $\sigma_{j,1}$ be the automorphism of $\boldsymbol{Q}(\zeta_{m'}, \zeta_p)$ such that

$$\sigma_{j,1}(\zeta_{m'}) = \zeta^j_{m'} \quad \text{and} \quad \sigma_{j,1}(\zeta_p) = \zeta_p.$$

Then $(\tilde{\vartheta}(\chi_{m'}, \chi_m))^{\sigma_{j,1}} = \tilde{\vartheta}(\chi^j_{m'}, \chi^j_m)$ and $\tilde{\vartheta}(\chi^j_{m'}, \chi^j_m) = \dfrac{g_F(\chi^j_{m'})}{p^{\frac{f'-f}{2}} g_K(\chi^j_m)}$.

We define cyclotomic classes,

$$S_i^{(h,p^d)} = \gamma_0^i \left\langle \gamma_0^h \right\rangle, \quad S_i^{(m,p^f)} = \gamma^i \left\langle \gamma^m \right\rangle.$$

Theorem 10.9 *(Momihara [138]):* *Let H be a subset of $\{0, 1, \ldots, h-1\}$ such that $\sum_{i \in H} \zeta_{p_1}^i = 0$. Assume that*

$$D_0 = \bigcup_{i \in H} S_i^{(h,p^d)}$$

is a skew Hadamard difference set or Paley type regular PDS in $\boldsymbol{F}_{p^d}$.

Then

$$D = \bigcup_{i_1=0}^{p_1^{e-1}-1} \bigcup_{i \in H} S_{2i_1+ip_1^{e-1}}^{(m,\, p^f)}$$

is a skew Hadamard difference set or Paley type regular PDS in $\boldsymbol{F}_{p^f}$.

Proof: We will show the theorem by induction on e. Assume D_0 is a skew Hadamard difference set (we can prove the case that D_0 is a Paley type regular PDS similarly). We have $|H| = p_1$ from $|D_0| = \frac{1}{2}(p^d - 1)$ and we see that if $i \in H$, then $i + p_1 \notin H$ from $-1 \in S_{p_1}^{(h,p^d)}$.

Assume that D is a skew Hadamard difference set in $K = \boldsymbol{F}_{p^f}$ and put $m' = mp_1 = 2p_1^{e+1}, f' = \varphi(m')/g = fp_1$. We will show the subset

$$D' = \bigcup_{i_1=0}^{p_1^{e}-1} \bigcup_{i \in H} S_{2i_1+ip_1^{e}}^{(m',\, p^{f'})}$$

is a skew Hadamard difference set in $F = \boldsymbol{F}_{p^{f'}}$.

Let χ_m be a primitive character of order m of K and $\chi_{m'}$ be a primitive character of order m' of F.

For any nontrivial additive character η_u of K for $\gamma^u \in K$, we have

$$\begin{aligned}\eta_u(D) &= \frac{1}{2p_1^e} \sum_{j=0}^{2p_1^e-1} \chi_{2p_1^e}^{-j}(\gamma^u) g_K\left(\chi_m^j\right) \sum_{i \in H} \zeta_{2p_1}^{-ij} \sum_{i_1=0}^{p_1^{e-1}-1} \zeta_{p_1^e}^{-i_1 j} \\ &= \frac{1}{2}\left(-1 \pm \sqrt{\delta p^f}\right) \end{aligned} \tag{10.9}$$

where $\delta = -1$. Notice that we identify the subset D with the element $\sum_{d \in D} d$ of $\boldsymbol{Z}[F]$. Let η'_a be an additive character of F and ξ be a generator of $F^\times$. We obtain

$$|D'| = |H|\, p_1^e \frac{p^{f'} - 1}{m'} = \frac{1}{2}\left(p^{f'} - 1\right).$$

For any nontrivial additive character η'_a, $\xi^a \in F$,

$$\eta'_a(D') = \frac{1}{2p_1^{e+1}} \sum_{j=0}^{2p_1^{e+1}-1} \chi_{m'}^{-j}\left(\xi^a\right) g_F\left(\chi_{m'}^j\right) \sum_{i \in H} \zeta_{2p_1}^{-ij} \sum_{i_1=0}^{p_1^e-1} \zeta_{2p_1^{e+1}}^{-2i_1 j}.$$

If $j \neq 0$ is even, then the order of $\chi_{m'}^j$ is a power of an odd prime p_1. Hence $\tilde{\vartheta}(\chi_{m'}^j, \chi_m^j) = 1$ from Theorem 10.8. If j is odd, then the order of $\chi_{m'}^j$ has a form $2p_1^v$. From Theorem 10.7, $\tilde{\vartheta}(\chi_{m'}^j, \chi_m^j) = (-1)^{\frac{(p-1)(f'-f)}{4}} = \epsilon$.

By substituting $p^{\frac{f'-f}{2}}\tilde{\vartheta}(\chi^j_{m'},\chi^j_m)g_K(\chi^j_m)$ for $g_F(\chi^j_{m'})$ with these values, we have

$$\begin{aligned}\eta'_a(D') &= \frac{p^{\frac{f'-f}{2}}\epsilon}{2p_1^{e+1}}\sum_{j=0}^{2p_1^{e+1}-1} g_K\left(\chi^j_m\right)\sum_{i\in H}\zeta_{2p_1}^{-ij}\sum_{i_1=0}^{p_1^e-1}\zeta_{2p_1^{e+1}}^{-(2i_1+a)j}\\ &+\frac{p^{\frac{f'-f}{2}}}{2p_1^{e+1}}(-\epsilon+1)\sum_{j=0}^{p^{e+1}-1} g_K\left(\chi^{2j}_m\right)\sum_{i\in H}\zeta_{2p_1}^{-2ij}\sum_{i_1=0}^{p_1^e-1}\zeta_{2p_1^{e+1}}^{-(2i_1+a)2j}\\ &-\frac{p^{\frac{f'-f}{2}}}{2p_1^{e+1}}\left(-p_1^{e+1}\right)+\frac{1}{2p_1^{e+1}}\left(-p_1^{e+1}\right).\end{aligned}$$

We adjust the value for $j=0$ with the third and fourth terms.

We put $j=2p_1^e x+y$ for $0\le x\le p_1-1$ and $0\le y\le 2p_1^e-1$. Then the first term is written as

$$\frac{p^{\frac{f'-f}{2}}\epsilon}{2p_1^{e+1}}\sum_{y=0}^{2p_1^e-1} g_K\left(\chi^y_m\right)\sum_{i\in H}\zeta_{2p_1}^{-iy}\sum_{i_1=0}^{p_1^e-1}\left(\sum_{x=0}^{p_1-1}\zeta_{p_1}^{-(a+2i_1)x}\right)\zeta_{2p_1^{e+1}}^{-(a+2i_1)y}.$$

For each $a\in\{0,1,\ldots,m'\}$, there exists a unique $i_a\in\{0,1,\ldots,p_1-1\}$ such that $a+2i_a\equiv 0\pmod{p_1}$. We write $a+2i_a=p_1z_a$. Every $i_1\in\{0,1,\ldots,p_1^e-1\}$ such that $i_a\equiv i_1\pmod{p_1}$ holds $a+2i_1\equiv 0\pmod{p_1}$. Thus

$$\sum_{i_1=0}^{p_1^e-1}\left(\sum_{x=0}^{p_1-1}\zeta_{p_1}^{-(a+2i_1)x}\right)\zeta_{2p_1^{e+1}}^{-(a+2i_1)y}=p_1\zeta_{2p_1^e}^{-z_a y}\sum_{\ell=0}^{p_1^{e-1}-1}\zeta_{p_1^e}^{-\ell y}.$$

It follows the first term is

$$\begin{aligned}&\frac{p^{\frac{f'-f}{2}}\epsilon}{2p_1^e}\sum_{y=0}^{2p_1^e-1}\zeta_{2p_1^e}^{-z_a y}g_K(\chi^y_m)\sum_{i\in H}\zeta_{2p_1}^{-iy}\sum_{\ell=0}^{p_1^{e-1}-1}\zeta_{p_1^e}^{-\ell y}\\ &=\frac{p^{\frac{f'-f}{2}}\epsilon}{2p_1^e}2p_1^e\eta_{z_a}(D)=\frac{\epsilon}{2}\left(-p^{\frac{f'-f}{2}}\pm\sqrt{\delta p^{f'}}\right)\end{aligned}$$

from Eq. (10.9).

From the assumption, $\sum_{i\in H}\zeta_{p_1}^{-ij}=0$ if $j\ne 0$, or p_1 if $j=0$. Thus the second term is $\frac{-p^{\frac{f'-f}{2}}}{2}(-\epsilon+1)$. Altogether we obtain

$$\begin{aligned}\eta'_a(D') &= \frac{\epsilon}{2}\left(-p^{\frac{f'-f}{2}}\pm\sqrt{\delta p^{f'}}\right)-\frac{1}{2}p^{\frac{f'-f}{2}}(-\epsilon+1)+\frac{1}{2}p^{\frac{f'-f}{2}}-\frac{1}{2}\\ &=\frac{1}{2}\left(-1\pm\epsilon\sqrt{\delta p^{f'}}\right).\end{aligned}$$

We easily verify D' is skew. We see $-1\in S_{p^e}^{(m,p^{f'})}$. Assume that $\alpha\in S_{2i_1+p_1^{e-1}i}^{(m,p^{f'})}$. Then $-\alpha\in S_{2i_1+p_1^{e-1}i+p^e}^{(m,p^{f'})}$. If $-\alpha\in D$, then $i\in H$ and $i+p_1\in H$. It contradicts $i+p_1\notin H$. □

Example 10.7 *(Momihara [138]):* *Let $p_1=13$, $p=3$. Let $h=2p_1=2\cdot 13$ and $m=2\cdot 13^e$. The subgroup $\langle 3\rangle$ is of index 4 of $(\mathbf{Z}/2\cdot 13\mathbf{Z})^\times$ and $f=\varphi(2\cdot 13)/4=3$. Let $Q=\langle 23\rangle$ be the subgroup of index 2 of $(\mathbf{Z}/2\cdot 13\mathbf{Z})^\times$ and let $H=Q\cup 2Q\cup\{0\}$. Then*

$$H=\{0,1,23,9,25,3,17,2,20,18,24,6,8\}.$$

We verify that if $i \in H$, then $i + 13 \notin H$ and $\sum_{i \in H} \zeta_{13}^i = 0$. The subset

$$D_0 = \bigcup_{i \in H} S_i^{(26,3^3)} = \{\gamma^i \mid i \in H\}$$

is a skew Hadamard difference set over $\mathbf{F}_{3^3}$.
The subgroup $\langle 3 \rangle$ is of index 4 of $(\mathbf{Z}/2 \cdot 13^e\mathbf{Z})^\times$ for every positive integer e. From Theorem 10.9, we obtain a family of skew Hadamard difference sets with parameters

$$\left(3^{\frac{\varphi(m)}{4}}, \frac{1}{2}\left(3^{\frac{\varphi(m)}{4}} - 1\right), \frac{1}{4}\left(3^{\frac{\varphi(m)}{4}} - 3\right)\right).$$

11

Asymptotic Existence of Hadamard Matrices

11.1 Notations

Table 11.1 gives the notations which are used in this chapter.

11.2 Introduction

11.2.1 de Launey's Theorem

While it is conjectured that Hadamard matrices exist for all orders $4t$ $t > 0$, sustained effort over five decades only yielded a theorem of the type "that for all odd natural numbers q, there exists an Hadamard matrix of order $q^{2(a+b\log_2 q)}$ where a and b are nonnegative constants." To prove the Hadamard conjecture it is necessary to show we may take $a = 2$ and $b = 0$. Seberry [213] showed that we may take $a = 0$ and $b = 2$. This was improved by Craigen [30], who showed that we may take $a = 0$ and $b = \frac{3}{8}$. Then astonishingly, de Launey [45] and de Launey and Kharaghani [48] showed, using a number theoretic argument of Erdös and Odlyzko [69], that there are enough Paley Hadamard matrices to ensure that for all $\epsilon > 0$, the set of odd numbers q for which there is an Hadamard matrix of order $q^{22+\lfloor \epsilon \log_2 k \rfloor}$ has positive density in the natural numbers. It is beyond the scope of this book to prove this result but it is so important we have chosen to report it.

11.3 Seberry's Theorem

Seberry (Wallis) [213] gave the first asymptotic theorem in 1976. She showed

Theorem 11.1: *Given an odd number q, there exists a* $t_0 > 0$, *so that for every* $t \geq t_0$, *there exists an Hadamard matrix of order* $2^t q$.

We do not give here the proof which can be found in [158].

We note that by careful counting in some cases there is a slightly stronger bound but the result proved was

Theorem 11.2 ***(Wallis [213]):*** *If* $q > 3$ *is an integer, then there exists an Hadamard matrix of order* $2^t q$, *where* $t = \lfloor 2\log_2(q-3) \rfloor + 1$.

We do not prove this result here (the proof can be found in [213] and [158]). We proceed to a stronger result of Craigen [30].

Hadamard Matrices: Constructions using Number Theory and Algebra, First Edition. Jennifer Seberry and Mieko Yamada.

Table 11.1 Notations used in this chapter.

$\underline{x}$	Represents $-x$
$[x]$	The largest integer not exceeding x
$\lceil x \rceil$	The integer truncating the fractional part of x
$\lfloor x \rfloor$	The smallest integer exceeding x
A^H	The adjoint of a matrix A
$OD(n; s_1, s_2, \dots, s_u)$	Orthogonal design of order n and type $(s_1, \dots, s_u)$
SP_m	The signed group of $m \times m$ permutation matrices
$(0, S)$-matrix	A matrix whose nonzero entries are in S
S-matrix	A matrix with no zero entries
$H = SH(n, S)$	The signed group Hadamard matrix
$W = SW(n, k, S)$	The signed weighing matrix of weight k

11.4 Craigen's Theorem

Craigen [30] in 1995 used groups containing a distinguished central involution and sequences with zero autocorrelation function was able to greatly improve Seberry's result. We strongly acknowledge Craigen's ingenious contribution to this section as sharpened by de Launey and Gordon [48] (W. de Launey. Private conversation. Seville, Spain, 2007), he showed

Theorem 11.3: *For any positive number t, there exists an Hadamard matrix*

1) *of order $2^{2b}t$, where b is the number of nonzero digits in the binary expansion of t, and*
2) *of order $2^s t$ for $s = 6\lfloor \frac{1}{16}((t-1)/2) \rfloor + 2$.*

Craigen's theorem implies that there is an Hadamard matrix of order 2^s whenever $2^s >\geq ct^a$, where we may take $a = \frac{3}{8}$ and $c = 2^{\frac{26}{16}}$.

11.4.1 Signed Groups and Their Representations

Definition 11.1: A **signed group** of order n is a group of order $2n$ written multiplicatively with a distinguished central element of order 2, denoted by -1.

Signed groups may be identified by their presentation, which is similar to the presentation of groups. The group of quaternions of order 8, but a signed group of order 4 is traditionally presented as follows:

$$Q = \langle i, j : i^2 = j^2 = -1, ij = -ji \rangle.$$

Other small signed groups of interest are the trivial signed group:

$$T = \langle -1 \rangle$$

which is the cyclic group of order 2, but has order 1 as a signed group. The complex signed group is

$$S_{\mathbf{C}} = \langle i : i^2 = -1 \rangle$$

which is the cyclic group of order 4, and the dihedral group of order 8 is

$$D = \langle i, j : i^2 = j^2 = 1, ij = -ji \rangle$$

which is to be a signed group of order 4. The most important signed groups to us are the direct analog of the symmetric groups, namely for each m, the signed group of $m \times m$ signed permutation matrices, which we denote by SP_m. Notice that $SP_2 \cong D$.

Definition 11.2: A **representation of degree** m **of a signed group** S is a homomorphism from S to the general linear group $GL(m, \mathbf{R})$ or $GL(m, \mathbf{C})$ where $\mathbf{R}$ is the real number field and $\mathbf{C}$ is the complex number field.

Definition 11.3: If a representation of a signed group is in the monomial matrices, then the representation is called a **monomial representation**. A real monomial representation (**remrep** is an abbreviation) is a subgroup of SP_m. Remreps have the important property that multiplicative inversion in the signed group corresponds to the transpose matrix.

Let N be a subgroup of a finite group G and $t_1, \dots, t_m$ be a complete system of representatives in G/N. Furthermore let σ be a representation of N. The induced representation σ_G of G is defined as

$$\sigma_G(g) = \left(\sigma(t_i g t_j^{-1})\right)_{1 \le i,j \le m}$$

where $\sigma_G(g) = 0$ if $g \notin N$.

We know if σ is a one-dimensional representation of N, $m = 1$, then the induced representation σ_G is monomial.

Example 11.1: *Let* $G = S_{\mathbf{C}}$ *and* $N = \langle -1 \rangle = \{1, -1\}$. *We define the representation* σ *of* N *as* $\sigma(1) = 1$ *and* $\sigma(-1) = -1$. *Then we have*

$$\sigma_G(i) = \begin{bmatrix} \sigma(1 \cdot i \cdot 1) & \sigma\left(1 \cdot i \cdot i^{-1}\right) \\ \sigma(i \cdot i \cdot 1) & \sigma\left(i \cdot i \cdot i^{-1}\right) \end{bmatrix} = \begin{bmatrix} 0 & 1 \\ -1 & 0 \end{bmatrix},$$

$$\sigma_G(1) = \begin{bmatrix} 1 & 0 \\ 0 & 1 \end{bmatrix} = I \text{ and } \sigma_G(-1) = \begin{bmatrix} -1 & 0 \\ 0 & -1 \end{bmatrix} = -I.$$

We see σ_G *is a monomial representation.*

Definition 11.4: The **signed group ring** $\mathbf{Z}[S]$ is defined in a similar to a group ring containing both $\mathbf{Z}$ and S by canonical injections, that preserve multiplication and map -1 to -1. That is, for $x \in \mathbf{Z}[S]$,

$$x = \sum_{s \in S} z_s s, \quad z_s \in \mathbf{Z}.$$

It has a canonical involution $x \to x^*$ obtained by extending multiplicative inverse in S linearly,

$$x^* = \sum_{s \in S} z_s s^{-1}.$$

The signed group ring associated with $S_{\mathbf{C}}$ is the ring of Gaussian integers. Conjugation in this ring is the usual complex conjugation.

Every representation σ_G of a signed group extends to a representation σ_R of the signed group ring, that is

$$\sigma_R\left(\sum_{s \in S} z_s s\right) = \sum_{s \in S} z_s \sigma_G(s), \quad z_s \in \mathbf{Z}.$$

The reader may verify that a degree m remrep of S extends to a representation of $\mathbf{Z}[S]$ as matrices of order m in which conjugation corresponds to matrix transpose.

Definition 11.5: The involution induces an **adjoint** for $\mathbf{Z}[S]$-matrices having entries in $\mathbf{Z}[S]$. For $A = (a_{ij})$, $a_{ij} \in \mathbf{Z}[S]$, we define the **adjoint** A^H **of** A as $A^H = (a_{ij}^*)$. A is **normal** if $AA^H = A^H A$.

Note that if $S = S_{\mathbb{C}}$, the adjoint is the usual Hermitian adjoint.
We denote a matrix whose nonzero entries are in S by a $(0, S)$-matrix and a S-matrix if they have no zero entries. We are most interested in $(0, S)$ matrices.

Definition 11.6: A S-matrix A of order n satisfying $AA^H = I$ is called a **signed group Hadamard matrix** and denoted by $SH(n, S)$. A $(0, S)$-matrix W of order n satisfying $WW^H = kI$ is called a **signed weighing matrix of weight** k and denoted by $SW(n, k, S)$.

Section 11.4 will be devoted to the construction of signed group Hadamard matrices from sequences. We will show how Hadamard matrices may be constructed in turn from these and thereby obtain powerful new results on the asymptotic existence of Hadamard matrices.

11.4.2 A Construction for Signed Group Hadamard Matrices

We shall repeatedly use the following simple generalization of lemma 11 of [31].

Definition 11.7: The **support of a matrix** is the set of coordinates where its entries are nonzero. A matrix is **quasi-symmetric** if it has the same support as its transpose. Two matrices M and N are said to be **disjoint** if they have disjoint support. It is equivalent that $M * N = 0$.
The **support of a sequence** is the set of coordinates where its entries are nonzero. A sequence of length ℓ is **quasi-symmetric** if the corresponding $\ell \times \ell$ circulant matrix is quasi-symmetric. Sequences of the same length are **disjoint** if the corresponding matrices are disjoint.

Lemma 11.1: *Let A, B be normal commuting, disjoint $(0, S)$-matrices of order n. So $A^HA = AA^H$ and $B^HB = BB^H$. If*

$$C = \begin{bmatrix} A+B & A-B \\ A^H - B^H & -A^H - B^H \end{bmatrix},$$

then $CC^H = C^HC = 2I_2 \times (AA^H + BB^H)$. Moreover, if A and B are both quasi-symmetric and S has a remrep of degree m, then there is a $(0, SP_{2m})$-matrix D of order n, having the same support as $A + B$, such that $DD^H = D^HD = AA^H + BB^H$. Further, if A and B are both circulant, so is D.

Proof: The claim concerning C may be verified directly. The matrix D is obtained as follows: Reorder the rows and columns of C so that the resulting matrix, D_0, is partitioned into 2×2 blocks whose entries are the (i,j), $(i+n,j)$, $(i,j+n)$, and $(i+n,j+n)$ entries of C, $1 \le i,j \le n$. Each nonzero block of D_0 will have one of the forms

$$\begin{bmatrix} a & a \\ b & -b \end{bmatrix} \text{ or } \begin{bmatrix} a & -a \\ b & b \end{bmatrix}, \quad \text{where } a, b \in S.$$

Multiplying D_0 on the right by $\frac{1}{2}I_n \times \left[\begin{smallmatrix} 1 & 1 \\ 1 & - \end{smallmatrix}\right]$, we get a $(0, S)$-matrix D_1, whose nonzero 2×2 blocks have one of the forms

$$\begin{bmatrix} a & 0 \\ 0 & b \end{bmatrix} \text{ or } \begin{bmatrix} 0 & a \\ b & 0 \end{bmatrix}, \quad \text{where } a, b \in S. \tag{11.1}$$

and such that $D_1D_1^H = D_1^HD_1 = (AA^H + BB^H) \times I_2$. The 2×2 matrices of the form (11.1) comprise a signed group which has a remrep of degree $2m$ (obtained by replacing $0, a, b$ with the $m \times m$ matrices representing them). Replacing each block of D_1, with the element of SP_{2m} it thus represents gives D. If A and B are circulant, then C consists of four circulant blocks, so D_0 and D_1 are block circulant (2×2 blocks) and consequently D is circulant. □

Theorem 11.4: *Suppose $X_1, \dots, X_n$, are disjoint, quasi-symmetric sequences having zero periodic autocorrelation of period q and weight w.*

1) If these are all $(0, \pm 1)$-sequences, then there is a circulant $SW(q, w, SP_{2^{n-1}})$.
2) If X_1, X_2 are $(0, \pm 1, \pm i)$-sequences and $X_3, \dots, X_n$ are $(0, \pm 1)$-sequences, then there is a circulant $SW(q, w, SP_{2^n})$.

Proof: We let $M_1, \dots, M_n$ be the circulant matrices having first rows $X_1, \dots, X_n$. The first item follows by iterating Lemma 11.1, $n-1$ times: The first iteration uses M_1 and M_2 to obtain a circulant $(0, SP_2)$-matrix N_1, the next uses N_1, and M_3 to obtain a circulant $(0, SP_4)$-matrix N_2, etc. At each step, N_i and M_{i+2} are disjoint, normal circulant $(0, S)$-matrices. Since M_{i+2} has real entries, they also commute, so Lemma 11.1 applies again. The desired matrix is N_{n-1}.

The second case is similar, except that the first iteration of Lemma 11.1 uses the complex circulant matrices M_1, M_2 to obtain a circulant $(0, SP_4)$-matrix. □

Now if a signed group Hadamard matrix H of order $q > 1$ is to exist, it is necessary that q be even, a fact (theorem 2.15 of [26]) which can be seen by observing that each $s \in S$ must appear as a term in the inner product of two rows of H as many times as $-s$. We consider a number of ways to use known sequences with zero autocorrelation to get sequences of length $q = 2p$ with the desired properties.

If U and V are sequences of length $\ell \leq \frac{(p-1)}{2}$, and V^*, the reverse of the elements of V, has the same support as U, then a and b may be chosen so that $a + b = p - \ell - 1$ and the quasi-symmetric sequences $X_U = (0_{a+1}, U, 0_{2b+1}, V, 0_a)$ and $X_V = (0_{b+1}, V, 0_{2a+1}, -U, 0_b)$ are disjoint. These have periodic autocorrelation equal to twice the periodic autocorrelation of the sequences $(U, 0_{2p-1})$ and $(V, 0_{2p-1})$.

For example, if $U = 00111-00$ and $V = 0011-100$ then U and V^*, where $V^* = 001-1100$, have the same support.

Suppose now that $U_1, \dots, U_t, V_1, \dots, V_t$ are sequences having zero autocorrelation and weight w, with total length no more than $p-1$, and V_j^* is the same length and has the same support as $U_j, j = 1, \dots, t$. Then we may perform the above construction for each j, varying the values of a and b so as to obtain $2t$ quasi-symmetric, disjoint sequences of length $2p$ having zero periodic autocorrelation and weight $2w$. The easiest way of guaranteeing the correct relationship between each U_j and V_j is to require them to be (± 1)-sequences of the same length. Fortunately, sets of 2 and 4 (± 1)-sequences with zero autocorrelation are known to exist in abundance (see, for example [166, 252]). Complex Golay sequences [28] are also natural to use (i.e. for X_1, X_2 in part 2 of Theorem 11.4). In fact, we can use any number of (± 1)-sequences gathered into pairs of the same length, with at most two complex sequences of equal length, as long as we have zero autocorrelation and the total of all the lengths is less than or equal to $p-1$.

Let us now consider how large the signed group must be in Theorem 11.4. The measure of the "size" of a signed group that we are most interested in is the degree of its smallest remrep. Our constructions give a signed subgroup of SP_{2^n}, so we know that this number is at most 2^n (2^{n-1} if only real sequences are used).

A most crude bound on n is obtained using only Golay sequences of length 2^t, which exist for all t. Let p be an odd positive integer. Henceforth, let us write $N = N(p)$ for the number of nonzero binary digits of p. If the jth nonzero binary digit of $\frac{(p-1)}{2}$ corresponds to 2^t, we form sequences X_{2j+1}, X_{2j+2} of length $2p$, using Golay sequences U, V of length 2^t in the above construction. We also use $X_1 = (1, 0_{2p-1})$, $X_2 = (0_p, 1, 0_{p-1})$ (or we could use the single sequence $(1, 0_{p-1}, i, 0_{p-1})$). Altogether this gives us $2N$ disjoint, quasi-symmetric sequences with zero periodic autocorrelation, period $2p$, and weight $2p$. The following result now follows by Theorem 11.4.

Theorem 11.5: *For any odd positive integer p, there exists a circulant $SH(2p, SP_{2^{2N-1}})$.*

In contrast to Theorem 11.4, we have established here a uniform bound (best possible, in fact) on the exponent of 2 necessary in the order of a signed group Hadamard matrix (although we do not restrict the size of the signed group to make this statement, we clearly have some control on it)!

That the signed group Hadamard matrices we construct are all circulant is interesting in light of the long-standing conjecture that there is no circulant Hadamard matrix of order >4 [128], and the difficulty of finding more than a handful of circulant complex Hadamard matrices [190]. Because of their strong algebraic structure it is likely that these matrices will be found to have applications involving the coding, transmission, storage, and retrieval of information.

11.4.3 A Construction for Hadamard Matrices

The following ingenious construction is due to Rob Craigen [31]. Lemma 9 of [31] gives the connection between orthogonal matrices whose nonzero entries lie in a signed group and those whose nonzero entries are ± 1, and it has the following immediate consequence.

Theorem 11.6: *Let m be the order of an Hadamard matrix. Suppose there exists a $SH(n,S)$ and S has a remrep of degree m, then there is an Hadamard matrix of order mn.*

Proof: [Sketch] Replace each entry s of the $SH(n,S)$ with $\pi(s)H$, where π is the remrep and H is an Hadamard matrix of order m. □

Using direct sum, we see that if S has a remrep of degree m, it has a remrep of any degree mq, q a positive integer. Now SP_m has a remrep of degree m (i.e. the natural representation), and so also of any degree mq. Using Theorems 11.4 and 11.6, we give here some consequences of the method of Section 11.4.2.

Theorem 11.7: *Suppose there are (± 1)-sequences $A_1, B_1, \dots, A_t, B_t$ with zero autocorrelation, A_i, B_i both having length λ_i, $i = 1, \dots, t$, where $\sum_{i=1}^{t} \lambda_i = \Lambda$. Then Hadamard matrices exists in orders*

i) $4^{t+1}(2\Lambda + 1)$;
ii) $2 \cdot 4^{t+1}(2\Lambda + \Gamma)$, $\Gamma = 3, 5, 11, 13$.
iii) $2 \cdot 4^{t+2}(2\Lambda + 2g + 1)$, *where g is a complex Golay number.*
Moreover
iv) if there is an Hadamard matrix of order $2 \cdot 4^t q$, then there is an Hadamard matrix of order $4^{t+1}(2\Lambda + 1)q$;
v) if there is an Hadamard matrix of order $4^{t+1}q$, then there are Hadamard matrices of order $2 \cdot 4^{t+1}(2\Lambda + \Gamma)q$, $\Gamma = 3, 5, 11, 13$;
vi) if there is an Hadamard matrix of order $4^{t+2}q$, then there is an Hadamard matrix of order $2 \cdot 4^{t+2}(2\Lambda + 2g + 1)q$.

Proof: [Sketch] Part 1 follows by applying Theorem 11.6 (using an Hadamard matrix of order 2^{2t+1}) to the signed group Hadamard matrices obtained in part 1 of Theorem 11.4 from $2t$ sequences, $X_1 = \left(1, 0_{2\Lambda+1}\right)$ and $X_2 \left(0_\Lambda, 1, 0_{\lambda+1}\right)$ constructed as in Section 11.4.2. For part 3, we use the two complex Golay sequences, U, V to form complex sequences X_U and X_V similar to the real ones obtained in Section 11.4.2, giving $2t + 4$ sequences to be used in part 2 of Theorem 11.4. For (*ii*) and $\Gamma = 3, 5$, we replace $(1, 0_{2\Lambda}, 0, 0_{2\Lambda})$ and $(0, 0_{2\Lambda}, 1, 0_{2\Lambda})$ with complex sequences

$$(1, -, 0_{2(\Lambda-1)}, 0, 0, 0, 0_{2(\Lambda-1)}, 1), \quad (0, 0, 0_{2(\Lambda-1)}, 1, i, 1, 0_{2(\Lambda-1)}, 0),$$

obtained from complex Golay sequences of lengths 3, and

$$\left(i, -i, i, 0_{2(\Lambda-2)}, 0, 0, 0, 0, 0, 0_{2(\Lambda-2)}, 1, 1\right),$$
$$\left(0, 0, 0, 0_{2(\Lambda-2)}, 1, i, i, 1, -i, 0_{2(\Lambda-2)}, 0, 0\right),$$

similarly obtained from complex Golay sequences of length 5. For $\Gamma = 11, 13$ we use the complex Golay sequences of lengths 11 and 13 [30]. Parts 4,5, and 6 follow from Theorem 11.6, with $m = 2 \cdot 4^t q, 4^{t+1} q$, and $4^{t+2} q$, respectively, using the signed group Hadamard matrices obtained in Theorem 11.4. □

Of course, we may replace (ii) and (v) in the statement of the theorem with any odd complex Golay number.

Corollary 11.1: *For any odd positive integer p, there exists a (block-circulant) Hadamard matrix of order* $2^{2N}p$.

This may be viewed as a corollary to either Theorem 11.7 (using Golay sequences of lengths 2^t), or Theorem 11.5 (via Theorem 11.6). The construction for Theorem 11.5 does not even come close to using the full power now available to us, considering the wide variety of sequences known. As a bound on the exponent of 2, however, it provides a strict improvement on Theorem 11.2 whenever $N(p) < \lfloor 2\log_2(p-3) \rfloor / 2$, which is the case for all $p > 5$ except $p = 11, 13, 23, 47$ (in which case equality is attained) and $p = 2^s - 1$ for some s (in which case the exponent given in Theorem 11.2 is one less). On average, $N(p)$ is about $\frac{1}{2}\log_2(p)$, so the exponent we obtain is on average about half the best previous one.

The possibilities for applying Theorem 11.4 seem endless, and this method will be subject to continual refinement as more is understood regarding sequences. For now, we shall be content with an even more convincing demonstration of the power of our method, and the construction of a few Hadamard matrices of relatively small order in Section 11.5.

The following lemma (essentially theorem 4 of [252]) will be helpful.

Lemma 11.2: *If there are base sequences of lengths* $m, m, m+1, m+1$ *and* $n, n, n+1, n+1$, *then there are four* (± 1)*-sequences of length* $(2m+1)(2n+1)$, *having zero autocorrelation. Consequently, there are also four such sequences of length* $2^s(2m+1)(2n+1)$, $s \geq 0$. *In particular, we may choose m and n to be any nonnegative integers up to 30, or any Golay number.*

Theorem 11.8: *If p is an odd positive integer, then there is an Hadamard matrix of order* $2^t p$, *where* $t = 4\lceil \frac{1}{6}\log_2((p-1)/2) \rceil + 2$.

Proof: Base sequences provide $4(\pm 1)$-sequences with zero autocorrelation, the sum of whose lengths is twice any odd number up to $61 (= 2^6 - 3)$. Since 100 and 26 are Golay numbers, there are four sequences whose total length is $252 (= 4 \cdot 63)$. Since we can double the length of any such set of sequences, we have four sequences, the sum of whose lengths is $2^t q$ for any $t > 0$, $q < 63$, and $t > 1$, $q = 63$. In this way we associate four sequences with every nonzero digit of the base 64 expansion of $(p-1)/2$, except when $p = 127 \mod 128$. The sum of the lengths of these sequences will be $p - 1 = 2\Lambda$, and there will be at most $4\lceil \frac{1}{6}\log_2((p-1)/2) \rceil$ sequences in the resulting list.

We now take care of the remaining case. If $p = 127$, the result is known. If $p > 127$, $p \equiv 127 \mod 128$, we argue that the last two digits of $(p-1)/2$ in base 64 represent a number that is half the sum of the lengths of $8\,(\pm 1)$-sequences with zero autocorrelation, as follows.

Write $(p-1)/2 \equiv r \mod 64^2$, $r = 64k + 63$, $0 \leq k \leq 63$. Now as long as $1 \leq k \leq 31$, Lemma 11.2 gives four sequences of length $33k$, and $63 - 2k$ is half the sum of the lengths of base sequences. This gives eight sequences altogether, with total length $4(33k) + 2(63 - 2k) = 2r$. On the other hand, if $34 \leq k \leq 63$ or $k = 32$, Lemma 11.2 gives four (± 1)-sequences of length $33(k-1)$, and $129 - 2k$ is half the sum of the lengths of base sequences, and we have eight sequences with total length $4 \cdot 33(k-1) + 2(129 - 2k) = 2r$.

There remain two cases:

a) $k = 0$, $r = 63 = 52 + 10 + 1$ – half the sum of the lengths of 6 Golay sequences – and
b) $k = 33$, $r = 2175 = 2080 + 64 + 31$ – half the sum of the lengths of 4 Golay sequences and 4 base sequences.

This takes care of the exceptional cases, showing that for $p \neq 127$, $L = (p-1)/2$ is the sum of the lengths of at most $4\lceil \frac{1}{6} \log_2((p-1)/2) \rceil$ (± 1)-sequences with zero autocorrelation. The result follows by an application of part 1 of Theorem 11.7. □

11.4.4 Comments on Orthogonal Matrices Over Signed Groups

Though the method of Theorem 11.8 lacks finesse, it gives an exponent t about $\frac{1}{3}$ the size of that given in Theorem 11.2 when p large. There is a qualitative difference between the two results: in Theorem 11.2 the order of magnitude of 2^t compares to p^2, while in Theorem 11.8, it compares to $p^{\frac{2}{3}}$. So for the first time, we have that the power of two is less significant than the odd factor in the order of Hadamard matrices.

Theorem 11.8 demonstrates that an Hadamard matrix of order $2^t p$ exists with $t \leq 10$ for $p < 8000$ (Theorem 11.2 gives up to $t = 25$ in this range) and with $t \leq 14$ for $p < 500000$ (Theorem 11.2 gives up to $t = 37$). We shall see in Section 11.4.5 that it is a simple matter to significantly improve even on Theorem 11.8 in these ranges with our method.

Seberry's and Craigen's asymptotic formulae for t in terms of q, versus the Hadamard conjecture is given in Figure 11.1.

In [31], it was conjectured that for every even positive integer n, there is a signed group S such that a $SW(n, n-1, S)$ exists. This conjecture can be proved by leaving out the sequence $(1, 0_{n-1})$ in the method of Section 11.4.2. Again we have a bound on the size of S, which is better by a factor of 2 than the bound obtained for $SH(n, S)$. The signed group weighing matrices we obtain are circulant while, in contrast, *ordinary* $W(n, n-1)$ are *never* circulant unless $n = 2$ [29]. There are some handy replication theorems [31] that give infinite classes of $SW(N, N-1, S)$ from these, which in turn give signed group Hadamard matrices $SH(N, S')$, where if S has a remrep of degree m, then S' has a remrep of degree $2m$.

Moreover, simply by making our sequences long enough, we see that it is possible to obtain $SW(n, w, S)$ for any even n and $w \leq n$, where S is a suitable signed group (depending only on w). These are also circulant, and give

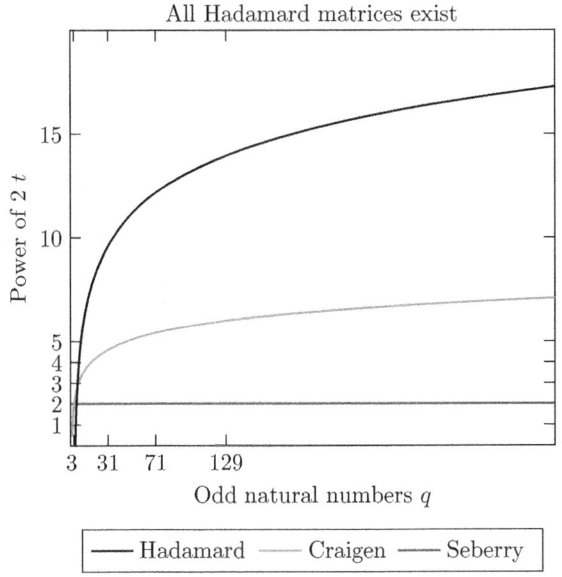

Figure 11.1 Asymptotic support for the Hadamard conjecture. Seberry's and Craigen's asymptotic formulae for t in terms of q, versus the Hadamard conjecture.

block-circulant (ordinary) weighing matrices. Therefore, if we ignore the size of the signed group involved, we have a best possible result for the existence of signed group weighing matrices in even orders: they always exists! What can be shown in odd orders remains to be seen. Such matters will be discussed further elsewhere.

11.4.5 Some Calculations

Here we construct the Hadamard matrix of order $2^6 \cdot 13$ given in Corollary 11.1. Of course this matrix is not new, but it will serve as an example of manageable size.

Now $\frac{(13-1)}{2} = 6$ has binary expansion 110. We therefore use Golay sequences $X = (111-), Y = (11-1)$ of length 4 and $U = (11), V = (1-)$ of length 2. As described in Section 11.4.2, these give disjoint, quasi-symmetric sequences

$$
\begin{aligned}
X_1 &= (10000000000000000000000000) = (1, 0_{12}, 0, 0_{12})\\
X_2 &= (00000000000001000000000000) = (0_{13}, 1, 0_{12})\\
X_3 &= (0111-0000000000000000011-1) = (0, X, 0_8, 0, 0_8, Y),\\
X_4 &= (00000000011-10---100000000) = (0, 0_8, Y, 0, -X, 0_8),\\
X_5 &= (0000011000000000000001-0000) = (0, 0_4, U, 0_6, 0, 0_6, V, 0_4),\\
X_6 &= (00000001-000000000--000000) = (0, 0_6, V, 0_4, 0, 0_4, -U, 0_6),
\end{aligned}
$$

having zero periodic autocorrelation with weight 26.

After one iteration of Lemma 11.1 in Theorem 11.4, we obtain the circulant matrix N_1, which has first row = $(1, 0_{12}, i, 0_{12})$. The remaining four iterations in turn produce $N_2, \dots, N_5$, which have first rows:

$$
\begin{aligned}
&(1juj\underline{u}00000000i00000000uj\underline{u}j), \quad (1juj\underline{u}0000kv\underline{k}vi\underline{vk}vk0000uj\underline{u}j),\\
&(1juj\underline{u}w\ell 00kv\underline{k}vi\underline{vk}vk00\ell\underline{w}uj\underline{u}j), \quad (1juj\underline{u}w\ell x\underline{m}kv\underline{k}vi\underline{vk}vk\underline{mx}\ell\underline{w}uj\underline{u}j),
\end{aligned}
$$

where $\underline{z}$ represents $-z$ in a sequence, and $i, j, k, \ell, m, u, v, w, x$, generate a subgroup S of SP_{2^5} having presentation

$$
\begin{aligned}
&\langle i, j, k, \ell, m, u, v, w, x : i^2 = j^2 = k^2 = \ell^2 = m^2 = u^2 = v^2 = w^2 = x^2 = 1,\\
&\qquad ab = -ba, a, b \in \{i, j, k, \ell, m, u, v, w, x\},\ a \neq b\rangle
\end{aligned}
$$

(these relations can be inferred by examining the construction of Lemma 11.1). Now $N_5 = SH(26, S)$, and S has a remrep of any degree $32q$. So by Theorem 11.6, if there is an Hadamard matrix of order $32q$, there is an Hadamard matrix of order $64 \cdot 13q$. Now one might wonder about the usefulness of this fact, since the existence of an Hadamard matrix of order $4 \cdot 13$ is known to carry this implication without the our method (see [27] or [37]). Suppose, however, that we did not have an Hadamard matrix of order $4 \cdot 13$. Even an Hadamard matrix of order $8 \cdot 13$ would produce no better than $128 \cdot 13q$ by known methods, which can multiply the odd parts of the orders of two Hadamard matrices only by incurring a "penalty" in the exponent of 2, so that the new exponent is one less than the sum of the exponents of the original matrices. But with our method, we would still be able to multiply by q with *no penalty*!

Now let us obtain some *new* Hadamard matrices. We see that $N(1319) = 6$. The stronger version of Seberry's bound [213] gave the best previous Hadamard matrix of order $2^t \cdot 1319$, namely $t = 18$ ($t = 20$ is given by Theorem 11.2), and Corollary 11.1 effortlessly improves this by a factor of 2^6, giving $2^{12} \cdot 1319$. Similarly, Theorem 11.8 gives $2^{10} \cdot 1319$. Now $1319 = 2 \cdot 659 + 1$, and $659 = 2 \cdot 329 + 1$. Since there are 4-complementary sequences of length $329 = 7 \cdot 47$ (these can be, obtained using Lemma 11.2) and 1 is a Golay number, we can take $t = 3$ and $\Lambda = 659$ in part 1 of Theorem 11.7, giving an Hadamard matrix of order $2^8 \cdot 1319$. Better yet: $1319 = 4 \cdot 329 + 3$, and so by doubling the length of the above complementary sequences, we can take $t = 2$ in part 2 of Theorem 11.7, further improving this to $2^7 \cdot 1319$.

Here is a construction that could have been included in [35], illustrating the utility of the multiplication in Theorem 11.6. The construction for theorem 12 of [31] gives $SH(2(g+3), SP_4)$, where g is a Golay number. Hence

$SH(2 \cdot 419, SP_4)$ exists. Now there is an Hadamard matrix of order $4. \cdot 3 = 12$, and so by Theorem 11.6 there is an Hadamard matrix of order $2 \cdot 4 \cdot 3 \cdot 419 = 2^3 \cdot 1257$, which is new.

Table 11.2 lists a few new exponents resulting from the development of signed groups, comparing these to Theorem 11.2 and the best previously known exponent [166] (J. Seberry. Table of Hadamard matrices of order 2'p, p < 4000. unpublished, 1990).

Table 11.2 Exponents for Hadamard matrices of order $2^t p$ from various sources.

	Previous results		Results using signed groups			
p	Theorem 11.9	Best previous	References [26, 28, 29]	(Base 2) Theorem 11.5	(Base 64) Theorem 11.8	t From Theorem 11.7
419	17	4	3	10	10	5
479	17	16	4	16	10	7
491	17	15	5	14	1	7
599	18	8	6	12	10	8
653	18	4	3	10	10	7
659	18	17	4	10	10	6
739	19	16	6	12	10	6
839	19	8	4	12	10	7
1223	20	8	6	12	10	7
1257	20	4	3	12	10	7
1319	20	18	4	12	10	7
1439	20	19	10	16	10	8
1447	20	19	7	14	10	7
1499	21	18	7	16	10	7
1567	21	19	6	14	10	7
1571	21	18	7	10	10	7
1783	21	7	6	18	10	7
1913	21	19	7	16	10	7
1987	21	16	9	14	10	7
2033	21	4	3	16	10	6
2039	21	20	7	20	10	8
2063	22	8	7	10	10	8
2287	22	20	12	16	10	7
2293	22	22	6	6	10	7
2371	22	9		10	10	7
2671	22	9	7	16	10	7
2677	22	9	6	14	10	8
2687	22	21		18	10	9
2699	22	21	7	12	10	9
2879	22	21	7	18	10	9

(Continued)

Table 11.2 (*Continued*)

	Previous results		Results using signed groups			
p	Theorem 11.9	Best previous	References [26, 28, 29]	(Base 2) Theorem 11.5	(Base 64) Theorem 11.8	*t* From Theorem 11.7
2913	23	7	6	12	10	6
2939	23	8	6	18	10	9
2995	23	9	6	16	10	7
2999	23	22		18	10	9
3119	23	21	7	14	10	7
3271	23	21	6	14	10	7
3295	23	18	6	18	10	8
3307	23	20	6	16	10	7
3343	23	22		14	10	8
3359	23	22	7	16	10	8
3371	23	20	7	14	10	7
3437	23	16	6	16	10	7
3539	23	21	6	16	10	7
3547	23	22		18	10	9
3571	23	7	6	18	10	9
3669	23	7	6	14	10	6
3719	23	19	6	14	10	9
3749	23	4	3	14	10	7
3907	23	21	6	14	10	7

Source: Adapted from Craigen [30, pp. 252–253], Reproduced with permission of Elsevier.

11.5 More Asymptotic Theorems

Many far reaching asymptotic theorems are due to Craigen, Kharaghani, Holzmann and Ghaderpour, de Launey, Flannery, and Horadam [32, 46, 48, 79].

These asymptotic theorems have been of the type "given an odd number q, there exists a $t_0 > 0$, so that for every $t \geq t_0$, there exists a cocyclic Hadamard, complex Hadamard, orthogonal design of type $OD(n; s_1, \ldots, s_u)$, regular symmetric Hadamard matrix of order $2^t q$."

There has as yet been only partial such theorem for skew Hadamard matrices or symmetric Hadamard matrices or other kinds of Hadamard matrices.

We give three important research questions posed by Warwick de Launey shortly before he left to work on Erdös' book in Appendix C.

11.6 Skew Hadamard and Regular Hadamard

This is taken from the original chapter of [166]. We do not prove them here.

The first asymptotic existence theorem, which was given by Seberry–Wallis in 1976, uses a "plug in" technique to give a result for every odd q:

Theorem 11.9 ***(Seberry [213]):*** *Let q be any odd natural number. Then there exists a $t(\leq [2\log_2(q-3)]+1)$ so that there is an Hadamard matrix of order $2^t q$. (The best known bounds are $t \leq [\log_2(q-3)(q-7)-1]$ for $q(\text{prime}) \equiv 3 \pmod 4$ and $t \leq [\log_2(q-1)(q-5)]+1$ for $p(\text{prime}) \equiv 1 \pmod 4$).*

The proof of this theorem allows a number of special cases of interest and stronger results in some cases where q is not prime (for a similar case example).

Corollary 11.2 ***(Seberry [153]):*** *Let q be any odd natural number. Then there exists a regular symmetric Hadamard matrix with constant diagonal of order $2^{2t}q^2$, $t \leq [2\log_2(q-3)]+1$.*

Corollary 11.3 ***(Seberry [153]):***

i) Let p and $p+2$ be twin prime powers. Then there exists a $t \leq [\log_2(p+3)(p-1)(p^2+2p-7)]-2$ so that there is an Hadamard matrix of order $2^t p(p+2)$.
ii) Let $p+1$ be the order of a symmetric Hadamard matrix. Then there exists a $t \leq [\log_2(p-3)(p-7)]-2$ so that there is an Hadamard matrix of order $2^t p$.

Corollary 11.4 ***(Seberry [153]):*** *Let $Q = pq$ be an odd natural number. Suppose all $OD(2^s p; 2^r a, 2^r b, 2^r c)$ exist, $s \geq s_0$, $2^{s-r}p = a+b+c$. Then there exists an Hadamard matrix of order $2^t \cdot p \cdot q, s \leq t \leq [2\ \log_e((q-3)/p)]+r+1$. (The best known bounds are $s \leq t \leq [\log_e((q-3)(q-7)/p)]-1+r$ for $q(prime) \equiv 3 \pmod 4$ and $st \leq [\log_e((q-1)(q-5)/p)]+r+1$ for $q(prime) \equiv 1 \pmod 4$.)*

Example 11.2: *Let $q = 3 \cdot 491$. We know there is an Hadamard matrix of order 12 and the best bound of the theorem gives an Hadamard matrix of order $2^{15} \cdot 491$. So there is an Hadamard matrix of order $2^{16} \cdot 3 \cdot 491$ using the multiplication theorem.*

The best value from the corollary gives an Hadamard matrix of order $2^{13} \cdot 3 \cdot 491$ from using the $OD(2^{12} \cdot 3; 22, 3, 2^{12} \cdot 3 - 25)$.

Other similar results are known. The Table A.17 gives an indication of the smallest t for each odd natural number q for which an Hadamard matrix is known.

A list of the construction methods used is given in Section A.1 of the Appendix A.

The results for skew Hadamard matrices are far less complete than for Hadamard matrices.

Theorem 11.10 ***(Seberry [153],[166, p. 497]):*** *Let $q \equiv 5 \pmod 8$ be a prime power and $p = \frac{1}{2}(q+1)$ be a prime. Then there is a skew Hadamard matrix of order $2^t p$ where $t \leq \lfloor 2\log_2(p-2) \rfloor$.*

12

More on Maximal Determinant Matrices

12.1 Notations

Table 12.1 gives the notations which are used in this chapter.

12.2 *E*-Equivalence: The Smith Normal Form

The following theorem is due to Smith [171] and has been reworded from the theorems and proofs in MacDuffee [131, p. 41] and Marcus and Minc [132, p. 44].

Definition 12.1: Let E be a Euclidean domain. Two matrices X and Y are **E-equivalent** or **Euclidean equivalent** if one can be obtained from the other by:

i) interchanging rows/columns;
ii) adding a multiple of a row/column to another row/column.

We denote by $X^{\mathbf{E}}Y$ that X is Euclidean equivalent to Y. When $\boldsymbol{E}$ is the rational integer ring $\boldsymbol{E}$, we denote by $X^{\mathbf{E}}Y$.

Theorem 12.1 : *If A is any integer-matrix of order n and rank r, then there is a unique matrix*

$$D = \text{diag}(a_1, a_2, \dots, a_r, 0, \dots, 0),$$

such that $A^{\mathbf{E}}D$ *and* $a_i | a_{i+1}$ *where the* a_i *are nonnegative. The greatest common divisor of the* $i \times i$ *sub-determinants of A is*

$$a_1 a_2 a_3 \dots a_i.$$

If $A^{\mathbf{E}}E$ *where*

$$E = \left[\begin{array}{ccccc|c} a_1 & & & & & \\ & a_2 & & & & \\ & & a_3 & & & \\ & & & \ddots & & \\ & & & & a_i & \\ \hline & & & & & F \end{array}\right]$$

then a_{i+1} *is the greatest common divisor of the nonzero elements of the remaining submatrix F.*

Hadamard Matrices: Constructions using Number Theory and Algebra, First Edition. Jennifer Seberry and Mieko Yamada.

Table 12.1 Notations used this chapter.

$X^E Y$	X is Euclidean equivalent to Y where $\boldsymbol{E}$ is a Euclidean domain
$A^E B$	A is $\boldsymbol{E}$-equivalent to B where $\boldsymbol{E}$ is the rational integer ring
SNF	Smith normal form
D_k	D-optimal matrix of order k
H_k	An Hadamard matrix of order n
M_k	The absolute value of $k \times k$ minor
$g(n, A)$	The growth factor on a matrix A
CP	Completely pivoted
p_j	Pivot
$A(j)$	The absolute value of $j \times j$ principal minor is a rational integer ring

Definition 12.2: The a_i of Theorem 12.1 are called the invariants of A, and the diagonal matrix D is called the **Smith normal form** or **SNF**.

Definition 12.3: Two Hadamard matrices are said to have **equivalent SNF**, if one can be obtained from the other by a series of operations

1) add an integer multiple of one row to another,
2) negate some row,
3) reorder the rows

and the corresponding column operations. The entries in the SNF are called **invariants**.

The following result of Jennifer Wallis and W.D. Wallis [215] for odd m has been reproved and strengthened by Newman, Spence, and W.D. Wallis.

Theorem 12.2: *Any Hadamard matrix of order* $4m$*, where* m *is square-free, is* $\boldsymbol{E}$*-equivalent to*

$$\operatorname{diag}\left(1, \underbrace{2, 2, \dots, 2}_{2m-1 \text{ times}}, \underbrace{2m, 2m, \dots, 2m}_{2m-1 \text{ times}}, 4m\right).$$

Proof: By Theorem 12.1 an Hadamard matrix, H of order $4m$, is $\boldsymbol{E}$-equivalent to $\operatorname{diag}(a_1, a_2, \dots, a_{4m})$ where $a_i | a_{i+1}$. Write

$$PHQ = \operatorname{diag}\left(a_1, a_2, \dots, a_{4m}\right). \tag{12.1}$$

Now

$$HH^\top = 4mI$$

So

$$(PHQ)\left(Q^{-1}H^\top P^{-1}\right) = 4mI,$$

which, by (12.1) implies that

$$Q^{-1}H^\top P^{-1} = \operatorname{diag}(4m/a_1, 4m/a_2, \dots, 4m/a_{4m}). \tag{12.2}$$

However, it is clear that H and $H^{\top}$ have the same invariant factors. Thus, since $4m/a_{i+1}|4m/a_i$ $(1 \le i \le 4m)$, it follows at once that the invariant factors of $H^{\top}$ are $(4m/a_{4m}, 4m/a_{4m-1}, \dots, 4m/a_1)$ which can be identified with $(a_1 a_2, \dots, a_{4m})$.

Consequently

$$a_{4m+1-i} a_i = 4m. \tag{12.3}$$

Using the fact that $a_i|a_{i+1}$ and $a_{2m}a_{2m+1} = 4m$ from (12.3), it is seen that

$$4m \equiv 0 \pmod{a_{2m}^2}.$$

Thus if m is square-free, $a_{2m} = 1$ or 2. However, since a_2 is the greatest common divisor of the two rowed minors of H, clearly $a_2 = 2$, for any 2×2 matrix with entries ± 1 has determinant 0 or ± 2. It follows that $a_{2m} = 2$ and the invariant factors of H are

$$1, \underbrace{2, 2, \dots, 2}_{2m-1 \text{ times}}, \underbrace{2m, 2m, \dots, 2m}_{2m-1 \text{ times}}, 4m.$$

□

Corollary 12.1: *Every Hadamard matrix of order 4m, m square-free, is **E**-equivalent to*

$$\mathrm{diag}(a_1, a_2, \dots, a_{4m})$$

where $a_1 = 1$, $a_2 = 2$, $a_{4m-1} = 2m$, $a_{4m} = 4m$, $a_i|a_{i+1}$, *and* $a_i a_{4m+1-i} = 4m$.

Definition 12.4: The diagonal matrix of order $4m$

$$\mathrm{diag}\left(1, \underbrace{2, 2, \dots, 2}_{2m-1 \text{ times}}, \underbrace{2m, 2m, \dots, 2m}_{2m-1 \text{ times}}, 4m\right)$$

will be called the **standard form**.

The following result can be obtained by noting that the rank of the incidence matrix of a (v, k, λ)-configuration is v (see Marcus and Minc [132, p. 46]):

Theorem 12.3: *The incidence matrix of a (v, k, λ)-configuration is equivalent to a diagonal matrix with entries*

$$\begin{cases} 1 & \frac{1}{2}(v+1) \text{ times,} \\ (k-\lambda) & \frac{1}{2}(v-3) \text{ times,} \\ k(k-\lambda) & \text{once,} \end{cases}$$

when $k - \lambda$ is square-free and $(k - \lambda, k) = 1$.

12.3 *E*-Equivalence: The Number of Small Invariants

We know from Corollary 12.1 that an Hadamard matrix has exactly one invariant equal to 1, and that the next invariant is 2. In this section, a lower limit for the number of invariants equal to 2 is found as a consequence of a general result of independent interest. We write $[x]$ for the largest integer not exceeding x.

Theorem 12.4 ***(J. Wallis [214]):*** *Suppose B is an $n \times n$ matrix with nonzero determinant whose entries are all 0 and 1, the zero and identity elements of a Euclidean domain* **E**. *Then the number of invariants of B under* **E***-equivalence (see Definition 12.1) which equal 1 is at least*

$$[\log_2 n] + 1.$$

Proof: Write t for $[\log_2 n]$; that is, t is the unique integer such that $2^t \leq n \leq 2^{t+1}$. We shall use a sequence of equivalence operations to transform B to

$$\begin{bmatrix} I_t & 0 \\ 0 & D \end{bmatrix},$$

where 1 is a greatest common divisor of the entries in D.

The first part of the process is to reorder the rows and columns of B so that, in the reordered matrix, columns 1 and 2 have different entries in the first row, columns 3 and 4 are identical in the first row but have different entries in the second row, and in general columns $2i - 1$ and $2i$ are identical in the first $i - 1$ rows but differ in row i for $i = 1, 2, \dots, t$. This is done using the following algorithm.

Step 1. Select two columns of B which have different entries in the first row. (If the first row of B has every entry 1, it will first be necessary to choose some other row as row 1.) Reorder the columns so that the two chosen columns become columns 1 and 2 (in either order).

Step 2. Select two columns of the matrix just formed, neither of them being columns 1 or 2, which have identical entries in the first row. Reorder the rows of the matrix other than row 1 so that the two columns chosen have different entries in the new second row. Reorder columns after column 2 so that the new pair become columns 3 and 4.

Step *k*. In the matrix resulting from step $k - 1$, select two columns to the right of column $2k - 2$ which are identical in rows 1 to $k - 1$. Reorder the rows after row $k - 1$ and the columns after column $2k - 2$ so that the chosen columns become columns $2k - 1$ and $2k$ and differ in their kth row.

It is always possible in step k to find a row in which the two chosen columns differ, since the matrix cannot have two identical columns. Therefore, step k only requires that we can choose two columns from the $n - 2k + 2$ available ones which are identical in their first $k - 1$ places. Since there are only $2k - 1$ different $(0, 1)$-vectors of length $k - 1$, this will be possible provided

$$2^{k-1} < n - 2k + 2,$$

and this is always true for $1 \leq k \leq t$ except when $k = t$ and $n = 4$ or $n = 8$. In the case $n = 4$, it is easy to check by hand that every $(0, 1)$ matrix of nonzero determinant is **E**-equivalent to a matrix on which t steps can be carried out. If $n = 8$, step 3 will be impossible if the first two rows of the last four columns contain all $(0, 1)$-vectors of length 2; typically the first two rows are

$$\begin{matrix} 1 & 0 & 1 & 1 & 1 & 1 & 0 & 0 \\ * & * & 1 & 0 & 1 & 0 & 1 & 0 \end{matrix}$$

and in this case we can proceed with step 3 if we first apply the column permutation (37)(48). Consequently t steps can always be carried out.

In the second stage, select if possible two columns (columns a and b say) which are identical in rows 1 to t; reorder the later rows so that columns a and b differ in row $t + 1$. If the selection was impossible then $n = 2^t$ and the first t rows of the matrix constitute the 2^t different column vectors, so one column (column a say) will start with t zeros; reorder the rows from $t + 1$ on so that there is a 1 in the $(a, t + 1)$ position. In either case, if column

a were in a pair chosen in stage 1, reorder the pair if necessary so that a is even; and similarly for b if two were chosen.

The third stage isolates certain entries ± 1 by carrying out t steps:

Step 1. Subtract column 2 from column 1, so that the $(1, 1)$ entry becomes ± 1. Then add a suitable multiple of the first column to every other column to ensure that row 1 has every entry 0 except the first, and similarly eliminate all entries except the first from column 1 by adding suitable multiples of row 1 to the other rows.

After step 1 the matrix has first row and column $(\pm 1, 0, 0, \dots, 0)$; whatever multiple of column 1 was added to column $2k - 1 (2 \leq k \leq t)$, the same multiple was added to column $2k$, so that the $(k, 2k - 1)$ and $(k, 2k)$ entries still differ by 1. It will be seen from the description of the general step that, after $k - 1$ steps, the matrix will have its first $k - 1$ rows zero except for entries ± 1 in the $(i, 2i - 1)$ positions, $1 \leq i \leq k - 1$; the first $k - 1$ odd-numbered columns will be zero except at those positions; and for $k \leq i \leq t$ the $(i, 2i - 1)$ and $(i, 2i)$ entries differ by 1 and the $(j, 2i - 1)$ and $(j, 2i)$ entries are equal when $j < i$. After step k this description can be extended by replacing $k - 1$ by k.

Step k. Subtract column $2k$ from column $2k - 1$, so that the $(k, 2k - 1)$ entry becomes ± 1. Add a suitable multiple of column $2k - 1$ to every subsequent column, and then add suitable multiples of row k to the later rows, so that row k and column $2k - 1$ become zero except at their intersection.

Observe that if $k < i \leq t$ the $(k, 2i - 1)$ and $(k, 2i)$ entries were equal before step k, so the same multiple of column $2k - 1$ was added to both columns $2i - 1$ and $2i$ and the differences between these columns is unchanged. If two columns, a and b, were chosen at stage 2, then the difference between those columns is unaltered in the t steps; if only one column was chosen then that column is unaltered in the t steps since it has never had a nonzero entry in its kth row to be eliminated.

Finally, reorder the columns so that the former columns $1, 3, \dots, 2t - 1$ become the first t columns. We obtain

$$\begin{bmatrix} I_t & 0 \\ 0 & D \end{bmatrix};$$

D contains either two entries which differ by 1 (corresponding to the former $(a, t + 1)$ and $(b, t + 1)$ entries) or has an entry 1 (the former $(a, t + 1)$ entry) depending on the course followed at stage 2, and in either case the greatest common divisor of entries of D is 1.

Therefore, B has at least $t + 1$ invariants equal to 1. □

Corollary 12.2: *An Hadamard matrix of order $4m$ has at least*

$$\lfloor \log_2(4m - 1) \rfloor + 1$$

invariants equal to 2, and by Theorem 12.4 it has at least this number of invariants equal to $2m$.

Proof: Let A be an Hadamard matrix of order $4m$; assume A to be normalized. Subtract row 1 from every other row and them column 1 from every other column; we obtain

$$A = \begin{bmatrix} 1 & 0 \\ 0 & -2B \end{bmatrix}$$

where B is an $(0, 1)$-matrix of size $4m - 1$ with nonzero determinant. The first invariant of A is 1; the others are double the invariants of B. The result follows from Theorem 12.4. □

Theorem 12.5 *(J. Wallis [214]):* *Suppose **E** is a Euclidean domain with characteristic not equal to 2, and suppose f and g are monotonic nondecreasing functions which satisfy:*

i) any Hadamard matrix of order N has at least f(N) invariants equal to 2;
*ii) any (0, 1)-matrix over **E** which has nonzero determinant and is of size r × r has at least g(r) invariants equal to 1.*

Then

$$f(N) \le \lceil \log_2(N-1) \rceil + 1 \tag{12.4}$$

*and, if 2 is a non-unit of **E**,*

$$g(r) \le \lceil \log_2 r \rceil + 1.$$

Proof: The function on the right hand side of (12.4) is a step-function which increases in value just after N takes as its value a power of 2. So, if (12.4) is false, we must have

$$f\left(2^t\right) > \left\lceil \log_2\left(2^t - 1\right) \right\rceil + 1 = t$$

for some t. However, for every t, there is an Hadamard matrix A of order 2^t which has precisely t invariants 2. Let

$$H = \begin{bmatrix} 1 & 1 \\ 1 & -1 \end{bmatrix}$$

which has invariants $\{1, 2\}$, and define A as the direct product of t copies of H.

Then $A \sim D$, where D is the direct product of t copies of diag(1, 2); D is a diagonal matrix whose entries are powers of 2, and 2^a occurs $\binom{t}{a}$ times. These must be the invariants of A. A has precisely t invariants equal to 2.

If we consider A as a matrix over $\mathbf{E}$, rather than an integer matrix, and pass to B as in the proof of Corollary 12.2, then B is a (0, 1)-matrix over $\mathbf{E}$ of size $r = 2^t - 1$, and has nonzero determinant (as the characteristic of $\mathbf{E}$ is not 2). The invariants of B are 2^a, $\binom{t}{a+1}$ times each, for $a = 0, 1, \dots, t-1$. 2^a and 2^b are the same invariant if and only if 2^{b-a} is a unit of $\mathbf{E}$, and this cannot occur when $a \ne b$ provided 2 is a non-unit. So, the matrix B can be used to prove the part of the theorem involving g. □

Theorem 12.5 shows that the results of Theorem 12.4 and Corollary 12.2 are best possible in a certain sense unless $\mathbf{E}$ has characteristic 2 or 2 is a unit of $\mathbf{E}$. If 2 were a unit then the matrix B has every invariant 2 (or 1, which is the same thing), and if $\mathbf{E}$ had characteristic 2 then we could not divide by 2 to get B.

12.4 *E*-Equivalence: Skew Hadamard and Symmetric Conference Matrices

Theorem 12.6 ***(J. Wallis [214]):*** *If there is a skew Hadamard matrix of order n then there is a skew Hadamard matrix of order 2n **E**-equivalent to the standard form.*

Proof: The theorem is easily proven when $n = 1$ or 2, so put $n = 4m$. Suppose A is a skew Hadamard matrix with canonical diagonal matrix D; suppose P and Q are unimodular integral matrices such that

$$D = PAQ.$$

Then

$$Q^{-1}A^{\top}P^{-1} = nD^{-1},$$

and nD^{-1} is the matrix D with the order of its entries reversed (12.2). For convenience write

$$\begin{bmatrix} 1 & & \\ & 2C & \\ & & 4m \end{bmatrix};$$

C is a diagonal integral matrix of order $n - 2$.

Consider the matrix

$$K = \begin{bmatrix} A & A \\ -A^{\top} & A^{\top} \end{bmatrix}$$

which is skew Hadamard of order $2n$. K is equivalent to

$$\begin{bmatrix} P & 0 \\ Q^{-1} & -Q^{-1} \end{bmatrix} \begin{bmatrix} A & A \\ -A^{\top} & A^{\top} \end{bmatrix} \begin{bmatrix} Q & P^{-1} \\ 0 & -P^{-1} \end{bmatrix}$$

$$= \begin{bmatrix} PAQ & 0 \\ Q^{-1}\left(A + A^{\top}\right) Q & 2Q^{-1}A^{\top}P^{-1} \end{bmatrix}$$

$$= \begin{bmatrix} D & 0 \\ 2I & 2nD^{-1} \end{bmatrix}$$

using the fact that $A + A^{\top} = 2I$. This last matrix is

$$\left[\begin{array}{c|c|c|c|c|c} 1 & & & & & \\ \hline & 2C & & & & \\ \hline & & n & & & \\ \hline 2 & & & 2n & & \\ \hline & 2I & & & nC^{-1} & \\ \hline & & 2 & & & 2 \end{array}\right].$$

Subtract twice row 1 from row $n + 1$; subtract column $2n$ from column n; then we can reorder the columns and rows to obtain

$$\left[\begin{array}{c|c|c|c|c|c} 1 & & & & & \\ \hline & 2 & & & & \\ \hline & & n & & & \\ \hline & & & 2n & & \\ \hline & & & & 2C & 0 \\ \hline & & & & 2I & nC^{-1} \end{array}\right].$$

Every entry of C divides $2n$, so $\frac{1}{2}nC^{-1}$ is integral. So the second direct summand is integrally equivalent to

$$\begin{bmatrix} -I & C \\ 0 & I \end{bmatrix} \begin{bmatrix} 2C & 0 \\ 2I & nC^{-1} \end{bmatrix} \begin{bmatrix} -\frac{1}{2}nC^{-1} & I \\ I & 0 \end{bmatrix} = \begin{bmatrix} nI & 0 \\ 0 & 2I \end{bmatrix}$$

and the invariants of K are as required. □

Lemma 12.1: *If there is an Hadamard matrix of order $n = 8m$, then there is an Hadamard matrix of order $2n$ with at least $12m - 1$ invariants divisible by 4.*

Proof: If A is Hadamard of order n and has canonical diagonal matrix D, then

$$H = \begin{bmatrix} A & A \\ -A & A \end{bmatrix}$$

is Hadamard of order $2n$ and is equivalent to the diagonal matrix

$$D' = \begin{bmatrix} D & 0 \\ 0 & 2D \end{bmatrix}.$$

Since $n = 8m$, the last $4m$ invariants of A are divisible by 4. Every entry of $2D$ except the first is divisible by 4. So D' has at least $12m - 1$ entries divisible by 4. Even if D' is not in canonical form, it is easy to deduce that D' (and consequently H) has at least $12m - 1$ invariants divisible by 4. □

We note

Theorem 12.7 ***(J. Wallis [214]):*** *If there is a symmetric conference matrix of order n then there exists an Hadamard matrix of order 2n which is E-equivalent to the standard form.*

12.5 Smith Normal Form for Powers of 2

We note that the SNF of the inverse of an Hadamard matrix is the same (up to a constant) as that of Hadamard matrix. In the proof they appear in reverse order but to satisfy the divisibility property they must be reordered. This was observed by Spence [173]. W.D. Wallis and Jennifer Seberry (Wallis) [215] showed that for an Hadamard matrix, H, of order $4t$, where t is square free, the SNF of H is in standard form. More recently, T.S. Michael and W.D.Wallis [136] have shown that all skew Hadamard matrices have SNF in standard form.

However, the powers of 2 are different. Marshall Hall [91] gave five in-equivalence classes of Hadamard matrices of order 16: HI, HII, $HIII$, HIV, $HV = HIV^T$. These have SNF

$$\text{diag}\,(1, \underbrace{2, 2, \dots, 2}_{4}, \underbrace{4, 4, \dots, 4}_{6}, \underbrace{8, 8, \dots, 8}_{4}, 16).$$

$$\text{diag}\,(1, \underbrace{2, 2, \dots, 2}_{5}, \underbrace{4, 4, \dots, 4}_{4}, \underbrace{8, 8, \dots, 8}_{5}, 16).$$

$$\text{diag}\,(1, \underbrace{2, 2, \dots, 2}_{6}, \underbrace{4, 4, \dots, 4}_{2}, \underbrace{8, 8, \dots, 8}_{6}, 16).$$

and for HIV and HV

$$\text{diag}\,(1, \underbrace{2, 2, \dots, 2}_{7}, \underbrace{8, 8, \dots, 8}_{7}, 16).$$

The Sylvester–Hadamard matrix of order 16 belongs to Marshall Hall's class HI. It is known that the number of 2s in the SNF of an Hadamard matrix of order $4t$ is $\geq \log_2 4t$.

In fact the Sylvester–Hadamard matrix of order 2^t always has exactly t 2's in its SNF.

Lemma 12.2: *Assume that the SNF of the matrix A is*

$$D = \text{diag}\,(a_1, a_2, \dots, a_r, 0, \dots 0).$$

Then the SNF of $\left[\begin{smallmatrix} A & A \\ A & -A \end{smallmatrix}\right]$ *is comprised of*

$$\text{diag}\,(a_1, a_2, \dots, a_r, 2a_1, 2a_2, \dots, 2a_r, 0, \dots 0).$$

(These may have to be reordered.)

Remark 12.1: The number of occurrences of 2^r in the SNF of the Sylvester matrix of order 2^k is $\binom{k}{r}$ for $0 \leq r \leq k$.

Example 12.1: *We show the steps to form the SNF of an Hadamard matrix by Gaussian elimination:*

$$\begin{bmatrix} 1 & 1 & 1 & 1 \\ - & 1 & 1 & - \\ - & - & 1 & 1 \\ - & 1 & - & 1 \end{bmatrix} \rightarrow \begin{bmatrix} 1 & 1 & 1 & 1 \\ 0 & 2 & 2 & 0 \\ 0 & 0 & 2 & 2 \\ 0 & 2 & 0 & 2 \end{bmatrix} \rightarrow \begin{bmatrix} 1 & 1 & 1 & 1 \\ 0 & 2 & 2 & 0 \\ 0 & 0 & 2 & 2 \\ 0 & 0 & \bar{2} & 2 \end{bmatrix} \rightarrow \begin{bmatrix} 1 & 1 & 1 & 1 \\ 0 & 2 & 2 & 0 \\ 0 & 0 & 2 & 2 \\ 0 & 0 & 0 & 4 \end{bmatrix}.$$

So the SNF is diag $(1, 2, 2, 4)$.

12.6 Matrices with Elements $(1, -1)$ and Maximal Determinant

We recall the original interest in Hadamard matrices was that an Hadamard matrix $H = (h_{ij})$ of order n satisfies the equality of Hadamard's inequality that

$$(\det H)^2 \le \prod_{j=1}^{n} \sum_{i=1}^{n} |h_{ij}|^2$$

for elements in the unit circle.

This has led to further studies into the maximum of the determinant of $(1, -1)$ matrices of any order. This problem was first brought to the attention of one of us (Seberry) by a 1970 report for the USAF by Stanley Payne at Dayton, Ohio.

Definition 12.5: A ***D*-optimal matrix** of order n is an $n \times n$ matrix with entries ± 1 having maximum determinant.

Like Hadamard matrices, we can always write the D-optimal matrix in a normalized form. It is well known that Hadamard matrices of order n have absolute value of determinant $n^{n/2}$ and thus are D-optimal matrices for $n \equiv 0 \pmod 4$.

Remark 12.2: Since some authors use the term D-optimal design for $(1, -1)$ matrices of order n and form

$$\begin{bmatrix} A & B \\ B^\top & -A^\top \end{bmatrix}$$

where $AA^\top + BB^\top = (2n-2)I + 2J$ [73] we have defined the D-optimal matrix as above, to be an $n \times n$ matrix of $(1, -1)$ elements with maximal determinant.

It is simple to show that the matrices

$$\begin{bmatrix} 1 & 1 \\ 1 & - \end{bmatrix} \quad \begin{bmatrix} 1 & 1 & 1 \\ 1 & - & 1 \\ - & 1 & 1 \end{bmatrix} \quad \begin{bmatrix} 1 & 1 & 1 & 1 \\ 1 & - & 1 & - \\ 1 & 1 & - & - \\ 1 & - & - & 1 \end{bmatrix}$$

have maximum determinant for n= 2, 3, 4. We call these matrices (or their Hadamard equivalent) D_2, D_3, and D_4. The following result ensures the existence of the D-maximal matrix of order 4 with determinant 16 in every Hadamard matrix.

Remark 12.3: Since D_2 and D_3 are embedded in D_4, Theorem 1.13 implies that every Hadamard matrix of order ≥ 4 contains a sub-matrix equivalent to D_2 and D_3.

Now we explore embedding D-optimal matrices of orders $m = 5, 6, 7$, and 8 in Hadamard matrices of order n.

Notation 12.6: We write H_j for an Hadamard matrix of order j, S_j for the Sylvester–Hadamard matrix of order j and D_j for a D-optimal matrix of order j. The notation $D_j \in H_n$ is used to say "D_j is embedded in some H_n." Whenever the word determinant or minor is mentioned in this work, we mean its absolute value.

12.7 *D*-Optimal Matrices Embedded in Hadamard Matrices

Using the research of H. Kharaghani and W. Orrick [107], J. Seberry and M. Mitrouli [162] we see combinatorial methods may be used to show that the unique, up to equivalence, 5×5 sub-matrix of elements ± 1 with determinant 48, the unique, up to equivalence, 6×6 sub-matrix of elements ± 1 with determinant 160, and the unique, up to equivalence, 7×7 sub-matrix of elements ± 1 with determinant 576 cannot be embedded in the Hadamard matrix of order 8. H. Kharaghani and W. Orrick [107] give the unique D-maximal matrices under Hadamard equivalence operations.

The 5×5 $(1, -1)$ matrix with maximal determinant 48 is

$$D_5 = \begin{bmatrix} 1 & 1 & 1 & 1 & 1 \\ 1 & - & 1 & - & - \\ 1 & 1 & - & - & - \\ 1 & - & - & 1 & - \\ 1 & - & - & - & 1 \end{bmatrix};$$

the 6×6 $(1, -1)$ matrix with maximal determinant 160 is

$$D_6 = \begin{bmatrix} 1 & 1 & 1 & 1 & 1 & 1 \\ 1 & - & 1 & - & - & 1 \\ 1 & 1 & - & - & - & 1 \\ 1 & - & - & 1 & - & 1 \\ 1 & 1 & 1 & 1 & - & - \\ 1 & - & - & - & 1 & - \end{bmatrix};$$

and the 7×7 $(1, -1)$ matrix with maximal determinant 576 is

$$D_7 = \begin{bmatrix} 1 & 1 & 1 & 1 & 1 & 1 & 1 \\ 1 & 1 & - & - & - & 1 & 1 \\ 1 & - & 1 & - & - & 1 & 1 \\ 1 & - & - & 1 & 1 & - & 1 \\ 1 & - & - & 1 & 1 & 1 & - \\ 1 & 1 & 1 & - & 1 & - & - \\ 1 & 1 & 1 & 1 & - & - & - \end{bmatrix}.$$

The $(1, -1)$ with maximal determinant 4096 is the Hadamard matrix of order 8.

12.7.1 Embedding of D_5 in H_8

Lemma 12.3: *The D-optimal matrix of order 5 (D_5) is not embedded into an Hadamard matrix of order 8 (H_8).*

Proof: We attempt to extend

$$\begin{bmatrix} 1 & 1 & 1 & 1 & 1 \\ 1 & - & 1 & - & - \\ 1 & 1 & - & - & - \\ 1 & - & - & 1 & - \\ 1 & - & - & - & 1 \end{bmatrix}$$

to H_8. Without loss of generality we choose $h_{16} = h_{17} = h_{18} = 1$. We note rows 2, 3, 4, 5 each contain three -1s and two 1s, so for them to be orthogonal with the first row each needs to be extended by one -1 and two 1s. We also note the mutual inner product of rows 2, 3, 4, 5 is $+1$. It is not possible to extend them by choosing the single -1 in individual columns as there are four rows and three columns. Hence by the pigeon hole principle this is impossible. Hence D_5 does not exist embedded into an Hadamard matrix of order 8. □

We note Edelman and Mascarenhas [66] and Seberry et al. [164] have shown $D_5 \in H_{12}$.

12.7.2 Embedding of D_6 in H_8

Lemma 12.4: *The D-optimal matrix of order 6 (D_6) is not embedded into an Hadamard matrix of order 8 (H_8).*

Proof: We extend partially the 6×6 matrix, D_6 by adding two columns. We note without any loss of generality we may choose $h_{17} = h_{18} = h_{27} = -h_{28} = 1 = h_{67} = h_{68}$ so we have

$$H_8^{\text{partial}} = \begin{bmatrix} 1 & 1 & 1 & 1 & 1 & 1 & 1 & 1 \\ 1 & - & 1 & - & - & 1 & 1 & - \\ 1 & 1 & - & - & - & 1 & h_{37} & h_{38} \\ 1 & - & - & 1 & - & 1 & h_{47} & h_{48} \\ 1 & 1 & 1 & 1 & - & - & h_{57} & h_{58} \\ 1 & - & - & - & 1 & - & 1 & 1 \end{bmatrix}.$$

Now the inner product of row 5 with row 1 gives $h_{57} + h_{58} = -2$ while the inner product of row 5 and row 2 gives $h_{57} - h_{58} = 0$. So $h_{57} = h_{58} = -1$. But now the inner product of rows 5 and 6 cannot be zero, so D_6 cannot be extended to H_8. □

12.7.3 Embedding of D_7 in H_8

Lemma 12.5: *The D-optimal matrix of order 7 (D_7) is not embedded into an Hadamard matrix of order 8 (H_8).*

Proof: To embed D_7 in H_8, it is merely necessary to note that for H_8 we can choose $h_{18} = 1$. We note that every other row of H_8 must have 4 ones and 4 -1s so we extend D_7 thus:

$$D_7^{\text{extended}} = \left[\begin{array}{ccccccc|c} 1 & 1 & 1 & 1 & 1 & 1 & 1 & 1 \\ 1 & 1 & - & - & - & 1 & 1 & - \\ 1 & - & 1 & - & - & 1 & 1 & - \\ 1 & - & - & 1 & 1 & - & 1 & - \\ 1 & - & - & 1 & 1 & 1 & - & - \\ 1 & 1 & 1 & - & 1 & - & - & - \\ 1 & 1 & 1 & 1 & - & - & - & - \end{array}\right]$$

But rows 2 and 3 are not orthogonal so the result is not possible. □

Remark 12.4: Since all the 7×7 minors of the 8×8 Hadamard matrix are equal to 512 (an immediate consequence of Hadamard definition), this also proves the Lemma 12.5.

12.7.4 Other Embeddings

M. Mitrouli and J. Seberry searched for *D*-optimal matrices of order $m = 5$, 6, 7, and 8 embedded into classes of Hadamard matrices. By selecting m rows and columns of the Hadamard matrices tested, they checked if the determinants of the $m \times m$ sub-matrices are equal to det(D_m). More specifically, they searched the full list of Hadamard matrices of orders $n = 12, 16, 20, 24$, and 28 for this purpose. There are exactly 1, 5, 3, 60, and 487 inequivalent Hadamard matrices respectively, in each order. Their findings are summarized in Table 12.2.

By examining Table 12.2 one notices that

- The D-optimal matrix of order $m = 5$ is embedded in all Hadamard matrices of the specific orders we study.
- The D-optimal matrix of order $m = 6$ is embedded in almost all Hadamard matrices of the specific orders we study. It is not embedded into one Hadamard matrix of order $n = 16$: the Sylvester-Hadamard matrix.
- The D-optimal matrix of order $m = 7$ is embedded in all Hadamard matrices of the specific orders we study. It is not embedded in two Hadamard matrices of order $n = 16$, one of which is the Sylvester-Hadamard matrix.
- The D-optimal matrix of order $m = 8$ (i.e. the Hadamard matrix of order 8) is embedded in all Hadamard matrices we study except for the Hadamard matrix of order $n = 12$.

In Table 12.3 we summarise the above results with some extensions and conjectures added in the last column.

The above results and the posed conjectures are indicative of the following Warwick de Launey's theorem (W. de Launey. Private conversation. Seville, Spain, 2007).

Theorem 12.8: *For every ± 1 sub-matrix there exists an n_0, large enough, so that the sub-matrix is $\in H_{n_0}$.*

Remark 12.5: Using the same combinatorial methods as used to show $D_7 \notin H_8$, we can show that $H_n \notin H_{n+4}, \dots, H_{2n-4}$.

This general area of research is still ongoing. See Appendix C for a research question on embedding sub-matrices.

Table 12.2 Hadamard matrices with embedded D-optimal matrix of order m.

	n				
m	12	16	20	24	28
5	1	5	3	60	487
6	1	4	3	60	487
7	1	3	3	60	487
8	0	5	3	60	487

Source: Seberry and Mitrouli [162, table 1, p. 297], Reproduced with permission of Springer Nature.

Table 12.3 Existence or not of D-optimal matrix in Hadamard matrices.

$D_2 \in D_4$				$D_2 \in H_{4t}\ \forall\, t$
$D_3 \in D_4$				$D_3 \in H_{4t}\ \forall\, t$
$D_4 \in H_8$	$D_4 \in H_{12}$	$D_4 \in H_{16}$	$D_4 \in H_{20}, H_{24}, H_{28}$	$D_4 \in H_{4t}\ \forall\, t$
$D_5 \notin H_8$	$D_5 \in H_{12}$	$D_5 \in H_{16}$	$D_5 \in H_{20}, H_{24}, H_{28}$	$D_5 \in H_{4t}\ \forall\, t > 2$
$D_6 \notin H_8$	$D_6 \in H_{12}$	$D_6 \in H_{16}$	$D_6 \in H_{20}, H_{24}, H_{28}$	$D_6 \in H_{4t}\ \forall\, t > 2$
		$D_6 \notin S_{16}$		$D_6 \in S_{32}, S_{64}$
$D_7 \notin H_8$	$D_7 \in H_{12}$	$D_7 \in H_{16}$	$D_7 \in H_{20}, H_{24}, H_{28}$	$D_7 \in H_{4t}\ \forall\, t > 2$
		$D_7 \notin S_{16}$		$D_7 \notin S_{32}, D_7 \in S_{64}$
$D_8 = H_8$	$D_8 \notin H_{12}$	$D_8 \in H_{16}$	$D_8 \in H_{20}, H_{24}, H_{28}$	

Source: Seberry and Mitrouli [162, table 2, p. 298], Reproduced with permission of Springer Nature.

12.8 Embedding of Hadamard Matrices within Hadamard Matrices

We acknowledge the work of Christou et al. [19] in the following material.

Frequently, in several applications it is useful to know if specific Hadamard matrices are embedded in other Hadamard matrices of higher order, i.e. if an Hadamard matrix of order m is a submatrix of an Hadamard matrix of order n, when $m < n$. We denote this by $H_m \in H_n$. Regarding the existing embedding properties of the Hadamard matrices [4,13], it is known that H_4 is embedded in H_n for any order $n > 4$ and H_n is embedded in H_{2n} due to the doubling construction of Sylvester, Lemma 1.9, which means that the matrix

$$H_{2n} = \begin{bmatrix} H_n & H_n \\ H_n & -H_n \end{bmatrix}$$

is always a Hadamard matrix of order $2n$ when H_n is a Hadamard matrix of order n. These properties can be expressed as

$$H_4 \in H_n \text{ and } H_n \in H_{2n}. \tag{12.5}$$

In 1965, Cohn [21] proved that H_n can have a Hadamard submatrix H_m when $m \leq \frac{1}{2}n$. Using matrix algebra, Vijayan [196] proved that $(n-k) \times n$ row-orthogonal matrices with ± 1 elements can be extended to $n \times n$ Hadamard matrices when $k \leq 4$. More recently, Evangelaras et al. [70] used the distance distribution from coding theory to search for normalized Hadamard matrices of order n embedded in normalized Hadamard matrices of order $m \geq 2n$, and Brent and Osborn [17] generalized Cohn's result to maximal determinant submatrices of Hadamard matrices showing that if H_n has a maximal determinant submatrix M of order m, then $m < (\frac{1}{2}(n) + 5 \ln n)$ or $m \geq n-2$. Several other researchers have dealt with this problem in the past employing mostly combinatorial methods and the approaches that have been developed so far are either constructive [162] or employ the Hadamard conjecture [78] providing partially inconclusive results.

We examine the conditions under which an Hadamard matrix of order $n-k$ can be embedded in an Hadamard matrix of order n, denoted by $H_{n-k} \in H_n$. The current approach is based on a relation between the minors of Hadamard matrices presented in [184] and, by employing differential calculus and elementary number theory, first, we show that $H_{n-4} \notin H_n$ and $H_{n-8} \notin H_n$. Then, for any positive integers n and k multiples of 4, we proceed with the generalization

$$H_{n-k} \notin H_n \text{ for } k < \frac{1}{2}n \tag{12.6}$$

which is equivalent to Cohn's result in [21]. The above relation (12.6) was also considered in [17] where it is proven using Szöllősi's result [184] about the minors of an Hadamard matrix and calculus techniques.

We first analyze this approach in more depth providing an alternative proof of Szöllősi's result and an analytic description of the steps of the proof of (12.6), starting from $H_{n-4} \notin H_n$. Then, we provide a new number theoretic proof for $H_{n-8} \notin H_n$. Finally, in Section 12.10, we study the problem whether H_{n-k} can exist embedded in H_n when $4 \leq k < n$, and the connection between the order of the matrix H_{n-k} and the values which form the spectrum of the determinant function for (± 1)-matrices [43, 143].

12.9 Embedding Properties Via Minors

The current study of the embedding and extension properties of Hadamard matrices is motivated by the results obtained from [184] which lead to a simple relation connecting the minors of a (± 1)-matrix. The proof of this result was based on the properties of the generalized matrix determinant. A simplified proof of the same result has recently been presented by Banica et al. [7]. A more elegant, direct proof employing only the Jacobi identity [74] is given next.

Proposition 12.1: Let H_n an Hadamard matrix of order $n \geq 4$. If M_k denotes the absolute value of a $k \times k$ minor of H_n where $k = d$ or $n - d$, then for any $1 < d < n$ it holds:

$$M_{n-d} = n^{\frac{1}{2}n-d} M_d \tag{12.7}$$

Proof: An Hadamard matrix of order n can be considered as a block-matrix of the form

$$H_n = \begin{bmatrix} A & B \\ C & D \end{bmatrix}$$

where A is a $(n-d) \times (n-d)$ (± 1)-matrix and D is a $d \times d$ (± 1)-matrix for $1 < d < n$. If $U = (\sqrt{n})^{-1} H_n$, then U is orthogonal, because

$$UU^{\top} = \frac{1}{\sqrt{n}} H_n \frac{1}{\sqrt{n}} H_n^{\top} = \frac{1}{n} H_n H_n^{\top} = \frac{1}{n} n I_n = I_n.$$

Consequently, U is invertible and its inverse has the form:

$$U^{-1} = U^{\top} = \frac{1}{\sqrt{n}} \begin{bmatrix} A^{\top} & C^{\top} \\ B^{\top} & D^{\top} \end{bmatrix}$$

Using Jacobi's determinant identity [74] for U, it follows:

$$\det\left(\frac{1}{\sqrt{n}} A\right) = \det U \cdot \det\left(\frac{1}{\sqrt{n}} D^{\top}\right) \overset{|\det U|=1}{\Longleftrightarrow}$$

$$\left|\det\left(n^{-\frac{1}{2}} A\right)\right| = \left|\det\left(n^{-\frac{1}{2}} D^{\top}\right)\right| \Leftrightarrow$$

$$n^{-\frac{1}{2}(n-d)} |\det A| = n^{\frac{-d}{2}} \left|\det D^{\top}\right| \overset{|\det D^{\top}|=|\det D|}{\Longleftrightarrow}$$

$$|\det A| = n^{\frac{n}{2}-d} |\det D|$$

Since the absolute determinant of a matrix remains invariant under row or column interchange, the last equation holds for any $(n-d) \times (n-d)$ and $d \times d$ minors of H_n. □

The next lemma specifies the values of the determinant of a square (± 1)-matrix of order $n \leq 6$ and gives a more general property for the determinant of order $n > 6$. These values will be useful in the following.

Lemma 12.6 *(Day and Peterson [43])*: *Let B be an $n \times n$ matrix with elements ± 1. Then*

i) $\det B$ is an integer and 2^{n-1} divides $\det B$,
ii) when $n \leq 6$, the only possible values for $\det B$ are given in Table 12.4, and they do all occur.

According to Lemma 12.6-(i), if M_k denotes the absolute value of a $k \times k$ minor of a (± 1)-matrix of order $n \geq k$, then

$$M_k = p2^{k-1} \tag{12.8}$$

where p is either a positive integer, or zero. Table 12.4 gives the possible absolute values of the determinant of $n \times n$ matrices with elements ± 1.

Definition 12.7: The **spectrum of the determinant function for** (± 1)**-matrices** is defined to be the set of values taken by $p = 2^{1-k} |\det R_k|$ as the matrix R_k ranges over all $k \times k(\pm 1)$-matrices.

Orrick and Solomon give a list of values for p in [143]. They instance all values for $k = 1, 2, \ldots, 11$, and 13. Also, conjectures have been formulated for $k = 12, 14, 15, 16$, and 17.

Table 12.4 Possible determinant absolute values.

n	1	2	3	4	5	6
det B	1	0,2	0,4	0,8,16	0,16,32,48	0,32,64,96,128,160

Source: Christou et al. [19, table 1, p. 157], Reproduced with De Gruyter open access licence.

12.10 Embeddability of Hadamard Matrices

Considering the above results, we begin the study of the embedding of Hadamard matrices of order $n-k$ when $n>8$ and $k=4$. The following proposition can be established.

Proposition 12.2: An Hadamard matrix of order $n-4$ cannot be embedded in an Hadamard matrix of order n for any $n=4t$ with integer $t>2$.

$$H_{n-4} \notin H_n,\ n>8 \tag{12.9}$$

Proof: If $t=1$, then $n=4$ and $H_{n-4}=H_0$ which does not exist. If $t=2$, then $n=8$ and $H_{n-4}=H_4 \in H_8 = H_n$ which is true according to (12.5). Therefore, an integer $t>2$ must be considered in the following.

Assuming that an Hadamard matrix of order $n-4$ can be embedded in an Hadamard matrix of order n, the relation (12.7) for $d=4$ implies that

$$|\det H_{n-4}| = n^{\frac{n}{2}-4} M_4 \tag{12.10}$$

where M_4 is the absolute value of a 4×4 minor of H_n. However, it is known that

$$|\det H_{n-4}| = (n-4)^{\frac{1}{2}(n-4)} \tag{12.11}$$

and, according to Lemma 12.6-(ii), the possible nonzero values that M_4 can take are 8 or 16. Considering both cases, the value of M_4 will be denoted by m in the following.

Combining (12.10) and (12.11), it follows that:

$$\begin{aligned}
(n-4)^{\frac{1}{2}(n-4)} &= n^{\frac{n}{2}-4}\cdot m \quad \Leftrightarrow \\
\frac{1}{2}(n-4)\ln(n-4) &= \frac{1}{2}(n-8)\ln(n)+\ln(m) \quad \overset{n=4t}{\Longleftrightarrow} \\
(t-1)\ln(4(t-1)) &= (t-2)\ln(4t)+\frac{1}{2}(\ln(m)) \quad \Leftrightarrow \\
(t-1)(\ln 4+\ln(t-1)) &= (t-2)(\ln 4+\ln(t))+\ln(\sqrt{m}) \quad \Leftrightarrow \\
(t-1)\ln 4+(t-1)\ln(t-1) &= (t-2)\ln 4+(t-2)\ln(t)+\ln(\sqrt{m}) \quad \Leftrightarrow
\end{aligned}$$

$$(t-1)\ln(t-1)-(t-2)\ln t = \ln\left(\frac{1}{4}\sqrt{m}\right) \tag{12.12}$$

Since $t>2$, every term in (12.12) can be divided by the nonzero algebraic expression $(t-1)(t-2)$. Then, (12.12)is transformed into

$$\frac{\ln(t-1)}{t-2}-\frac{\ln(t)}{t-1} = \frac{\ln\left(\frac{1}{4}\sqrt{m}\right)}{(t-1)(t-2)} \tag{12.13}$$

Using the function $f(x)=\frac{\ln(x)}{x-1}$, the above Eq. (12.13) can be expressed in the form:

$$f(t-1)-f(t) = \frac{\ln\left(\frac{1}{4}\sqrt{m}\right)}{(t-1)(t-2)} \tag{12.14}$$

The real function f is well defined and differentiable in the interval $(2, +\infty)$. Moreover, for every $x \in (2, +\infty)$ we have

$$f(x) > 0 \text{ and } \frac{df}{dx} = \frac{\left(1 - \frac{1}{x}\right) - \ln(x)}{(x-1)^2} < 0.$$

The latter shows that f is a strictly decreasing function in the interval $(2, +\infty)$. Hence, for any $t > 2$ it follows that

$$t - 1 < t \Leftrightarrow f(t-1) > f(t) \Leftrightarrow f(t-1) - f(t) > 0 \tag{12.15}$$

Consequently, the left part of the Eq. (12.14) is always positive whereas its right part is either negative (for $m = 8$) or zero (for $m = 16$). Therefore, the assumption that was made in the derivation of the Eq. (12.10) is invalid for any $n = 4t$ with integer $t > 2$. Thus, H_{n-4} cannot be embedded in H_n. □

12.11 Embeddability of Hadamard Matrices of Order $n - 8$

In the case of H_{n-k} with $k = 4$, the possible nonzero values that M_k can take on are only two. Conversely, when $k = 8$, the determinant spectrum [143] includes more than two values which depend on a specific integer p. Hence, a different approach must be followed for H_{n-k} when $n > 16$ and $k = 8$. The next proposition illustrates this approach and its proof is based on elementary number theory.

Proposition 12.3: An Hadamard matrix of order $n - 8$ cannot be embedded in an Hadamard matrix of order n for any $n = 4t$ with integer $t > 4$.

$$H_{n-8} \notin H_n,\ n > 16 \tag{12.16}$$

Proof: For $t = 1, 2$ no matrix H_{n-8} can be determined and an integer $t > 2$ will be considered in the following.

Assuming that an Hadamard matrix of order $n - 8$ can be embedded in an Hadamard matrix of order n, the relation (12.7) for $d = 8$ implies that

$$|\det H_{n-8}| = n^{\frac{n}{2}-8} M_8 \tag{12.17}$$

where M_8 is the absolute value of a 8×8 minor of H_n. Moreover, it is known that

$$|\det H_{n-8}| = (n-8)^{\frac{1}{2}(n-8)} \tag{12.18}$$

and, according to (12.8), $M_8 = p \cdot 2^7$, where p is a positive integer. For the 8×8 case it has been confirmed that the possible existing values for the integer p are $1, 2, \ldots, 18, 20, 24$, and 32 [143].

Combining (12.17) and (12.18), it follows that:

$$(n-8)^{\frac{1}{2}(n-8)} = n^{\frac{n}{2}-8} p 2^7 \overset{n=4t\ t>2}{\Longleftrightarrow} p = 2t^4 \left(\frac{t-2}{t}\right)^{2(t-2)}. \tag{12.19}$$

The above Eq. 12.19 is satisfied by the pairs of values $(t, p) = (3, 18)$ and $(t, p) = (4, 32)$ which correspond to the valid cases of $H_4 \in H_{12}$ and $H_8 \in H_{16}$, respectively. However, for integers $t > 4$, the term p in (12.19) cannot be an integer. A proof of this statement based on elementary number theory is presented below.

Assuming that p is an integer for any integer $t > 4$, the Eq. (12.19) is written equivalently in the form of an equality between two integers:

$$pt^{2t} = 2t^8(t-2)^{2(t-2)}. \tag{12.20}$$

The next two cases are considered:

i) $t = 2^m$ for integer $m \geq 3$
The prime factorization of the left-hand side of (12.20) shows that the least power of 2 is 2^{2mt}, whereas the prime factorization of the right-hand side of (12.20) shows that the least power of 2 is exactly $2^{1+8m+2t-4}$. Therefore, considering that p might also be a power of 2, it follows that:

$$2mt \leq 8m + 2t - 3 \Leftrightarrow t \leq 4 + \frac{5}{2(m-1)}.$$

The above inequality implies that $t \leq 5$ which contradicts the hypothesis for the integer t in this case, i.e. $t \geq 2^3 = 8$.

ii) t is divisible by a prime integer $r \geq 3$
If $m \geq 1$ is an integer such that r^m is the maximum power of r that divides t, then the prime factorization of the left-hand side of (12.20) shows that the least power of r is r^{2mt}, whereas the prime factorization of the right-hand side of (12.20) shows that the least power of r is exactly r^{8m}. Considering again that p might also be a power of r, it follows that:

$$2mt \leq 8m \Leftrightarrow t \leq 4.$$

The above inequality contradicts the general hypothesis for the integer t, i.e. $t > 4$.

Consequently, there is no integer $t > 4$ which can give a valid p for the minor M_8. Therefore, the assumption that was made in the derivation of the Eq. (12.17) is invalid for any $n = 4t$ with integer $t > 4$ and as a result H_{n-8} cannot be embedded in H_n. □

Remark 12.6: The problem of the nonexistence of integers p satisfying (12.19) can also be investigated using tools from calculus. Specifically, if p in (12.19) is regarded as a real function of t, then $p(t)$ is differentiable in the interval $(4, +\infty)$ with first derivative:

$$\frac{dp}{dt} = \frac{4t^7}{(t-2)^4}\left(\frac{t-2}{t}\right)^{2t}\left(t\ln\left(\frac{t-2}{t}\right)+4\right)$$

It can easily be proven that $\frac{dp}{dt} > 0$ for every $t > 4$, which implies that $p(t)$ is a strictly increasing function in the interval $(4, +\infty)$. As a result, for every $t > 4$, it holds that

$$p(t) > p(4) \Leftrightarrow p > 32$$

which contradicts the fact that $p \leq 32$ [143].

12.12 Embeddability of Hadamard Matrices of Order $n - k$

In the proof of Proposition 12.3, the parameter p plays a key role in the study of the embedding properties of Hadamard matrices. Given a positive integer k, by the Hadamard conjecture, the absolute value of the maximal determinant of a (± 1)-matrix of order k is always less than or equal to $k^{\frac{1}{2}k}$ [89]. Therefore, for any minor M_k we have

$$\left.\begin{array}{l} M_k = p2^{k-1} \\ M_k \leq k^{\frac{k}{2}} \end{array}\right\} \Leftrightarrow p2^{k-1} \leq k^{\frac{k}{2}} \Leftrightarrow p \leq 2\left(\frac{k}{4}\right)^{\frac{k}{2}} \tag{12.21}$$

If $\hat{p}$ is used to denote the maximum value of p, then (12.21) implies that

$$\hat{p} := max(p) = 2\left(\frac{k}{4}\right)^{\frac{k}{2}} \overset{k=4r}{\Longleftrightarrow} \hat{p} = 2r^{2r}. \tag{12.22}$$

The relation (12.21) forms a necessary condition for the embeddability of Hadamard matrices. Hence, by studying the range of values that p can take, the results obtained from Propositions 12.2 and 12.3 can be generalized for an Hadamard matrix of order $n - k$.

Let $n = 4t$ and $k = 4r$ where t, r are positive integers. Generally, $0 < k < n$ and thus, $0 < r < t$. The cases of $\{t > 2,\ r = 1\}$ and $\{t > 4,\ r = 2\}$ have been examined in Propositions 12.2 and 12.3. The case of $\{t > 2,\ r > 2;\ t > r\}$ will be considered in the following.

Assuming that an Hadamard matrix of order $n - k$ can be embedded in an Hadamard matrix of order n, the relation (12.7) for $d = k$ implies that

$$|\det H_{n-k}| = n^{\frac{n}{2}-k} M_k \tag{12.23}$$

Furthermore, we have

$$|\det H_{n-k}| = (n-k)^{\frac{1}{2}(n-k)} \tag{12.24}$$

If we combine (12.8), (12.23), and (12.24), we get the next important algebraic relation which connects p, and consequently the spectrum of the determinant function, with the order of the matrix H_{n-k}:

$$p = 2t^{2r}\left(\frac{t-r}{t}\right)^{2(t-r)} \tag{12.25}$$

If $t = 2r$, or equivalently $n = 2k$, the integer p attains its maximum value $\hat{p}$, thus $p = \hat{p}$. Then, $H_{n-k} = H_k \in H_{2k} = H_n$ as mentioned in (12.5). Therefore, in the following, we shall examine the existence of positive integers t, r satisfying the inequalities:

$$p < \hat{p},\ t > r \tag{12.26}$$

Let $\theta = \frac{r}{t}$. Since $t > r$, it follows that $0 < \theta < 1$ and $0 < 1 - \theta < 1$. Then,

$$p < \hat{p} \Leftrightarrow 2t^{2r}\left(1 - \frac{r}{t}\right)^{2t\left(1-\frac{r}{t}\right)} < 2r^{2r} \Leftrightarrow$$
$$(1-\theta)\ln(1-\theta) - \theta\ln\theta < 0 \tag{12.27}$$

Now, for every $\theta \in (0, 1)$ we consider the real function:

$$h(\theta) = (1-\theta)\ln(1-\theta) - \theta\ln\theta \tag{12.28}$$

Using calculus, it can be proven that $h(\theta) \geq 0$, if $\theta \in (0, \frac{1}{2})$, and $h(\theta) < 0$, if $\theta \in (\frac{1}{2}, 1)$. The graph of the function $h(\theta)$ is illustrated in Figure 12.1

Studying the sign of the function $h(\theta)$, where $\theta = \frac{r}{t} = \frac{k}{n} \in (0, 1)$, provides important information about the behavior of p for the various values of the integers n and k when $n > k$.

12.12.1 Embeddability–Extendability of Hadamard Matrices

The preceding analysis provides conclusive results on the embedding problem of Hadamard matrices H_{n-k} which form the next theorem.

Theorem 12.9: *An Hadamard matrix of order $n - k$ cannot be embedded in an Hadamard matrix of order n for any positive integers n and k multiples of 4 when $k < \frac{n}{2}$. That is*

$$H_{n-k} \notin H_n,\ 4 \leq k < \frac{n}{2}. \tag{12.29}$$

Proof: If $n > 2k$, then $4t > 8r$ and $\theta < \frac{1}{2}$. Consequently, it is $h(\theta) > 0$, which implies that the inequality (12.25) cannot be satisfied by the specific values of θ. Therefore, for $n > 2k$ there are no integers p satisfying the conditions (12.26) and the Eq. (12.23), which supports the embedding property of the matrix H_{n-k}, is not valid. □

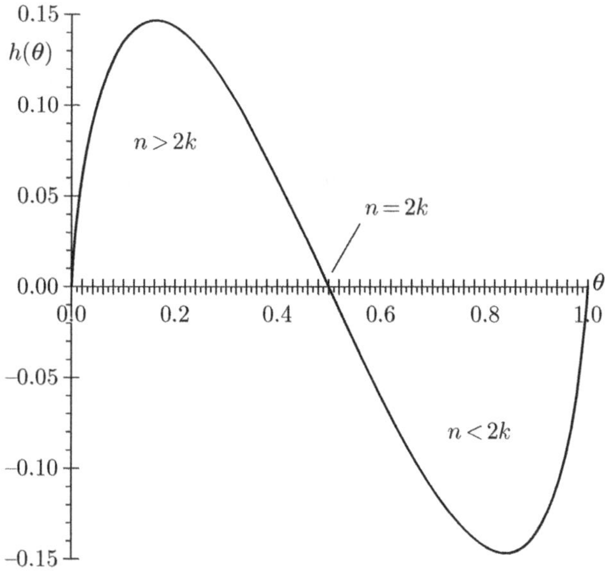

Figure 12.1 Graph of the function $h(\theta) = (1-\theta)\ln(1-\theta) - \theta \ln \theta$. *Source:* Christou et al. [19], De Gruyter open access licence.

However, the sign analysis of the function $h(\theta)$ also reveals that for $n \leq 2k < 2n$ there are positive integers p which satisfy $p \leq \hat{p}$.

If $n = 2k$, then $h(\theta) = 0 \Leftrightarrow p = \hat{p}$. This is the case of $H_k \in H_{2k}$ which is true according to (12.5). Conversely, if $n < 2k < 2n$, there are values of p such that $p < \hat{p}$. Considering (12.25) for $t = \frac{n}{4}$ and $r = \frac{k}{4}$, the discrete function

$$\mathcal{P}(n,k) = 2\left(\frac{n}{4}\right)^{\frac{k}{2}}\left(\frac{n-k}{n}\right)^{\frac{n-k}{2}} \quad \text{for } \frac{n}{2} \leq k < n \text{ and } \begin{cases} n = 8, 12, 16\ldots \\ k = 4, 8, 12, \ldots \end{cases} \tag{12.30}$$

provides values for the parameter p which satisfy (12.23), expressed as

$$|\det H_{n-k}| = 2^{(n-k)-1}\left(\frac{n}{4}\right)^{\frac{n}{2}-k} p. \tag{12.31}$$

The above results form the basis to pose the following conjecture.

Conjecture 12.8: Consider a Hadamard matrix H_n. If $H_n^{(k)}$ is a $k \times k$ submatrix of H_n, where $n \geq 8$ and $k \geq 4$ are integers multiples of 4 such that $\frac{n}{2} \leq k < n$, and $|\det H_n^{(k)}| = p2^{k-1}$ with $p = \mathcal{P}(n,k)$, then an Hadamard matrix of order $n-k$ may exist embedded in the Hadamard matrix of order n, i.e.

$$H_{n-k} \in H_n, \quad 4 \leq \frac{n}{2} \leq k < n. \tag{12.32}$$

12.12.2 Available Determinant Spectrum and Verification

Both results (12.29) and (12.32) reveal a characteristic embedding and extension pattern for Hadamard matrices and their proof is based on the properties of the minors of Hadamard matrices. The key element in this study is the range of the values of the parameter p. Theorem 12.9 provides conclusive results for every $p > \hat{p}$. Conversely, (12.32) holds for specific values of p which can be obtained from (12.30).

For a fixed order $n \geq 8$, the relation (12.32) holds for every $k = \frac{n}{2}$ and $k = n - 4$, since they are linked to the already known cases $H_n \in H_{2n}$ and $H_4 \in H_n$, respectively. Furthermore, for $4 \leq k \leq 16$ the computed values of $p = \mathcal{P}(n, k)$ are already included in the available and confirmed spectrum for $k = 4, 8$, and in the conjectured spectrum for $k = 12, 16$ given by Orrick and Solomon [143].

We also examined some cases where $k > \frac{n}{2}$ using the Hadamard matrices of order 20, 24, and 28 (Paley-type) given by Sloane [170].[1] The following results, obtained by using Matlab and a quad-core AMD-A10/8Gb-Ram machine, verify Conjecture 12.8 for $n = 20, 24, 28$ and $k = 12, 16, 20$.

i) $n = 20$, $k = 12$, and $p = \mathcal{P}(20, 12) = 800$.
There exists a 12×12 submatrix $H_{20}^{(12)} = [a_{ij}]$ of H_{20} where

$$i \in \{1, 2, 3, 4, 5, 6, 7, 8, 9, 11, 14, 20\},$$
$$j \in \{1, 2, 3, 4, 7, 8, 13, 14, 17, 18, 19, 20\}$$

such that

$$|\det H_{20}^{(12)}| = 800 \cdot 2^{11} = 1638400.$$

The above result implies that H_8 may exist embedded in H_{20}. We can confirm that there is an 8×8 submatrix $A = [a_{ij}]$ of H_{20} where

$$i \in \{1, 2, 3, 4, 5, 6, 9, 12\},$$
$$j \in \{1, 2, 3, 4, 8, 14, 15, 18\}$$

which satisfies Definition 1.12, Lemma 1.6, and Eq. (12.31). That is

$$AA^{\top} = A^{\top}A = 8I \text{ and}$$
$$|\det A| = 2^{(20-12)-1} \cdot \left(\frac{20}{4}\right)^{\frac{20}{2}-12} \cdot 800 = 4096 = |\det H_8|$$

ii) $n = 24$, $k = 16$, and $p = \mathcal{P}(24, 16) = 41472$.
There exists a 16×16 submatrix $H_{24}^{(16)} = [a_{ij}]$ of H_{24} where

$$i \in \{1, 2, 3, 4, 5, 6, 7, 8, 9, 10, 11, 12, 15, 18, 22, 23\},$$
$$j \in \{1, 3, 6, 7, 9, 10, 12, 13, 14, 15, 16, 18, 19, 21, 23, 24\}$$

such that

$$|\det H_{24}^{(16)}| = 41472 \cdot 2^{15} = 1358954496$$

The above result implies that H_8 may exist embedded in H_{24}. We can confirm that there is an 8×8 submatrix $A = [a_{ij}]$ of H_{24} where

$$i \in \{1, 2, 3, 4, 5, 6, 12, 21\},$$
$$j \in \{2, 5, 7, 10, 11, 17, 18, 19\}$$

which satisfies Definition 1.12, Lemma 1.6, and Eq. (12.31). That is

$$AA^{\top} = A^{\top}A = 8I \text{ and}$$
$$|\det A| = 2^{(24-16)-1} \cdot \left(\frac{24}{4}\right)^{\frac{24}{2}-16} \cdot 41472 = 4096 = |\det H_8|$$

1 had.20.pal, had.24.pal, had.28.pal2

iii) $n = 28$, $k = 20$, and $p = \mathcal{P}(28, 20) = 3764768$.
There exists a 20×20 submatrix $H_{28}^{(20)} = [a_{ij}]$ of H_{28} where

$$i \in \{1, 2, 3, 4, 5, 6, 7, 8, 9, 10, 11, 12, 13, 14, 15, 16, 17, 19, 23, 27\},$$
$$j \in \{1, 2, 3, 4, 5, 7, 9, 11, 13, 14, 15, 16, 17, 19, 20, 22, 23, 24, 26, 27\}$$

such that

$$|\det H_{28}^{(20)}| = 3764768 \cdot 2^{19} = 1973822685184$$

The above result implies that H_8 may exist embedded in H_{28}. We can confirm that there is an 8×8 submatrix $A = [a_{ij}]$ of H_{28} where

$$i \in \{1, 2, 3, 4, 5, 6, 16, 17\},$$
$$j \in \{4, 6, 9, 14, 15, 19, 21, 26\}$$

which satisfies Definition 1.12, Lemma 1.6, and Eq. (12.31). That is

$$\begin{aligned} AA^\top &= A^\top A = 8I \text{ and} \\ |\det A| &= 2^{(28-20)-1} \cdot \left(\frac{28}{4}\right)^{\frac{28}{2}-20} \cdot 3764768 = 4096 = |\det H_8| \end{aligned}$$

iv) $n = 28$, $k = 16$, and $p = \mathcal{P}(28, 16) = 71442$.
There exists a 16×16 submatrix $H_{28}^{(16)} = [a_{ij}]$ of H_{28} where

$$i \in \{1, 2, 3, 4, 6, 8, 9, 11, 15, 16, 17, 18, 20, 22, 23, 25\},$$
$$j \in \{1, 2, 3, 4, 6, 8, 9, 11, 15, 16, 17, 18, 20, 22, 23, 25\}$$

such that

$$|\det H_{28}^{(16)}| = 71442 \cdot 2^{15} = 2341011456.$$

The above result implies that H_{12} may exist embedded in H_{28}. We can confirm that there is an 12×12 submatrix $A = [a_{ij}]$ of H_{28} where

$$i \in \{1, 2, 3, 4, 8, 11, 15, 16, 17, 18, 22, 25\},$$
$$j \in \{1, 2, 3, 4, 8, 11, 15, 16, 17, 18, 22, 25\}$$

which satisfies Definition 1.12, Lemma 1.6, and Eq. (12.31). That is

$$\begin{aligned} AA^\top &= A^\top A = 12I \text{ and} \\ |\det A| &= 2^{(28-16)-1} \cdot \left(\frac{28}{4}\right)^{\frac{28}{2}-16} \cdot 71442 = 2985984 = |\det H_{12}| \end{aligned}$$

The aforementioned submatrices are not unique. There are several different row and column arrangements (i, j) which also satisfy the conditions for (12.32) to hold. In Table 12.5, we summarize the confirmed results obtained from (12.29) and (12.32) for $n = 8, 12, \ldots, 28$ and $k = 4, 8, \ldots, 24$.

12.13 Growth Problem for Hadamard Matrices

During the process of Gaussian Elimination to solve linear equations or to invert a matrix the pivots can become very large and so, with rounding errors included, unstable. This is the origin of the "growth problem."

Table 12.5 Embeddability of Hadamard matrices H_{n-k}.

n	k=4	k=8	k=12	k=16	k=20	k=24
8	$H_4 \in H_8$					
12	$H_8 \in H_{12}$	$H_4 \in H_{12}$				
16	$H_{12} \in H_{16}$	$H_8 \in H_{16}$	$H_4 \in H_{16}$			
20	$H_{16} \in H_{20}$	$H_{12} \in H_{20}$	$H_8 \in H_{20}$	$H_4 \in H_{20}$		
24	$H_{20} \in H_{24}$	$H_{16} \in H_{24}$	$H_{12} \in H_{24}$	$H_8 \in H_{24}$	$H_4 \in H_{24}$	
28	$H_{24} \in H_{28}$	$H_{20} \in H_{28}$	$H_{16} \in H_{28}$	$H_{12} \in H_{28}$	$H_8 \in H_{28}$	$H_4 \in H_{28}$

Source: Christou et al. [19, table 2, p. 164], Reproduced with De Gruyter open access licence.

We give some results concerning the pivot values that appear when Gaussian Elimination with complete pivoting is applied to Hadamard matrices. Using the pivot patterns which have been evaluated, we prove the existence or nonexistence of specific D-maximal matrices in them.

Hadamard matrices are now investigated relative to the well-known growth problem. Traditionally, backward error analysis for Gaussian Elimination (GE), see, e.g. [43], on a matrix $A = (a_{ij}^{(1)})$ is expressed in terms of the **growth factor**

$$g(n, A) = \frac{max_{i,j,k}|a_{ij}^{(k)}|}{max_{i,j}|a_{ij}^{(1)}|},$$

which involves all the elements $a_{ij}^{(k)}$, $k = 1, 2, \dots, n$, that occur during the elimination. Matrices with the property that no row and column exchanges are needed during GE with complete pivoting are called **completely pivoted** (CP). In other words, at each step of the elimination, the element of largest magnitude (the "pivot") is located at the top left position of every sub-matrix appearing during the process. For a CP matrix A we have

$$g(n, A) = \frac{max\{p_1, p_2, \dots, p_n\}}{|a_{11}^{(1)}|},$$

where $p_1, p_2, \dots, p_n$ are the pivots of A.

The following lemma gives a useful relation between pivots and minors and a characteristic property for CP matrices.

Lemma 12.7: [38] *Let A be a CP matrix.*

i) The magnitude of the pivots which appear after application of GE operations on A is given by

$$p_j = \frac{A(j)}{A(j-1)}, \quad j = 1, 2, \dots, n, \quad A(0) = 1. \tag{12.33}$$

where $A(j)$ denotes the absolute value of $j \times j$ principal minor.

ii) The maximum $j \times j$ leading principal minor of A, when the first $j - 1$ rows and columns are fixed, is $A(j)$.

From the above Lemma, we see that the calculation of minors is important in order to study pivot structures. Moreover, the maximum $j \times j$ minor appears in the upper left $j \times j$ corner of A. So, if the existence of a matrix with maximal determinant, i.e. a D-optimal design, is proved for a CP matrix A, we can indeed assume that it always appears in its upper left corner.

H-equivalent operations do not preserve pivots, i.e. the pivot pattern is not invariant under H-equivalence, and many pivot patterns can be observed. So, H-equivalent matrices do not necessarily have the same pivot pattern.

Pivots can also be given from the relation [66]

$$p_{n+1-k} = \frac{nA[k-1]}{A[k]}, \quad k = 1, 2, \ldots, n, \quad A[0] = 1. \tag{12.34}$$

where $A[k]$ denotes the absolute value of the determinant of the lower right $k \times k$ principal sub-matrix.

In 1968 Cryer [38] conjectured that the maximum growth at each stage of Gaussian Elimination was less than or equal to the order of the matrix and equalled the order only if the matrix was Hadamard.

Gould [87] proved that the first part of the conjecture was not true. He found matrices as follows

n	Growth
13	13.02
18	20.45
20	24.25
25	32.99

Thus, the following conjecture remains open.

Conjecture 12.9 (Cryer): The growth of an Hadamard matrix is its order.

Concerning progress on this conjecture, there were specified pivot patterns of Sylvester–Hadamard matrices [43] and of Hadamard matrices of orders 12 and 16 [66, 123]. Next we connect, for the first time, the values of the pivots appearing to the patterns with the existence or not of D-optimal matrices in these Hadamard matrices. Thus, through the pivots we will provide another proof for Lemmas 12.3–12.5 and a reconfirmation of the results of Table 12.2.

The Pivot Structure of H_8

The growth of H_8 is 8 [43] and its unique pattern is:

$$1\ 2\ 2\ 4\ 2\ 4\ 4\ 8$$

Day and Peterson [43] specify that $H_8(5) = 32$, $H_8(6) = 128$, and $H_8(7) = 512$. From this we can reconfirm Lemmas 12.3, 12.4, and 12.5, i.e. $D_5,\ D_6,\ D_7 \notin H_8$.

The Pivot Structure of H_{12}

In [66] it was proved that the growth of H_{12} is 12 and its unique pattern was determined as:

$$1\ 2\ 2\ 4\ 3\ \tfrac{10}{3}\ \tfrac{18}{5}\ 4\ 3\ 6\ 6\ 12$$

Lemma 12.8: *The D-optimal matrix of order 5 appears in the Hadamard matrix of order 12.*

Proof: From the unique pivot pattern of H_{12} we see that p_5 is 3. From Lemma 12.7 we have that:

$$p_5 = \frac{H_{12}(5)}{H_{12}(4)}$$

According to Theorem 1.13 and Lemma 12.7, $H_{12}(4)$ always takes the value 16 thus the value of $H_{12}(5)$ will be 48. This means that the D-optimal matrix of order 5 always appears in the Hadamard matrix of order 12. □

Lemma 12.9: *The D-optimal matrix of order 6 appears in the Hadamard matrix of order 12.*

Proof: From the unique pivot pattern of H_{12} we see that p_6 is $\frac{10}{3}$. From Lemma 12.7 we have that:

$$p_6 = \frac{H_{12}(6)}{H_{12}(5)}$$

According to Lemmas 12.7 and 12.8, $H_{12}(5)$ takes the value 48 thus the value of $H_{12}(6)$ will be 160. This means that the D-optimal matrix of order 6 always appears in the Hadamard matrix of order 12. □

Lemma 12.10: *The D-optimal matrix of order* 7 *appears in the Hadamard matrix of order* 12.

Proof: From the unique pivot pattern of H_{12} we see that p_7 is $\frac{18}{5}$. From Lemma 12.7 we have that:

$$p_7 = \frac{H_{12}(7)}{H_{12}(6)}$$

According to Lemmas 12.7 and 12.9, $H_{12}(6)$ takes the value 160 thus the value of $H_{12}(7)$ will be 576. This means that the D-optimal matrix of order 7 always appears in the Hadamard matrix of order 12. □

Lemma 12.11: *The D-optimal matrix of order* 8 *does not appear in the Hadamard matrix of order* 12.

Table 12.6 The 34 pivot patterns of the Hadamard matrix of order 16.

Pivot	First class (Sylvester–Hadamard)	Second class	Third class	Fourth class
1	1	1	1	1
2	2	2	2	2
3	2	2	2	2
4	4	4	4	4
5	2,3	2,3	2,3	2,3
6	$4,\frac{8}{3}$	$4,\frac{10}{3}$	$4,\frac{8}{3},\frac{10}{3}$	$4,\frac{10}{3}$
7	2,4	$4,\frac{8}{10/3},\frac{16}{5}$	$4,\frac{18}{5}$	$4,\frac{18}{5}$
8	4,6,8	4,5,6	$4,\frac{9}{2},5,6$	4,5,6
9	$2,4,\frac{8}{3}$	$2,4,\frac{8}{3},\frac{16}{3},\frac{16}{5}$	$2,4,\frac{9}{2},\frac{8}{3},\frac{16}{5}$	$2,4,\frac{9}{2},\frac{8}{3}\ \frac{16}{5}$
10	4,8	4,5	$4,5,\frac{16}{18/5}$	$4,5,\frac{16}{18/5}$
11	4,6,8	$4,6,\frac{16}{10/3}$	$4,6,\frac{16}{10/3}$	$4,6,\frac{16}{10/3}$
12	$8,\frac{16}{3}$	$8,\frac{16}{3}$	$8,\frac{16}{3}$	$8,\frac{16}{3}$
13	4,8	4	4	4
14	8	8	8	8
15	8	8	8	8
16	16	16	16	16

Source: Seberry and Mitrouli [162, table 3, p. 304], Reproduced with permission of Springer Nature.

Proof: From the unique pivot pattern of H_{12} we see that p_8 is 4. From Lemma 12.7 we have that:

$$p_8 = \frac{H_{12}(8)}{H_{12}(7)}$$

According to Lemmas 12.7 and 12.10, $H_{12}(7)$ takes the value 576 thus the value of $H_{12}(8)$ will be 2304 and not 4096 which is the value of the D-optimal case. This means that the D-optimal matrix of order 8 does not appear in the Hadamard matrix of order 12. □

The Pivot Structure of H_{16}

In [123] it was proved that the growth of H_{16} is 16 and it was determined that there were exactly 34 pivot patterns up to H-equivalence given in Table 12.6.

The interesting fact is that **8** occurred as pivot p_{13} only in the Sylvester–Hadamard matrices. Since $p_{13} = \frac{nH_{16}[3]}{H_{16}[4]}$ and as $H_{16}[3]$ equals 4 (the only possible nonzero value for the determinant of a 3×3 matrix with elements ± 1), in order $p_{13} = 8$ we must have that $H_{16}[4]$ takes the value 8 (possible nonzero values for the determinants of a 4×4 matrix with elements ± 1 are 8 and 16). There are further research questions on the pivot structure of CP Hadamard matrices in Appendix C.

The next open case is the specification of the growth factor of the Hadamard matrix of order 20.

A

Hadamard Matrices

A.1 Hadamard Matrices

One of us (Seberry) has a table containing odd integers $q < 40{,}000$ for which Hadamard matrices orders $2^t q$ exist compiled in 1992. Another compilation for $q < 10{,}000$ is given in [22]. In Section A.3, we give the update of the table in [166] for $q \leq 3000$. The key for the methods of construction follow: Note that not all construction methods appear, only those that, in the opinion of the authors, enabled us to compile the tables efficiently.

A.1.1 Amicable Hadamard Matrices

Table A.1 gives the amicable Hadamard matrices key methods of construction.

Table A.1 Amicable Hadamard matrices key methods of construction.

Key	Method	Explanation
$a1$	$p^r + 1$	$p^r \equiv 3 \pmod 4$, is a prime power [201]
$a2$	$2(q+1)$	$2q+1$ is a prime power, $q \equiv 1 \pmod 4$, is a prime [204]
$a5$	nh	n, h, are amicable Hadamard matrices [201]

Source: Adapted from Seberry and Yamada [166, p. 535], Wiley.

A.1.2 Skew Hadamard Matrices

Table A.2 gives the skew Hadamard matrices key construction methods.

Hadamard Matrices: Constructions using Number Theory and Algebra, First Edition. Jennifer Seberry and Mieko Yamada.

Table A.2 Skew Hadamard matrices key methods of construction.

Key	Method	Explanation
$s1$	$2^t \prod k_i$	t all positive integers, $k_i - 1 \equiv 3 \pmod 4$ a prime power [144]
$s2$	$(p-1)^u + 1$	p is a skew Hadamard matrix, $u > 0$ is an odd integer [191]
$s3$	$2(q+1)$	$q \equiv 5 \pmod 8$ is a prime power [181]
$s4$	$2(q+1)$	$q = p^t$ is a prime power where $p \equiv 5 \pmod 8$ and $t \equiv 2 \pmod 4$ [182, 219]
$s5$	$4m$	$3 \le m \le 33, 127$ [58, 99, 183] $m \in \{37, 43, 67, 113, 127, 157, 163, 181, 241\}$ [54, 56]
$s6$	$4(q+1)$	$q \equiv 9 \pmod{16}$ is a prime power [216]
$s7$	$(\vert t\vert + 1)(q+1)$	$q = s^2 + 4t^2 \equiv 5 \pmod 8$ is a prime power and $\vert t\vert + 1$ is a skew Hadamard matrix [209]
$s8$	$4(q^2+q+1)$	q is a prime power, $q^2 + q + 1 \equiv 3, 5, 7 \pmod 8$ a prime or $2(q^2+q+1)+1$ is a prime power [177]
$s0$	hm	h is a skew Hadamard matrix and m is an amicable Hadamard matrix [204]

Source: Adapted from Seberry and Yamada [166, p. 535], Wiley.

A.1.3 Spence Hadamard Matrices

Table A.3 gives the Spence Hadamard matrices key construction methods.

Table A.3 Spence Hadamard matrices key methods of construction.

Key	Method	Explanation
$p1$	$4(q^2+q+1)$	$q^2+q+1 \equiv 1 \pmod 8$ is a prime [177]
$p2$	$4n$ or $8n$	$n, n-2$ are prime powers, if $n \equiv 1 \pmod 4$ there exists a Hadamard matrix of order $4n$, if $n \equiv 3 \pmod 4$ there exists a Hadamard matrix of order $8n$ [174]
$p3$	$4m$	m is an odd prime power for which an integer $s \ge 0$ such that $\frac{(m-(2^{s+1}+1))}{2^{s+1}}$ is an odd prime power [174]

Source: Adapted from Seberry and Yamada [166, p. 538], Wiley.

A.1.4 Conference Matrices Give Symmetric Hadamard Matrices

The methods in Table A.4 give symmetric Hadamard matrices of order $2n$ and conference matrices of order n with the exception of $c6$ which produces an Hadamard matrix. The order of the Hadamard matrix is given in the "Method" column.

Table A.4 Conference Hadamard matrices key methods of construction.

Key	Method	Explanation
c1	$2(p^r+1)$	$p^r \equiv 1 \pmod 4$ is a prime power [80, 144]
c2	$2((h-1)^2+1)$	h is a skew Hadamard matrix [11]
c3	$2(q^2(q-2)+1)$	$q \equiv 3 \pmod 4$ is a prime power $q-2$ is a prime power [133]
c4	$2(5.9^{2t+1}+1)$	$t \ge 0$ [163]
c5	$2((n-1)^s+1)$	n is a conference matrix $s \ge 2$ [191]
c6	nh	n is a conference matrix h is a Hadamard matrix [80]

Source: Adapted from Seberry and Yamada [166, pp. 538–539], Wiley.

Note: a conference matrix of order n exists only if $n-1$ is the sum of two squares.

A.1.5 Hadamard Matrices from Williamson Matrices

If a Williamson matrix of order $2^t q$ exists then there is a Hadamard matrix of order $2^{t+2}q$, the same key as in the Index of Williamson Matrices is used to index the Hadamard matrices produced from them.

A.1.6 OD Hadamard Matrices

Table A.5 give the OD Hadamard matrices key construction methods.

Table A.5 Orthogonal design Hadamard matrices key methods of construction.

Key	Method	Explanation
$o1$	$2^{t+2}q$	If a T-matrix of order $2^t q$ exists then there is a Hadamard matrix of order $2^{t+2}q$ [23, 194]
$o2$	ow	o is an $OD(4n; n, n, n, n)$ and w is a Williamson matrix [9, 23, 206]
$o3$	$8pw$	An $OD(8p; p; p; p; p; p; p; p; p)$ exists for $p = 1, 3$ and there exist 8-Williamson matrices of order w [145]
$o4$	$71w$	T-sequences [119]

Source: Adapted from Seberry and Yamada [166, p. 539], Wiley.

A.1.7 Yamada Hadamard Matrices

Table A.6 gives the Yamada Hadamard matrices key construction methods.

Table A.6 Yamada Hadamard matrices key methods of construction.

Key	Method	Explanation
$y1$	$4q$	$q \equiv 1 \pmod 8$ is a prime power $\frac{1}{2}(q-1)$ is a Hadamard matrix [237]
$y2$	$4(q+2)$	$q \equiv 5 \pmod 8$ is a prime power $\frac{1}{2}(q+3)$ is a skew Hadamard matrix [237]
$y3$	$4(q+2)$	$q \equiv 1 \pmod 8$ is a prime power $\frac{1}{2}(q+3)$ is a conference matrix [237]

Source: Adapted from Seberry and Yamada [166, p. 539], Wiley.

A.1.8 Miyamoto Hadamard Matrices

Table A.7 gives the Miyamoto Hadamard matrices key construction methods.

Table A.7 Miyamoto Hadamard matrices key methods of construction.

Key	Method	Explanation
$m1$	$4q$	$q \equiv 1 \pmod 4$ is a prime power $q-1$ is a Hadamard matrix [137]
$m2$	$8q$	$q \equiv 3 \pmod 4$ is a prime power $2q-3$ is a prime power [137]

Source: Adapted from Seberry and Yamada [166, p. 540], Wiley.

A.1.9 Koukouvinos and Kounias

Table A.8 gives the Koukouvinos and Kounias key construction method. All the orders given by this construction are of order divisible by 8.

Table A.8 Koukouvinos and Kounias key method of construction.

Key	Method	Explanation
$k1$	$2^t q$	$2^t q = g_1 + g_2$ where g_1 and g_2 are the lengths of Golay sequences [113]

Source: Adapted from Seberry and Yamada [166, p. 540], Wiley.

A.1.10 Yang Numbers

Four suitable sequences of length t can be multiplied by y to get four suitable sequences yt where $y = 1, \dots, 31, 51, 59, 2g + 1$ where g is the length of the Golay sequence.

A.1.11 Agaian Multiplication

Table A.9 gives the Agaian multiplication key construction method.

Table A.9 Agaian multiplication key method of construction.

Key	Method	Explanation
$d1$	$2^{t+s-1}pq$	Where $2^t p$ and $2^s q$ are the orders of Hadamard matrices [1]

Source: Adapted from Seberry and Yamada [166, p. 540], Wiley.

A.1.12 Craigen–Seberry–Zhang

Table A.10 gives the Craigen–Seberry–Zhang key construction method.

Table A.10 Craigen–Seberry–Zhang key method of construction.

Key	Method	Explanation
cz	$2^{t+s+u+w-4}$	Where $4a, 4b, 4c, 4d$ are the orders of Hadamard matrices [37]

Source: Adapted from Seberry and Yamada [166, p. 540], Wiley.

A.1.13 de Launey

Table A.11 gives the de Launey key construction.

Table A.11 de Launey key method of construction.

Key	Method	Explanation
$d\ell$	$2^7 ab \dots k\ell$	Where $4a, 4b, \dots, 4k, 4\ell$ are the orders of Hadamard matrices [44]

A.1.14 Seberry/Craigen Asymptotic Theorems

Table A.12 gives the Seberry/Craigen key construction method.

Table A.12 Seberry/Craigen key methods of construction.

Key	Method	Explanation
se	$2^t q$	Where t is the smallest integer such that for given odd q, $a(q+1)+b(q-3)=2^t$ has a solution for a, b nonnegative integers [213]
*cr*1	$2^t q$	Of order $2^{2b}t$, where t is the number of nonzero digits in the binary expansion of t, and
*cr*2	$2^t q$	Of order $2^s t$ for $s = 6\left\lfloor \frac{1}{16}((t-1)/2)\right\rfloor + 2$, t a positive integer [30]

A.1.15 Yang's Theorems and Đoković Updates

Table A.13 uses the powerful theorems of Yang [252] to give the updates and corrections to previous published tables [22, 166].

Table A.13 Yang key methods of construction.

Key	Method	Explanation
yy	$4yh(r+s)w$	Where y is a Yang number, and there exist $WL(h)$, $BS(r,s)$, and Williamson type matrices of order w [62]

A.1.16 Computation by Đoković

In Table A.14 Đoković [62] gives a few examples and lists the acceptable choices for the parameters $y, h, (r,s), w$ (see Table A.13). In some cases, there are several such choices, which may give different constructions for $HM(4n)$.

A.2 Index of Williamson Matrices

Table A.15 gives a list of constructions for these matrices, the methods used, and the discoverer – with apologies to any one excluded.

Good matrices exist for $S = \{1, \dots, 39\}$.

Note: The fact that if there is a Williamson matrix of order n then there is a Williamson matrix of order $2n$, is used in the calculation of *wh*.

We now give in Table A.16 known Williamson-type matrices of orders < 2000. The order in which the algorithms were applied was $w_1, w_2, w_3, w_4, w_5, w_6$, *wi*, *wj*, *wk*, *wl*, *wn*, *w#q*, *w#r*, and then others if it appeared they might give a new order. To interpret the results in the table, we note that if there is an Hadamard matrix of order $4q$, then it can be a Williamson-type matrix, but this was not included. A notation *w#x* means that 8-Williamson matrices are known, but not four, so an $OD(8s; s, s, s, s, s, s, s, s)$ is needed to get an Hadamard matrix. The notation 47, 3, *w#p* means that there are 8-Williamson matrices of order 47, and thus an Hadamard matrix of order $8 \cdot 47 = 2^3 \cdot 47$. A notation with *wn* indicates that there are four Williamson-type matrices but they are of even order. The notation 35, 3, *wn* means that there are four Williamson-type matrices of order 70 and an Hadamard matrix of order 280.

Table A.14 Parameters for the construction of $HM(4n)$.

n	y	h	(r,s)	w	n	y	h	(r,s)	w
2773	59	1	(24,23)	1	4495	31	5	(15,14)	1
3953	59	1	(34,33)	1		31	5	(1,0)	29
4097	4097	1	(1,0)	1		31	1	(15,14)	5
	1	1	(2049,2048)	1		31	1	(3,2)	29
4389	19	1	(17,16)	7		29	5	(16,15)	1
	19	1	(11,10)	11		29	5	(1,0)	31
	19	1	(6,5)	21		29	1	(3,2)	31
	19	1	(4,3)	33		29	1	(16,15)	5
	11	1	(29,28)	7	5201	5201	1	(1,0)	1
	11	1	(11,10)	19		1	1	(2601,2600)	1
	11	1	(10,9)	21	5875	25	5	(24,23)	1
	11	1	(4,3)	57		25	1	(24,23)	5
	11	1	(1,0)	399		5	1	(24,23)	25
	7	1	(29,28)	11		5	5	(24,23)	5
	7	1	(17,16)	19		1	5	(24,23)	25
	7	1	(10,9)	33	5913	1	1	(41,40)	73
	7	1	(6,5)	57	7373	1	1	(100,1)	73
	7	1	(1,0)	627	9065	49	5	(19,18)	1
	1	1	(6,5)	399		49	5	(1,0)	37
4453	1	1	(31,30)	73		49	1	(19,18)	5
						49	1	(3,2)	37
						37	5	(25,24)	1
						37	5	(1,0)	49
						37	1	(25,24)	5
						37	1	(3,2)	49

Source: Đoković [62, table 1, p. 257], Reproduced with permission of Wiley & Sons.

A.3 Tables of Hadamard Matrices

Table A.17 gives the orders of known Hadamard matrices. The table gives the odd part q of an order, the smallest power of 2, t, for which the Hadamard matrix is known and a construction method. If there is no entry in the t column the power is 2. Thus, there are Hadamard matrices known of orders $2^2 \cdot 105$ and $2^3 \cdot 179$. Order 107 was found by Kharaghani and Tayfeh-Rezaie [109]. We see at a glance, therefore, that the smallest order for which an Hadamard matrix is not yet known is $4 \cdot 167$. Since the theorems of Seberry and Craigen ensure that a t exists for every q, there is either a t entry for each q, or $t = 2$ is implied.

With the exception of order $4 \cdot 163$, marked dj, which was announced by Đoković [54], the method of construction used is indicated. The order in which the algorithms were applied reflects the fact that other tables were being constructed at the same time. Hence, the "Amicable Hadamard," "Skew Hadamard," "Conference Matrix," "Williamson Matrix," direct "Complex Hadamard" were implemented first (in that order). The tables reflect this and not the priority in time of a construction or its discoverer.

Table A.15 Williamson matrices key method of construction.

Key	Method	Explanation
$w1$	$\{1,\ldots,33,37,41,43\}$	[57, 112, 233]
$w2$	$\frac{1}{2}(p+1)$	$p \equiv 1 \pmod 4$ a prime power [82, 192, 221]
$w3$	9^d	d a natural number [139, 195]
$w4$	$\frac{1}{2}(p(p+1))$	$p \equiv 1 \pmod 4$ a prime power [208, 222]
$w5$	$s(4s+3), s(4s-1)$	$s \in \{1,3,5,\ldots,31\}$ [211]
$w6$	93	[211]
$w7$	$\frac{1}{4}((f-1)(4f+1))$	$p = 4f+1, f$ odd, is a prime power of the form $1+4t^2$, $\frac{1}{8}(f-1)$ is the order of a good matrix [210]
$w8$	$\frac{1}{4}((f+1)(4f+1))$	$p = 4f+1, f$ odd, is a prime power of the form $25+4t^2$, $\frac{1}{8}(f+1)$ is the order of a good matrix [210]
$w9$	$\frac{1}{2}(p(p-1))$	$p = 4f+1$ is a prime power and $\frac{1}{4}(p-1)$ is the order of a good matrix [210]
$w0$	$(p+2)(p+1)$	$p \equiv 1 \pmod 4$ a prime power, $p+3$ is the order of a symmetric Hadamard matrix [210]
wa	$\frac{1}{2}((f+1)(4f+1))$	$p = 4f+1, f$ odd, is a prime power of the form $9+4t^2$, $\frac{1}{2}(f-1) \equiv 1 \pmod 4$ a prime power [210]
wb	$\frac{1}{2}((f-1)(4f+1))$	$p = 4f+1, f$ odd, is a prime power of the form $49+4t^2$, $\frac{1}{2}(f-3) \equiv 1 \pmod 4$ a prime power [210]
wc	$2p+1$	$q = 2p-1$ is a prime power p is a prime [137, 165]
wd	7.3^i	$i \geq 0$ [139]
$w\#e$	$7^{i+1}, 11.7^i$	$i \geq 0$ (Gives 8-Williamson matrices) [154]
wf	$\frac{1}{2}(q^d(q+1))$	$q \equiv 1 \pmod 4$ is a prime $d \geq 2$ [139, 176]
wg	$\frac{1}{2}(p^2(p+1))$	$p \equiv 1 \pmod 4$ is a prime power [155]
wh	$\frac{1}{4}(p^2(p+1))$	$p \equiv 3 \pmod 4$ is a prime power and $\frac{1}{4}(p+1)$ is the order of a Williamson type matrix [155]
wi	q^2	$q \equiv 1 \pmod 4$ is a prime power [230]
wi	$q+2$	$q \equiv 1 \pmod 4$, is a prime power and $\frac{1}{2}(q+1)$ is a prime power [137]
wj	$q+2$	$q \equiv 1 \pmod 4$, is a prime power; $\frac{1}{2}(q+3)$ is the order of a symmetric conference matrix [137]
wk	q	$q \equiv 1 \pmod 4$ is a prime power; $\frac{1}{2}(q-1)$ is the order of a symmetric conference matrix or the order of a symmetric Hadamard matrix [137]
wl	q	$q \equiv 1 \pmod 4$, is a prime power and $\frac{1}{4}(q-1)$ is the order of a Williamson type matrix [137]
wm	q	$q \equiv 1 \pmod 4$, is a prime power and $\frac{1}{2}(q-1)$ is the order of a Hadamard matrix [165]
wn	wn	w is the order of a Williamson type matrix n is the order of a symmetric conference matrix [165]
wx	q^2	$q \equiv 1 \pmod 4$ a prime power Xia and Lui [227], Xiang [231]
wo	$2wu$	w and u are the orders of Williamson type matrices [165]
$w\#p$	$2q+1$	$q+1$ is the order of an amicable Hadamard matrix and q is the order of a Williamson type matrix [165]
$w\#q$	q	q is a prime power and $\frac{1}{2}(q-1)$ is the order of a Williamson type matrix [165]
$w\#r$	$2q+1$	$q \equiv 1 \pmod 4$ is a prime power or $q+1$ is the order of a symmetric conference matrix and q is the order of a Williamson type matrix [165]
$w\#s$	2.9^t+1	$t > 0$ [165]

Table A.16 Williamson and Williamson-type matrices.

q	t	How	q	t	How	q	t	How	q	t	How
1		w1	83		wi	165	3	wn	247	3	wn
3		w1	85		w2	167	3	w#p	249	3	wn
5		w1	87		w2	169		w2	251	3	w#q
7		w1	89		wl	171	3	wn	253	3	wn
9		w1	91		w2	173		wl	255		w2
11		w1	93		w5	175		w2	257		wl
13		w1	95		w5	177		w2	259	3	wn
15		w1	97		w2	179	3	w#q	261		w2
17		w1	99		w2	181		w2	263		
19		w1	101		wk	183	3	wn	265		w2
21		w1	103	3	w#q	185	3	wn	267	3	wn
23		w1	105	3	wn	187		w2	269		
25		w1	107	3	w#q	189		w5	271		w2
27		w1	109		wk	191	3	w#p	273	3	wn
29		w1	111	3	wn	193		wk	275	3	wn
31		w1	113		wk	195		w2	277		wk
33		w1	115		w2	197		wk	279		w2
35	3	wn	117		w2	199		w2	281		wl
37		w1	119	3	wn	201		w2	283	3	w#q
39		wi	121		w2	203	3	w9	285		w2
41		w1	123		wi	205		w2	287	3	wn
43		w1	125		wk	207	3	wn	289		w2
45		w2	127	3	w#p	209	3	wn	291	3	wn
47	3	w#p	129		w2	211		w2	293		wl
49		w2	131			213			295		
51		w2	133	3	wn	215	3	wn	297		w2
53		wk	135		w2	217		w2	299	3	wn
55		w2	137		wl	219	3	wn	301		w2
57		w2	139		w2	221	3	wn	303	3	w7
59	3	w#q	141		w2	223			305	3	wn
61		w2	143	3	wn	225		w2	307		w2
63		w2	145		w2	227	3	w#q	309		w2
65	3	wn	147		w2	229		w2	311		
67	3	w#q	149		wk	231		w2	313		w2
69		w2	151	3	w#q	233		wl	315		w5
71			153		w4	235			317		wk
73		wk	155	3	wn	237	3	wn	319	3	wo
75		w2	157		w2	239			321		w2
77	3	wn	159		w2	241		wk	323	3	wn
79		w2	161	3	wn	243		wj	325		w4
81		w3	163	3	w#q	245	3	wn	327		w2

(Continued)

Table A.16 (*Continued*)

q	t	How	q	t	How	q	t	How	q	t	How
329			411		w2	543		wi	625		w2
331		w2	413			545	3	wn	627		wi
333	3	w9	415		w2	547		w2	629	3	wn
335			417	3	wn	549		w2	631	3	w#q
337		w2	419			551	3	wn	633	3	wn
339		w2	421		w2	553	3	wn	635	3	w#r
341	3	wn	423		wi	555		w2	637	3	wn
343	3	wn	425	3	wn	557		wk	639		w2
345	3	wn	427		w2	559		w2	641		wk
347	3	w#q	429		w2	561	3	wn	643	3	w#q
349		wk	431			563	3	w#q	645		w2
351		w2	433		wk	565		w2	647		
353		wl	435		w4	567	3	wn	649		w2
355		w2	437	3	wn	569		wm	651		w2
357	3	wn	439		w2	571	3	w#q	653		
359			441		w2	573			655	3	w#p
361		wk	443			575	3	wn	657	3	wn
363		wi	445	3	wn	577		w2	659		
365		w2	447	3	wn	579		wj	661		w2
367		w2	449		wk	581	3	wn	663		w5
369	3	wn	501			583	3	wo	665	3	wn
371	3	wn	503			585	3	wn	667	3	wn
373		wl	505		w2	587	3	w#q	669		
375		wf	507		w2	589	3	wn	671	3	wn
377	3	wn	509			591		w2	673		wk
379		w2	511		w2	593		wk	675		wi
381		w2	513	3	wn	595	3	wn	677		wk
383			515	3	w#r	597		w2	679	3	wn
385		w2	517		w2	599			681		w2
387		w2	519	3	wn	601		w2	683		
389		wk	521		wl	603	3	wn	685		w2
391	3	wn	523	3	w#q	605	3	wn	687		w2
393			525		w2	607		w2	689	3	w9
395	3	wn	527	3	wn	609		w2	691		w2
397		wk	529		wl	611			693	3	wn
399		w2	531		w2	613		wl	695	3	wn
401		wk	533	3	wn	615		w2	697	3	wn
403	3	wn	535		w2	617		wl	699	3	wn
405		w2	537			619		w2	701		wk
407	3	wn	539	3	wn	621	3	wn	703		w4
409		wk	541		wk	623	3	wn	705		w2

(*Continued*)

Table A.16 (*Continued*)

q	t	How	q	t	How	q	t	How	q	t	How
707	3	*wn*	789			871		*w*2	953		*wl*
709		*wk*	791	3	*wn*	873	3	*wn*	955		
711	3	*wn*	793	3	*wn*	875	3	*wn*	957		*w*2
713	3	*wn*	795	3	*wn*	877		*w*2	959	3	*wn*
715		*w*2	797		*wk*	879		*wi*	961		*wk*
717		*w*2	799		*w*2	881		*wk*	963	3	*wn*
719			801		*w*2	883	3	*w*#*q*	965	3	*wn*
721			803	3	*wo*	885		*w*5	967		*w*2
723	3	*wn*	805		*w*2	887			969	3	*wn*
725	3	*wn*	807		*w*2	889		*w*2	971		
727		*w*2	809		*wl*	891	3	*wn*	973	3	*wn*
729		*w*3	811		*w*2	893			975		*w*2
731	3	*wo*	813	3	*wn*	895		*w*2	977		*wk*
733	3	*w*#*q*	815			897	3	*wn*	979	3	*wo*
735		*wi*	817	3	*wn*	899	3	*wn*	981	3	*wn*
737			819		*w*2	901		*w*2	983		
739			821		*wk*	903	3	*wn*	985	3	*wn*
741		*w*2	823	3	*w*#*q*	905	3	*wn*	987		*w*2
743			825	3	*wn*	907			989	3	*wn*
745		*w*2	827			909	3	*wn*	991		
747		*w*2	829		*w*2	911			993	3	*wn*
749			831	3	*wn*	913	3	*wo*	995	3	*wn*
751	3	*w*#*q*	833	3	*wn*	915	3	*w*9	997		*w*2
753			835		*w*2	917			999		*w*2
755			837	3	*wn*	919	3	*w*#*q*	1001	3	*wn*
757		*wl*	839			921	3	*wn*	1003		
759		*wi*	841		*w*2	923	3	*w*#*r*	1005	3	*wn*
761		*wl*	843		*wi*	925		*w*2	1007	3	*wn*
763			845	3	*wn*	927	3	*wn*	1009		*w*2
765	3	*wn*	847		*w*2	929		*wl*	1011	3	*wn*
767			849		*w*2	931		*w*2	1013	3	*wn*
769		*wk*	851	3	*wn*	933			1015		*w*2
771	3	*wn*	853			935	3	*wn*	1017	3	*wn*
773		*wl*	855		*w*2	937		*w*2	1019	3	*w*#*r*
775		*w*2	857			939		*w*2	1021		*wk*
777		*w*2	859	3	*w*#*q*	941			1023	3	*wn*
779	3	*wn*	861		*w*2	943	3	*wn*	1025	3	*wn*
781			863			945		*w*2	1027		*w*2
783	3	*wn*	865	3	*wn*	947	3	*w*#*q*	1029	3	*wn*
785	3	*wn*	867		*w*2	949	3	*wn*	1031		
787			869	3	*wn*	951		*w*2	1033		*wk*

(*Continued*)

Table A.16 (*Continued*)

q	t	How	q	t	How	q	t	How	q	t	How
1035		w2	1115	3	w#r	1195		w2	1275		w2
1037	3	wn	1117		wk	1197		w2	1277	3	w#q
1039			1119		w2	1199	3	wo	1279		w2
1041		w2	1121			1201		w2	1281	3	wn
1043	3	wn	1123			1203		wi	1283	3	w#q
1045		w2	1125	3	wn	1205	3	wn	1285	3	wn
1047	3	wn	1127	3	wn	1207		w5	1287	3	wn
1049		wm	1129		wk	1209		w2	1289		wl
1051	3	w#q	1131	3	wn	1211	3	wn	1291	3	w#q
1053	3	wn	1133			1213	3	w#q	1293		
1055	3	wn	1135		w2	1215		wi	1295	3	wn
1057		w2	1137		w2	1217		wk	1297		w2
1059	3	wn	1139		w5	1219		w2	1299	3	wn
1061		wk	1141		w2	1221		w2	1301		wl
1063	3	w#q	1143	3	wn	1223			1303	3	w#q
1065		w2	1145	3	wn	1225		w4	1305		w2
1067	3	wn	1147		w2	1227	3	wn	1307	3	w#r
1069		w2	1149		w2	1229		wk	1309		w2
1071		w2	1151			1231	3	w#p	1311		w2
1073	3	wn	1153		wk	1233	3	wn	1313	3	wn
1075	3	wn	1155		w2	1235	3	wn	1315		
1077		w2	1157	3	wn	1237		w2	1317		w2
1079	3	wn	1159	3	wn	1239		w2	1319		
1081		w2	1161	3	wn	1241	3	wo	1321		wl
1083	3	wn	1163			1243	3	wn	1323		wi
1085	3	wn	1165	3	wn	1245	3	wn	1325	3	wn
1087	3	w#p	1167		w2	1247	3	wo	1327	3	w#p
1089	3	wn	1169			1249		wl	1329		w2
1091			1171		w2	1251		wi	1331	3	wn
1093	3	w#q	1173	3	wn	1253			1333	3	wn
1095		wi	1175			1255			1335	3	wn
1097		wl	1177			1257			1337		
1099		w2	1179		w2	1259			1339		w2
1101	3	wn	1181			1261		w2	1341		wb
1103			1183		wf	1263	3	wn	1343	3	wn
1105		w2	1185	3	wn	1265	3	wn	1345		w2
1107		w2	1187	3	w#q	1267	3	wn	1347		w2
1109		wl	1189		w2	1269	3	wn	1349		
1111		w2	1191		w2	1271	3	wn	1351	3	wn
1113	3	wn	1193		wl	1273			1353	3	wn

(*Continued*)

Table A.16 (*Continued*)

q	t	How	q	t	How	q	t	How	q	t	How
1355	3	wn	1435	3	wn	1515	3	wn	1595	3	wn
1357		w2	1437			1517	3	wn	1597		wk
1359			1439			1519		w2	1599	3	wn
1361		wk	1441			1521		w2	1601		wk
1363			1443	3	wn	1523	3	w#q	1603	3	wn
1365		w2	1445	3	wn	1525		w2	1605		w2
1367			1447			1527			1607		
1369		wl	1449		w2	1529	3	wn	1609		w2
1371		w2	1451			1531		w2	1611		w2
1373	3	w#q	1453		wl	1533	3	wn	1613	3	w#q
1375		w2	1455		w2	1535	3	wn	1615		w2
1377		w2	1457			1537	3	wo	1617	3	wn
1379	3	wn	1459		w2	1539	3	wn	1619	3	w#q
1381	3	w#q	1461			1541			1621		wk
1383		wi	1463	3	wn	1543			1623		wi
1385	3	wn	1465	3	wn	1545		w2	1625	3	wn
1387	3	wn	1467	3	wn	1547	3	wn	1627		w2
1389		w2	1469	3	wn	1549		wk	1629		w2
1391			1471	3	w#p	1551	3	wn	1631	3	wn
1393	3	wn	1473			1553		wk	1633		
1395		w2	1475			1555		w2	1635	3	wn
1397			1477		w2	1557	3	wn	1637		wl
1399		w2	1479		w2	1559			1639	3	wo
1401		w2	1481		wl	1561		w2	1641	3	wn
1403	3	wn	1483	3	w#q	1563		w2	1643	3	wn
1405		w2	1485		w2	1565	3	wn	1645		
1407	3	wn	1487			1567			1647	3	wn
1409		wl	1489		wl	1569		w2	1649	3	wn
1411	3	wo	1491			1571			1651		w2
1413	3	wn	1493		wl	1573	3	wn	1653	3	wn
1415			1495	3	wn	1575	3	wn	1655	3	wn
1417		w2	1497	3	wn	1577	3	wn	1657		w2
1419		w2	1499			1579			1659		wi
1421	3	wn	1501		w2	1581	3	wn	1661		
1423			1503			1583			1663		
1425		w5	1505	3	wn	1585		w2	1665		w2
1427			1507	3	wo	1587		wh	1667		
1429		w2	1509			1589			1669	3	w#q
1431		w2	1511			1591		w2	1671	3	wn
1433			1513	3	wn	1593	3	wn	1673		

(*Continued*)

Table A.16 *(Continued)*

q	t	How	q	t	How	q	t	How	q	t	How
1675			1757			1839		w2	1921	3	wn
1677	3	wn	1759		w2	1841			1923	3	wn
1679	3	wn	1761			1843	3	wn	1925	3	wn
1681		w2	1763	3	wn	1845	3	wn	1927		w2
1683		wj	1765		w2	1847			1929		
1685	3	wn	1767		w2	1849		w2	1931		
1687		w2	1769	3	wn	1851		w2	1933	3	w#q
1689			1771		w2	1853	3	wo	1935		wi
1691	3	wn	1773	3	wn	1855		w2	1937	3	wn
1693		wl	1775	3	wn	1857	3	wn	1939		w2
1695		n	1777		wl	1859	3	wn	1941		w2
1697		wl	1779		w2	1861		w2	1943		
1699	3	w#q	1781	3	wn	1863	3	wn	1945		w2
1701	3	wn	1783			1865	3	wn	1947	3	wn
1703			1785	3	wn	1867		w2	1949		
1705	3	wn	1787			1869	3	wn	1951	3	w#p
1707		w2	1789	3	w#q	1871			1953	3	wn
1709		wk	1791		w2	1873		wk	1955	3	wn
1711			1793			1875		wf	1957		
1713			1795			1877		wk	1959		w2
1715	3	w#r	1797		w2	1879	3	w#q	1961	3	wn
1717		w2	1799	3	wn	1881		w2	1963		
1719			1801		wk	1883	3	w#r	1965		w2
1721		wl	1803	3	wn	1885		w2	1967	3	wn
1723	3	w#q	1805		wh	1887	3	wn	1969		
1725		w2	1807		w2	1889		wm	1971	3	wn
1727	3	wn	1809		w2	1891		w4	1973	3	w#q
1729		w2	1811			1893			1975	3	wn
1731		w2	1813	3	wn	1895	3	wn	1977		
1733		wl	1815	3	wn	1897		w2	1979		
1735		w2	1817	3	wn	1899		w2	1981		
1737	3	wn	1819		w2	1901	3	w#q	1983	3	wn
1739			1821	3	wn	1903	3	wo	1985	3	wn
1741		w2	1823	3	wn	1905	3	wn	1987		
1743		w5	1825	3	wn	1907	3	w#q	1989	3	wn
1745	3	wn	1827		w5	1909	3	wn	1991	3	wn
1747			1829			1911		w2	1993		wl
1749	3	wn	1831			1913			1995		w2
1751			1833	3	wn	1915			1997		wk
1753		wl	1835	3	wn	1917		w2	1999	3	w#p
1755		wi	1837		w2	1919	3	wn			

Source: Adapted from Seberry and Yamada [166, table 9.1, pp. 514–517 and table A.1, pp. 543–547], Wiley & Sons.

Table A.17 Orders of known Hadamard matrices.

q	*t*	How	*q*	*t*	How	*q*	*t*	How	*q*	*t*	How
1		*a*1	77		*a*1	153		*w*4	229		*c*1
3		*a*1	79		*c*1	155		*a*1	231		*c*1
5		*a*1	81		*w*3	157		*c*1	233		*a*2
7		*c*1	83		*a*1	159		*c*1	235		*o*1
9		*c*1	85		*c*1	161		*a*1	237		*a*1
11		*a*1	87		*a*1	163		*a*1	239		*a*1
13		*c*1	89		*a*2	165		*a*1	241		*wk*
15		*a*1	91		*c*1	167	3	*ok*	243		*a*1
17		*a*1	93		*w*5	169		*c*1	245		*o*1
19		*c*1	95		*a*1	171		*a*1	247		*o*1
21		*a*1	97		*c*1	173		*a*1	249		*o*2
23		*w*1	99		*c*1	175		*c*1	251		*yy*
25		*c*1	101		*wk*	177		*c*1	253		*o*1
27		*a*1	103		*y*2	179		*w#q*	255		*a*1
29		*a*2	105		*a*1	181		*c*1	257		*wl*
31		*c*1	107		*ktr*	183		*o*2	259		*o*1
33		*a*1	109		*wk*	185		*a*1	261		*c*1
35		*a*1	111		*a*1	187		*c*1	263		*a*1
37		*c*1	113		*a*2	189		*w*5	265		*c*1
39		*wi*	115		*c*1	191	3	*dj*	267		*o*2
41		*a*1	117		*a*1	193		*wk*	269		*m*2
43		*w*1	119		*o*1	195		*c*1	271		*c*1
45		*a*1	121		*c*1	197		*a*1	273		*a*1
47		*o*1	123		*a*1	199		*c*1	275		*o*1
49		*c*1	125		*a*1	201		*c*1	277		*wk*
51		*c*1	127		*y*2	203		*a*1	279		*c*1
53		*a*1	129		*c*1	205		*c*1	281		*a*1
55		*c*1	131		*a*1	207		*a*1	283	3	*w#q*
57		*a*1	133		*o*1	209		*o*1	285		*c*1
59		*o*1	135		*c*1	211		*c*1	287		*o*1
61		*c*1	137		*a*1	213		*yy*	289		*c*1
63		*a*1	139		*c*1	215		*a*1	291		*a*1
65		*o*1	141		*a*1	217		*c*1	293		*a*1
67		*o*1	143		*a*1	219		*o*2	295		*o*1
69		*c*1	145		*c*1	221		*a*1	297		*a*1
71		*a*1	147		*a*1	223	3	*a*1	299		*o*1
73		*wk*	149		*wk*	225		*c*1	301		*c*1
75		*c*1	151		*y*2	227		*a*1	303		*w*7

(Continued)

Table A.17 (Continued)

q	t	How	q	t	How	q	t	How	q	t	How
305		o1	381		a1	457		wk	533		a1
307		c1	383		a1	459		wi	535		c1
309		c1	385		c1	461		wk	537	3	o3
311	3	m3	387		c1	463		yy	539		o2
313		c1	389		wk	465		c1	541		wk
315		a1	391		o1	467		a1	543		wi
317		wk	393		a1	469		c1	545		a1
319		o1	395		a1	471		c1	547		c1
321		a1	397		wk	473		w5	549		c1
323		a1	399		c1	475		o1	551		a1
325		w4	401		wk	477		a1	553		o2
327		a1	403		o1	479	4	cr1	555		c1
329		o1	405		a1	481		c1	557		wk
331		c1	407		a1	483		a1	559		c1
333		w9	409		wk	485		o2	561		a1
335		o1	411		c1	487	3	w#p	563		a1
337		c1	413		o1	489		c1	565		c1
339		c1	415		c1	491	5	cr1	567		a1
341		o1	417		a1	493		o1	569		wm
343		o2	419	3	cr1	495		a1	571	3	a1
345		o1	421		c1	497		a1	573	3	a1
347	3	w#q	423		wi	499		c1	575		o1
349		wk	425		a1	501		a1	577		c1
351		c1	427		c1	503		a1	579		wj
353		wl	429		c1	505		c1	581		o2
355		c1	431		a1	507		a1	583		o1
357		a1	433		wk	509		a2	585		a1
359	4	a1	435		w4	511		c1	587		a1
361		wk	437		a1	513		o1	589		o2
363		a1	439		c1	515	3	w#r	591		c1
365		a1	441		c1	517		c1	593		a1
367		c1	443	3	m3	519		o2	595		o1
369		o1	445		o2	521		a1	597		c1
371		a1	447		a1	523	3	w#q	599	6	cr1
373		wl	449		wk	525		a1	601		c1
375		a1	451		wj	527		o1	603		a1
377		o1	453		a1	529		wl	605		o2
379		c1	455		o1	531		c1	607		c1

(Continued)

Table A.17 (*Continued*)

q	t	How	q	t	How	q	t	How	q	t	How
609		c1	685		c1	761		a2	837		a1
611		o1	687		c1	763		o2	839	4	cr1
613		wl	689		w9	765		o1	841		c1
615		a1	691		c1	767		a1	843		a1
617		a1	693		o1	769		wk	845		o1
619		c1	695		o2	771		a1	847		c1
621		o1	697		o1	773		wl	849		c1
623		o2	699		o2	775		c1	851		o2
625		c1	701		a1	777		c1	853	3	a1
627		wi	703		w4	779		o1	855		c1
629		o1	705		a1	781	3	a1	857		m2
631		yy	707		o2	783		o1	859	3	a1
633		a1	709		wk	785		o2	861		c1
635		a1	711		a1	787	3	m3	863	3	m3
637		o2	713		a1	789	3	a1	865		o2
639		c1	715		c1	791		a1	867		a1
641		a2	717		c1	793		o2	869		o2
643	3	w#q	719	4	a1	795		o1	871		c1
645		a1	721	3	d1	797		a1	873		a1
647	3	m3	723		o2	799		c1	875		a1
649		c1	725		o1	801		a1	877		c1
651		c1	727		c1	803		o2	879		wi
653		a2	729		w3	805		c1	881		wk
655		y2	731		o2	807		c1	883	3	w#q
657		o2	733		m2	809		a2	885		a1
659	4	cr2	735		a1	811		c1	887		a1
661		c1	737		o2	813		a1	889		c1
663		w5	739	6	cr2	815		a1	891		o1
665		a1	741		a1	817		o2	893		a1
667		o2	743		a1	819		c1	895		c1
669	3	a1	745		c1	821		wk	897		o2
671		a1	747		c1	823	3	w#q	899		o2
673		wk	749	4	d1	825		a1	901		c1
675		a1	751	3	a1	827		a1	903		o1
677		a1	753		a1	829		c1	905		o2
679		o2	755		a1	831		a1	907	3	m3
681		c1	757		wl	833		a1	909		o1
683		a1	759		wi	835		c1	911		a1

(*Continued*)

Table A.17 (*Continued*)

q	t	How	q	t	How	q	t	How	q	t	How
913		o2	989		o2	1065		a1	1141		c1
915		a1	991	3	a1	1067		o2	1143		o2
917	3	o2	993		o2	1069		c1	1145		o2
919	3	a1	995		o2	1071		a1	1147		c1
921		o2	997		c1	1073		o2	1149		c1
923		a1	999		c1	1075		o2	1151		a1
925		c1	1001		a1	1077		c1	1153		wk
927		o2	1003		o1	1079		o2	1155		c1
929		wl	1005		a1	1081		c1	1157		o2
931		c1	1007		a1	1083		o2	1159		o2
933	4	d1	1009		c1	1085		a1	1161		a1
935		a1	1011		o2	1087	3	w#p	1163		a1
937		c1	1013		a1	1089		o1	1165		o2
939		c1	1015		c1	1091		a1	1167		c1
941		m2	1017		o2	1093	3	w#q	1169	4	k1
943		o1	1019	3	w#r	1095		wi	1171		c1
945		a1	1021		wk	1097		wl	1173		a1
947	3	w#q	1023		a1	1099		w2	1175		o1
949		o2	1025		a1	1101		o2	1177	4	d1
951		a1	1027		c1	1103	3	m3	1179		c1
953		a2	1029		o2	1105		c1	1181		a1
955	3	a1	1031	6	a1	1107		c1	1183		wf
957		c1	1033		wk	1109		wl	1185		o2
959		o2	1035		a1	1111		c1	1187	3	w#q
961		wk	1037		o2	1113		a1	1189		c1
963		a1	1039	3	a1	1115	3	w#r	1191		c1
965		o2	1041		c1	1117		wk	1193		wl
967		c1	1043		o2	1119		c1	1195		c1
969		o2	1045		c1	1121		a1	1197		a1
971	6	a1	1047		o2	1123	3	m3	1199		o2
973		o2	1049		a2	1125		o1	1201		c1
975		c1	1051	3	w#q	1127		a1	1203		wi
977		a1	1053		a1	1129		wk	1205		o2
979		o2	1055		a1	1131		a1	1207		w5
981		a1	1057		c1	1133	3	d1	1209		c1
983		a1	1059		o2	1135		c1	1211		o2
985		o2	1061		a1	1137		a1	1213		m2
987		a1	1063	3	w#q	1139		w5	1215		wi

(*Continued*)

Table A.17 (*Continued*)

q	*t*	How	*q*	*t*	How	*q*	*t*	How	*q*	*t*	How
1217		*wk*	1293		*a*1	1369		*wl*	1445		*a*1
1219		*c*1	1295		*a*1	1371		*a*1	1447	7	*cr*1
1221		*c*1	1297		*c*1	1373		*m*2	1449		*c*1
1223	6	*cr*1	1299		*o*2	1375		*c*1	1451	6	*a*1
1225		*w*4	1301		*wl*	1377		*a*1	1453		*wl*
1227		*o*2	1303	3	*w#q*	1379		*o*2	1455		*c*1
1229		*a*2	1305		*c*1	1381		*m*2	1457		*a*1
1231		*y*2	1307		*a*1	1383		*a*1	1459		*c*1
1233		*a*1	1309		*c*1	1385		*o*2	1461		*a*1
1235		*o*1	1311		*c*1	1387		*o*2	1463		*a*1
1237		*c*1	1313		*o*2	1389		*c*1	1465		*o*2
1239		*c*1	1315	3	*d*1	1391		*a*1	1467		*a*1
1241		*o*2	1317		*c*1	1393		*o*2	1469		*o*2
1243		*o*2	1319	4	*cr*1	1395		*c*1	1471	3	*w#p*
1245		*o*2	1321		*wl*	1397		*dj*	1473	3	*a*1
1247		*a*1	1323		*wi*	1399		*c*1	1475		*o*1
1249		*wl*	1325		*o*1	1401		*c*1	1477		*c*1
1251		*a*1	1327	3	*w#p*	1403		*o*2	1479		*c*1
1253		*a*1	1329		*c*1	1405		*c*1	1481		*a*1
1255	3	*a*1	1331		*a*1	1407		*o*1	1483	3	*a*1
1257	3	*cr*1	1333		*o*2	1409		*a*2	1485		*a*1
1259	4	*a*1	1335		*o*2	1411		*o*2	1487	3	*m*3
1261		*c*1	1337		*a*1	1413		*a*1	1489		*wl*
1263		*a*1	1339		*c*1	1415		*a*1	1491	3	*a*1
1265		*a*1	1341		*wb*	1417		*c*1	1493		*wl*
1267		*o*2	1343		*o*2	1419		*c*1	1495		*o*1
1269		*o*1	1345		*c*1	1421		*a*1	1497		*a*1
1271		*o*1	1347		*a*1	1423	3	*a*1	1499	7	*cr*1
1273		*o*2	1349	3	*d*1	1425		*w*5	1501		*c*1
1275		*a*1	1351		*o*2	1427	3	*m*3	1503		*a*1
1277		*a*1	1353		*o*1	1429		*c*1	1505		*o*2
1279		*c*1	1355		*a*1	1431		*c*1	1507		*o*2
1281		*o*1	1357		*c*1	1433		*m*2	1509	3	*a*1
1283	3	*w#q*	1359	3	*d*1	1435		*o*1	1511		*a*1
1285		*o*1	1361		*a*1	1437	6	*a*1	1513		*o*2
1287		*a*1	1363		*o*2	1439	8	*cr*2	1515		*o*2
1289		*a*2	1365		*c*1	1441	3	*a*1	1517		*a*1
1291	3	*w#q*	1367	3	*m*3	1443		*o*2	1519		*c*1

(*Continued*)

Table A.17 (Continued)

q	t	How	q	t	How	q	t	How	q	t	How
1521		c1	1595		a1	1669	3	w#q	1743		a1
1523		a1	1597		wk	1671		o2	1745		o2
1525		c1	1599		o2	1673		a1	1747	3	m3
1527	3	d1	1601		a2	1675		o2	1749		o1
1529		o2	1603		o2	1677		o1	1751	3	d1
1531		c1	1605		c1	1679		o2	1753		wl
1533		a1	1607		a1	1681		c1	1755		a1
1535		o2	1609		c1	1683		wj	1757		a1
1537		o1	1611		c1	1685		o2	1759		c1
1539		o1	1613		a1	1687		c1	1761		a1
1541		a1	1615		c1	1689	3	o3	1763		o2
1543	3	a1	1617		o2	1691		a1	1765		c1
1545		c1	1619	3	w#q	1693		wl	1767		c1
1547		o2	1621		wk	1695		a1	1769		o2
1549		wk	1623		a1	1697		wl	1771		c1
1551		a1	1625		o1	1699	3	a1	1773		o2
1553		a1	1627		c1	1701		a1	1775		o2
1555		c1	1629		c1	1703	3	o2	1777		wl
1557		o2	1631		o2	1705		o2	1779		c1
1559	4	a1	1633	3	a1	1707		a1	1781		o2
1561		c1	1635		o2	1709		wk	1783	6	cr1
1563		w2	1637		a1	1711		o2	1785		o1
1565		o2	1639		o2	1713	3	o3	1787	3	m3
1567	6	cr1	1641		a1	1715		o2	1789	3	w#q
1569		c1	1643		a1	1717		c1	1791		c1
1571	7	cr2	1645		o1	1719	3	a1	1793	4	a1
1573		o2	1647		o2	1721		a1	1795	5	d1
1575		a1	1649		o2	1723	3	w#q	1797		a1
1577		o2	1651		c1	1725		a1	1799		o1
1579	5	a1	1653		o2	1727		a1	1801		wk
1581		a1	1655		a1	1729		c1	1803		a1
1583	3	m3	1657		c1	1731		c1	1805		a1
1585		c1	1659		wi	1733		a2	1807		c1
1587		wh	1661	3	d1	1735		c1	1809		c1
1589	3	d1	1663	3	m3	1737		a1	1811		a1
1591		c1	1665		a1	1739		o2	1813		o2
1593		o1	1667	3	m3	1741		c1	1815		o1

(Continued)

Table A.17 (Continued)

q	t	How	q	t	How	q	t	How	q	t	How
1817		o2	1893	3	o3	1969	4	o2	2045		a1
1819		c1	1895		o2	1971		a1	2047		c1
1821		a1	1897		c1	1973		a2	2049		o1
1823	3	wn	1899		c1	1975		o2	2051		o2
1825		o2	1901		a1	1977		a1	2053	3	w#q
1827		a1	1903		o2	1979	4	a1	2055		a1
1829		o2	1905		o2	1981	4	d1	2057		o2
1831	3	m3	1907	3	w#q	1983		o2	2059	3	d1
1833		a1	1909		o2	1985		o2	2061		a1
1835		o2	1911		a1	1987	7	cr2	2063	7	cr1
1837		c1	1913	7	cr2	1989		o1	2065		c1
1839		c1	1915	3	a1	1991		a1	2067		c1
1841	3	d1	1917		c1	1993		wl	2069		a2
1843		o2	1919		o2	1995		c1	2071		o2
1845		o1	1921		o2	1997		wk	2073		a1
1847	3	m3	1923		a1	1999		y2	2075		o2
1849		c1	1925		a1	2001		c1	2077		c1
1851		c1	1927		c1	2003		a1	2079		c1
1853		a1	1929	3	o3	2005		o1	2081		wl
1855		c1	1931		a1	2007		c1	2083	3	w#q
1857		o2	1933	3	w#q	2009		o2	2085		o1
1859		o2	1935		wi	2011		c1	2087	4	a1
1861		c1	1937		o2	2013		o2	2089		c1
1863		a1	1939		c1	2015		a1	2091		a1
1865		a1	1941		c1	2017		wk	2093		w5
1867		c1	1943		o2	2019		wi	2095	3	a1
1869		o2	1945		c1	2021		o2	2097		a1
1871	3	m3	1947		o1	2023		o2	2099	3	w#r
1873		wk	1949	4	a1	2025		c1	2101		c1
1875		a1	1951		y2	2027	3	w#r	2103		o2
1877		a1	1953		o1	2029		c1	2105		a1
1879	3	a1	1955		o1	2031		a1	2107		o2
1881		a1	1957	3	d1	2033	3	cr2	2109		c1
1883	3	w#r	1959		c1	2035		o2	2111		a1
1885		c1	1961		o1	2037		a1	2113		wl
1887		a1	1963	3	d1	2039	7	cr1	2115		c1
1889		a2	1965		c1	2041		o2	2117		a1
1891		w4	1967		a1	2043		a1	2119	4	d1

(Continued)

Table A.17 *(Continued)*

q	t	How	q	t	How	q	t	How	q	t	How
2121		*c*1	2197		*wk*	2273		*a*1	2349		*o*1
2123		*o*2	2199		*c*1	2275		*c*1	2351		*a*1
2125		*o*1	2201		*a*1	2277		*o*2	2353		*o*2
2127		*c*1	2203	3	*a*1	2279		*o*2	2355		*a*1
2129		*a*2	2205		*a*1	2281		*c*1	2357		*m*2
2131		*c*1	2207	4	*a*1	2283		*o*2	2359		*o*2
2133		*o*2	2209		*wk*	2285		*o*2	2361		*c*1
2135		*a*1	2211		*c*1	2287	7	*cr*1	2363		*o*2
2137		*c*1	2213		*m*2	2289		*o*2	2365		*c*1
2139		*wi*	2215	4	*d*1	2291		*o*2	2367		*a*1
2141		*a*1	2217		*a*1	2293	6	*cr*1	2369	3	*d*1
2143	3	*w#q*	2219		*o*2	2295		*o*1	2371	6	*cr*2
2145		*c*1	2221		*c*1	2297		*a*1	2373		*a*1
2147		*o*2	2223		*o*1	2299		*c*1	2375		*o*2
2149		*c*1	2225		*o*2	2301		*a*1	2377		*wk*
2151		*o*2	2227	3	*o*2	2303		*o*2	2379		*o*2
2153		*wm*	2229		*c*1	2305		*o*2	2381		*m*2
2155	3	*a*1	2231		*a*1	2307		*a*1	2383	3	*w#q*
2157		*a*1	2233		*o*2	2309		*wk*	2385		*a*1
2159		*dj*	2235		*o*2	2311		*c*1	2387		*a*1
2161		*wk*	2237		*wk*	2313		*o*1	2389		*wk*
2163		*o*2	2239		*y*2	2315	4	*a*1	2391		*o*2
2165		*o*2	2241		*a*1	2317		*o*2	2393		*a*2
2167		*o*2	2243		*a*1	2319		*c*1	2395		*c*1
2169		*c*1	2245		*c*1	2321		*a*1	2397		*a*1
2171	4	*d*1	2247		*c*1	2323		*o*2	2399	8	*a*1
2173		*o*1	2249		*o*2	2325		*c*1	2401		*c*1
2175		*a*1	2251	5	*a*1	2327	4	*o*2	2403		*wi*
2177		*a*1	2253		*a*1	2329		*c*1	2405		*a*1
2179		*c*1	2255		*o*1	2331		*a*1	2407		*c*1
2181		*o*2	2257		*c*1	2333		*m*2	2409		*c*1
2183		*a*1	2259		*c*1	2335	3	*a*1	2411		*a*1
2185		*w*5	2261		*a*1	2337		*c*1	2413		*dj*
2187		*a*1	2263		*o*2	2339	4	*a*1	2415		*o*1
2189		*o*2	2265		*a*1	2341	3	*w#q*	2417		*wm*
2191		*o*2	2267		*a*1	2343		*a*1	2419		*o*1
2193		*o*1	2269	3	*w#q*	2345		*o*2	2421		*o*2
2195		*a*1	2271		*o*2	2347	3	*m*3	2423	4	*a*1

(Continued)

Table A.17 (*Continued*)

q	*t*	How	*q*	*t*	How	*q*	*t*	How	*q*	*t*	How
2425		o2	2501		o2	2577		c1	2653		o2
2427		o2	2503	3	a1	2579	3	w#q	2655		c1
2429	4	d1	2505		c1	2581		o2	2657		a1
2431		c1	2507		o2	2583		a1	2659	3	w#q
2433		o2	2509		o2	2585		o2	2661	3	d1
2435		a1	2511		c1	2587		o2	2663		a1
2437		wk	2513	4	k1	2589	4	d1	2665		c1
2439		c1	2515	3	d1	2591	3	m3	2667		a1
2441		wl	2517		a1	2593		wl	2669		o2
2443		o2	2519		o2	2595		c1	2671	7	cr1
2445		c1	2521		c1	2597		o2	2673		a1
2447		a1	2523		a1	2599		c1	2675		o2
2449		o2	2525		a1	2601		wf	2677	6	cr1
2451		a1	2527		o2	2603		o2	2679		o2
2453		a1	2529		o2	2605		c1	2681		a1
2455		c1	2531	3	m3	2607		a1	2683	7	a1
2457		w2	2533		o2	2609		wm	2685		a1
2459	3	w#q	2535		a1	2611		o2	2687	9	cr2
2461	3	a1	2537		o2	2613		o1	2689		wk
2463		a1	2539		c1	2615		a1	2691		c1
2465		a1	2541		a1	2617		c1	2693		a1
2467		c1	2543	6	a1	2619		c1	2695		o2
2469		c1	2545	3	a1	2621		m2	2697		c1
2471		a1	2547		o2	2623		o2	2699	7	cr2
2473		wk	2549		a2	2625		a1	2701		w4
2475		w5	2551		c1	2627	3	d1	2703		o1
2477		a1	2553		a1	2629	3	a1	2705		o1
2479		c1	2555		o2	2631		c1	2707		c1
2481		a1	2557		c1	2633		a1	2709		c1
2483		a1	2559		wi	2635		o1	2711	3	m3
2485		c1	2561		a1	2637		c1	2713		wk
2487		c1	2563		o2	2639		o2	2715		a1
2489	3	o2	2565		a1	2641		c1	2717		a1
2491		o1	2567		a1	2643		o2	2719		c1
2493		o2	2569		o2	2645		o2	2721		a1
2495		o2	2571	3	d1	2647	3	w#q	2723		a1
2497		c1	2573		o2	2649		c1	2725		c1
2499		o1	2575		w5	2651		o2	2727		o2

(*Continued*)

Table A.17 (*Continued*)

q	*t*	How	*q*	*t*	How	*q*	*t*	How	*q*	*t*	How
2729		*wl*	2797		*m*2	2865	3	*d*1	2933		*a*1
2731	3	*w#q*	2799		*wi*	2867		*a*1	2935		*c*1
2733	3	*a*1	2801		*wk*	2869		*c*1	2937		*o*2
2735		*a*1	2803	3	*w#q*	2871		*a*1	2939	6	*cr*2
2737		*o*1	2805		*o*1	2873		*a*1	2941		*c*1
2739		*c*1	2807		*o*1	2875		*c*1	2943		*o*2
2741		*a*2	2809		*wk*	2877		*o*2	2945		*a*1
2743		*o*2	2811		*a*1	2879	7	*cr*1	2947		*o*2
2745		*a*1	2813		*a*1	2881		*o*2	2949		*c*1
2747		*a*1	2815	3	*d*1	2883		*o*2	2951	3	*d*1
2749		*wk*	2817		*o*2	2885		*o*2	2953		*wl*
2751		*a*1	2819	3	*w#q*	2887	5	*a*1	2955		*o*2
2753		*a*2	2821		*c*1	2889		*w*5	2957		*a*1
2755		*o*2	2823	3	*d*1	2891		*o*2	2959		*y*2
2757		*a*1	2825		*a*1	2893	3	*a*1	2961		*o*1
2759		*o*2	2827		*c*1	2895		*a*1	2963	3	*w#q*
2761		*c*1	2829		*c*1	2897		*a*1	2965		*o*2
2763		*o*2	2831		*o*2	2899	4	*d*1	2967		*a*1
2765		*a*1	2833		*wk*	2901		*c*1	2969		*a*2
2767	3	*w#p*	2835		*c*1	2903	4	*a*1	2971	3	*a*1
2769	3	*o*2	2837		*wl*	2905		*o*2	2973	4	*d*1
2771		*a*1	2839		*y*2	2907		*c*1	2975		*o*1
2773		*dj*	2841	3	*a*1	2909		*wk*	2977		*c*1
2775		*o*2	2843	3	*m*3	2911		*c*1	2979		*o*2
2777		*m*2	2845		*c*1	2913	6	*cr*2	2981		*a*1
2779		*c*1	2847		*c*1	2915		*o*1	2983		*o*2
2781		*o*2	2849		*o*2	2917		*wl*	2985		*a*1
2783		*a*1	2851		*c*1	2919		*wi*	2987	3	*w#r*
2785		*c*1	2853		*a*1	2921		*dj*	2989		*o*2
2787		*c*1	2855	4	*d*1	2923		*o*2	2991		*c*1
2789		*m*2	2857		*wl*	2925		*a*1	2993		*a*1
2791		*c*1	2859		*c*1	2927	3	*m*3	2995	6	*cr*1
2793		*a*1	2861		*a*1	2929		*c*1	2997		*a*1
2795		*o*2	2863		*o*2	2931		*c*1	2999	9	*cr*1

Source: Adapted from Seberry and Yamada [166, table 11.2, pp. 536–538 and table A.2, pp. 548–554], Wiley & Sons.

Next the "Spence," "Miyamoto," and "Yamada" direct constructions were applied because they were noticed to fill places in the table. The methods *o*1 and of Koukouvinos and Kounias were now applied as lists of ODs were constructed. These were then used to "plug-in" the Williamson-type matrices implementing methods *o*2 and *o*3.

Dietrich Burde, in a post on https://math.stackexchange.com/questions/1733695/known-classes-of-had on 8 April 2016, indicates orders 213, 239, 311, and 463 are known but gives no method.

Finally, the multiplication theorems of Agaian, Seberry, and Zhang were applied. The Craigen, Seberry, and Zhang theorem was applied to the table that one of us (Seberry), had in the computer. The method and order of application was by personal choice to improve the efficiency of implementation. This means that some authors, for example, Baumert, Hall, Turyn, and Whiteman, who have priority of construction are not mentioned by name in the final table.

B

List of sds from Cyclotomy

B.1 Introduction

The combinations of supplementary difference sets and the union of supplementary difference sets are also supplementary difference sets. If $D_0, \dots, D_{n-1}$ are sds, and if any D_i is replaced by its complement $D_i^c = F - D_i$, then $D_0, \dots, D_i^c, \dots, D_{n-1}$ are also sds. Let $C_i, 0 \le i \le e-1$ be cyclotomic classes of order e. If the transformation $f : C_i \to C_{i+a}$ where a is an integer, acts on the supplementary difference sets $D_0, \dots, D_{n-1}$ simultaneously, then $D_0^f, \dots, D_{n-1}^f$ are also sds. Thus, we define $D_0^f, \dots, D_{n-1}^f$ and $D_0, \dots, D_i^c, \dots, D_{n-1}$ are equivalent to the supplementary difference sets $D_0, \dots, D_{n-1}$. We give the representatives of equivalent classes of sds in the following table.

This appendix describes Yamada's list of $n - \{q; k_1, \dots, k_n : \lambda\}$ *sds* [239]. Table B.1 gives the existence of a $(q, \frac{1}{2}(q-1), \frac{1}{4}(q-3))$ difference set and $2 - \{q; \frac{1}{2}(q-1); \frac{1}{2}(q-3)\}$ *sds* with the reference. Thereafter, each table gives the existence of $n - \{q; k_1, \dots, k_n; \lambda\}$ *sds*.

B.2 List of $n - \{q; k_1, \dots, k_n : \lambda\}$ sds

Table B.1 $e = 2, n = 1, 2$.

n	q		λ	Reference
$n = 1$	$q \equiv 3 \pmod 4$	C_0	$\frac{1}{4}(q-3)$	Section 8.3
$n = 2$	$q \equiv 1 \pmod 4$	$C_0 \quad C_1$	$\frac{1}{2}(q-3)$	Lemma 1.17

Source: Yamada [239, figure 1, p. 84]. Reproduced with permission of Elsevier.

Table B.2 $e = 4, n = 1$.

q		λ
$q = 1 + 4b^2 \equiv 5 \pmod{16}$	C_0	$\frac{1}{16}(q-5)$
$q = 9 + 4b^2 \equiv 13 \pmod{16}$	$C_0 \cup C_1 \cup C_2$	$\frac{3}{16}(3q-7)$

Source: Yamada [239, figure 1, p. 84], Reproduced with permission of Elsevier.

Hadamard Matrices: Constructions using Number Theory and Algebra, First Edition. Jennifer Seberry and Mieko Yamada.

Table B.3 $e = 4, n = 2$.

q		λ
$q \equiv 5 \pmod 8$	$C_0 \quad C_1$	$\frac{1}{8}(q-5)$
$q = \frac{1}{5}(1+z^2) \equiv 13 \pmod{16}$, $z \equiv 2 \pmod 5$	$C_0 \quad C_0 \cup C_3$ $C_0 \quad C_1 \cup C_2$	$\frac{1}{16}(5q-17)$
$q = 25 + 4b^2 \equiv 13 \pmod{16}$	$C_0 \quad C_0 \cup C_2$	$\frac{1}{16}(5q-17)$
$q = 9 + 4b^2 \equiv 13 \pmod{16}$	$C_0 \quad C_1 \cup C_3$	$\frac{1}{16}(5q-17)$
$q = a^2 + 4b^2 \equiv 5 \pmod 8$, $b = 1$	$C_0 \cup C_1 \quad C_0 \cup C_2$	$\frac{1}{2}(q-3)$
$q \equiv 5 \pmod 8$	$C_0 \cup C_1 \quad C_0 \cup C_3$	$\frac{1}{2}(q-3)$
$q = a^2 \equiv 1 \pmod 8$	$C_0 \cup C_1 \quad C_2 \cup C_3$	$\frac{1}{2}(q-3)$
$q = \frac{1}{5}(9+z^2) \equiv 5 \pmod 8$, $z \equiv 1 \pmod 5$	$C_0 \cup C_1 \quad C_0 \cup C_1 \cup C_3$ $C_0 \cup C_1 \quad C_1 \cup C_2 \cup C_3$	$\frac{1}{16}(13q-33)$
$q = 49 + 4b^2 \equiv 5 \pmod{16}$	$C_0 \cup C_2 \quad C_0 \cup C_1 \cup C_2$	$\frac{1}{16}(13q-33)$
$q = 1 + 4b^2 \equiv 5 \pmod{16}$	$C_0 \cup C_2 \quad C_1 \cup C_2 \cup C_3$	$\frac{1}{16}(13q-33)$
$q \equiv 5 \pmod 8$	$C_0 \cup C_1 \cup C_2 \quad C_0 \cup C_1 \cup C_3$	$\frac{1}{8}(9q-21)$

Source: Yamada [239, figure 2, p. 84]. Reproduced with permission of Elsevier.

Table B.4 $e = 4, n = 3$.

q		λ
$q = 9 + 4b^2 \equiv 1 \pmod 8$	$C_0 \quad C_2 \quad C_0 \cup C_2$	$\frac{1}{8}(3q-11)$
$q = 1 + 4b^2 \equiv 1 \pmod 8$	$C_0 \quad C_2 \quad C_1 \cup C_3$	$\frac{1}{8}(3q-11)$
$q = 25 + 4b^2 \equiv 13 \pmod{16}$	$C_0 \quad C_0 \quad C_0 \cup C_1 \cup C_2$ $C_0 \quad C_2 \quad C_0 \cup C_1 \cup C_2$	$\frac{1}{16}(11q-31)$
$q = \frac{1}{17}(1+z^2) \equiv 5 \pmod{16}$, $z \equiv 4 \pmod{13}$	$C_0 \quad C_0 \cup C_3 \quad C_0 \cup C_3$ $C_0 \quad C_0 \cup C_3 \quad C_1 \cup C_2$ $C_0 \quad C_1 \cup C_2 \quad C_1 \cup C_2$	$\frac{1}{16}(9q-29)$
$q = \frac{1}{5}(9+z^2) \equiv 5 \pmod{16}$, $z \equiv 1 \pmod 5$	$C_0 \quad C_0 \cup C_1 \quad C_1 \cup C_3$ $C_0 \quad C_1 \cup C_3 \quad C_2 \cup C_3$	$\frac{1}{16}(9q-29)$
$q = 81 + 4b^2 \equiv 5 \pmod{16}$	$C_0 \quad C_0 \cup C_2 \quad C_0 \cup C_2$	$\frac{1}{16}(9q-29)$
$q = 49 + 4b^2 \equiv 5 \pmod{16}$	$C_0 \quad C_1 \cup C_3 \quad C_1 \cup C_3$	$\frac{1}{16}(9q-29)$

(Continued)

Table B.4 (*Continued*)

q		λ
$q=a^2+4b^2\equiv 5 \pmod 8$, $b=1$	$C_0\quad C_0\cup C_1\quad C_0\cup C_1\cup C_2$ $C_0\quad C_2\cup C_3\quad C_0\cup C_1\cup C_2$ $C_0\quad C_0\cup C_1\quad C_0\cup C_2\cup C_3$ $C_0\quad C_2\cup C_3\quad C_0\cup C_2\cup C_3$	$\frac{1}{8}(7q-19)$
$q=\frac{1}{2}(1+z^2)\equiv 5 \pmod 8$, $z\equiv 1 \pmod 4$	$C_0\quad C_0\cup C_1\quad C_0\cup C_1\cup C_3$ $C_0\quad C_2\cup C_3\quad C_0\cup C_1\cup C_3$ $C_0\quad C_0\cup C_1\quad C_1\cup C_2\cup C_3$ $C_0\quad C_2\cup C_3\quad C_1\cup C_2\cup C_3$	$\frac{1}{8}(7q-19)$
$q\equiv 5 \pmod 8$	$C_0\quad C_1\cup C_3\quad C_0\cup C_1\cup C_2$	$\frac{1}{8}(7q-19)$
$q=49+4b^2\equiv 5 \pmod{16}$	$C_0\quad C_0\cup C_1\cup C_2\quad C_0\cup C_1\cup C_2$ $C_0\quad C_0\cup C_1\cup C_2\quad C_0\cup C_2\cup C_3$	$\frac{1}{16}(19q-47)$
$q=\frac{1}{17}(9+z^2)\equiv 13 \pmod{16}$, $z\equiv 12 \pmod{17}$	$C_0\cup C_3\quad C_0\cup C_3\quad C_0\cup C_1\cup C_2$ $C_0\cup C_3\quad C_1\cup C_2\quad C_0\cup C_1\cup C_2$ $C_1\cup C_2\quad C_1\cup C_2\quad C_0\cup C_1\cup C_2$	$\frac{1}{16}(17q-45)$
$q=\frac{1}{5}(49+z^2)\equiv 13 \pmod{16}$, $z\equiv 1 \pmod 5$	$C_0\cup C_1\quad C_0\cup C_2\quad C_0\cup C_1\cup C_2$ $C_0\cup C_2\quad C_2\cup C_3\quad C_0\cup C_1\cup C_2$	$\frac{1}{16}(17q-45)$
$q=\frac{1}{5}(1+z^2)\equiv 13 \pmod{16}$, $z\equiv 2 \pmod 5$	$C_0\cup C_1\quad C_1\cup C_3\quad C_0\cup C_1\cup C_2$ $C_1\cup C_3\quad C_2\cup C_3\quad C_0\cup C_1\cup C_2$	$\frac{1}{16}(17q-45)$
$q=11^2+4b^2\equiv 13 \pmod{16}$,	$C_0\cup C_2\quad C_0\cup C_2\quad C_0\cup C_1\cup C_2$	$\frac{1}{16}(17q-45)$
$q=25+4b^2\equiv 13 \pmod{16}$,	$C_1\cup C_3\quad C_1\cup C_3\quad C_0\cup C_1\cup C_2$	$\frac{1}{16}(17q-45)$
$q=25+4b^2\equiv 1 \pmod 8$,	$C_0\cup C_2\quad C_0\cup C_1\cup C_2\quad C_0\cup C_2\cup C_3$	$\frac{1}{8}(11q-27)$
$q=1+4b^2\equiv 1 \pmod 8$,	$C_1\cup C_3\quad C_0\cup C_1\cup C_2\quad C_0\cup C_2\cup C_3$	$\frac{1}{8}(11q-27)$
$q=9$	$C_0\quad C_1\quad C_0\cup C_1$	2

Source: Adapted from Yamada [239, figures 3 and 4, p. 85]. Reproduced with permission of Elsevier.

Table B.5 $e = 4, n = 4$.

q		λ
$q \equiv 1 \pmod 8$	$C_0 \quad C_1 \quad C_2 \quad C_3$	$\frac{1}{4}(q-5)$
$q = \frac{1}{13}(9+z^2) \equiv 13 \pmod{16}$, $z \equiv 11 \pmod{13}$	$C_0 \quad C_0 \quad C_0 \quad C_0 \cup C_1$ $C_0 \quad C_0 \quad C_0 \quad C_2 \cup C_3$ $C_0 \quad C_0 \quad C_2 \quad C_0 \cup C_1$ $C_0 \quad C_0 \quad C_2 \quad C_2 \cup C_3$	$\frac{1}{16}(7q-27)$
$q = \frac{1}{2}(9+z^2) \equiv 5 \pmod 8$, $z \equiv 1 \pmod 4$	$C_0 \quad C_0 \quad C_0 \cup C_1 \quad C_0 \cup C_2$ $C_0 \quad C_0 \quad C_0 \cup C_2 \quad C_2 \cup C_3$ $C_0 \quad C_2 \quad C_0 \cup C_1 \quad C_0 \cup C_2$	$\frac{1}{8}(5q-17)$
$q = \frac{1}{2}(1+z^2) \equiv 5 \pmod 8$ $z \equiv 1 \pmod 4$	$C_0 \quad C_0 \quad C_0 \cup C_1 \quad C_1 \cup C_3$ $C_0 \quad C_0 \quad C_1 \cup C_3 \quad C_2 \cup C_3$ $C_0 \quad C_2 \quad C_0 \cup C_1 \quad C_1 \cup C_3$	$\frac{1}{8}(5q-17)$
$q = \frac{1}{13}(1+z^2) \equiv 5 \pmod{16}$, $z \equiv 5 \pmod{13}$	$C_0 \quad C_0 \quad C_0 \cup C_1 \quad C_0 \cup C_1 \cup C_3$ $C_0 \quad C_0 \quad C_0 \cup C_1 \quad C_1 \cup C_2 \cup C_3$ $C_0 \quad C_0 \quad C_2 \cup C_3 \quad C_0 \cup C_1 \cup C_3$ $C_0 \quad C_0 \quad C_2 \cup C_3 \quad C_1 \cup C_2 \cup C_3$ $C_0 \quad C_2 \quad C_0 \cup C_1 \quad C_0 \cup C_1 \cup C_3$ $C_0 \quad C_2 \quad C_0 \cup C_1 \quad C_1 \cup C_2 \cup C_3$	$\frac{1}{16}(15q-43)$
$q = 81 + 4b^2 \equiv 5 \pmod{16}$	$C_0 \quad C_0 \quad C_0 \cup C_2 \quad C_0 \cup C_1 \cup C_2$ $C_0 \quad C_2 \quad C_0 \cup C_2 \quad C_0 \cup C_1 \cup C_2$	$\frac{1}{16}(15q-43)$
$q = 1 + 4b^2 \equiv 1 \pmod 8$	$C_0 \quad C_2 \quad C_0 \cup C_1 \cup C_3 \quad C_1 \cup C_2 \cup C_3$	$\frac{1}{4}(5q-13)$
$q = \frac{1}{37}(1+z^2) \equiv 13 \pmod{16}$, $z \equiv 6 \pmod{37}$	$C_0 \quad C_0 \cup C_3 \quad C_0 \cup C_3 \quad C_0 \cup C_3$ $C_0 \quad C_0 \cup C_3 \quad C_0 \cup C_3 \quad C_1 \cup C_2$ $C_0 \quad C_0 \cup C_3 \quad C_1 \cup C_2 \quad C_1 \cup C_2$ $C_0 \quad C_1 \cup C_2 \quad C_1 \cup C_2 \quad C_1 \cup C_2$	$\frac{1}{16}(13q-41)$
$q = \frac{1}{17}(25+z^2) \equiv 13 \pmod{16}$, $z \equiv 3 \pmod{17}$	$C_0 \quad C_0 \cup C_2 \quad C_0 \cup C_3 \quad C_0 \cup C_3$ $C_0 \quad C_0 \cup C_2 \quad C_0 \cup C_3 \quad C_1 \cup C_2$ $C_0 \quad C_0 \cup C_2 \quad C_1 \cup C_2 \quad C_1 \cup C_2$	$\frac{1}{16}(13q-41)$
$q = \frac{1}{17}(9+z^2) \equiv 13 \pmod{16}$, $z \equiv 5 \pmod{17}$	$C_0 \quad C_0 \cup C_3 \quad C_0 \cup C_3 \quad C_1 \cup C_3$ $C_0 \quad C_0 \cup C_3 \quad C_1 \cup C_2 \quad C_1 \cup C_3$ $C_0 \quad C_1 \cup C_2 \quad C_1 \cup C_2 \quad C_1 \cup C_3$	$\frac{1}{16}(13q-41)$
$q = \frac{1}{5}(81+z^2) \equiv 13 \pmod{16}$, $z \equiv 3 \pmod 5$	$C_0 \quad C_0 \cup C_2 \quad C_0 \cup C_2 \quad C_0 \cup C_3$ $C_0 \quad C_0 \cup C_2 \quad C_0 \cup C_2 \quad C_1 \cup C_2$	$\frac{1}{16}(13q-41)$

(Continued)

Table B.5 (*Continued*)

q		λ
$q=\frac{1}{5}(49+z^2)\equiv 13 \pmod{16}$, $z\equiv 1 \pmod 5$	C_0 $C_0\cup C_3$ $C_1\cup C_3$ $C_1\cup C_3$ C_0 $C_1\cup C_2$ $C_1\cup C_3$ $C_1\cup C_3$	$\frac{1}{16}(13q-41)$
$q=13^2+4b^2\equiv 13 \pmod{16}$	C_0 $C_0\cup C_2$ $C_0\cup C_2$ $C_0\cup C_2$	$\frac{1}{16}(13q-41)$
$q=11^2+4b^2\equiv 13 \pmod{16}$	C_0 $C_1\cup C_3$ $C_1\cup C_3$ $C_1\cup C_3$	$\frac{1}{16}(13q-41)$
$q=1+4b^2\equiv 1 \pmod 8$	C_1 $C_0\cup C_3$ $C_2\cup C_3$ $C_0\cup C_1\cup C_2$	$\frac{1}{3}(9q-25)$
$q=\frac{1}{5}(49+z^2)\equiv 13 \pmod{16}$, $z\equiv 4 \pmod 5$	C_0 $C_0\cup C_1$ $C_0\cup C_1\cup C_2$ $C_0\cup C_1\cup C_2$ C_0 $C_0\cup C_3$ $C_0\cup C_1\cup C_2$ $C_0\cup C_2\cup C_3$ C_0 $C_1\cup C_2$ $C_0\cup C_1\cup C_2$ $C_0\cup C_1\cup C_2$ C_0 $C_1\cup C_2$ $C_0\cup C_1\cup C_2$ $C_0\cup C_2\cup C_3$ C_2 $C_0\cup C_3$ $C_0\cup C_1\cup C_2$ $C_0\cup C_1\cup C_2$ C_2 $C_1\cup C_2$ $C_0\cup C_1\cup C_2$ $C_0\cup C_1\cup C_2$	$\frac{1}{16}(23q-59)$
$q=11^2+4b^2\equiv 13 \pmod{16}$	C_0 $C_0\cup C_2$ $C_0\cup C_1\cup C_2$ $C_0\cup C_1\cup C_2$ C_0 $C_0\cup C_2$ $C_0\cup C_1\cup C_2$ $C_0\cup C_2\cup C_3$	$\frac{1}{16}(23q-59)$
$q=\frac{1}{13}(25+z^2)\equiv 13 \pmod{16}$, $z\equiv 1 \pmod{13}$	C_1 $C_0\cup C_1$ $C_0\cup C_1\cup C_2$ $C_0\cup C_1\cup C_2$ C_1 $C_0\cup C_1$ $C_0\cup C_1\cup C_2$ $C_0\cup C_2\cup C_3$ C_1 $C_2\cup C_3$ $C_0\cup C_1\cup C_2$ $C_0\cup C_1\cup C_2$ C_1 $C_2\cup C_3$ $C_0\cup C_1\cup C_2$ $C_0\cup C_2\cup C_3$ C_3 $C_0\cup C_1$ $C_0\cup C_1\cup C_2$ $C_0\cup C_1\cup C_2$ C_3 $C_2\cup C_3$ $C_0\cup C_1\cup C_2$ $C_0\cup C_1\cup C_2$	$\frac{1}{16}(23q-59)$
$q=a^2+36\equiv 5 \pmod 8$	$C_0\cup C_1$ $C_0\cup C_2$ $C_0\cup C_2$ $C_0\cup C_2$	$q-3$
$q=\frac{1}{37}(9+z^2)\equiv 5 \pmod{16}$, $z\equiv 19 \pmod{37}$	$C_0\cup C_1$ $C_0\cup C_1$ $C_0\cup C_1$ $C_0\cup C_1\cup C_2$ $C_0\cup C_1$ $C_0\cup C_1$ $C_2\cup C_3$ $C_0\cup C_1\cup C_2$ $C_0\cup C_1$ $C_2\cup C_3$ $C_2\cup C_3$ $C_0\cup C_1\cup C_2$ $C_2\cup C_3$ $C_2\cup C_3$ $C_2\cup C_3$ $C_0\cup C_1\cup C_2$	$\frac{1}{16}(21q-57)$
$q=\frac{1}{17}(49+z^2)\equiv 5 \pmod{16}$, $z\equiv 6 \pmod{17}$	$C_0\cup C_1$ $C_0\cup C_1$ $C_0\cup C_2$ $C_0\cup C_1\cup C_2$ $C_0\cup C_1$ $C_0\cup C_2$ $C_2\cup C_3$ $C_0\cup C_1\cup C_2$ $C_0\cup C_2$ $C_2\cup C_3$ $C_2\cup C_3$ $C_0\cup C_1\cup C_2$	$\frac{1}{16}(21q-57)$
$q=\frac{1}{17}(1+z^2)\equiv 5 \pmod{16}$, $z\equiv 4 \pmod{17}$	$C_0\cup C_1$ $C_0\cup C_1$ $C_1\cup C_3$ $C_0\cup C_1\cup C_2$ $C_0\cup C_1$ $C_1\cup C_3$ $C_2\cup C_3$ $C_0\cup C_1\cup C_2$ $C_1\cup C_3$ $C_2\cup C_3$ $C_2\cup C_3$ $C_0\cup C_1\cup C_2$	$\frac{1}{16}(21q-57)$
$q=\frac{1}{5}(11^2+z^2)\equiv 5 \pmod{16}$, $z\equiv 3 \pmod 5$	$C_0\cup C_1$ $C_0\cup C_2$ $C_0\cup C_2$ $C_0\cup C_1\cup C_2$ $C_0\cup C_2$ $C_0\cup C_2$ $C_2\cup C_3$ $C_0\cup C_1\cup C_2$	$\frac{1}{16}(21q-57)$

(*Continued*)

Table B.5 (*Continued*)

q		λ
$q = 15^2 + 4b^2 \equiv 5 \pmod{16}$	$C_0 \cup C_2$ $C_0 \cup C_2$ $C_0 \cup C_2$ $C_0 \cup C_1 \cup C_2$	$\frac{1}{16}(21q - 57)$
$q = 81 + 4b^2 \equiv 5 \pmod{16}$	$C_1 \cup C_3$ $C_1 \cup C_3$ $C_1 \cup C_3$ $C_0 \cup C_1 \cup C_2$	$\frac{1}{16}(21q - 57)$
$q = \frac{1}{2}(25 + z^2) \equiv 5 \pmod{8}$, $z \equiv 3 \pmod{4}$	$C_0 \cup C_1$ $C_0 \cup C_2$ $C_0 \cup C_1 \cup C_2$ $C_0 \cup C_1 \cup C_2$ $C_0 \cup C_1$ $C_0 \cup C_2$ $C_0 \cup C_1 \cup C_2$ $C_0 \cup C_2 \cup C_3$ $C_0 \cup C_2$ $C_2 \cup C_3$ $C_0 \cup C_1 \cup C_2$ $C_0 \cup C_1 \cup C_2$	$\frac{1}{8}(13q - 33)$
$q = \frac{1}{2}(1 + z^2) \equiv 5 \pmod{8}$, $z \equiv 3 \pmod{4}$	$C_0 \cup C_1$ $C_1 \cup C_3$ $C_0 \cup C_1 \cup C_2$ $C_0 \cup C_1 \cup C_2$ $C_0 \cup C_1$ $C_1 \cup C_3$ $C_0 \cup C_1 \cup C_2$ $C_0 \cup C_2 \cup C_3$ $C_1 \cup C_3$ $C_2 \cup C_3$ $C_0 \cup C_1 \cup C_2$ $C_0 \cup C_1 \cup C_2$	$\frac{1}{8}(13q - 33)$
$q = \frac{1}{13}(81 + z^2) \equiv 5 \pmod{16}$, $z \equiv 7 \pmod{13}$	$C_0 \cup C_1$ $C_0 \cup C_1 \cup C_2$ $C_0 \cup C_1 \cup C_2$ $C_0 \cup C_1 \cup C_2$ $C_0 \cup C_1$ $C_0 \cup C_1 \cup C_2$ $C_0 \cup C_1 \cup C_2$ $C_0 \cup C_2 \cup C_3$ $C_0 \cup C_1$ $C_0 \cup C_1 \cup C_2$ $C_0 \cup C_2 \cup C_3$ $C_0 \cup C_2 \cup C_3$ $C_2 \cup C_3$ $C_0 \cup C_1 \cup C_2$ $C_0 \cup C_1 \cup C_2$ $C_0 \cup C_1 \cup C_2$	$\frac{1}{16}(31q - 75)$
$q \equiv 1 \pmod{8}$	$C_0 \cup C_1 \cup C_2$ $C_1 \cup C_2 \cup C_3$ $C_0 \cup C_2 \cup C_3$ $C_0 \cup C_1 \cup C_3$	$\frac{3}{4}(3q - 7)$
$q = 17$	C_2 $C_0 \cup C_2$ $C_1 \cup C_2$ $C_0 \cup C_1 \cup C_2$	16

Source: Adapted from Yamada [239, figures 5–11, pp. 85–89], Reproduced with permission of Elsevier.

C

Further Research Questions

C.1 Research Questions for Future Investigation

There are many research problems that arise from this book. We list here a few key questions that current and future researchers may wish to pursue.

C.1.1 Matrices

Research Question C.1 (Williamson matrices): Let $q \equiv 1 \pmod 4$ be a prime power. For which (if any) orders $n \geq 63$, $n \neq \frac{1}{2}(q+1)$ do Williamson matrices exist?

Research Question C.2 (Balonin–Seberry conjecture): Prove the Balonin–Seberry conjecture that all luchshie matrices exist.

C.1.2 Base Sequences

Research Question C.3: Base sequences $BS(m, n)$ are four complementary or suitable sequences of length m, m, n, n.

i) Do all base sequences $(2m+1, m)$ exist?
ii) For which m and n do base sequences $B(m, n)$ exist?

Research Question C.4 (Yang/Đoković conjecture): Prove that the Yang/Đoković conjecture $BS(n+1, n)$ exist.

C.1.3 Partial Difference Sets

Research Question C.5 (Ma on Partial Difference Sets [130]): Suppose G is an abelian group of order v such that $v \equiv 1 \pmod 4$.

i) If v is not a prime power, does there exist a Paley type PDS?
ii) If v is a prime power, does G need to be an elementary abelian?

C.1.4 de Launey's Four Questions

Research Question C.6 (de Launey's extension of Agaian's theorem): Find more results of the types of Agaian's theorem 6.2 (also called the Agaian–Sarukhanyan theorem), see de Launey's 1993 paper [44], the Craigen–Seberry–Xianmo Zhang's theorem [37], and the de Launey's theorem 6.4.

Hadamard Matrices: Constructions using Number Theory and Algebra, First Edition. Jennifer Seberry and Mieko Yamada.

Research Question C.7 (de Launey's Problem 1): Improve the known results on the density of Hadamard matrices in the set of natural numbers. See Warwick de Launey and Daniel M. Gordon [47]

Research Question C.8 (de Launey's Problem 2): Improve the known bounds on the existence of partial Hadamard matrices, See Warwick de Launey and David A. Levin [49].

Research Question C.9 (de Launey's Problem 3): Improve the de Launey bound on the order of the power of 2 for the existence of an Hadamard matrix of order $2^t q$, q an odd natural number. See Warwick de Launey [45].

C.1.5 Embedding Sub-matrices

Research Question C.10: Can we give a bound on the size of D_m that will embed into an S_n? (Note that $D_6 \notin S_{16}$ but $D_6 \in S_{32}, S_{64}$.)

C.1.6 Pivot Structures

Research Question C.11: For a CP Hadamard matrix of order 16, can the determinant of its lower right 4×4 principal sub-matrix only take the value 8 if the matrix is in the Sylvester Hadamard equivalence class?

Research Question C.12: Find the pivot structure of CP matrices of Hadamard matrices of orders >16.

C.1.7 Trimming and Bordering

Research Question C.13: [Trimming and bordering via M-constructions] We can see Yamada's matrix [236] or the J. Wallis–Whiteman [216] matrix with a border embodied in constructions for M-structures. Miyamoto has done further work using the quaternions rather than the complex numbers to build bigger M-structures [137]. Extend this work further.

C.1.8 Arrays

Research Question C.14: Does an $OD(4t; t, t, t, t)$ exist for every positive integer t?

Research Question C.15: WL arrays are only known for t odd integers $t = 5, 9$. Find WL arrays for other odd t.

References

1 S. S. Agaian. *Hadamard Matrices and their Applications*, volume 1168 of *Lecture Notes in Mathematics*. Springer-Verlag, Berlin-Heidelberg, 1985.

2 T. H. Andres. *Some combinatorial properties of complementary sequences*. M.Sc. Thesis, University of Manitoba, Winnipeg, 1977.

3 Gene Awyzio and Jennifer Seberry. On good matrices and skew Hadamard matrices. In *Algebraic Design Theory and Hadamard Matrices*, volume 133 of *Springer Proceedings in Mathematics & Statistics*, pages 13–28. Springer International Publishing, Cham, 2015.

4 N. A. Balonin, Y. N. Balonin, D. Ž Đoković, D. A. Karbovskiy, and M. B. Sergeev. Construction of symmetric Hadamard matrices. *Informatsionno-upravliaiushchie sistemy*, 5: 2–11, 2017.

5 N. A. Balonin and D. Ž Đoković. Symmetric Hadamard matrices of orders 268, 412, 436, and 604. *Informatsionno-upravliaiushchie sistemy*, 4: 2–8, 2018.

6 N. A. Balonin, D. Ž Đoković, and D. A. Karbovskiy. Construction of symmetric Hadamard matrices of order $4v$ for $v = 47, 73, 113$. *Spec. Matr.*, 6: 11–22, 2018.

7 T. Banica, I. Nechita, and M. M. Schlenker. Submatrices of Hadamard matrices: complementation results. *Electron. J. Linear Algebra*, 27: 197–212, 2014.

8 L. D. Baumert, S. W. Golomb, and M. Hall Jr. Discovery of an Hadamard matrix of order 92. *Bull. Am. Math. Soc.*, 68: 237–238, 1962.

9 L. D. Baumert and M. Hall Jr. A new construction for Hadamard matrices. *Bull. Am. Math. Soc.*, 71: 169–170, 1965.

10 V. Belevitch. Theory of $2n$-terminal networks with applications to conference telephony. *Electr. Commun.*, 27(3):231–244, 1950.

11 V. Belevitch. Conference networks and Hadamard matrices. *Ann. Soc. Scientifique Brux.*, T82:13–32, 1968.

12 J. Bell and D. Ž. Đoković. Construction of Baumert-Hall-Welch array and T-matrices. *Australas. J. Combin.*, 14: 93–107, 1996.

13 B. C. Berndt, R. J. Evans, and K. S. Williams. *Gauss and Jacobi Sums*. John Wiley & Sons, New York, 1998.

14 M. R. Best. The excess of a Hadamard matrix. *Indag. Math.*, 39: 357–361, 1977.

15 T. Beth, D. Jungnickel, and H. Lenz. *Design Theory*. Cambridge University Press, Cambridge, 2006.

16 P. B. Borwein and R. A. Ferguson. A complete description of Golay pairs for lengths up to 100. *Math. Comput.*, 73(246):967–985 (electronic), 2004.

17 R. P. Brent and J.-A. H. Osborn. On minors of maximal determinant matrices. *J. Integer Seq.*, 16: 1–30, 2013.

18 Y. Q. Chen. On the existence of abelian Hadamard difference sets and a new family of difference sets. *Finite Fields Appl.*, 3: 234–256, 1997.

19 D. Christou, M. Mitrouli, and J. Seberry. Embedding and extension properties of Hadamard matrices revisited. *Spec. Matr.*, 6(1):155–165, 2018.

Hadamard Matrices: Constructions using Number Theory and Algebra, First Edition. Jennifer Seberry and Mieko Yamada.

20 G. Cohen, D. Rubie, C. Koukouvinos, S. Kounias, J. Seberry, and M. Yamada. A survey of base sequences, disjoint complementary sequences and *OD*($4t; t, t, t, t$). *J. Combin. Math. Combin. Comput.*, 5: 69–103, 1989.

21 J. N. Cohn. Hadamard matrices and some generalizations. *Am. Math. Mon.*, 72: 515–518, 1965.

22 C. J. Colbourn and J. H. Dinitz, editors. *Handbook of Combinatorial Designs*. Discrete Mathematics and Its Applications. Chapman & Hall/CRC, New York, 2nd edition, 2007.

23 J. Cooper and Jennifer Seberry Wallis. A construction for Hadamard arrays. *Bull. Aust. Math. Soc.*, 7: 269–278, 1972.

24 R. S. Coulter and R. W. Matthews. Planar functions and planes of Lenz–Barlotti class II. *Des. Codes Crypt.*, 10: 167–184, 1997.

25 R. Craigen. Constructing Hadamard matrices with orthogonal pairs. *Ars Combin.*, 33: 57–64, 1990.

26 R. Craigen. *Constructions for orthogonal matrices*. Ph.D. Thesis, University of Waterloo, Waterloo, Ontario, 1991.

27 R. Craigen. Constructing Hadamard matrices with orthogonal pairs. *Ars Combin.*, 33: 57–64, 1992.

28 R. Craigen. Complex Golay sequences. *J. Combin. Math. Combin. Comput.*, 15: 161–169, 1994.

29 R. Craigen. Trace, symmetry and orthogonality. *Can. Math. Bull.*, 37: 461–467, 1994.

30 R. Craigen. Signed groups, sequences and the asymptotic existence of Hadamard matrices. *J. Combin. Theory*, 71: 241–254, 1995.

31 R. Craigen. The structure of weighing matrices having large weights. *Des., Codes Crypt.*, 5: 199–216, 1995.

32 R. Craigen, W. Holzmann, and H. Kharaghani. On the asymptotic existence of complex Hadamard matrices. *J. Combin. Des.*, 5: 319–327, 1996.

33 R. Craigen, W. Holzmann, and H. Kharaghani. Complex Golay sequences: structure and applications. *Discrete Math.*, 252 (1): 73–89, 2002.

34 R. Craigen and J. Jedwab. Comment on revised version of "The Hadamard Circulant Conjecture." *ArXiv e-prints*, arXiv:1111.3437v1 [math.CO], Nov 2011.

35 R. Craigen and H. Kharaghani. Hadamard matrices from weighing matrices via signed groups. *Des., Codes Crypt.*, 12 (1): 49–58, 1997.

36 R. Craigen and C. Koukouvinos. A theory of ternary complementary pairs. *J. Combin. Theory Ser. A*, 96(2):358–375, 2001.

37 R. Craigen, J. Seberry, and X. M. Zhang. Product of four Hadamard matrices. *J. Combin. Theory Ser. A*, 59 (2): 318–320, 1992.

38 C. W. Cryer. Pivot size in gaussian elimination. *Numer. Math*, 12 (4): 335–345, 1968.

39 J. A. Davis. Difference set in abelian 2-groups. *J. Combin. Theory (A)*, 57: 262–286, 1991.

40 J. A. Davis. Partial difference sets in *p*-groups. *Arch. Math.*, 63: 103–110, 1994.

41 J. A. Davis and J. Jedwab. A survey of Hadamard difference sets. In *Groups, Difference Sets, and the Monster: Proceedings of a Special Research Quarter at The Ohio State University, Spring 1993* (K. T. Arasu, J. F. Dillon, K. Harada, S. Sehgal, and R. Solomon, Eds.), pages 145=156 Ohio State University Mathematical Research Institute Publications. Walter De Gruyter, New York, 1996.

42 J. A. Davis and J. Jedwab. A unifying construction for difference sets. *J. Combin. Theory A*, 80: 13–78, 1997.

43 Jane Day and Brian Peterson. Growth in Gaussian elimination. *Am. Math. Mon.*, 95 (6): 489–513, 1988.

44 W. de Launey. A product of twelve Hadamard matrices. *Australas. J. Combin.*, 7: 123–127, 1993.

45 W. de Launey. On the asymptotic existence of Hadamard matrices. *ArXiv e-prints*, arXiv.1003.4001 [math CO], 2010. Also published in *J. Combin. Theory Ser. A*, 116: 1002–1008, 2009.

46 W. de Launey, D. L. Flannery, and K. J. Horadam. Cocyclic Hadamard matrices and difference sets. *Discrete Appl. Math.*, 102: 47–61, 2000.

47 W. de Launey and D. M. Gordon. On the density of the set of known Hadamard orders. *ArXiv e-prints*, arXiv.1004.4872 [math.CO], 2010.

48 W. de Launey and H. Kharaghani. On the asymptotic existence of cocyclic Hadamard matrices. *J. Combin. Theory*, 116 (6): 1140–1153, 2009.
49 W. de Launey and D. A. Levin. A Fourier-analytic approach to counting partial Hadamard matrices. *ArXiv e-prints*, arXiv.1003.4003 [math.PR], 2010.
50 P. Delsarte, J. M. Goethals, and J. J. Seidel. Orthogonal matrices with zero diagonal II. *Can. J. Math.*, 23: 816–832, 1971.
51 P. Dembowski and T. G. Ostrom. Planes of order n with collimeation groups of order n^2. *Math. Z.*, 193: 239–258, 1968.
52 O. Di Matteo, D. Ž. Đoković, and I. S. Kotsireas. Symmetric Hadamard matrices of order 116 and 172 exist. *Spec. Matr.*, 3: 227–234, 2015.
53 C. Ding and J. Yuan. A family of skew Hadamard difference sets. *J. Combin. Theory Ser. A*, 113: 1526–1535, 2006.
54 D. Ž. Đoković. Construction of some new Hadamard matrices. *Bull. Aust. Math. Soc.*, 45: 327–332, 1992.
55 D. Ž. Đoković. Williamson matrices of orders 4.29 and 4.31. *J. Combin. Theory*, 59: 309–311, 1992.
56 D. Ž. Đoković. Skew-Hadamard matrices of order 4.37 and 4.43. *J. Combin. Theory*, 61: 319–321, 1992.
57 D. Ž. Đoković. Williamson matrices of order $4n$ for $n = 33, 35, 39$. *Discrete Math.*, 115: 267–271, 1993.
58 D. Ž. Đoković. Good matrices of order 33, 35 and 127 exist. *J. Combin. Math. Combin. Comput.*, 14: 145–152, 1993.
59 D. Ž. Đoković. Note on Williamson matrices of orders 25 and 37. *J. Combin. Math. Combin. Comput.*, 18: 171–175, 1995.
60 D. Ž. Đoković. Equivalence classes and representatives of Golay sequences. *Discrete Math.*, 189 (1–3): 79–93, 1998.
61 D. Ž. Đoković. On the base sequence conjecture. *Discrete Math.*, 310 (13): 1956–1964, 2010.
62 D. Ž. Đoković. Hadamard matrices of small order and Yang conjecture. *J. Combin. Des.*, 18 (4): 254–259, 2010.
63 D. Ž. Đoković. Small orders of Hadamard matrices and base sequences. *Int. Math. Forum*, 6 (62): 3061–3067, 2011.
64 D. Ž. Đoković, O. Golubitsky, and I. Kotsireas. Some new orders of Hadamard and skew-Hadamard matrices. *J. Combin. Des.*, 22 (6): 270–277, 2014.
65 D. Ž. Đoković and I. S. Kotsireas. Some new periodic Golay pairs. *Numer. Algorithms*, 69 (3): 523–530, 2015.
66 A. Edelman and W. Mascarenhas. On the complete pivoting conjecture for a Hadamard matrix of order 12. *Linear Multilinear Algebra*, 38 (3): 181–187, 1995.
67 Genet M. Edmondson, Jennifer Seberry, and Malcolm R. Anderson. On the existence of Turyn sequences of length less than 43. *Math. Comput.*, 62: 351–362, 1994.
68 S. Eliahou, M. Kervaire, and B. Saffari. A new restriction on the lengths of Golay complementary sequences. *J. Combin. Theory A*, 55: 49–59, 1990.
69 P. Erdös and A. M. Odlyzko. On the density of odd integers of the form $(p-1)2^{-n}$ and related questions. *J. Number Theory*, 11: 257–263, 1979.
70 H. Evangelaras, C. Koukouvinos, and K. Mylona. On Hadamard embeddability. *J. Discrete Math. Sci. Crypt.*, 9 (3): 503–512, 2006.
71 T. Feng, K. Momihara, and Q. Xiang. Constructions of strongly regular Cayley graphs and skew Hadamard difference sets from cyclotomic classes. *Combinatorica*, 35: 413–434, 2015.
72 T. Feng and Q. Xiang. Cyclotomic constructions of skew Hadamard difference sets. *J. Combin. Theory A*, 119: 245–256, 2012.
73 Roderick J. Fletcher, Christos Koukouvinos, and Jennifer Seberry. New skew-Hadamard matrices of order $4 \cdot 59$ and new D-optimal designs of order $2 \cdot 59$. *Discrete Math.*, 286 (3): 251–253, 2004.
74 F. R. Gantmacher. *The Theory of Matrices*, volume 1. Chelsea, New York, 1959.

75 S. Georgiou, C. Koukouvinos, and S. Stylianou. On good matrices, skew Hadamard matrices and optimal designs. *Comput. Stat. Data Anal.*, 41 (1): 171–184, 2002.

76 Anthony V. Geramita, Joan Murphy Geramita, and Jennifer Seberry Wallis. Orthogonal designs. *Linear Multilinear Algebra*, 3: 281–306, 1975.

77 A. V. Geramita, N. J. Pullman, and J. S. Wallis. Families of weighing matrices. *Bull. Aust. Math. Soc.*, 10: 119–122, 1974.

78 A. V. Geramita and J. Seberry. *Orthogonal Designs: Quadratic Forms and Hadamard Matrices*, volume 45 of *Lecture Notes in Pure and Applied Mathematics*. Marcel Dekker, New York-Basel, 1st edition, 1979.

79 E. Ghaderpour and H Kharaghani. The asymptotic existence of orthogonal designs. *Australas. J. Combin.*, 58: 333–346, 2014.

80 J. M. Goethals and J. J. Seidel. Orthogonal matrices with zero diagonal. *Can. J. Math.*, 19: 1001–1010, 1967.

81 J. M. Goethals and J. J. Seidel. A skew-Hadamard matrix of order 36. *J. Aust. Math. Soc.*, 11: 343–344, 1970.

82 J. M. Goethals and J. J. Seidel. Quasi-symmetric block designs. In *Combinatorial Structures and Their Application: Proceedings of the Calgary International Conference*. Gordon and Breach, New York, 1970.

83 M. J. E. Golay. Complementary series. *IRE Trans. Inf. Theory*, 6: 400–408, 1960.

84 M. J. E. Golay. Note on complementary series. *Proc. IRE*, 50: 84, 1962.

85 K. Goldberg. Hadamard matrices of order cube plus one. *Proc. Am. Math. Soc.*, 17: 744–746, 1966.

86 B. Gordon, W. H. Mills, and L. R. Welch. Some new difference sets. *Can. J. Math.*, 14: 614–625, 1962.

87 N. Gould. On growth in Gaussian elimination with pivoting. *SIAM J. Matrix Anal. Appl.*, 12: 354-361, 1991.

88 M. Griffin. There are no Golay sequences of length 2.9^t. *Aequationwa Math.*, 15: 73–77, 1977.

89 Jacques Hadamard. Résolution d'une question relative aux déterminants. *Bull. des Sci. Math.*, 17: 240–246, 1893.

90 W. H. Haemers and Q. Xiang. Strongly regular graphs with parameters $(4m^4, 2m^2 + m^2, m^4 + m^2, m^4 + m^2)$ exist for all $m > 1$. *Eur. J. Combin.*, 31: 1553–1559, 2010.

91 Marshall Hall Jr. Hadamard matrices of order 16. Research summary 36-10, Jet Propulsion Laboratory, 1961. pages 21–26.

92 Marshall Hall Jr. *Combinatorial Theory*. John Wiley & Sons, New York, 2nd edition, 1986.

93 J. Hammer, R. Levingston, and J. Seberry. A remark on the excess of Hadamard matrices and orthogonal designs. *Ars Combin.*, 5: 237–254, 1978.

94 A. Hedayat and W. D. Wallis. Hadamard matrices and their applications. *Ann. Stat.*, 6: 1184–1238, 1978.

95 Wolf H. Holzmann and Hadi Kharaghani. A computer search for complex Golay sequences. *Australas. J. Combin.*, 10: 251–258, 1994.

96 Wolf H. Holzmann and Hadi Kharaghani. On the Plotkin arrays. *Australas. J. Combin.*, 22: 287–299, 2000.

97 Wolf H. Holzmann, Hadi Kharaghani, and Behruz Tayfeh-Rezaie. Williamson matrices up to order 59. *Des., Codes Crypt.*, 46 (3): 343–352, 2008.

98 K. J. Horadam. *Hadamard Matrices and Their Applications*. Princeton University Press, Princeton NJ, 2006.

99 David C. Hunt. Skew-Hadamard matrices of order less than 100. In *Proceedings of First Australian Conference on Combinatorial Mathematics*, pages 55–59, TUNRA, Newcastle, Australia, 1972.

100 David C. Hunt and Jennifer Wallis. Cyclotomy, Hadamard arrays and supplementary difference sets. In *Proceedings of the Second Manitoba Conference on Numerical Mathematics*, volume 7 of *Congressus Numerantium*, pages 351–381 University of Manitoba, Winnipeg, Canada, 1973.

101 Barry Hurley, Paul Hurley, and Ted Hurley. The Hadamard circulant conjecture. *ArXiv e-prints*, arXiv:1109.0748v1 [math.RA], Sep 2011. Also published in *Bull. Lond. Math. Soc.* 44 (1):206, 2012 (retracted).

102 N. Ito. On Hadamard groups. *J. Algebra*, 168: 981–987, 1994.

103 N. Ito. On Hadamard groups III. *Kyushu J. Math.*, 51: 369–379, 1997.

104 M. James. *Golay sequences*. Honours Thesis, University of Sydney, Australia, 1987.

105 S. Jauregui. Complementary sequences of length 26. *IRE Trans. Inf. Theory*, IT-7 (4): 323, 1962.

106 Hadi Kharaghani. Arrays for orthogonal designs. *J. Combin. Des.*, 8 (3): 166–173, 2000.

107 Hadi Kharaghani and William Orrick. *D*-optimal matrices, In *Handbook of Combinatorial Designs* (C. J. Colbourn and J. H. Dinitz, Eds.) pages 296–298. Chapman & Hall CRC Press, Bocas Racon, FL. 2nd edition, 2007.

108 Hadi Kharaghani and Behruz Tayfeh-Rezaie. Some new orthogonal designs in orders 32 and 40. *Discrete Math.*, 279 (1–3): 317–324, 2004.

109 Hadi Kharaghani and Behruz Tayfeh-Rezaie. A Hadamard matrix of order 428. *J. Combin. Des.*, 13 (6): 435–440, 2005.

110 Z. Kiyasu. *An Hadamard matrix and its applications.* Denshi-Tsushin Gakkai, Tokyo, 1980. (in Japanese.)

111 C. Koukouvinos. Base sequences $BS(n+1, n)$. http://www.math.ntua.gr/ ckoukouv/baseseq.htm, 2018.

112 C. Koukouvinos and S. Kounias. Hadamard matrices of the Williamson type of order $4m$, $m = pq$. An exhaustive search for $m = 33$. *Discrete Math.*, 68: 47–57, 1988.

113 C. Koukouvinos and S. Kounias. An infinite class of Hadamard matrices. *J. Aust. Math. Soc.*, 46: 384–394, 1989.

114 C. Koukouvinos and S. Kounias. Construction of some Hadamard matrices with maximal excess. *Discrete Math.*, 85 (3): 295–300, 1990.

115 C. Koukouvinos, S. Kounias, and J. Seberry. Further results on base sequences, disjoint complementary sequences, $OD(4t; t, t, t, t)$ and the excess of Hadamard matrices. *Ars Combin.*, 30: 241–256, 1990.

116 C. Koukouvinos, S. Kounias, J. Seberry, C. H. Yang, and J. Yang. Multiplication of sequences with zero autocorrelation. *Australas. J. Combin.*, 10: 5–15, 1994.

117 C. Koukouvinos, S. Kounias, J. Seberry, C. H. Yang, and J. Yang. On sequences with zero autocorrelation. *Des., Codes Crypt.*, 4: 327–340, 1994.

118 C. Koukouvinos, S. Kounias, and K. Sotirakoglou. On base and Turyn sequences. *Math. Comput.*, 55: 825–837, 1990.

119 Christos Koukouvinos and Jennifer Seberry. Constructing Hadamard matrices from orthogonal designs. *Australas. J. Combin.*, 6: 267–278, 1992.

120 S. Kounias, C. Koukouvinos and K. Sotirakoglou. On Golay sequences. *Discrete Math.*, 92 (1): 177–185, 1991

121 S. Kounias and K. Sotirakoglou. Construction of orthogonal sequences. In *Proceedings of the 14th Greek Statistical Conference 2001*, Skiathos, Greece, pages 267–278 (in Greek), 2001.

122 R. G. Kraemer. Proof of a conjecture on Hadamard 2-groups. *J. Combin. Theory Ser. A*, 63: 1–10, 1993.

123 Christos Kravvaritis and Marilena Mitrouli. The growth factor of a Hadamard matrix of order 16 is 16. *Numer. Linear Algebra Appl.*, 16 (9): 715–743, 2009.

124 S. Lang. *Cyclotomic Fields I and II.* Springer-Verlag, Berlin-Heidelberg, 1990.

125 Ka Hin Leung and Siu Lun Ma. Partial difference sets with Paley parameters. *Bull. Lond. Math. Soc.*, 27 (6): 553–564, 1995.

126 R. Lidl and H. Niederreiter. *Introduction to Finite Fields.* Cambridge University Press, Cambridge, 1994.

127 R. Lidl and H. Niederreiter. *Finite Fields.* Cambridge University Press, Cambridge, 1997.

128 Cantian Lin and W. D. Wallis. On the circulant Hadamard matrix conjecture. In *Proceedings of the Marshall Hall Conference on Coding Theory, Design Theory, Group Theory* (D. Jungnickel and S. A. Vanstone, Eds.), Burlington, Vermont, USA, pages 213–217. John Wiley & Sons, New York, 1993.

129 S. L. Ma. Partial difference sets. *Discrete Math.*, 52: 75–89, 1984.

130 S. L. Ma. A survey of partial difference sets. *Des. Codes Crypt.*, 4: 221–261, 1994.

131 C. C. MacDuffee. *The Theory of Matrices.* Chelsea, 1964. Reprint of First Edition.

132 M. Marcus and H. Minc. *A Survey of Matrix Theory and Matrix Inequalities.* Allyn and Bacon Series in Advanced Mathematics. Allyn and Bacon, Boston, MA, 1964.

133 R. Mathon. Symmetric conference matrices of order $pq^2 + 1$. *Can. J. Math.*, 30: 321–331, 1978.

134 B. R. McDonald. *Finite Rings with Identity*, volume 28 of *Pure and Applied Mathematics*. Marcel Dekker, New York, 1974.

135 R. L. McFarland. Difference sets in abelian groups of order $4p^2$. *Mitt. Math. Sem. Giessen.*, 192: 1–70, 1989.

136 T. S. Michael and W. D. Wallis. Skew-Hadamard matrices and the Smith normal form. *Des., Codes Crypt.*, 13 (2): 173–176, 1998.

137 M. Miyamoto. A construction of Hadamard matrices. *J. Combin. Theory*, 57: 86–108, 1991.

138 K. Momihara. Strongly regular Cayley graphs, skew Hadamard difference sets, and rationality of relative Gauss sums. *Eur. J. Comb.*, 34: 706–723, 2013.

139 A. C. Mukhopadhyay. Some infinite classes of Hadamard matrices. *J. Combin. Theory Ser. A*, 25: 128–141, 1978.

140 M. E. Muzychuk. On skew Hadamard difference sets. *ArXiv e-prints*, arXiv.1012.2089 [math.CO], Dec 2010.

141 G. Myerson. Period polynomials and Gauss sums. *Acta Arithmetica*, 39: 251–264, 1981.

142 Tamio Ono and Kazue Sawade. Whiteman's solution of Hadamard matrices. *Bull. Nagoya Inst. Tech.*, 35: 73–80, 1983. (In Japanese with English summary.)

143 W. P. Orrick and B. Solomon. Spectrum of the determinant function, 2018. http://www.indiana.edu/~maxdet/spectrum.html.

144 R. E. A. C. Paley. On orthogonal matrices. *J. Math. Phys.*, 12: 311–320, 1933.

145 M. Plotkin. Decomposition of Hadamard matrices. *J. Combin. Theory Ser. A*, (13): 127–130, 1972.

146 J. Polhill. Paley type partial difference sets in non-p-groups. *Des. Codes Crypt.*, 52: 163–169, 2009.

147 J. Polhill. Paley type partial difference sets in groups of order n^4 and $9n^4$ for any odd $n > 1$. *J. Combin. Theory A*, 117: 1027–1036, 2010.

148 Herbert John Ryser. *Combinatorial Mathematics*, volume 14 of *Carus Mathematical Monographs*. Mathematical Association of America, Buffalo, NY, distributed by Wiley, New York, 1963.

149 K. Sawade. A Hadamard matrix of order 268. *Graphs Combin.*, 1: 185–187, 1985.

150 U. Scarpis. Sui determinanti di valore massimoe. *Rend. R. Inst. Lombardo Sci. e Lett.*, 2 (31): 1441–1446, 1898.

151 B. Schmidt. Cyclotomic integers and finite geometry. *J. Am. Math. Soc.*, 12 (4): 929–952, 1999.

152 B. Schmidt. Williamson matrices and a conjecture of Ito's. *Des., Codes Crypt.*, 17 (1–3): 61–68, 1999.

153 J. Seberry. On skew Hadamard matrices. *Ars Combin.*, 6: 255–275, 1978.

154 J. Seberry. Some infinite families of Hadamard matrices. *J. Aust. Math. Soc. Ser. A*, 29: 235–242, 1980.

155 J. Seberry. A new construction for Williamson-type matrices. *Graphs Combin.*, 2: 81–87, 1986.

156 J. Seberry. SBIBD($4k^2$, $2k^2 + k$, $k^2 + k$) and Hadamard matrices of order $4k^2$ with maximal excess are equivalent. *Graphs Combin.*, 5: 373–383, 1989. Also appeared in *Proceedings of the Symposium on Hadamard Matrices* (K. Yamamoto, Ed.), pages 19–29, Tokyo, 1988. Tokyo Women's University.

157 J. Seberry. New families of amicable Hadamard matrices. *J. Stat. Theory Pract.*, 7 (4): 650–657, 2013. (In memory of Jagdish N. Srivastava.)

158 J. Seberry. *Orthogonal Designs: Hadamard Matrices, Quadratic Forms and Algebras*. Springer Nature, Cham, 1st edition, 2017.

159 J. Seberry and N. A. Balonin. Two infinite families of symmetric Hadamard matrices. Conference presentation, 16 December 2016. 40th Australasian Conference on Combinatorial Mathematics and Combinatorial Computing, Newcastle, NSW, Australia.

160 J. Seberry and N. A. Balonin. Two infinite families of symmetric Hadamard matrices. *Australas. J. Combin.*, 69 (3): 349–357, 2017. Also appeared as ArXiv e-print 1512.01732 "The propus construction for symmetric Hadamard matrices."

161 J. Seberry and R. Craigen. Orthogonal designs. In *CRC Handbook of Combinatorial Designs* (Charles J. Colbourn and Jeffrey H. Dinitz, Eds.), pages 400–406. CRC Press Series on Discrete Mathematics and Its Applications. CRC Press, Boca Raton, FL, 1st edition, 1996.

162 J. Seberry and M. Mitrouli. Some remarks on Hadamard matrices. *Crypt. Commun.*, 2 (2): 293–306, 2010. Dedicated with great respect to Warwick de Launey.
163 J. Seberry and A. L. Whiteman. New Hadamard matrices and conference matrices obtained via Mathon's construction. *Graphs Combin.*, 4: 355–377, 1988.
164 J. Seberry, T. Xia, C. Koukouvinos, and M. Mitrouli. The maximal determinant and subdeterminants of ± 1 matrices. *Linear Algebra Appl.*, 373 (Supplement C): 297–310, 2003. Combinatorial Matrix Theory Conference (Pohang, 2002).
165 J. Seberry and M. Yamada. On the products of Hadamard, Williamson and other orthogonal matrices using M-structures. *J. Combin. Math. Combin. Comput.*, 7: 97–137, 1990.
166 J. Seberry and M. Yamada. Hadamard matrices, sequences and block designs in *Contemporary Design Theory A Collection of Surveys* (D. J. Stinson and J. Dinitz, Eds.), pages 431–560. John Wiley & Sons, New York, 1992.
167 J. Seberry and X. M. Zhang. Some orthogonal designs and complex Hadamard matrices by using two Hadamard matrices. *Australas. J. Combin.*, 4: 93–102, 1991.
168 S. S. Shrikhande. On a two parameter family of balanced incomplete block designs. *Sankhya Ser. A*, 24: 33–40, 1962.
169 J. Singer. A theorem in finite projective geometry and some applications in number theory. *Trans. Am. Math. Soc.*, 43: 377–385, 1938.
170 Neil J. A. Sloane. A library of Hadamard matrices, 2018. http://neilsloane.com/hadamard/index.html.
171 H. J. S. Smith. Arithmetical notes. *Proc. Lond. Math. Soc.*, 4: 236–253, 1873. The three papers which form these Notes were read on Jan. 9 and Feb. 13, 1873.
172 Edward Spence. A new class of Hadamard matrices. *Glasgow J.*, 8: 59–62, 1967.
173 Edward Spence. A note on the equivalence of Hadamard matrices. *Not. Am. Math. Soc.*, 18: 624, 1971.
174 Edward Spence. Hadamard matrices from relative difference sets. *J. Combin. Theory Ser. A*, (19): 287–300, 1975.
175 Edward Spence. Skew-Hadamard matrices of Goethals–Seidel type. *Can. J. Math.*, 27: 555–560, 1975.
176 Edward Spence. An infinite family of Williamson matrices. *J. Aust. Math. Soc. Ser. A*, 24: 252–256, 1977.
177 Edward Spence. Skew-Hadamard matrices of order $2(q + 1)$. *Discrete Math.*, 18 (1): 79–85, 1977.
178 R. G. Stanton and D. A. Sprott. A family of difference sets. *Can. J. Math.*, 10: 73–77, 1958.
179 T. Storer. *Cyclotomy and Difference Sets*, volume 2 of *Lectures in Advanced Mathematics*. Markham Publishing Company, Chicago, IL, 1967.
180 J. J. Sylvester. Thoughts on inverse orthogonal matrices, simultaneous sign successions, and tesselated pavements in two or more colours, with applications to Newton's rule, ornamental tile-work, and the theory of numbers. *Phil. Mag.*, 34 (4): 461–475, 1867.
181 G. Szekeres. Tournaments and Hadamard matrices. *Enseign. Math.*, 15: 269–278, 1969.
182 G. Szekeres. Cyclotomy and complementary difference sets. *Acta. Arith.*, 18: 349–353, 1971.
183 G. Szekeres. A note on skew type orthogonal ± 1 matrices. *Colloq. Math. Soc. Janos Bolyai*, 52: 489–498, 1987. Presented at the Combinatorics Conference, Eger (Hungary).
184 F. Szöllősi. Exotic complex Hadamard matrices and their equivalence. *Crypt. Commun.*, 2: 187–198, 2010.
185 C. C. Tseng. Signal multiplexing in surface-wave delay lines using orthogonal pairs of Golay's complementary sequences. *IEEE Trans. Sonics Ultrasonics*, SU-18 (2): 103–107, 1971.
186 C. C. Tseng and C. L. Liu. Complementary sets of sequences. *IEEE Trans. Inf. Theorey*, IT-18: 644–652, 1972.
187 T. Tsuzuku. *Finite Groups and Finite Geometries*. Cambridge University Press, Cambridge, 1982.
188 Richard J. Turyn. Ambiguity functions of complementary sequences. *IEEE Trans. Inf. Theory*, IT-9 (1): 46–47, 1963.
189 Richard J. Turyn. Character sums and difference sets. *Pac. J. Math.*, 15: 319–346, 1965.
190 Richard J. Turyn. Complex Hadamard matrices. In *Combinatorial Structures and Their Application*, pages 435–437. Gordon and Breach, London, 1970.

191 Richard J. Turyn. On C-matrices of arbitrary powers. *Bull. Can. Math. Soc.*, 23: 531–535, 1971.

192 Richard J. Turyn. An infinite class of Williamson matrices. *J. Combin. Theory*, 12: 319–321, 1972.

193 Richard. J. Turyn. Four-phase Barker codes. *IEEE Trans. Inf. Theory*, IT-20: 366–371, 1974.

194 Richard J. Turyn. Hadamard matrices, Baumert–Hall units, four-symbol sequences, pulse compressions and surface wave encodings. *J. Combin. Theory Ser. A*, (16): 313–333, 1974.

195 Richard J. Turyn. A special class of Williamson matrices and difference sets. *J. Combin. Theory Ser. A*, 36: 111–115, 1984.

196 K. Vijayan. Hadamard matrices and submatrices. *J. Aust. Math. Soc.*, 22: 469–475, 1976.

197 Jennifer Wallis. A class of Hadamard matrices. *J. Combin. Theory*, 6: 40–44, 1969.

198 Jennifer Wallis. A note of a class of Hadamard matrices. *J. Combin. Theory*, 6: 222–223, 1969.

199 Jennifer Wallis. Hadamard designs. *Bull. Aust. Math. Soc.*, 2: 45–54, 1970.

200 Jennifer Wallis. (v, k, λ)-Configurations and Hadamard matrices. *J. Aust. Math. Soc.*, 11: 297–309, 1970.

201 Jennifer Wallis. Amicable Hadamard matrices. *J. Combin. Theory eory Ser. A*, 11: 296–298, 1971.

202 Jennifer Wallis. A skew-Hadamard matrix of order 92. *Bull. Aust. Math. Soc.*, 5: 203–204, 1971.

203 Jennifer Wallis. *Combinatorial matrices.* Ph.D. Thesis, La Trobe University, Melbourne, Australia, 1971.

204 Jennifer Wallis. Hadamard matrices. In *Combinatorics: Room Squares, Sum Free Sets, Hadamard Matrices* (W. D. Wallis, Anne Penfold Street, and Jennifer Seberry Wallis, Eds.) pages 273–489. Springer-Verlag, Berlin-Heidelberg-New York, 1972.

205 Jennifer Wallis. On supplementary difference sets. *Aequationes Math.*, 8: 242–257, 1972.

206 Jennifer Wallis. Hadamard matrices of order $28m$, $36m$, and $44m$. *J. Combin. Theory Ser. A*, (15): 323–328, 1973.

207 Jennifer Wallis. A note on amicable Hadamard matrices. *Util. Math.*, 3: 119–125, 1973.

208 Jennifer Wallis. Some matrices of Williamson type. *Util. Math.*, 4: 147–154, 1973.

209 Jennifer Wallis. Some remarks on supplementary difference sets. In *Infinite and finite sets*, pages 1503–1526. Colloq. Math. Soc. Janos bolyai, 1973.

210 Jennifer Wallis. Williamson matrices of even order. In *Combinatorial Mathematics: Proceedings of the Second Australian Conference* (D. A. Holton, Ed.), volume 403 of *Lecture Notes in Mathematics*, pages 132–142. Springer-Verlag, Berlin-Heidelberg-New York, 1974.

211 Jennifer Wallis. Construction of Williamson type matrices. *J. Linear Multilinear Algebra*, 3: 197–207, 1975.

212 Jennifer Wallis. Orthogonal designs V: orders divisible by eight. *Utili. Math.*, 9: 263–281, 1976.

213 Jennifer Wallis. On the existence of Hadamard matrices. *J. Combin. Theory Ser. A*, 21: 444–451, 1976.

214 W. D. Wallis, Anne Penfold Street, and Jennifer Seberry Wallis. *Combinatorics: Room Squares, Sum-Free Sets, Hadamard Matrices*, volume 292 of *Lecture Notes in Mathematics*. Springer-Verlag, Berlin-New York, 1972.

215 W. D. Wallis and Jennifer (Seberry) Wallis. Equivalence of Hadamard matrices. *Israel J. Math.*, 7: 122–128, 1969.

216 Jennifer Seberry Wallis and Albert Leon Whiteman. Some classes of Hadamard matrices with constant diagonal. *Bull. Aust. Math. Soc.*, 7: 233–249, 1972.

217 G. R. Welti. Quaternary codes for pulsed radar. *IRE Trans. Inf. Theory*, IT-6 (3): 400–408, 1960.

218 G. Weng and L. Hu. Some results on skew Hadamard difference sets. *Des. Codes Crypt.*, 50: 93–105, 2009.

219 Albert Leon Whiteman. An infinite family of skew-Hadamard matrices. *Pac. J. Math.*, 38: 817–822, 1971.

220 Albert Leon Whiteman. Skew-Hadamard matrices of Goethals–Seidel type. *Discrete Math.*, 2: 397–405, 1972.

221 Albert Leon Whiteman. An infinite family of Hadamard matrices of Williamson type. *J. Combin. Theory Ser. A*, 14: 334–340, 1973.

222 Albert Leon Whiteman. Hadamard matrices of Williamson type. *J. Aust. Math. Soc.*, 21 (21): 481–486, 1976.

223 John Williamson. Hadamard's determinant theorem and the sum of four squares. *Duke Math. J.*, 11: 65–81, 1944.

224 John Williamson. Note on Hadamard's determinant theorem. *Bull. Am Math. Soc.*, 53: 608–613, 1947.

225 R. M. Wilson and Q. Xiang. Constructions of Hadamard difference sets. *J. Combin. Theory Ser. A*, 77: 148–160, 1997.

226 Ming-Yuan Xia. Some infinite classes of special Williamson matrices and difference sets. *J. Combin. Theory Ser. A*, 61: 230–242, 1992.

227 M. Xia and G. Liu. An infinite class of supplementary difference sets and Williamson matrices. *J. Combin. A*, 58: 310–317, 1991.

228 M. Xia and G. Liu. A new family of supplementary difference sets and Hadamard matrices. *J. Statist. Plann. Inference*, 51: 283–291, 1996.

229 M. Xia, T. Xia, and J. Seberry. A new method for constructing Williamson matrices. *Des. Codes Crypt.*, 35: 191–209, 2005.

230 M. Xia, T. Xia, Jennifer Seberry, and H. Qin. Construction of T-matrices of order $6m + 1$. *Far East J. Math. Sci.*, 101 (8): 1731–1749, 2017.

231 Q. Xiang. Difference families from lines and half lines. *Eur. J. Combin.*, 19: 395–400, 1989.

232 Q. Xiang and Y. Q. Chen. On Xia's construction of Hadamard difference sets. *Finite Fields Appl.*, 2: 87–95, 1996.

233 M. Yamada. On the Williamson type j matrices of order $4 \cdot 29, 4 \cdot 41$ and $4 \cdot 37$. *J. Combin. Theory Ser. A*, 27: 378–381, 1979.

234 M. Yamada. Hadamard matrices generated by an adaption of the generalized quaternion type array. *Graphs Combin.*, 2: 179–187, 1986.

235 M. Yamada. On a relation between a cyclic relative difference set associated with the quadratic extensions of a finite field and the Szekeres difference sets. *Combinatorica*, 8: 207–216, 1988.

236 M. Yamada. On a series of Hadamard matrices of order 2^t and the maximal excess of Hadamard matrices of order 2^{2t}. *Graphs Combin.*, 4: 297–301, 1988.

237 M. Yamada. Some new series of Hadamard matrices. *J. Aust. Math. Soc. Ser. A*, 46: 371–383, 1989.

238 M. Yamada. Hadamard matrices of generalized quaternion type. *Discrete Math.*, 87: 187–196, 1991.

239 M. Yamada. Supplementary difference sets and Jacobi sums. *Discrete Math.*, 103 (1): 75–90, 1992.

240 M. Yamada. Difference sets over the Galois ring $GR(2^n, 2)$. *Eur. J. Combin.*, 23: 239–252, 2002.

241 M. Yamada. Supplementary difference sets constructed from $q + 1^{st}$ cyclotomic classes in $GF(q^2)$. *Australas. J. Combin.*, 39: 73–87, 2007.

242 M. Yamada. Difference sets over Galois rings with odd extension degrees and characteristic of even power of 2. *Des. Codes, Crypt.*, 67 (1): 37–57, 2013.

243 M. Yamada. Menon-Hadamard difference sets obtained from a local field by natural projections. In *Algebraic Design Theory and Hadamard Matrices* (C. Colbourn, Ed.), pages 235–249, *Springer Proceedings in Mathematics & Statistics*, volume 133. Springer, Cham, 2015.

244 K. Yamamoto. Decomposition fields of difference sets. *Pac. J. Math.*, 13: 337–352, 1963.

245 K. Yamamoto. On a generalized Williamson equation. *Colloq. Math. Soc. Janos Bolyai*, 37: 839–850, 1981.

246 K. Yamamoto. *Kumiawasesugaku (Combinatorics)*. Asakurashoten, Tokyo, 1989 (in Japanese).

247 K. Yamamoto and M. Yamada. Williamson Hadamard matrices and Gauss sums. *J. Math. Soc. Japan*, 37 (4): 703–717, 1985.

248 K. Yamamoto and M. Yamada. Hadamad difference sets over an extension of $\boldsymbol{Z}/4\boldsymbol{Z}$. *Util. Math.*, 34: 169–178, 1988.

249 C. H. Yang. Hadamard matrices and δ-codes of length $3n$. *Proc. Am. Math. Soc.*, 85: 480–482, 1982.

250 C. H. Yang. A composition theorem for δ-codes. *Proc. Am. Math. Soc.*, 89: 375–378, 1983.

251 C. H. Yang. Lagrange identities for polynomials and δ-codes of lengths $7t$ and $13t$. *Proc. Am. Math. Soc.*, 88: 746–750, 1983.

252 C. H. Yang. On composition of four symbol δ-codes and Hadamard matrices. *Proc. Am. Math. Soc.*, 107: 763–776, 1989.

253 J. Yang and L. Xia. Complete solving of explicit evaluation of Gauss sums in the index 2 case. *Sci. China Ser. A*, 53: 2525–2542, 2010.

254 Guoxin Zuo, Mingyuan Xia, and Tianbing Xia. Constructions of composite T-matrices. *Linear Algebra Appl.*, 438 (3): 1223–1228, 2013.

Index

h

r

s

Printed and bound by CPI Group (UK) Ltd, Croydon, CR0 4YY

16/04/2025

14658370-0005